Applied Statistics

Models and Intuition

Applied Statistics

Models and Intuition

ARNOLD I. BARNETT

George Eastman Professor of Management Science
Professor of Statistics
Sloan School of Management
Massachusetts Institute of Technology

Dynamic Ideas LLC
Belmont, Massachusetts

Dynamic Ideas LLC
43 Lantern Road
Belmont, Mass. 02478
USA

Email: info@dynamic-ideas.com
WWW Information and Orders: http://www.dynamic-ideas.com

Cover Design: Patrick Ciano

Publisher's Cataloging-in-Publication Data
Barnett, Arnold I.
Applied Statistics: Models and Intuition
Includes bibliographical references and index
1. Applied Statistics. 2. Data. 3. Applications. Title.

T56.B44 2015

ISBN: 978-0-9899108-8-0

Dedication

With deep gratitude, I dedicate this book to my wife Harriet, my children Andrew and Lauren (two legged) and Duncan and Lilly (four legged), my mother Doris Barnett, my sister Elaine Gell, my late father Leon Barnett and my late in-laws, Bebe and Shepard Schwartz.

CONTENTS

PREFACE

"Statistics is about numbers. It is also about common sense. The numbers (or data) constitute clues that help us answer questions we care about. But, like all good detectives, we have to identify the primary clues, combine them in intelligent ways, and recognize that a particular clue can have more than one interpretation. Good statistical analysis can bring insight and joy; poor statistical analysis can bring suffering and pain."

The quotation above is a shameless exercise in self-plagiarism, for it is the introductory paragraph of Chapter 2 of this book. But it conveys my deep beliefs about Statistics: that the field is important, that it is eminently logical, and that, when used properly, it can be outright fun.

In my experience, many people would concede (however grudgingly) that Statistics is important, but far fewer have an intuitive "feel" for Statistics and fewer still enjoy their experiences with it. This book aspires to introduce the subject in a way that conveys its power and elegance. It would be preposterous to suggest that all readers of this book will come away loving Statistics. But I hope that some of them will, and that most will conclude that Statistics is a trustworthy, congenial and valuable traveling companion.

This introductory textbook in Statistics is written from the viewpoint of Applied Mathematics. Under that perspective, theoretical rigor is not denigrated, but the book puts a high premium on offering an intuitive overview of key methods as well as clear evidence of their usefulness. Beyond teaching particular techniques, the book strives to create real comfort with the underlying notions of statistical thinking. Grasping such ideas is essential if they are to be used with competence and confidence.

Obviously, there are already many textbooks for a first course in Statistics. What might differentiate this book from the others are:

An Exceptionally Large Inventory of Real-Life Examples Drawn from Many Fields

Consistent with the theme of *Applied* Statistics, the book contains hundreds of worked examples, analyses of actual data, and practice exercises that serve as key components of the pedagogy. For example, we will explore with data which couples are the best candidates for *in vitro* fertilization, whether temperatures in New York City have been rising in a matter consistent with climate change, whether the Space Shuttle Challenger disaster was all but inevitable, when bystanders are least likely to help people in trouble, and what statistical evidence is considered compelling in job-discrimination cases. The aim is to have examples that are consistently stimulating and that will attract the attention of all readers. It would not be helpful if Engineering students lose interest whenever applications from Biology are raised, or vice-versa.

An Unusually Strong Emphasis on Achieving an Intuitive Understanding of the Material

Given my background in Physics and Applied Mathematics, I am very uncomfortable with accepting an answer that, while clearly correct, just doesn't seem to make sense. Thus, I see little value in presenting a key formula without trying to make clear why it is plausible. Indeed, in my experience as an applied statistician, people who lack an intuitive understanding of why a statistical method works will often apply it incorrectly. Striving for such understanding is therefore more a necessity than a luxury.

My treatment of how to estimate the dispersion of a series of numbers across a population illustrates my priorities for this book. A mathematical derivation shows that, given a random data sample from the population, the most natural formula to estimate the population variance suffers a downward bias. But if that finding is to be more illuminating than unnerving, it is important to conduct a "post mortem" to make clear with pictures and words why the bias arises and why it tends to wither away as the sample size gets larger. Once readers grasp these points, they are in a position to appreciate why a surprisingly simple adjustment of the formula eliminates the bias. And, lest the discussion seem overly abstract, the book applies the adjusted formula to several real-life data sets.

A Conversational Style of Presentation

I have tried to write the book in an informal and occasionally light-hearted style, though I placed far higher priority on lucidity than entertainment. The reader should *want* to pick up the book and read it, and should sense that the author is there with her as she tries to think the material through. Make no mistake: good statistical analysis is anything but a mechanical process, and the study of Statistics requires concentration and discipline. Beyond mathematics *per se*, the discussions and practice exercises are meant to make it easier to focus on the material, and to achieve the calm and self-confidence that are prerequisites to real mastery of Applied Statistics.

The Intended Audience

The book is intended to a support a one-semester course in Applied Statistics, though it could undergird a one-quarter course if some topics are skipped. The book strives to be suitable for both undergraduates and graduate students who concentrate in Engineering, Business, the Natural Sciences, and the Social Sciences, as well as for Math majors with an interest in applications. As suggested earlier, I am not among those who believe that engineers need learn from a book about

"Statistics for Engineers" or that business students need a volume about "Statistics for Business." The underlying principles of statistical analysis are the same regardless of the domain of application, and the thinking about (say) a regression model in Finance can benefit the Chemistry major as well as the economist.

The book does not minimize the importance of derivations of key results (the author got his doctorate in Mathematics), but it is possible for instructors to deemphasize these derivations and move directly to the results when it seems desirable. The reality is that there is a finite amount of time in the introductory Statistics course, and a tradeoff must be made between an emphasis on mathematical rigor and a zestful treatment of applications. This volume clearly tilts towards the latter, but proofs of most primary results appear either in the text or at the website that accompanies this book.

In the 21st century, the study of Statistics routinely assumes the availability of statistical computing packages, or at least access to statistical programs and calculators that are abundant on the Internet. This book is not geared to a particular statistical package (e.g., R, JUMP, SPSS) for the simple reason that all the usual packages provide adequate support for the book's analyses and data explorations. I believe that requiring any single package would be no more sensible than insisting, in days gone by, that students taking notes in class all use the same brand of pen.

Of course, one does not become a statistical expert based solely on an introductory course. Although this book contains two chapters on statistical sampling (including the modern method of bootstrapping) and two on regression analysis, some readers may find it important to take more advanced courses on these and other methodologies. The intent is to immerse the reader in enough statistical thinking that he can readily progress to such subsequent courses. Indeed, it would be immensely gratifying if this book served to inspire students (perhaps to their surprise) to continue their study of Statistics.

The book assumes that the reader has taken a first course in Calculus. Calculus is fundamental to such important concepts as probability density functions and maximum likelihood parameter estimation. But part of the reason for a Calculus requirement goes beyond its immediate utility. Completion of a Calculus course suggests a level of mathematical mastery and maturity that I believe appropriate to achieve comfort with the book's treatment of Applied Statistics. Even though Calculus is of little direct relevance to many topics covered in the book, the reasoning skills associated with instruction in that subject are often important.

Solution Manual

A solution manual that includes solutions to all exercises in the book is available to qualified instructors. Interested parties should contact the publisher at *info@dynamic-ideas.com* to obtain it.

Some Acknowledgements

Perhaps the hardest part of writing this book is preparing this section of acknowledgements, because I am fearful of accidentally leaving people out. Over the past several decades, I have taught thousands of students at MIT and worked with nearly 100 teaching assistants, and a huge number of these people have influenced my thinking about what material is important and how to explain it effectively. I have also benefitted greatly from my experience outside the academic world in research and consulting, which has given me a strong sense of the value of Applied Statistics in coping with real-world problems (many of which are reflected in the book). The lawyers, medical doctors, engineers, economists, and business and government leaders whom I've been lucky enough to work with are literally too numerous to mention, but my debt to them is enormous.

However, I should single out some of the remarkable individuals from the academic world who have encouraged me and influenced me in important ways. Among faculty, a partial list would include Alfred Blumstein (Carnegie Mellon), Jonathan Caulkins (Carnegie Mellon), David Czerwinski (San Jose State), the late Alvin Drake (MIT), Robert Freund (MIT), Mark Hansen (Berkeley), Kenneth Huang (Singapore Management University), the late Paul Gray (USC), Linda Green (Columbia), Edward Kaplan (Yale), Richard C. Larson (MIT), Gordon Kaufman (MIT), Daniel J. Kleitman (MIT), the late Polykarp Kusch (Columbia), John D. C. Little (MIT), the late Robert Machol (Northwestern University and the Federal Aviation Administration), Thomas Magnanti (MIT and Singapore Institute of Technology and Design), Susan Martonosi (Harvey Mudd), Donald Morrison (UCLA), Scott Neslin (Dartmouth), the late Gordon Newell (Berkeley), Amedeo Odoni (MIT), Cynthia Rudin (MIT), the late Abraham Siegel (MIT), Robert Shumsky (Dartmouth), and Roy Welsch (MIT). I am especially grateful to the brilliant publisher of Dynamic Ideas Press, Dimitris Bertsimas (MIT). Among former MIT students, I owe much to Robin Bose, Brian Chang, Lincoln Chandler, Jesse Goranson, Bob Jordan, Mark Kamal, Sean Leow, Steve Miller, Arvind Nagarajan, C. S. Venkatakishan, and in particular Anthony Lofaso, who prepared most of the solutions to exercises in the book. I am of course grateful to my wife and children, who make my life better with statistical certainty

A Final Thought

It is unfortunately the case that many data analyses are performed now by people who do not even know the assumptions of the statistical models they are using, let alone whether those assumptions are satisfied in the matter at hand. There are too many "walking wounded professionals" who apply statistical techniques that they understand only hazily and regard rather warily. I have tried to write a book that will keep readers out of these categories, and that will lead them to consider data analysis not the least interesting aspect of their applied work but among the most stimulating.

To state the obvious: I hope that readers and instructors find this a valuable book, and that they will let me know about changes that could make it better.

Arnold Barnett
Cambridge, Massachusetts
December 2014

CHAPTER 1

Fundamentals of Applied Probability

Double Trouble (Based on a True Story)

The operations manager of the air-freight company was mightily upset. The company flew Boeing-727 freighters with eight compartments for containers. Now the government was considering a rule that would bar loading two heavy containers (each weighing more than 9600 pounds) into adjacent compartments. The fear was that, if the plane encountered an extreme wind gust while in a "double heavy" situation, it could suffer a catastrophic loss of control.

The manager accosted the statistician and began a bitter lament. "If we have to weigh each container before loading it, the effect on our efficiency will be disastrous. And the thing is, only 7% of our containers are heavy. So we rarely have two heavies on the same plane, let alone next to one another. Even if we loaded at random without weighing anything, the chance of being in "double heavy" mode is infinitesimal. Right?"

"I'm not sure it's infinitesimal," the statistician responded. More than a little annoyed, the manager snapped "you're the statistician. Why don't you work it out?"

To Be Continued

Introduction

This author believes strongly that one can't really understand Statistics without some background in Probability. Statistics is basically the quantitative analysis of data, and nearly all data suffer uncertainty of one form or another. Probability offers a precise description of uncertainty, and a framework for clarifying what can and cannot be said on the basis of the data at hand. Under the circumstances, this treatment of Applied Statistics starts with an abbreviated treatment of Applied Probability.

In an ideal world, the reader will already have studied Probability for at least one quarter or semester. (The author has written a book titled *Applied Probability: Models and Intuition* as a basis for such a course.) In the real world, however, not everyone who wants to learn Statistics has the time for a prior course. This chapter presents a brisk survey of important topics in Applied Probability and, if the reader can follow the material and do the exercises, then nothing related to Probability should present any problems in subsequent chapters. If, however, the treatment seems too terse at some points, then turning to the corresponding section of an introductory text on Probability could be wise.

1.1 Some Basic Probability

Because basic probability often invokes the ideas of set theory, we do well to start our work by reviewing those ideas. In simplest terms, a set S is a collection of elements, where the word "elements" is defined broadly enough to include people, objects, events, or bankrupt corporations. Some definitions:

A **subset** A of set S contains some (but generally not all) elements in S.

The **intersection** of two subsets A and B is the set of elements that are included in both A and B. The intersection is commonly assigned the notation $A \cap B$.

The **union** of two subsets A and B is the set of items included in one or both of A and B. The union is usually assigned the notation $A \cup B$.

A **empty set**, usually given the symbol $\varnothing$, is a set that has no elements from S in it. For example, if S is the set of integers between 65 and 70, then the set of integers within S that are perfect squares (i.e., integers of the form x^2, where x is an integer) has no members. Thus, that set is an empty set. So is the set of integers within S that are multiples of 16.

Two subsets A and B within S are said to be **disjoint** if $A \cap B = \varnothing$. For example, if A is the set of even numbers in the range 65–70 and B is the set of odd numbers in that range, then A and B are disjoint.

Two subsets A and B within S are said to be **complementary** if two conditions are met: $A \cap B = \varnothing$ and $A \cup B = S$. Complementary sets arise when the elements in S are partitioned into two disjoint sets.

Subset B is **contained** in Subset A if all elements of B are also elements of A. This relationship is usually written as $B \subset A$.

Probability starts out with the notion of an *experiment*, which is some process that leads to a single outcome that (generally) is not known in advance with certainty. For example, a coin toss might be the experiment, and the outcome might be either heads or tails. A few more definitions:

The **sample space** S is the set of all possible outcomes of an experiment (plus the empty set $\varnothing$).

An **event** A is a subset of S.

Two events A and B are said to be **mutually exclusive** if the occurrence of one of them precludes the occurrence of the other. They do not share any common elements in the sample space (or, in the notation of set theory, $A \cap B = \varnothing$). Mutually exclusive events are also described as **disjoint**.

An **elementary event** cannot be expressed as the union of two or more disjoint events.

Following the terminology of set theory, two events A and B are **complementary** if $A \cap B = \varnothing$ and $A \cup B = S$. In general, we will use the notation A^c for the complement of event A.

For two events A and B, $A \cup B$ is the event that ***at least one*** of A and B occurs (and possibly both).

EXAMPLE 1.1

Double Header?

A coin is tossed twice in a row. Several events associated with the outcomes are:

A: *both tosses lead to heads*

B: *the second toss yields a different outcome than the first*

C: *both tosses lead to tails*

D: *the first toss yields heads*

(i) What is the sample space here?
(ii) Among these events, which pairs are mutually exclusive?
(iii) Which pairs are not mutually exclusive?
(iv) Are any of these pairs complementary?
(v) Are all possible outcomes of the experiment covered by these events?

Solution

(i) The sample space S consists of all possible outcomes of the double toss, which in simplest form are the four outcomes:

(Heads, Heads) (Heads then Tails) (Tails then Heads) (Tails, Tails).

(ii) A and B are mutually exclusive, as are A and C. B and C are also mutually exclusive, as well as C and D.

(iii) A and D and B and D are not mutually exclusive.

(iv) No pairs of events above are complementary, because the union of no two of them equals the full sample space.
(v) Yes: because $A \cup B \cup C = S$, they cover all possibilities.

EXERCISE 1.1

(i) In words, what are events $B \cap D$, $B \cup C$, and $A \cup D$ in Example 1.1?
(ii) For what pairs of events is the intersection the empty set?
(iii) In words, what is the complement of D?
(iv) If E is the event that the second toss yields tails, is $B = E \cup D$? Briefly discuss.

Kolmogorov's Axioms

Probabilities are numbers assigned to individual events, and are chosen to correspond to the intuitive concept of likelihood. Higher probabilities and greater likelihood are meant to go hand in hand. Kolmogorov brought greater precision to the quantification of probability by advancing several axioms that a legitimate assignment of probabilities to events must satisfy:

Kolmogorov's Axioms

Suppose that S is the sample space consisting of all possible events that can arise in an experiment, and that there are n events labeled A_1 through A_n that, collectively, include all elements in the sample space. The probabilities assigned to these events must have the following properties:

- $P(A_i) \geq 0$ for all i (1-1)
- $P(\varnothing) = 0$ (1-2)
- $P(S) = P(A_1 \cup A_2 \ldots \cup A_n) = 1$ (1-3)
- If $A_i \cap A_j = \varnothing$, then $P(A_i \cup A_j) = P(A_i) + P(A_j)$ (1-4)

These axioms imply some other rules about probabilities, which we present below.

If $A_1 \cap A_2 = \varnothing$ and $A_1 \cup A_2 = S$, then:

$$P(A_2) = 1 - P(A_1) \tag{1-5}$$

As noted, A_1 and A_2 are **complementary** events.

In words, if there are only two possible events and they are mutually exclusive, (1-5) says that the probability of one of them is one minus the probability of the other. That is true because $P(S) = 1$ and, given that $S = A_1 \cup A_2$ and $A_i \cap A_j = \varnothing$, we have:

$$P(S) = 1 = P(A_1 \cup A_2) = P(A_1) + P(A_2) \Rightarrow P(A_2) = 1 - P(A_1)$$

The additive rule in (1-4) can be generalized to several mutually exclusive events.

If events (A_1, A_2, A_k) are pairwise mutually exclusive, (i.e. events for which $A_i \cap A_j = \varnothing$ for all $1 \le i \ne j \le n$ then:

$$P(A_1 \cup A_2 \ldots A_k) = \sum_{i=1}^{k} P(A_i) \qquad (1\text{-}6)$$

Furthermore:

If A is an event and $B_1, \ldots B_k$ is a series of mutually exclusive (disjoint) events for which $B_1 \cup B_2 \ldots \cup B_n = S$, then:

$$P(A) = P(A \cap B_1) + P(A \cap B_2) \ \ldots \ + P(A \cap B_k) = \sum_{i=1}^{k} P(A \cap B_i) \qquad (1\text{-}7)$$

(This rule is called the Law of Alternatives.)

Because the B_i's are mutually exclusive, the events $A \cap B_i$ and $A \cap B_j$ must be mutually exclusive if $i \ne j$. Furthermore, the fact that $B_1 \cup B_2 \ldots \cup B_n = S$ means that every point in the sample space contained in A must also be contained in one of the B_i's. Therefore, the $A \cap B_j$'s are mutually exclusive events the union of which must be A itself. We can therefore apply (1-6) to reach (1-7).

Another rule arises quickly. Suppose that $A \subset B$. (Recall that $\subset$ means "is contained in.") Then $A \cap B = A$, meaning that:

$$\text{If } A \subset B \text{ then } P(A \cap B) = P(A) \qquad (1\text{-}8)$$

Kolmogorov's axioms (1-1) → (1-4) impose requirements that any viable assignment of probabilities must meet. But they do not imply what particular assignment is appropriate in a given situation. For example, if the sample space consists of exactly two disjoint points 1 and 2, then assigning 70% probability to point 1 and 30% probability to 2 would satisfy Kolmogorov. But so would assigning 43% probability to 1 and 57% to 2. We need to supplement the Kolmogorov axioms in some defensible way that favors some assignments over others. But *how* to supplement those axioms is a subject that has provoked disagreement.

Three Perspectives on Setting Probabilities

As is ably discussed in deGroot and Schervish* (2002), there are three major interpretations of probability that have been proposed: subjective, classical, and frequency. The authors note that "no single scientific interpretation of probability is accepted by all statisticians, philosophers, and other authorities (because) ... each interpretation of probability that has been proposed by some authorities has been criticized by others." For these reasons, the authors conclude that "the true meaning of probability is still a highly controversial subject." Here we make no attempt to resolve the controversy, but instead offer something of a mixture of the three interpretations to assign probabilities in particular circumstances.

Subjective Probability and Equally Likely Events

The *subjective* concept of probability entails the use of a numerical scale to reflect varying degrees of likelihood. On a scale from 0 to 1, under which 0 reflects impossibility and 1 reflects certainty, a person chooses a number that reflects her belief about whether event A will occur. The number selected is considered her *subjective probability* of event A. Asked the question "which would surprise you more: if A occurred or if A didn't occur?" she would be told to choose 1/2 if her answer was "neither outcome would surprise me more than the other." (In that case, she deems *A equally likely* to occur as not to occur.) The choice of 3/4 would suggest considerable surprise if A doesn't happen, while the choice of 1/8 would imply even greater surprise if A did happen. Of course, different people might assign very different subjective probabilities to the same event: one person might estimate that there is a 60% chance that Hillary Clinton will be elected President of the United States in 2016, while someone else assigns probability 25% to her election.

The notion of equal likelihood in subjective probabilities is assumed to extend to any two mutually exclusive events. Two mutually exclusive events A and B are considered *equally likely* if, when asked "which would surprise your more: having A occur or having B occur?," the person assigning probabilities would again answer "neither outcome would surprise me more than the other." (We will not engage in deep probing of what we mean by "surprise," but will rather assume that we have an intuitive understanding of the concept.)

Of course, subjective probability assessments are required to be consistent with Kolmogorov's axioms. It simply would not do, for example, if the probabilities assigned to three mutually exclusive events added up to 1.5.

A Classical Axiom about Equally Probable Events

In classical probability, some prior understanding of the words "equally likely" is assumed to exist, perhaps tied to the concept of "surprise" in connection with

---

* M. deGroot and M. Schervish, (2002) *Probability and Statistics*, Addison-Wesley

subjective probability. It is postulated that equally likely events will share the same numerical probability:

Equal Probability Axiom:

If two mutually exclusive events A and B are equally likely, then

$$P(A) = P(B) \tag{1-9}$$

Coupled with Kolmogorov's axioms, (1-9) allows the assignment of probabilities in some situations. For example, we would presumably treats heads (event A) and tails (event B) as equally likely if a fair coin were tossed, meaning that $P(A) = P(B)$ under (1-9). But because A and B are mutually exclusive and because one of them must occur, they are complementary events. Therefore (1-3) and (1-4) tell us that:

$$P(A \cap B = P(A) + P(B) = 1.$$

But then $P(A) = P(B)$ means that $P(A) + P(B) = 2P(A) = 1 \Rightarrow P(A) = P(B) = 1/2$.

EXAMPLE 1.2

Sixth Sense

If $P(A_j)$ is the chance of getting the number i on the toss of a fair six-sided die for $j = 1,2, \ldots,6$, then what is $P(A_i)$?

Solution

Here the sample space S is the integers from 1 to 6, the only possible outcomes in such a die toss. Because $A_1 \cup A_2 \ldots \cup A_6 = S$, we know from (1-3) that $P(A_1 \cup A_2 \ldots \cup A_6) = 1$, and from (1-6) that $P(A_1 \cup A_2 \ldots A_6) = \sum_{i=1}^{6} P(A_i)$. On a fair die, $P(A_i)$ is the same for all i; if we call the common value p, then $\sum_{i=1}^{6} p = 6p = 1$. *Thus,* $P(A_I) = \frac{1}{6}$ *for* $i = 1, 2, \ldots ,6$.

EXERCISE 1.2

Did you know there is such a thing as a 20-sided die? Take a look. On such a die, what is the probability of a getting a 3?

While the equal probability axiom is stated above for mutually exclusive events, we assume that it can be extended to cases in which A and B are not mutually exclusive. For example, two problems that might arise tomorrow morning are that my car will not start (event A) or that the newspaper will not arrive (event B). These events are not mutually exclusive, but suppose that I am told by a soothsayer that exactly one of these events will occur tomorrow. If I would be at a complete loss to guess which one (i.e., if neither resolution of the soothsayer's prophecy would surprise me more than the other), then I am in effect declaring the two events equally likely and the axiom can be taken to require that $P(A) = P(B)$.

The equal-probability axiom has a limitation: it does not directly tell us what probabilities to assign to two complementary events A and B when we do not think A and B equally likely. Here, however, we might consider something of a synthesis of the equal probability axiom and the subjective concept of probability. Suppose again that A is the event that my car will *not* start the next time I try to use it, and B the complementary event that it will start. Suppose further that I view having the car fail to start as equally likely with getting a 1 when I pick a ball at random from an urn that has 1000 balls, each labeled with a different integer from 1 to 1000. As we know, the equal probability axiom for the urn choice yields a 1/1000 probability of getting a one. And a second application of the equal probability axiom—involving the car and the urn—then obliges me to assign 1/1000 probability to the car failing to start. Given that $P(A \cap B) = P(A) + P(B) = 1$, the choice $P(A) = 1/1000$ implies that $P(B) = 999/1000$. In this way, we might indirectly benefit from the equal probability principle even in cases where its direct use for A and B would be flagrantly inappropriate.

Relative Frequency

Our subjective assignments of probability are sometimes disciplined by the third interpretation of probability, namely, the concept of *relative frequency*. If the experiment that can result in the event can be conducted a large number of times under independent and identical conditions, the *fraction* of times that event A occurs should converge to $P(A)$. In other words, probability should correspond to *the fraction of time the event occurs "in the long run."*

Because of the relative frequency principle, a subjective estimate of probability can sometimes be shown to be incorrect. For example, suppose that someone estimates that a coin will come up heads with probability 7/10. Then if the coin is tossed a million times and it comes up heads 50% of the time, the estimate of 7/10 is contradicted by long-term experience. (More precisely, either one disavows the original estimate or winds up disavowing some of the axioms of probability.) Of course, it is not always possible to conduct lots of experiments to determine long-term outcomes. If the issue is the probability that Hillary Clinton will be elected the US President in 2016, the idea of arranging ten million independent elections and observing the fraction in which Hillary Clinton wins does not exude realism.

EXAMPLE 1.3 Rain Man

To assess the accuracy of a weatherman's predictions, five years of his forecasts are assembled. It is discovered that there were 120 days on which he said that there was a 40% probability of rain the next day. In fact, it rained on the following day on 26 occasions. Does that outcome reduce confidence in his 40% forecasts? Discuss briefly.

Solution

Well, yes. 40% of 120 is 48, so the relative frequency principle suggests that the number of subsequent days with rain should have been about twice as large as the 26 observed. Because 26/120 = .217, he would have had better overall accuracy had he had estimated the chance of rain as 20% on each of those 120 occasions.

EXERCISE 1.3 Let's pursue the two tosses of a fair coin mentioned in Example 1.1.

(i) Taking account of what elementary events are possible (return to Example 1.1 if your memory needs refreshing), what is the probability of getting exactly one head?
(ii) Exactly two?
(iii) Exactly zero?
(iv) If you had to guess how many heads would come up, would guess would you make?
(v) What is the probability your guess would be correct?
(vi) Does this example suggest that relative frequency is not a reliable guide to the outcome in the short run? Briefly discuss.

Not everyone would accept this particular approach to mixing various interpretations of probability; as noted, the topic evokes a controversy that has gone on far longer than our lives. Once probabilities have been assigned to elementary events, however, there is no controversy about how to use those probabilities to determine the probabilities of more complex events.

We need to introduce one more fundamental probability concept: the idea of conditional probability.

1.2 Conditional Probability

The ***conditional probability of* B *given* A** is denoted $P(B|A)$ and is defined by the rule:

$$P(B|A) = \frac{P(A \cap B)}{P(A)} \tag{1-10}$$

The intuitive interpretation of $P(B|A)$ is as the probability we assign to B knowing that A has occurred. Event B originally had probability $P(B)$, but the information that event A occurred in the experiment could justify revising the probability of event B, and (1-10) indicates how we should do so.

For example, suppose a fair six-sided die is tossed, and event A and B follow:

A: *getting an even number*

B: *getting a four*

If we learn that the toss resulted in an even number (but nothing more about the outcome), how should we adjust the probability of event B? The initial probability of B (i.e., $P(B)$) is 1/6. Because $P(A \cap B) = \frac{1}{6}$ and $P(A) = \frac{1}{2}$, (1-10) tells us that $P(B|A) = \frac{1}{3}$. That revision of the probability makes sense: Given that A occurred (an even number), the outcome of the toss must be a 2, 4, or 6. Because these three outcomes initially were equally likely—and because knowing that the toss came up even does not change that equality—the probability assigned to B should go from 1/6 to 1/3. Under (1-10), it does.

We can derive (1-10) under the relative frequency concept of Probability. Over a very large number of experiments (N), suppose that N_A of them result in event A and $N_{A\cap B}$ result in both A and B. Then, when A arises, the fraction of time that B arises in the long run is approximated by $N_{A\cap B}/N_A$. For example, if N = 100,000, N_A = 60,000 and $N_{A\cap B}$ = 20,000, then $P(B|A) \approx N_{A\cap B}/N = 1/3$. But the frequency interpretation of probability also entails the approximations $P(A) \approx N_A/N$ and $P(A \cap B) \approx N_{A\cap B}/N$. Therefore, $P(B|A) \approx N_{A\cap B}/N = (N_A/N)/\ (N_{A\cap B}/N)$, which becomes $P(B|A) = \frac{P(A \cap B)}{P(A)}$.

EXAMPLE 1.4

Another Die Is Cast

A fair six-sided die is tossed twice. Among the events that might arise are:

A: *the first toss yields an even number*

B: *the second toss yields an odd number*

C: *the sum of the two outcomes is even*

Find $P(A)$, $P(B)$, $P(C)$, $P(B \cap C)$, $P(C|B)$, $P(C|A \cap B)$, and $P(B|C)$

Solution

The long way to do this is to list the $6 * 6 = 36$ possible (i,j) pairs that can result from the tosses, assign probability 1/36 to each, and then add the probabilities for all the mutually exclusive pairings that lead to each event. But we can save time if we recognize that the successive outcomes fall into four categories on the odd/even dimension:

O_{12}: (Both Odd) O_1E_2:(Odd then Even) E_1O_2:(Even then Odd) E_{12}: (Both Even)

All these categories are equally likely: there are three odd and three even possibilities on each toss (and we are assuming that the first outcome does not influence the second). Thus $P(O_{12}) = P(O_1E_2) = P(E_1O_2) = P(E_{12}) = 1/4$

As for the events of interest:

$P(A) = P(E_1O_2 \cup E_{12}) = P(E_1O_2) + P(E_{12}) = 1/2$ (because E_1O_2 and E_{12} are mutually exclusive)

$P(B) = P(E_1O_2) + P(O_{12}) = 1/2$

$P(C) = P(O_{12}) + P(E_{12}) = 1/2$

$P(B \cap C) = P(O_{12}) = 1/4$

$P(C|B) = P(B \cap C)/P(C) = 1/2$

$P(C|A \cap B) = P(C \cap (A \cap B))/P(A \cap B) = 0$ because C and $A \cap B$ are disjoint subsets of the sample space, meaning that $P(C \cap (A \cap B)) = 0$.

$P(B|C) = P(B \cap C)/P(B) = 1/2$

All these probabilities make sense, and we could have worked them out in our heads without formal use of conditional probability. But it is comforting that use of (1-10) yields the answers we knew were correct on other grounds. Rest assured, however, that we will be using (1-10) in lots of cases where we cannot work out answers in our heads.

EXERCISE 1.4 In Example 1.4, work out $P(A \cap B)$, $P((A \cap B)|C)$ and $P((A \cap B)|C^c)$ where C^c is the complement of C.

Independent Events

Two events A and B are said to be ***independent*** if:

$$P(B|A) = P(B) \tag{1-11}$$

In other words, knowing that A has occurred does not change our assessment of the likelihood of B.

When A and B are independent, (1-11) implies a simple product rule:

$$P(B|A) = \frac{P(A \cap B)}{P(A)} \rightarrow P(B) = \frac{P(A \cap B)}{P(A)} \rightarrow P(A \cap B) = P(A)P(B)$$

If A and B are independent events, then

$$P(A \cap B) = P(A)P(B) \tag{1-12}$$

If A is independent of B, does that mean B is independent of A? Yes. We know that:

$$P(A|B) = \frac{P(A \cap B)}{P(B)}. \text{ But if } P(A \cap B) = P(A)P(B), \text{ then } P(A|B) = P(A).$$

In consequence, *independence is a two-way street.*

EXAMPLE 1.5

Backed Up

Many kinds of equipment have the feature of **redundancy**, under which backup systems take over certain functions when the primary systems fail. Such arrangements are common with respect to a hospital's electrical supply: if the usual power source fails, secondary generators are supposed to come on-line at once. (Obviously, it is hoped that whatever caused the primary system to fail does not simultaneously cripple the backup system.) Consider two events about a piece of "redundant" equipment for which both the primary and backup systems are up and running now:

A_1: *the primary system fails in the next hour*

A_2: *the backup system fails in the next hour*

Let $p_i = P(A_i)$ for $i = 1,2$.

Suppose that a study of historical data about the equipment leads one to hypothesize that:

(i) p_1 and p_2 are exactly the same, both of them sharing the common value p.
(ii) A_1 and A_2 are independent events.

Statements (i) and (ii) constitute a *probabilistic model* about the equipment's reliability. Assume that this model is valid and consider five events:

B: *both systems fail in the next hour*

C: *the backup system does not fail over the next hour*

D: *the primary system fails in the next hour but not the backup*

E: *exactly one of the two systems fails in the next hour*

F: *at least one system fails in the next hour*

What is the probability of each of these events?

Solution

$P(B)$ (both systems fail)

B is the same as the event $A_1 \cap A_2$. Because the A_i's are independent, we can use (1-11) to write:

$$P(B) = P(A_1 \cap A_2) = P(A_1)P(A_2)$$

Exploiting the fact that $P(A_i) = p$ for each i, we simplify the last equation to $\mathbf{P(B) = p^2}$.

$P(C)$ (backup system does not fail)

As defined, the events C and A_2 are complementary. Thus, we can use (1-5) to write:

$$\mathbf{P(C)} = 1 - p$$

$P(D)$ (the primary system fails but not the backup)

D is the joint event A_1 and C. To find $P(D) = P(A_1 \cap C)$, we must face a question: While we know that A_1 and A_2 are independent, can the same be said of A_1 and C? The answer is yes because, if the failure of the primary system does not affect the chance that the backup fails, it must likewise leave untouched the chance that it does *not* fail. Therefore:

$$P(D) = P(A_1 \cap C) = P(A_1)P(C) = \mathbf{p(1-p)}$$

$P(E)$ (exactly one of the systems fails)

Event D is only one way to get exactly one failure: it is also possible that that the backup system fails but not the primary one. If we denote this latter event by Y, then $P(E) = P(D \cup Y)$. Clearly, D and Y are mutually exclusive, so $P(E) = P(D \cup Y) = P(D) + P(Y)$. We know that $P(D) = p(1 - p)$; identical arguments establish that $P(Y)$ also equals $p(1 - p)$, which means that:

$$P(E) = P(D) + P(Y) = p(1-p) + p(1-p) = \mathbf{2p(1-p)}$$

$P(F)$ (at least one system fails)

Here we can use the earlier findings about B and E to write:

$$P(F) = P(B \cup E) = P(B) + P(E) = p^2 + 2p(1-p)$$

EXERCISE 1.5 Suppose that p is the probability that the primary system in Example 1.11 fails in the next hour, and q is the corresponding probability for the backup system ($p \neq q$). Again, failures of the two systems are independent events. Find the probabilities of events $B \Rightarrow F$ in terms of p and q.

Bayes' Theorem

In using (1-10), we assume that $P(A) > 0$. If $P(A) = 0$, after all, the concept of $P(B|A)$ is meaningless. (If there are no such things as unicorns, we should not worry about what fraction of them drives Toyota Corollas.) When $P(A) > 0$, we know from (1-10) that:

For any two events A and B:

$$P(B|A) = \frac{P(A \cap B)}{P(A)} \tag{1-13}$$

But another use of (1-10) assuming $P(B) > 0$ yields:

$$P(A|B) = \frac{P(A \cap B)}{P(B)} \tag{1-14}$$

(1-14) can be rewritten as:

$$P(A \cap B) = P(B)P(A|B) \tag{1-15}$$

Substituting (1-15) into (1-13) yields:

Bayes' Theorem:

$$P(B|A) = \frac{P(B)P(A|B)}{P(A)} \tag{1-16}$$

The practical value of that formula might seem dubious on the grounds that, unless one knew $P(B|A)$ already, one might seem unlikely to know all three of the quantities needed to deduce it from (1-16). But things are not always what they seem: It turns out that Bayes' Theorem is one of the most powerful results in Probability and Statistics.

The great advantage of the Bayesian formula (1-16) is that it allows the revision of probabilities in the light of new information (i.e. from $P(B)$ to $P(B|A)$). Quite often, a variety of indicators about the likelihood of an event (e.g. that a particular geological basin contains oil) arise at different times. One wants to include the latest findings in the probability estimate while not diminishing too much the weight accorded earlier developments. Bayes' Theorem provides for a natural and accurate synthesis of all of the facts already amassed.

EXAMPLE 1.6

Bayes' Hit?

(Sorry) If $P(A)$ and $P(B) > 0$, does $P(B|A) = P(A|B)$ mean that $P(B)$ must equal $P(A)$?

Solution:

Well, yes and no. First, suppose that $P(B|A) > 0$. From Bayes Theorem, we have $P(B|A) = \frac{P(B)P(A|B)}{P(A)}$. Therefore, the equality of $P(B|A) = P(A|B)$ implies that the ratio of $P(B)$ to $P(A)$ is one. Hence, $P(A)$ and $P(B)$ must be equal.

But suppose that A and B are mutually exclusive, in which case $P(B|A) = P(A|B) = 0$. There is no requirement that $P(A)$ and $P(B)$ be equal just because the events are mutually exclusive. And this outcome does not contradict Bayes' Theorem, which simply becomes $0 = 0$ in this case.

EXERCISE 1.6

If $P(B|A) = 1$ and $P(A) = 3/5$, do we have enough information to determine $P(B)$? Either explain why we do or give two examples that show that $P(B)$ can take on different values even with this information.
Also, is it possible that $P(B) = 1/2$ with this information? Discuss briefly.

The Joint Probability Law

Equation (1-15) can be generalized beyond two events A and B. If we have three events A, B, and C, then we can write $D = A \cap B$ and then write:

$$P(A \cap B \cap C) = P(D \cap C) = P(D)P(C|D)$$

But (1-15) also implies that $P(D) = P(A \cap B) = P(A)P(B|A)$

Therefore:

$$P(A \cap B \cap C) = P(A)P(B|A)P(C|A \cap B) \tag{1-17}$$

We have in (1-17) a three-event generalization of (1-15). And we can use this reasoning again to get a four-event generalization and, through applying the reasoning recursively, to get a rule that applies for k events $A_1, A_2, \ldots A_k$, which we call the joint probability law:

Joint Probability Law:

$$P(A_1 \cap A_2 \ldots \cap A_k) = P(A_1)P(A_2|A_1)P(A_3|A_1 \cap A_2) \ldots P(A_k|A_1 \cap A_2 \ldots \cap A_{k-1}) \tag{1-18}$$

The joint probability law is basically a "sudden death" rule: To get all events $A_1 \rightarrow A_n$, first A_1 must occur: otherwise we have already failed. Given that A_1 occurred, we require that A_2 also occurred; if not, the k-event sequence is again doomed. Then we must ask about A_3 given A_1 and A_2, and so on. Equation (1-18) is the probability of surviving all n stages of this discovery process.

If the A_j's are independent—in which case $P\left(A_j \middle| A_1 \cap A_2 \ \dots \ \cap A_{j-1}\right) = P\left(A_j\right)$ then (1-18) becomes a simple product rule:

If $A_1 \rightarrow A_n$ are independent events, then:

$$P\left(A_1 \cap A_2 \ \dots \ \cap A_k\right) = P\left(A_1\right)P\left(A_2\right) \ \dots \ P\left(A_k\right) \tag{1-19}$$

We have now stated all the basic laws of Probability that we will need in this book. There can be debate about which specific formulas should be called "laws." Kolmogorov's Axioms, the equal probability axiom, and the conditional probability rule (1-10) might be viewed as the fundamental laws, because all the others we presented are derived from them. As noted, however, people speak of the Law of Alternatives and the Joint Probability Law. But there is no point engaging in semantic quibbles, for we have work to do to gain greater familiarity with the principles we have discussed. To move towards that familiarity, we now employ the principles in some practical exercises.

1.3 Some Probability Problems

EXAMPLE 1.7 Length of Life

According to the *US National Center for Health Statistics*, a baby born in the US in 2004 would have an 88.0% chance of living until age 60 and a 53.9% chance of living until age 80. Based on these data:

- If a baby born in 2004 actually lives until age 60 (in 2064), what is the probability that baby will live until age 80?
- What is the probability that an American baby born in 2004 will have a lifespan between 60 and 80 years?

Solution

To answer the first question, we note that it really is an exercise in conditional probability. Suppose we define events A and B by:

A: *the baby lives until age 60*

B: *the baby lives until age 80*

The question is asking for $P(B|A)$.

As always, we can cite (1-10) and write: $P(B|A) = P(A \cap B)/P(A)$. $P(A) = .88$ given the data. As for $P(A \cap B)$, we note that everyone who reaches age 80 has previously reached age 60 (i.e. that $B \subset A$), meaning that $P(A \cap B) = P(B)$ under (1-8). But $P(B) = .539$ given the data. Hence $P(B|A) = P(A \cap B)/P(A) = (.539)/(.88) = \mathbf{.613}$.

Is this last outcome surprising? At birth, the baby has a 53.9% chance of reaching age 80. But if the person is still alive in 2064, the estimated chance of reaching age 80 only increases to 61.3%. One might have expected a greater increase than that: 60 years is 3/4 of 80 years, and making it 3/4 of the way might seem to increase sharply the chance of making it all the way. The problem with that reasoning is that, because $P(A) = .88$, the first sixty years that have passed are years of low death risk: indeed, the twenty years between ages 60 and 80 entail far higher risk than the first 60 years combined. A person who reaches 60, therefore, has not avoided 3/4 of the risk of not reaching 80, but instead a far lesser proportion.

Which brings us to the second question: what is the chance the baby's lifespan will be between 60 and 80? The temptation is to take the drop between the chance of reaching 60 and the chance of reaching 80, meaning $.88 - .539 = .341$. But, in Probability, we should be wary of yielding at once to our temptations. Instead, we might define event C by:

C: *lifespan between 60 and 80*

Let A^c and B^c be the *complements* of events A and B respectively. Thus, A^c is the chance of *not* reaching 60 and B^c of not reaching 80.

We can then write:

$$P(B^c) = P(A^c) + P(C)$$

There are, after all, two mutually exclusive ways of *not* reaching age 80: dying before age 60, or dying between 60 and 80.

We have the equivalent equation:

$$P(C) = P(B^c) - P(A^c)$$

We know that $P(B^c) = 1 - P(B) = 1 - .539 = .461$, while $P(A^c) = 1 - .88 = .12$. It follows that $P(C) = .461 - .12 = .341$. In this instance, therefore, the simple subtraction rule was correct! No one is saying that our gut instincts always lead us astray; they do so often enough, however, that we do well to work out the answer rather than proclaim that it is obvious.

Of course, these statistics are based on longevity patterns that prevailed in 2004. The chance that someone who reached 50 would reach 51, for example, might be estimated by the fraction of Americans who turned 50 in 2003 who were still alive one year later. Advances in medicine over the next half century could increase the likelihood of reaching 80 considerably, meaning that the 59.3% estimate could well be pessimistic. Any probabilistic forecast involves assumptions, and one does well to find out what they are before taking the forecast too seriously.

EXERCISE 1.7 Data suggest a 20.8% probability that an American baby born in 2004 will be alive 90 years later. Given that information and the data in Example 1.7, what is the conditional probability that a 2004 baby who lives to age 80 will make it to age 90?

EXAMPLE 1.8

The Birthday Problem

One of the classic probability problems asks: if M people are assembled in a room, what is the chance no two of them share the same birthday? Approximately what is the answer?

Solution

This calculation starts off in a routine enough way but it ends on a note of great surprise.

To solve this problem, one must make some assumptions. Here we make the two standard ones:

- There is no prior relationship among the various birthdays (i.e., that we are not (say) at a convention of twins).
- Birthdays are uniformly distributed over the year. (This assumption means that, neglecting Leap Year, the chance of being born on any particular date is 1 in 365.)

Suppose that one went around the room asking each of the people his or her birthday. Let A_i be the event that the ith person questioned mentions a date that has *not* come up before, and let C be the event that all M people have different birthdays. Then

$$P(C) = P(A_1 \cap A_2 \ldots \cap A_M) = P(A_1)P(A_2|A_1) \ldots P(A_M|A_1 \cap A_2 \ldots \cap A_{M-1}) \qquad (1\text{-}20)$$

The quantities on the right-hand side of (1-20) can readily be deduced. Obviously, $P(A_1) = 1$, because the first person asked cannot repeat a date that came up previously. The second respondent will announce a new date *unless* he was born the same day as the first; under uniformity, the chance of such overlap is 1/365 and thus $P\left(A_2|A_1\right)1 = 1 - \frac{1}{365} = \frac{364}{365}$.

To find $P(A_3|A_1 \text{ and } A_2)$, we note that if A_1 and A_2 have both occurred, there are *two* dates the third person must avoid to prevent repetition. The chance of doing so is 363/365. To find $P(A_k|A_1 \cap A_2 \ldots \cap A_{k-1})$, we note that the kth person must avoid the birthdays of the previous $k - 1$, which are all assumed to be different. We therefore reach the formula:

$$P\left(A_k|A_1 \cap A_2 \ldots \cap A_{k-1}\right) = \frac{365 - (k-1)}{365}$$

We can therefore express $P(C)$ in (1-20) by:

$$P(C) = 1 * \left(\frac{364}{365}\right) * \left(\frac{363}{365}\right) \dots \frac{365 - (M-1)}{365}$$
$$= \left(\frac{365}{365}\right) * \left(\frac{364}{365}\right) * \left(\frac{363}{365}\right) \dots \frac{366 - M}{365}$$
$$\text{or } P(C) = \frac{365 * 364 * 363 * \dots * (366 - M)}{365^M}$$

This is an opportune time to remind readers about the concept of factorials. For any positive integer k, the quantity written as k! and spoken as k-factorial is the product of all the integers from 1 to k. Thus 3! = 6 and 4! = 24. With this definition, the quotient $j!/k!$ for $j > k$—which includes all whole numbers from 1 to k in both numerator and denominator—is simply the product of all the integers between $k + 1$ and j. Bearing this last statement in mind, we can rewrite (1-21) as:

$$P(C) = \frac{(365!/(365 - M!))}{365^M} \tag{1-22}$$

From the viewpoint of calculation (1-22) is often friendlier than it seems, for there are superb simple approximations for $k!$ when k is very large.

The birthday problem attains its zest when (1-22) is evaluated at particular values of M. When M is as small as 25, the probability of *no* shared birthdays is only .48. When $M = 60$, the probability drops to 1 in 170. And for 100 individuals, the chance that all have different birthdays drops to 1 in 3 *million*.

Most people find such statistics astounding. When $M = 60$, there are six times as many dates as individuals, leaving plenty of room for the people to spread out their birthdays. (Indeed, each one could take up nearly a week.) But while birthday overlap is thus wholly unnecessary, the calculation tells us it is not just likely but almost inevitable.

Nor is the outcome some peculiar consequence of the uniform-birthdays assumption. One can show rigorously that the chance of birthday overlap is actually *lower* under this hypothesis than any other. The reason is that an alternate assumption must involve clumping of birthdays at certain times of the year, a circumstance that would make duplication even more probable.

A Post Mortem

As we think about the problem a bit more, however, the answer starts to appear less preposterous. Suppose that someone enters a room where 59 others are present (i.e., $M = 60$) and wonders whether any were born on the same day as she. Because, collectively, their birthdays cover roughly a sixth of the year, the chance that her birthday is shared by someone else is something like 1 in 6.

Likewise, each of the other 60 people in the rooms has roughly a 1 in 6 chance of sharing a birthday with someone. But that circumstance reminds us of a analogous situation. One in six is also the chance that a fair die comes up three. If such a die is cast 60 times, the probability that it *never* comes up three is extremely low. (We would expect roughly 60 ∗ (1/6) = 10 threes to arise.) In the same way, the chance that all 60 people avoid the 1 in 6 risk of birthday overlap should also be very low.

The comparison just made is not perfect: different tosses of a die are independent, while the birthday overlaps of different people are not (e.g. if one person shares his birthday with 12 others, each of them must be in the same situation). But the similarity with the die-toss example is more powerful than the difference. The point is that for each of 60 people to avoid a "collision" with all 59 of the others, *thousands* of potential collisions must be avoided. Doing so takes a lot of luck. The actual answer is telling us that, somewhere along the line, this luck is likely to run out.

EXERCISE 1.8 In a group of 80 people, what is the chance of no birthday overlap? Numerical answer!

EXAMPLE 1.9

Strong Stuff?

When testing the effectiveness of a new drug, pharmaceutical companies do not simply give the drug to some disease-sufferers and observe the proportion that improves. For one thing, some of these sufferers might have gotten better on their own; for another, the very act of taking medicine can lead a patient to perceive improvements whether or not they truly occur. (This phenomenon—known as the *placebo effect*—may be especially common if the malady is in part psychological.)

For these reasons, a typical drug-effectiveness study will divide the disease-sufferers at random into two groups. Half are given the new pill, while the other half is given a harmless but useless substitute (called the *placebo*). None of those taking part in the experiment know what type of pill they are receiving. Then, the patient-reported success rate for the new drug is compared with that for the placebo, allowing an assessment of the drug's true potency.

The results of such clinical trials are sometimes given wide publicity. A *Time* magazine ad for Viagra, for example, stated that 80% of those in an experiment who used Viagra reported improvement, as compared to 25% among those who had the placebo.

What do these statistics imply about the effectiveness of Viagra?

Solution

The temptation is to subtract 25 from 80, and thus to estimate Viagra's true effectiveness as 55%. Such a calculation, however, is harsh on Viagra.

Among the 80% of Viagra users who reported improvement, some physically benefited from the drug (the genuine beneficiaries), while the remainder reported improvement for other reasons (such as the placebo effect). Thus, the genuine effectiveness of Viagra is less than 80%. But how much less? If q is the fraction of users genuinely helped by the drug, then 80% is roughly the sum of two quantities:

- q
- 25% of $1 - q$

Why the second quantity? We know that 25% of those who got the placebo felt better. It seems reasonable to posit that among those who used Viagra but did *not* physically benefit—namely, those for whom Viagra was *in effect* a placebo—roughly 25% would likewise report improvements. The 80% of Viagra users who gained, therefore, included all of those directly helped plus about 25% of the rest.

These observations imply the equation:

$$q + .25(1 - q) = .8 \tag{1-23}$$

From simple algebra on (1-23), we see that $q \approx .73$. The 80% of Viagra patients who reported gains included the 73% with genuine benefits (User → Gain → Report Gain) and about 1/4 of the remaining 27% without such benefits (User → No Gain → Report Gain), a path with probability $.27 * .25 \approx .07$. Of course, 73% is considerably higher than the 55% success rate readers of *Time* might have inferred from subtracting 25% from 80%.

Some readers might wonder whether the Viagra success rate should be estimated as 80% rather than 73%, on the grounds that the 7% of users who erroneously believed that they gained because of the drug (i.e., took the path User → No Gain → Report Gain) did derive *some* benefit, whatever its origin. The problem with this view is that placebo effects are often temporary, meaning that the 7% boost because of "mind over matter" might not persist.

EXERCISE 1.9 Among those who have a meal that Mendel cooked, all those who found it delicious will tell him so, as will 40% of those who did not find it delicious (out of politeness). If 76% of the guests tell Mendel that the meal was delicious, what percentage actually thought it was?

1.4 Discrete Probability Distributions

Some applications of the laws of probability are of such widespread practical importance that, rather than "reinvent the wheel" every time they come up, it is worth exploring them in detail and devising general formulas. In this section, we introduce the potent concepts of probability distribution and random variable, paying special attention to several widely-used distributions.

Random Variables

We start with a definition:

A **random variable** is a mathematical function that assigns a number to each elementary event in the sample space.

(To put it more mathematically, a random variable is a "mapping" from the sample space to the real numbers.) Different elements of the sample space are not necessarily assigned different numbers: all that is required is that each elementary event receives a unique number. The specific number that the random variable yields—based on which elementary event occurred—is called the **realization** of the random variable.

Typically, a random variable is designated by a capital letter like X, while a particular realization is denoted by a small letter like x. The quantity $P(X = x)$ is the probability that the realization of X is x.

EXAMPLE 1.10 Fit to Eat?

There are 12 sealed cups of strawberry yogurt at a supermarket, and Minerva is about to pick two at random to buy. Unbeknownst to her and the store owner, three of the cups contain yogurt that is sour and inedible. Random variable X assigns to each selection that Minerva might make the number of sour cups that she buys.

(i) What are the elementary events associated with Minerva's purchase?
(ii) How many such events are there?
(iii) How many such events involve two sour cups? Exactly one sour cup?
(iv) What are the realizations of X, and what is the probability associated with each one?

Solution

(i) To make things easier, we number the 12 cups from 1 to 12 based on their locations in the display (e.g., left to right). The elementary events can denoted by (i,j), where i and j are the two cups that Minerva selects $(j > i)$.
(ii) Readers might recall from high school that the total number of distinct "committees" of size k that can be created from a population of size n is given by $\binom{n}{k} = \frac{n!}{k!(n-k)!}$. Here we need the number of "committees" of two yogurt cups that can arise from a total of 12 cups or, more formally, the total number of (i,j) combinations when $1 \le i < j \le 12$. That number is $\binom{12}{2} = \frac{12 * 11}{2} = 66$.

Given that Minerva chose two cups at random, all 66 combinations are equally likely.

(iii) Suppose that i_1, i_2, and i_3 are the numbers of the three cups with sour yogurt ($i_3 > i_2 > i_1$). There are only three combinations of two cups that are both sour: (i_1, i_2), (i_1, i_3), and (i_2, i_3). But there are 27 combinations in which exactly one of the two cups is sour, namely, i_1 and one of the nine cups that are OK, i_2 and one of the nine, or i_3 and one of the nine.

(iv) Random variable X has three possible realizations: zero, one, or two. The number of (*i*,*j*) pairs in which both cups are OK is 66 – 3 – 27 = 36. Each of the distinct (*i*,*j*) combinations has a 1/66 chance of arising. We therefore have the chart:

Realization of X (x)	Probability of Occurrence
0	36/66
1	27/66
2	3/66

EXERCISE 1.10 Suppose that there was only one cup with sour yogurt in Example 1.10. Answer parts (i)–(iv) of Example 1.11 under this new condition.

Shorthand for a Random Variable

Often, in a situation such as Example 1.10, we will see wording like "let random variable X be the number of sour cups that Minerva buys." As we know, however, the random variable is the *function* that assigns to her selection the number of sour cups it includes, while the actual number she buys is the *realization* of the random variable. In other words, the shortened description conflates the realization of the random variable with the random variable itself. Yet the description does achieve brevity, and it makes clear how the random variable in question was defined. For these reasons, *we will adopt this shorthand in the remainder of this chapter and beyond.*

Random variables can generally be classified as **continuous** or **discrete**.

Discrete random variable
A discrete random variable *X* assigns the elementary events in the sample space to a discrete set of (separate) points on the real-number line.

Continuous random variable
A continuous random variable *X* assigns the elementary events to an entire interval (or intervals) of real numbers.

The random variable discussed in Example 1.10 is discrete. A random variable that reflects the future temperature at some time and place (i.e., which maps the temperature t to the real line under the rule $X = t$) is continuous, because the temperature itself could be not only (say) 32°F or 33°F but any real number between 32 and 33. We focus on discrete random variables in the next few sections and then turn to continuous random variables

Discrete Probability Distributions

A listing of each possible realization of a discrete random variable and the probability that it will arise is called the **probability mass function** of the random variable.

A discrete mass function distribution takes the form:

$$X = \begin{cases} a_1 & \text{w.p. } p_1 \\ a_2 & \text{w.p. } p_2 \\ & \cdots \\ a_M & \text{w.p. } p_M \end{cases} \qquad \text{where w.p. = with probability} \tag{1-24}$$

We will follow the convention that the a_i's are listed in *ascending* order. In (1-24), it must be the case that:

$$p_j > 0 \text{ for all } j, \text{ and that } \sum_{j=1}^{M} p_j = 1$$

The number of possible realizations of a discrete random variable need not be finite (i.e., M could be ∞). For example, suppose that a fair coin is to be tossed, and let random variable X be the number of tosses until "heads" first occurs. (Again, we are using shorthand: X maps each possible sequence into the number of tosses until the first head comes up, while the actual number of tosses is the realization of X.) There is no finite upper limit on X here: for any positive integer k, the probability is $(1/2)^k$ that the first k tosses are all "tails," meaning that X will exceed k. The quantity $(1/2)^k$ obviously gets very small as k gets very large, but it never reaches zero when k is finite.

In many instances, one can express the probability mass function more succinctly than with a huge chart like (1-24). For example, it might be the case that $a_k = k$ for $k \geq 1$ while p_k is some simple function $g(k)$. When that happens, the situation can be described in a single line: $P(X = k) = g(k)$ for nonnegative integer values of k.

Summary Statistics for a Discrete Random Variable

The probability mass function for a discrete random variable tells the variable's full story, but that can be a problem. If the variable can take on (say) 300 values with 300 different probabilities, there is the danger of "information overload" that creates more confusion than enlightenment. To circumvent this problem, it has become customary to compute a few **summary statistics** about random variables that provide useful first impressions about their behavior.

The Mean

Perhaps the most common such statistic is the **mean**, which is also called the *average*, the *expected value*, and the *expectation*.

For discrete random variable X with a probability mass function of form (1-24), the **mean** μ is defined by:

$$\mu = \sum_{j=1}^{M} a_j p_j \tag{1-25}$$

The mean is also written as $E(X)$ or sometimes μ_X (to make clear that X is the random variable under discussion).

Thus defined, the mean is a weighted average of the values that X can take on, with the weight of each value being the probability assigned to it. The mean is often considered a measure about the middle of the distribution. By its very definition, the mean must fall between the lowest X-value of a_1 and the highest a_M. But the mean need not be halfway between a_1 and a_M: it can tilt towards one or the other depending on the locations and probabilities of the intermediate a_j's.

EXAMPLE 1.11 **Mean Feat**

Find the mean for X, which is the outcome for the toss of a fair six-sided die.

Solution

For a fair six-sided die, Equation (1-25) yields a mean μ that follows:

$$\mu = (1/6) * 1 + (1/6) * 2 + \ldots (1/6) * 6 = (1/6) * (1 + 2 + 3 + 4 + 5 + 6) = 3.5.$$

EXERCISE 1.11 Remember the 20-sided fair die? Find the mean outcome when such a die is tossed.

An Intuitive Interpretation of the Mean

We will get to the Law of Large Numbers shortly, but should now make clear what the law implies. Suppose that the experiment that generates random variable X can be repeated, independently and under identical conditions. Then the arithmetic average of the individual outcomes for X will approach μ as the number of repetitions gets very large. In other words, the mean corresponds to the long-term average of x-outcomes (i.e., of realizations of X).

We can see the correspondence between mean and long-term average if we turn to the fair six-sided die. We have calculated its mean outcome as 3.5. If the die is tossed a very large number of times (with its molecular structure unaffected by the repeated tossing), then we would anticipate about equal numbers of 1's and 6's, and these outcomes will average out to 3.5. We would likewise expect equal numbers of 3's and 4's, which average 3.5, and of 2's and 5's, again averaging 3.5. Since the tosses yield three groups of outcomes all of which average to 3.5, it is no surprise that the overall long-term average is itself 3.5 (i.e., the mean).

The Variance and Standard Deviation

While the mean conveys some sense about the middle of a distribution, it does not reveal how much the individual outcomes spread around the middle. Suppose that X and Z represent the outcomes of two different investments, and that X must equal 60 (i.e., $X = 60$ w.p. 1), while Z is equally likely to be 0 or 120. Both these random variables have mean 60 under (1-25), but they are far from identical. Random variable X represents a sure thing, while Z reflects a "double or nothing" bet with 60 as the baseline value. Opinions can differ about which of the opportunities is more attractive, but few investors would be indifferent between X and Z.

To reflect the spread of a discrete distribution around its mean μ, it is common to calculate the **variance** σ^2 of the random variable.

The **Variance** σ^2 of Random Variable X that has probability mass function given by (1-24) follows:

$$\sigma^2 = \sum_{j=1}^{M}\left(a_j - \mu\right)^2 p_j \qquad (1\text{-}26)$$

The variance squares the difference between each a_j and the mean μ, and then weights that squared quantity by the probability that X equals a_j. It is no surprise that variance is described as "mean squared spread around the mean."

A difficulty with the variance is that it is not measured in the same units as the mean. If x (and thus μ) is expressed in dollars, for example, then σ^2 is expressed in "squared" dollars; if the mean is in degrees Centigrade, then the variance is in "squared" degrees Centigrade. Neither of these squared statistics has a reasonable interpretation. To make the measure about the spread of X both more plausible and more compatible with the measure of its "middle" value, people have taken to emphasizing the **standard deviation** of x, which is denoted σ and is simply the **positive square root** of the variance.

The **Standard Deviation** of a random variable is the positive square root of its variance:

$$\sigma = +\sqrt{\sigma^2} \tag{1-27}$$

The standard deviation is measured in the same units as the mean, and suggests the typical distance between *individual* outcomes of the random process and the mean of such outcomes. Note that σ is very small when the a_j's all huddle closely around μ, but that it grows rapidly as the a_j's spread out.

Assuming that μ is a reasonable measure of the middle of the distribution, then σ is a reasonable measure of the distribution's *spread around the middle.*

It has become traditional to report the mean and variance (or standard deviation) of a discrete random variable at the same time that one presents its probability mass function. The two summary statistics give a good overview of how the variable behaves, while the mass function allows for whatever fine-tuning seems important.

EXAMPLE 1.12

A Toss and a Spread

For a six-sided fair die, find the variance and standard deviation of the outcome X of a single toss.

Solution

For the toss of a fair six-sided die, the variance of the outcome is given under (1-26) by:

$$\sigma^2 = \sum_{j=1}^{M}\left(a_j - \mu\right)^2 p_j = (1/6) * (1 - 3.5)^2 + (1/6) * (2 - 3.5)^2 + \ \ldots \ (1/6)(6 - 3.5)^2 = 2.92$$

- Under (1-27), the standard deviation σ of the outcome is $\sqrt{2.92} = 1.71$

Because 1.71 is roughly half of 3.5, this last result means that it is not uncommon for variations around the mean to be about half as large as the mean itself. That is, of course, the case: when the die yields a 1 or a 6 (i.e., 1/3 of the time), the outcome is more half a mean away from the mean of 3.5.

EXERCISE 1.12

You guessed it: find the variance and standard deviation for one outcome of the toss of a fair 20-sided die.

1.5 The Binomial Distribution

In applications that range from industrial quality control to medical clinical trials to political opinion polling, **binomial** random variables and the binomial probability distribution are crucially important.

A probabilistic process is said to be **binomial** if it satisfies four basic properties:

(i) It consists of a sequence of **n** trials for some $n > 1$.
(ii) Each trial has exactly two possible outcomes, A and B, which are complementary.
(iii) $P(A)$, the probability of A, takes the same value on all n trials. $P(B)$ is likewise fixed at $1 - p$.
(iv) The n trials are independent of one another. (For example, what happens on the ith trial has nothing to do with what happened on the $i - 1^{st}$ trial or any previous trials.)

In this setting, the individual trials are called *Bernoulli trials*.

Though binomial processes are abundant, so are processes that may seem to be binomial but are not. Several repeated flips of a fair coin create a binomial process, with n = number of flips and $p = 1/2$. But if we classify several successive days by the dichotomy "rain/no rain," then, even if the probability of rain is the same on all days in a given month, a tendency of rainy days to come in streaks would lead us to revise the chance that it rains tomorrow based on the weather today. Such revision would violate the independence assumption stated in (iv).

In a binomial process, the elementary events are the different sequences of A's and B's that could arise over the n trials (e.g., *ABABBAB*). Suppose that random variable X assigns to each such sequence the number of A's that it contains. In shorthand, however, we say "let random variable X be the number of A's that come up over the n trials."

In practical situations, it is often important to know about X. This discrete random variable X is said to follow a *binomial distribution*. Clearly, X must be an integer that can neither exceed n nor fall below 0. But we want to move beyond that trivial observation to find the chance that X can take on any particular integer value. Put differently, we want the probability mass function of X, namely, $P(X = k)$ for each k from 0 to n. The quantity $P(X = k)$ varies with both n and p, which are called the *parameters* of the binomial process. We use the following notation:

If random variable X is the number of A's that arise over a binomial process that has n trials and for which $P(A) = p$ on each one, then we say that X has a ***binomial distribution with parameters n and p***. We use the notation **X is $B(n,p)$**.

The Probability Mass Function of a Binomial Random Variable

Here we specify the formula for $P(X = k)$ when X is $B(n,p)$. To do so, we first define events A_i and B_i as follows:

A_i: *the ith trial of the binomial process leads to outcome* A. (i = 1,2, … ,n)

B_i: *the ith trial of the binomial process leads to outcome* B.

We determine $P(X = k)$, starting at the extreme values for k and working our way inwards.

Case 1. $X = n$ and $X = 0$

These two cases are quite straightforward. The outcome $X = n$ can only arise if event A arises on every one of the n trials. We can therefore write:

$$P(X = n) = P(A\text{'s come up on all } n \text{ trials}) = P(A_1 \cap A_2 \ldots \cap A_n)$$

Because the different trials of a binomial process are independent, we have:

$$P(A_1 \cap A_2 \ldots \cap A_n) = P(A_1)\, P(A_2)\, P(A_n)$$

Moreover, because $P(A_i) = p$ for all i, we can go on to write:

$$P(X = n) = p^n$$

The case $X = 0$ is analogous, the only difference being that, instead of getting A's on all n trials, we get all B's. Since each of these B's has a probability $1 - p$ of arising, a modification of the above analysis yields:

$$P(X = 0) = (1 - p)^n$$

Case 2. $X = n - 1$ and $X = 1$

To get $X = n - 1$, we must get A on $n - 1$ of the trials and B on the remaining one. There are n mutually exclusive ways of getting all A's but one, depending on where the lone B appears within the n trials:

B	A	A	…	A
A	B	A	…	A
A	A	B	…	A
		…		
A	A	A	…	B

FIGURE 1.1 The n different elementary events for which $X = n - 1$ in a binomial process with n trials.

Each of these sequences has probability $p^{n-1}(1-p)$, because there are $n-1$ factors of p in the multiplicative expression for the probability and one factor of $1-p$. Adding the probabilities of these mutually exclusive events, we get:

$$P(X = n - 1) = np^{n-1}(1 - p)$$

By analogous reasoning, we reach:

$$P(X = 1) = np(1 - p)^{n-1}$$

Case 3. $X = k$

Finding the probability that X takes a particular value k gets more complicated when k moves towards interior values further away from 0 and n. Even for $k = 5$, where do the five A's fall among the many trials? The first five trials only? The last five? The 3rd, 9th, 27th, 36th, and 88th?

The answer is easier than we might think. For any particular outcome of the n trials that yields A exactly k times, we could list the trial numbers on which A arose (in ascending order). The number of different ways of getting exactly k A's in n trials is equal to the number of distinct combinations of k integers in the range from 1 to n. (Right?) But, as you might recall from high school, this number is simply $\binom{n}{k}$, which is $\frac{n!}{k!(n-k)!}$.

Furthermore, any particular outcome with k A's and n-k B's in n trials has probability $p^k(1-p)^{n-k}$. Different sequences that meet this criterion are mutually exclusive events: for any two such sequences, there must be at least two individual trials in which one saw an A and the other a B. Putting the pieces together, we can write:

The binomial probability mass function:

$$P(X = k) = \binom{n}{k} p^k (1 - p)^{n-k} \quad (1\text{-}28)$$

$$\text{where } \binom{n}{k} = \frac{n!}{k!(n-k)!}, \quad 0! = 1, \text{ and } 0 \le k \le n$$

EXAMPLE 1.13 **One Day Flu?**

Suppose that a pharmaceutical company claims that it has a new cure for the flu, and that 96% of flu sufferers will fully recover within 24 hours of taking the drug. As a first test on its claim, a clinic will give the drug to 25 people diagnosed with the flu and then check on their condition 24 hours later. Suppose too that the company's claim is *perfectly accurate*.

(i) Let random variable X be the number of people tested who will recover from the flu within 24 hours. What can we say about the probability distribution of X?
(ii) What is the chance that all 25 people tested will recover within 24 hours?
(iii) What is the chance that the percentage tested who recover will exactly match the claimed percentage?
(iv) What is the probability that only 80% of the people tested recover within 24 hours?

Solution

(i) If we treat different flu sufferers as independent in terms of responsiveness to the drug (which seems reasonable), then X is B(25, .96). Each patient-test constitutes a trial of the drug with a 96% chance of success (as defined by the 24-hour standard) and 4% chance of failure. With independence, all the conditions for a binomial process are met. (We assume that the 96% statistic includes the possibility that the patient would be cured even without the drug.)
(ii) Here we want $P(X = 25)$, which is given under (1-28) by:

$$P(X = 25) = .96^{25} = .360$$

(iii) If X = 24, then the cure rate is 96%, which exactly matches the claim. We write:

$$P(X = 24) = \binom{25}{24}(.96)^{24}(.04) = \frac{25!}{24! * 1!} \cdot 96^{24} * .04 = .96^{24} = .375$$

(iv) The cure rate is 80% if X = 20. We write:

$$P(X = 20) = \binom{25}{20}(.96)^{20}(.04)^{5} = \frac{25!}{20! * 5!} \cdot 96^{20} * .04^{5} = .0024$$

We see that the chance is about 3 in 4 that the observed cure rate will be 96% or higher, and that the chance of falling far short of the true rate of 96% because of "bad luck" with the patients tested is very small. Of course, these calculations are based on treating the 96% figure as correct. If the true cure rate is well below 96%, then $P(X = 25)$, $P(X = 24)$, and $P(X = 20)$ could be drastically different.

EXERCISE 1.13 Find the probability in Example 1.13 that at least 90% of the flu sufferers are cured within a day.

Binomial Table

Working with (1-28) can get tedious in many situations. Suppose, for example, that $n = 33$, and we want to find the $P(X \le 14)$. Finding $\binom{n}{k}$ can be unpleasant when (say) $n = 33$, and $k = 14$: who wants to work with 14! and the product of integers between 20 and 33? And, whatever p is, we have to multiply $\binom{33}{14}$ by $p^{14}(1 - p)^{19}$. Furthermore, finding $P(X = 14)$ is just the start of our work: we also have to calculate $P(X = k)$ for every integer k from 0 to 13, and then add all the various probabilities together. Wouldn't it be nice if there were electronic calculators that did the work for us?

Well, there are (which should be no surprise, given the practical importance of binomial calculations).

Your instructor will tell you which particular binomial calculators are at your disposal, depending on which statistical package you are using. (Excel works binomial probabilities, as do many hand calculators.) Necessarily, the calculator will require that you specify n and p, so that it knows which binomial distribution is appropriate. Then, you have to make clear what you want to know (e.g., $P(X = 3)$.) Such calculators routinely tell you $P(X = k)$ and $P(X \le k)$ for a specified value of k.

If you can find $P(X \le k)$ for any k, then you can deduce any probability that interests you:

- Because $P(X \le k) + P(X > k) = 1$, it follows that:

$$P(X > k) = 1 - P(X \le k) \tag{1-29}$$

- If we want $P(c \le X \le d)$ for two integers c and d with $d > c$, then we can write:

$$P(X \le d) = P(c \le X \le d) + P(X \le c - 1) \rightarrow P(c \le X \le d) = P(X \le d) - P(X \le c - 1) \tag{1-30}$$

To understand (1-30), note that there are two mutually exclusive ways to get $X \le d$: X can be between c and d, or X can fall below c. (Because X is an integer, its highest value below c is $c - 1$).

Mean and Variance of a Binomial Random Variable

The mean and variance of a binomial process follow simple formulas:

If X is B(n,p), then:

$$\begin{aligned} E(X) &= np \\ \sigma^2(X) &= np(1 - p) \end{aligned} \tag{1-31}$$

The (brief) derivation of (1-31) appears at the course website.

EXAMPLE 1.14 **Double Trouble (Based on a True Story)**

Now we return to the drama at the beginning of this chapter, which arises from a true story. To recapitulate: An air-freight company had eight adjacent positions on its Boeing-727 aircraft for freight containers (Figure 1.2). The weights of these containers depend on what they are carrying, and company statistics indicated that 7% of the containers should be classified as "heavy." While heavy containers are not inherently hazardous, having two such containers next to one another is considered dangerous should the plane encounter an extreme wind gust.

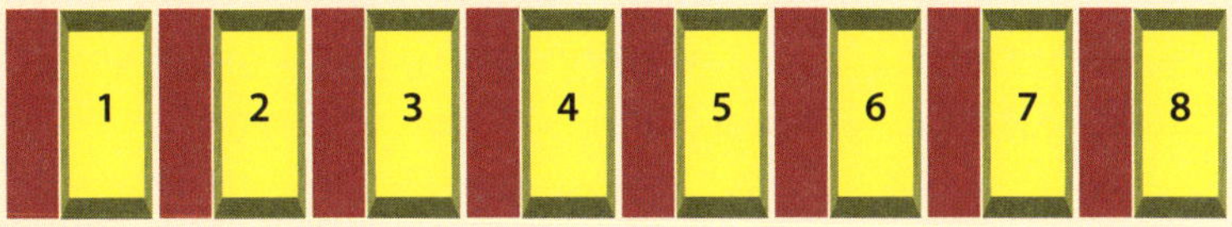

FIGURE 1.2 Eight positions for containers on a 727-freighter aircraft

Understandably, the company wanted to know how often it will be in a "double-heavy" situation if it loads containers at random. It assumed that the weights of different containers can be treated as independent (i.e., that heavy containers do not come in streaks). With that assumption and a 7% chance that a given container is heavy, what is the probability a randomly-loaded 727 will have two (or more) heavy containers next to one another?

Solution

To work out the probability that a given plane would be "double-heavy," we define event *DH* and random variable *X* by:

DH: the plane has at least two heavy containers in a row among its eight loading positions

X: the total number of heavy containers on the plane

We want to estimate *P*(DH) given the information at hand.

We know that from the Law of Alternatives that:

$$P(DH) = P(X = 0 \cap DH) + P(X = 1 \cap DH) + \ldots P(X = 8 \cap DH)$$

which we can in turn restate under the joint-probability rule as:

$$P(DH) = P(X = 0)P(DH | X = 0) + P(X = 1)P(DH | X = 1) + \ldots P(X = 8)P(DH | X = 8)$$

or

$$P(DH) = \sum_{k=0}^{8} P(X = k) * P(DH|X = k) \quad (1\text{-}32)$$

If each container independently has a 7% chance of being heavy, then X follows a *binomial* distribution, namely, B(8,.07). The binomial table in this case informs us that:

$$P(X = 0) = .56,\ P(X = 1) = .34, \text{ and } P(X = 2) = .09$$

These numbers add up to 99%. For a first approximation, therefore, we will *not pay attention* to the possibility of having three or more heavies on the plane. (In effect, we will assume that $P(X = k) \approx 0$ for $k \geq 3$.)

What about $P(DH|X = k)$? We see at once that both $P(DH|X = 0)$ and $P(DH|X = 1)$ are zero (how can there be two heavies together when there aren't two on the plane?). Thus, assuming that $P(X = k) \approx 0$ for $k \geq 3$, we need only concentrate on finding $P(DH|X = 2)$ in (1-32).

$P(DH|X = 2)$ is straightforward to find if we are patient. Because $\binom{8}{2} = 28$, there are 28 possible positions of the two heavy containers among the eight container positions. Seven of these create the unwanted double-heavy situation: (1,2), (2,3), (3,4), (4,5), (5,6), (6,7), (7,8). All 28 possibilities are equally likely under a binomial process (how could it be otherwise, when $p = .07$ for all containers?), from which it follows that $P(DH|X = 2) = 7/28 = 1/4$.

Neglecting the possibility that $X > 2$, we would therefore approximate the probability of a double-heavy by $P(DH) \approx P(X = 2) * P(DH|X = 2) = .09 * (.25) \approx .022$. Though small, this outcome is not so "infinitesimal" as the operations manager believed.

EXERCISE 1.14 Let's not neglect the possibility that $X = 3$, and let's amend the $P(DH)$ accordingly.

(i) What is $P(X = 3)$ here?
(ii) What is $P(DH|X = 3)$?
(iii) What is the revised estimate of $P(DH)$?
(iv) Is it easier to defend ignoring the possibility that $X = 4$ than that $X = 3$? Briefly discuss.

1.6 Some Further Mean and Variance Formulas

For discussing discrete probability distributions other than the binomial, we need some additional formulas about mean and variance. We therefore present these rules in this section, with their derivations at the website.

Linear Functions of a Discrete Random Variable

Suppose that random variable V is a linear function of discrete random variable X, of the form $V = c + bX$, where c and b are constants. How do the mean and variance of V compare to those of X? The answer is: In a straightforward way.

Rule 1:

Mean of a linear function of a discrete random variable:
If $V = cX + b$ where c and b are constants, then

$$E(V) = cE(X) + b \tag{1-33}$$

Rule 2:

Variance of a linear function of a discrete random variable:
If $V = cX + b$ where c and b are constants, then

$$\sigma^2(V) = c^2\sigma^2(X) \tag{1-34}$$

Rule 3:

Standard deviation of a linear function of a discrete random variable:
If $V = cx + b$ where c and b are constants, then

$$\sigma(V) = |c|\sigma(V) \tag{1-35}$$

where $|c|$ = absolute value of c

Rule 4:

Another formula for the variance $\sigma^2(X)$

$$\sigma^2(X) = \sum_{i=1}^{M} a_j^2 p_j - \mu_X^2 = E\left(X^2\right) - \mu_X^2 \tag{1-36}$$

The Conditional Expectation Rule

Suppose that there are *n mutually exclusive and exhaustive events* ($A_1, A_2, \ldots A_n$). If a fair die is tossed, for example, A_1 could be the circumstance that the toss comes up even and A_2 that it comes up odd. The conditional expectation rule states that:

Rule 5:

Conditional expectation rule:
If $A_1 \rightarrow A_n$ are mutually exclusive and $\sum_{i=1}^{n} P(A_i) = 1$, then:

$$E(X) = \sum_{i=1}^{n} E(X|A_i)P(A_i) \tag{1-37}$$

where $E(X|A_i)$ = the mean of X given that A_i occurred

In words (1-37) says that the mean of X is a weighted average of its conditional means over the various A_i's, with the weight assigned to each conditional mean equal to the probability that the associated event A_i occurs.

Derivations of these various rules appear at the website.

1.7 The Geometric Distribution

Like a binomial process, a geometric process involves a series of independent Bernoulli trials, each of which can end in either of two complementary outcomes, A or B. Once again, $P(A)$ takes the same value p throughout the trials, which we assume satisfies the modest condition: $0 < p < 1$. But while a binomial process involves a fixed number of trials (n), a geometric process typically continues until the first trial at which A comes up, at which point the process stops (or, at a minimum, we lose interest in what happens thereafter).

Geometric process
A series of Bernoulli trials that ends when the first "success" arises.

Let random variable X be the number of trials until a geometric process ends. X must be a positive integer, and $P(X = 1) = P(A \text{ on first trial}) = p$. Unlike the binomial distribution which has two parameters—n and p—the geometric only has p.

To find $P(X = k)$ for $k > 1$, we write:

$$P(X = k) = P\left(A \text{ on } k^{\text{th}} \text{ trial and } B \text{ on previous } k-1 \text{ trials}\right) = P\left(B_1 \cap B_2 \ldots B_{n-1} \cap A_n\right)$$
$$= (1-p)(1-p)(1-p)p = p(1-p)^{k-1}$$

where $P(B_j) = P(B \text{ arises on } j\text{th trial})$, etc.

To repeat:

Geometric distribution:
If X is geometric with parameter p, then

$$P(X = k) = p(1-p)k - 1 \qquad (1\text{-}38)$$

Note that $P(X = 1) = p$ and that, for $k \geq 2$, $P(X = k) = (1-p)P(X = k-1)$. Because $1 - p < 1$, it follows that $P(X = k) < P(X = k-1)$. We therefore reach two striking conclusions:

- ***Regardless of the (nonzero) value of p***, the geometric process is more likely to end on the first trial than on any other particular trial (e.g., the fourth).
- ***Regardless of the value of p***, $P(X = k)$ declines steadily as k increases.

We note further that:

If X is geometric with parameter p, then:

$$P(X > k) = P(\text{event } B \text{ on first } k \text{ trials}) = (1 - p)^k \tag{1-39}$$

$$P(X \leq k) = 1 - P(X > k) = 1 - (1 - p)^k \tag{1-40}$$

$E(X)$ and $\sigma^2(X)$, the mean and variance, respectively, of the length of the geometric process, follow the simple rules:

The mean and variance of the a geometric random variable
If X is geometric with parameter p (*i.e., if* X *is* G(p)), then:

$$E(X) = 1/p \tag{1-41}$$

$$\sigma^2(X) = (1 - p)/p$$

At the website, we derive (1-41) using the conditional expectation rule.

EXERCISE 1.15 Let random variable X be the number of times a fair die is tossed until the first three comes up. What are $P(X = 4)$, $P(X \leq 4)$, $E(X)$, and $\sigma(X)$?

1.8 The Hypergeometric Distribution

The hypergeometric distribution arises from the question:

Suppose that a population is composed of two non-overlapping groups, one of size m and the other of size n. If j members of the population are chosen **at random**, what is the probability that exactly k of them will come from group 1 (and $j - k$ from group 2)?

"Chosen at random" means that all $m + n$ members of the full population have an equal chance of being selected.

We define the hypergeometric distribution as the answer to that question:

Hypergeometric distribution:
A hypergeometric random variable is the number of selections from one group in a two-group population when a fixed number of members of the population are chosen at random (without replacement).

If random variable X is the number of members from group 1 in the random drawing of size j, then the calculation of $P(X = k)$ depends on familiar ideas about counting. We assume that group 1 is of size m, that group 2 is of size n, and that j members will be selected. For simplicity, we also assume that the individual members of the population are numbered 1 through $m + n$, with $1 \Rightarrow m$ being from group 1 and $m + 1 \Rightarrow m+n$ being from group 2.

The key considerations are these:

- The number of distinct combinations of j members from the population of size $m + n$ is $\binom{m+n}{j}$. In a single random drawing, any given combination has probability $1/\binom{m+n}{j}$ of arising.
- Out of the m members of Group 1, the number of distinct subgroups of size k that can be formed is $\binom{m}{k}$. Similarly, $\binom{n}{j-k}$ subgroups of size $j - k$ can be formed from Group 2.
- Thus, the number of "double subgroups" with total size j that can be formed, consisting of one subgroup of size k from Group 1 and the other of size $m - k$ from Group 2, is $\binom{m}{k}\binom{n}{j-k}$.
- Because the different "double subgroups" that achieve a $(k, j - k)$ split between Groups 1 and 2 reflect mutually exclusive outcomes of the random drawings, we can add their probabilities to get the overall chance of such a split. As noted, each outcome that yields a $(k,j - k)$ split—like all outcomes—has probability $1/\binom{m+n}{j}$.
- Thus, the probability of a $(k,j - k)$ split is $\binom{m}{k}\binom{n}{j-k}/\binom{m+n}{j}$.

Putting it all together, we get:

Let random variable X be the number of Group 1 members selected when j members are selected at random from a population consisting of m members of Group 1 and n of Group 2. The probability mass function for X satisfies:

$$P(X = k) = \binom{m}{k}\binom{n}{j-k}/\binom{m+n}{j} \tag{1-42}$$

where $k = 0, 1, \ldots, \min(j,m)$

The distribution of X is said to be **hypergeometric** with parameters (j,m,n).

The formula for $P(X = k)$ might initially seem alarming, for it requires computing lots of factorials. Yet fear is unwarranted because electronic tables for the hypergeometric distribution are abundant.

EXAMPLE 1.15

Under Fire

The borough of Manhattan in New York City has over 21,000 miles of electrical cable under its streets. To get access to those cables, workers can use 51,219 separate *manholes* from street level, which are covered by circular or rectangular plates. Sometimes, insulation breakdowns in the cables cause smoke, fire, or explosions at the nearest manhole, which is never pleasant for nearby pedestrians.

Because the cables differ in age and in other respects, some manholes are more vulnerable to trouble than others. Dr. Cynthia Rudin and a team of experts from Consolidated Edison (the electric company that serves Manhattan) and Columbia University devised a system for ranking the 51,219 manhole covers by the risk of major difficulty.* Before using the system to make decisions about preemptive repairs of cables, the investigators sought evidence that it had predictive power.

In a one-year period after the system was devised, 44 manholes in Manhattan suffered major trouble. Five of these manholes were in the top 1000 in the risk ranking, meaning that roughly 2% of the manholes (1000/51,219) experienced 11% (5/44) of the difficulties. Not bad, but does that outcome show the power of the predictive system or might it just reflect lucky guesses?

Solution

A hypergeometric calculation helps us choose between these possibilities. Suppose that the system was no better than guesswork, and the chance that a manhole would suffer trouble bore no relation to its ranking. Then the 44 problematic manholes would behave like a random selection of 44 different manholes, and the probability that exactly five of them would be in the top 1000 in the ranking would be governed by a hypergeometric distribution:

$$P(\text{exactly five events in top 1000 manholes}) = \frac{\binom{1000}{5}\binom{50,219}{39}}{\binom{51,219}{44}}$$

Note that the outcome here involves five events in the manholes ranked in the top 1000 for risk, and 39 events in the remaining 50,219 manholes.

Even though the calculation involves such frightful numbers as 51,219!, hypergeometric calculators can handle it. (Actually, the 51,219! is divided by 51175!, so the computer is multiplying 44 factors rather than 51,219). We learn from the calculator:

$$P(\text{exactly five events in top 1000}) = 0.0014$$

Moreover, the calculator shows that *P*(*at least* five events in top 1000) is only 0.0016. In other words, the outcome "five events" is in the top 1/6 of 1% of outcomes under

* C. Rudin, R. J. Passonneau, A. Radeva, H. Dutta, S. Ierome, and D. Issac, "A Process for Predicting Manhole Events in Manhattan," *Journal of Machine Learning*, 80, pp. 1–31, (2010).

the assumption of random guessing. If the system were worthless, the chance is only 1 in 600 that it would have done so well. For that reason, it seems implausible to dismiss the model's performance as due to sheer luck.

EXERCISE 1.16 In their paper, Rudin and her co-authors also made the following statement:

> "Out of the top 5000 structures in our ranked list, 9 of them had events in 2007. This means that the top 10% of structures contained 20% of the events (out of the 44 in total). The probability of a result this good or better via random guessing is approximately 2.4%."

Perform the hypergeometric calculation that leads to this 2.4% outcome.

1.9 The Poisson Process

For the last discrete distribution we discuss in detail, we turn to one of the most important. In a *Poisson* process, an unknown number of "items" are randomly scattered over some type of "region." Most commonly, events are spread over a period of time, and we introduce the Poisson distribution here in that setting. In thinking about a Poisson process, we first assume that the time period is divided into ***extremely small increments*** each of size Δ. If T is the length of the entire period, then there are a total of T/Δ such increments. A Poisson distribution has one key parameter, generally labeled λ. Within any particular increment, the following axioms hold *to first order*:

- The probability that exactly one event occurs during the increment is $\lambda\Delta$.
- The probability that no events occur during the increment is $1 - \lambda\Delta$.
- The probability of two or more events during the increment is zero.

Note that we said that these assumptions hold "to first order." This means that the chance of two or more events is not literally zero; instead, it can be proportional to Δ^2 or some higher power of Δ. Likewise, the exact probabilities of zero events or one event might need to be adjusted by some higher power of Δ. But if Δ is exceedingly small, then Δ^2 is pathetically so: (If Δ^2 is 1 in 10,000, for example, then Δ^2 is only

FIGURE 1.2 A timeline broken into small increments of length Δ, during two of which (°) one event occurred while the rest had no events.

1 in 100 million.) Thus, any revision required of the first-order probabilities to make them exact is treated as utterly negligible.

There is one further axiom in a Poisson process:

- Different increments (Δ's) are ***independent*** in terms of how many events they contain.

In summary, a Poisson process essentially allows for at most one event in an extremely short period (Figure 1.2). The chance of an event is the same in all periods. What happens in one period has nothing to do with what happened in any other, before or after. This is a prescription for events that arise *randomly and unpredictably* over time, except for the information implied by the numerical value of λ.

Processes that are exactly or approximately Poisson arise in highly diverse settings. Radioactive decay can be modeled as a time-related Poisson process. In many business contexts, the arrivals of customer demands over time is approximately Poisson-distributed, a circumstance of major importance in scheduling, inventory control, and queuing theory. Individual health related events arising in a large population (e.g. births or heart attacks) are often essentially Poisson in their timing. Poisson processes occur over space if, when the space is broken up into small increments, items are randomly distributed across these increments (i.e. the chance of an item is the same in all increments). Molecular biologists approximate the number of DNA molecules in a given stretch of a human genome (the part of a cell that transmits genetic instructions) as following a Poisson distribution.

The Poisson Distribution

Of considerable interest in a Poisson process is the random variable X, the **total** number of items or events that occur an interval of length T.

> In a time-related Poisson process with parameter λ, the number of events in a period of length T is a discrete random variable on the nonnegative integers that follows a **Poisson probability distribution**.

P(X = 0) in a Poisson process

If an interval of length T is broken into T/Δ segments each of length Δ, then the probability of no events at all over the full period (i.e., that $X = 0$)would be given under the first-order Poisson assumptions by:

$$P(X = 0 \text{ over period } T) \approx (1 - \lambda\Delta)^{T/\Delta} \quad (1\text{-}43)$$

We get (1-43) by multiplying the probabilities of zero events during each of the T/Δ individual Δ's which is appropriate because of the independence assumption.

Indeed, we can work out an approximation for $P(X = k)$ for $k > 0$, because the accumulation of events over a period of length T follows a binomial process. Each Δ can be construed as offering a trial in that process, in which outcome A (an event) has probability $\lambda\Delta$, while the complementary outcome B is having no event. With $X \approx B(T/\Delta, \lambda\Delta)$, we have:

$$P(X = k \text{ over period } T) \approx \binom{T/\Delta}{k}(\lambda\Delta)^k(1 - \lambda\Delta)^{T/\Delta - k} \tag{1-44}$$

Formula (1-44) is valuable because it points out that a Poisson process can be viewed as a binomial process with a very large number of trials. But the very fact that T/Δ can get extremely large can be problematic (e.g, if T = one day and Δ = one second, then $T/\Delta = 60 * 60 * 24 = 86{,}400$). And, in any case, (1-44) is an approximation, for it ignores the adjustments essentially proportional to Δ^2 tied to two or more events over increment Δ. It would be nice to have an expression for $P(X = k)$ that is both simpler and exact. Fortunately, such an expression can be derived.

As is shown at the website, the more general expression for $P(X = k$ over period $T)$ is:

In a Poisson process with parameter λ, the probability of exactly k events over a period of length T follows:

$$P(X = k \text{ over period } T) = (\lambda T)^k e^{-\lambda T}/k! \tag{1-45}$$

where k is a nonnegative integer

This formula is interesting for two reasons. Because it yields a positive probability for any nonnegative integer k, it means that a Poisson random variable in principle can take on an infinite number of possible values. Of course, the probabilities drop off dramatically as k increases beyond a certain point, and the sum of the $P(X = k)$'s for all k's equals one. Furthermore, (1-45) uses λ and T only through their product λT. Thus, the probability of (say) three events is the same when $\lambda = 3$ and $T = 1$ as it is when $\lambda = 1$ and $T = 3$.

EXERCISE 1.17 If $P(X = 1) = P(X = 0)$ in a Poisson process with parameter λ observed for a period of length T, what is λ in terms of T?

EXAMPLE 1.17 **Born Again?**

Suppose that births in a large city follow a Poisson process with rate λ per minute, and that the last blessed event occurred z minutes ago. What is the probability that the next birth will occur during the next five minutes?

Solution

To answer that question, we define four relevant events:

A: The last birth occurred z minutes ago

B: The next birth will occur during the next five minutes

C: At least one birth occurs over the next five minutes

D: No births occur within the next five minutes

In terms of these four events, the question is asking us to find $P(B|A)$. But C and D are worth contemplating because they make the problem easier to solve.

Because of assumption (ii) about the Poisson process, the value of z cannot reveal *anything* about the timing of the next event; thus, $P(B|A) = P(B)$. That past events in a Poisson process say nothing about future ones is the ***memoryless*** property of the process. Memoryless is sometimes viewed as an intriguing attribute of the Poisson process, but it is embodied in the process by definition.

Because B and C are two different descriptions of the same event, it follows that $P(B) = P(C)$. And because C and D are complementary, we have: $P(C) = 1 - P(D)$. To find $P(D)$, we set $T = 5$ and use (1-45) to determine $P(X = 0)$. Substitution into the formula yields:

$$P(D) = e^{-5\lambda}$$

Hence $P(B) = P(C) = 1 - P(D) = 1 - e^{-5\lambda}$

Thus $P(B|A) = 1 - e^{-5\lambda}$ for *any* value of z. If $\lambda = .02$ per minute, for example, then the probability the next birth occurs within five minutes is $1 - e^{-.1} = .095$.

EXERCISE 1.18 Under the Poisson process of Example 1.17, what is the probability that the next birth is between 5 and 10 minutes from now?

EXAMPLE 1.18 **Too Close for Comfort (Based on a true story)**

Now, a problem about air traffic control. Suppose that two aerial routes—one Eastbound and one Northbound—cross at an altitude of 35,000 feet at junction J (Figure 1.3). In the absence of air-traffic control, suppose that the times at which eastbound planes would arrive at the junction would reflect a Poisson process with parameter λ_E (per minute). Likewise, northbound planes would arrive under an independent Poisson process with parameter λ_N. All planes are jets that move at a speed of 600 miles per hour along their routes.

The US Federal Aviation Administration (FAA) thinks it dangerous if two planes cruising at the same altitude get within 5 miles of one another (in which case they are said to *conflict*). The idea is that, if a conflict arises, the planes are traveling so

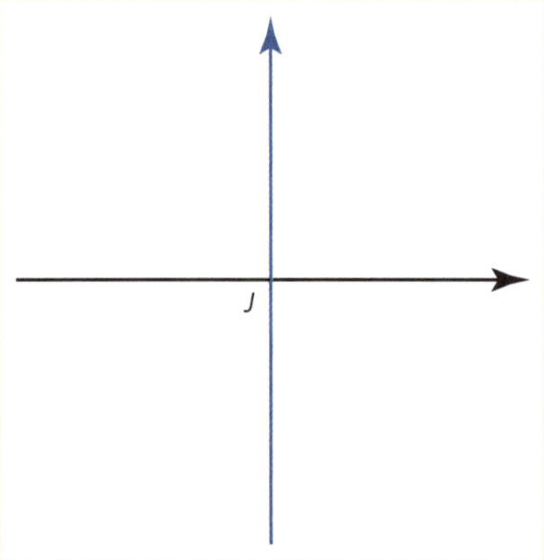

FIGURE 1.3 Eastbound and Northbound Air Routes Cross at J

fast that they might be within seconds of a collision. With the FAA standard in mind, find the probabilities of two interesting events:

(i) E: the chance that an eastbound plane that has just reached J is in conflict *at that moment* with a northbound plane
(ii) N: the chance that a northbound plane that has just reached J is at that moment in conflict with an eastbound plane

Solution

(i) To find $P(E)$, we note that the conflict occurs if, at the time the eastbound plane reaches J, there is a northbound plane within five miles of J. If E^c is the complement of E, then E^c requires that there be no northbound plane within five miles of the junction. It is easier to find $P(E^c)$ than $P(E)$, so we will do so and then invoke the rule $P(E) = 1 - P(E^c)$.
We aren't told anything about planes that are not at the junction, so how can we determine whether a Northbound aircraft is within five miles of J? Well, we can exploit the clue that planes travel at 600 miles per hour (which works out to ten miles per minute, or one mile every six seconds). Suppose that a plane is north of J and within five miles of it. Then, the plane must have passed through J *within the last thirty seconds.* Similarly, if a northbound plane is still south of J but less than five miles away, it will reach J within the *next* thirty seconds.
Thus, if an eastbound plane reaches J at time t, there will be a conflict at t if any northbound planes pass through J between $t - 0.5$ (in minutes) and $t + 0.5$. And there will be *no* conflict if no northbound plane reaches t over the interval $(t - 0.5, t + 0.5)$.
We can therefore write:

$$P(E^C) = P(\text{no northbound arrivals at } J \text{ over } (t - 0.5, t + 0.5)) = e^{-\lambda_N}$$

and thus that $P(E) = 1 - e^{-\lambda_N}$

(ii) The reasoning is the same as for $P(E)$, so we can write:

$$P(N) = 1 - e^{-\lambda_E}$$

It might seem surprising that $P(E)$ and $P(N)$ differ, given that each conflict we are considering involves one eastbound and one northbound plane. If, however, $\lambda_N > \lambda_E$ (for example), then more northbound planes reach J per hour than do eastbound planes. Thus, if equal *numbers* of northbound and eastbound planes face conflicts, the *percentage* of conflicts is lower for northbound planes passing through J than eastbound ones. And $P(E)$ and $P(N)$ reflect these percentages.

This example assumed the absence of air-traffic control, and random arrivals at J under Poisson processes. In reality, aircraft arrival times at junctions would never be left to chance alone. What these calculations suggest is the frequency at which *potentially* hazardous situations would arise based on activity levels and operational randomness in the air-traffic system. They therefore indicate the magnitude of the challenge facing air-traffic controllers. The magnificence with which the controllers meet this challenge is suggested by a statistic: between 1990 and 2013, over 11 billion passengers travelled in scheduled commercial aircraft in the United States. The number killed in midair collisions was zero.

EXERCISE 1.19 Suppose that Eastbound jets passing through J travel at speeds of 600 miles per hour, while Northbound planes have speeds of 500 miles per hour. (The discrepancy could arise because Eastbound planes benefit from West-to-East tailwinds, while South-to-North planes receive no comparable benefit.) How would the probabilities $P(E)$ and $P(N)$ in Example 1.18 change (if at all) because of this change in Northbound speeds?

1.10 Continuous Random Variables

Cumulative Distribution Functions

A continuous random variable can take on an infinite number of values, so creating a probability mass function like (1-24) is out of the question. Instead, we focus on probabilities associated with ranges. For a continuous random variable X and a given number y, we say:

For continuous random variable x, its **cumulative distribution function** $F_x(y)$ is defined by:

$$F_X(y) = P(X \leq y) \qquad (1\text{-}46)$$

(For short, $F_x(y)$ is often called the *CDF* of x.) In words, $F_x(y)$ is the chance that X does not exceed y. The quantity y is not a variable: it is a *numerical constant*. For the mid-August Boston temperature, for example, we might be told that $F(92) = .85$ while $F(60) = .02$. Those statements would mean that the temperature has an 85% chance of not exceeding 92° but only a 2% chance of not exceeding 60° (and hence a 98% chance of doing so).

Why are we interested in $F_x(y)$? The reason is that, if we are told $F_x(y)$ for all values of y, we can deduce *any* probabilistic quantities about X that we desire. We can learn the chance that X exceeds any number, falls below any number, or falls between any two numbers. If we know $F_x(y)$, we know as much as someone who possesses the probability mass function of a discrete random variable.

To justify this last statement, we start by relating $F_x(y)$—the probability that X is at or below y—to $P(X < y)$. If X is at or below y, one of two mutually exclusive possibilities has arisen: $X < y$ or $X = y$. Therefore:

$$P(X \le y) = P(X = y) + P(X < y) \tag{1-47}$$

But for a continuous random variable, $P(X = y)$ is treated as zero. (What is the probability that the temperature is 85° to one hundred trillion zillion places?) Thus, (1-47) becomes:

$$F_x(y) = P(X \le y) = P(X < y) \tag{1-48}$$

Equation (1-48) shows that $F(y)$ tells us $P(X < y)$ as well as $P(X \le y)$, and means that the symbols $\le$ and $<$ can be treated as identical in connection with continuous random variables (as can $\ge$ and $>$). (Because $P(X = y)$ can exceed zero for discrete random variables, these statements do not apply to them.)

Random variable X must take on *some* value; it can either exceed y or fail to do so. Hence:

$$P(X \le y) + P(X > y) = 1, \text{ or:}$$

$$P(X > y) = 1 - P(X \le y) = 1 - F_x(y) \tag{1-49}$$

Thus, $F_x(y)$ can be used to determine the chance that x exceeds any given number.

Now consider two numbers c and d, where $d > c$. The event that $X \le d$ can arise in two mutually exclusive ways: X falls below c, or X falls in the range (c,d). In consequence:

$$P(X \le d) = P(X < c) + P(c \le X \le d) \tag{1-50}$$

But because $P(X \le d) = F_X(d)$ by definition and $P(X < c) = F_X(c)$ by (1-48), we can restate (1-50) as:

If $d > c$, then:

$$P(c \le X \le d) = F_X(d) - F_X(c) \quad (1\text{-}51)$$

(This rule also applies if ≤ is replaced by <.)

Under (1-51), a simple subtraction using $F_X(y)$ at two y-values **yields the chance that X falls between those two values**.

It would be absurd to say that there is a 50% chance that X falls below 20 but a 30% chance that it falls below 40. As y goes up, after all, the chance X falls below y cannot go down. It follows that $F_X(y)$ is a *nondecreasing* function of y. But there can be ranges of y in which $F_X(y)$ does not increase: if X cannot fall between 5 and 6, for example, then $F_X(5.5) = F_X(5.2)$ (Right?). A plot of $F_X(y)$ vs. y will typically be horizontal in some regions and increasing in others (Figure 1.4). However, $F_X(y)$ cannot make any vertical shifts, which would imply a positive probability of being at a particular value y. Vertical shifts are associated with discrete random variables.

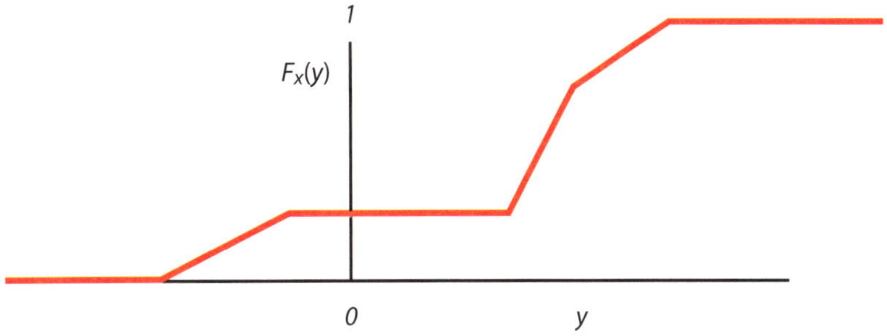

FIGURE 1.4 For a continuous random variable X, $F_x(y)$ cannot decrease as y increases, but it can be flat in some regions. It cannot make any vertical jumps.

Now an important definition:

The number C_A, the **A^{th} percentile** of continuous random variable X, has the property that:

$$P(X \le C_A) = F_X(C_A) = A\% \quad (1\text{-}52)$$

For example, if $C_{40} = 76°$ F for the noon temperature in Boston on August 15, then there is a 40% chance that the temperature then will not exceed 76° F.

Because $P(X = C_A) = 0$, it is also true that $P(X < C_A) = A\%$. It follows from (1-52) that $P(X > C_A) = (100 - A)\%$.

Perhaps the most famous percentile is the 50th, which is called the **median** and which competes with the mean as a measure of the "middle" of the distribution. The

95th percentile of X is often used to epitomize a high but not preposterous outcome: X is nineteen times as likely to fall below C_{95} as to exceed it (95/5), yet it is not the highest value possible. Likewise, the 5th percentile C_5 is often treated as a low but not bizarre outcome. The distance between the 25th and 75th percentiles of X (i.e., $C_{75} - C_{25}$) is called the **interquartile range** of X-values: it has a 50% chance of including X and represents something of a midsection of the distribution.

We will work numerical examples momentarily, but first we should meet an important relative of the cumulative distribution function.

The Probability Density Function

While the CDF tells us the chance that X doesn't exceed y, it doesn't tell us whether X falls only slightly below y or several miles below. To understand the local behavior of X for a continuous random variable, it is useful to work with its probability density function.

The *probability density function* of X, denoted $f_X(y)$ and often shortened to the *pdf*, plays a role similar to λ in the Poisson process:

If $f_X(y)$ is the **probability density function** of continuous random variable X, then:

$$P(y \le X \le y + \Delta) \approx f_X(y)\Delta \text{ when } \Delta \text{ is extremely small} \quad (1\text{-}53)$$

Note that $f_X(y)$ is *not* itself a probability: it must be multiplied by the length of a very small interval starting at y to yield the chance that X falls in that interval. The quantity $f_X(y)$ must be nonnegative (though it can exceed one).

If we plot $f_X(y)$ vs. y, we see from (1-53) that the chance that X falls within a very small Δ starting at y is *proportional to the height* of the density curve at y. Thus, the peaks of the curve indicate the neighborhoods where X is most likely to arise (Figure 1.5). The place where the density function is maximized is called the **mode** of X.

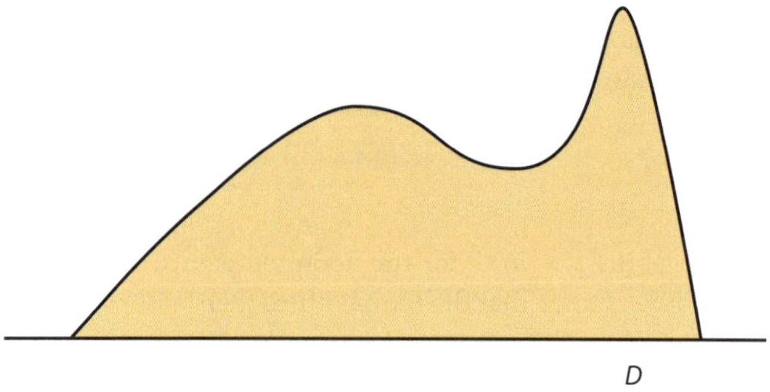

FIGURE 1.5 A Density Function $f_X(y)$ in which X Shows Up Most Often Near D, Its Mode

If we break the real number line into adjacent segments of length Δ, and then add the probabilities that X falls in each one, the sum must be *one*. (After all, X must be somewhere, and we are adding the probabilities for all the mutually-exclusive places it can be.) But we might recall that integration in Calculus is a process of summation, which means that, if we use the notation 'dy' rather than Δ for a miniscule increment that approaches zero, we can write:

For *any* probability density function $f_X(y)$:

$$\int_{-\infty}^{\infty} f_X(y)dy = 1 \tag{1-54}$$

When X is restricted to the finite range (a,b), we can restate (1-54) as follows:

If X is limited to the finite range (a,b), then:

- $f_X(y) = 0$ outside that range
- $\int_a^b f_X(y)dy = 1$ (1-55)

A major advantage of working with the density function is that it immediately yields formulas for the mean and variance of X. As defined for discrete random variables, the mean is a summation of the form $\mu = \sum_{j=1}^{m} a_j p_j$. Analogously, we define μ for continuous random variable X by multiplying each y-value by the chance that X falls extremely close to it (i.e., between y and $y + dy$), and then summing up over all possibilities:

When random variable X is continuous, then its mean follows:

$$E(X) = \mu_X = \int_{-\infty}^{\infty} y f_X(y)dy \tag{1-56}$$

Again in analogy with the discrete-variable formula, the variance $\sigma^2(X)$ is given by:

$$\sigma^2(X) = \int_{-\infty}^{\infty} (y - \mu_X)^2 f_X(y)dy \tag{1-57}$$

The Probability Density Function and the Cumulative Distribution Function

Given the definition of $f_X(y)$, we can write: $P(c \le X \le d) = \int_c^d f_X(y)dy$ (for $d > c$). But we have also established in (1-51) that $P(c \le X \le d) = F_X(d) - F_X(c)$, where $F_X(y)$ is the CDF of X. It follows that:

For continuous random variable X:

$$\int_c^d f_X(y)dy = F_x(d) - F_x(c) \tag{1-58}$$

We could also obtain this equation using the definitions of pdf and CDF and a connection between them that arises from Calculus. We note that:

$$P(y \le X \le y + dy) = F_X(y + dy) - F_X(y) \approx f_X(y)dy \text{ (for } dy \text{ very small)}$$

Dividing both sides of this relationship by dy yields:

$$F_X(y) \approx (F_X(y + dy) - F_X(y))/dy$$

As $dy \to 0$, the expression on the right become the derivative of $F_x(y)$ with respect to y. In short, *the density function at a particular value is the derivative of the cumulative distribution function at that value.* We might recall from Calculus that the CDF is therefore the indefinite integral of the pdf.

For a continuous random variable X, the **derivative** of its CDF $F_x(y)$ with respect to y at any y-value is its probability density function $f_X(y)$ at that value.

Under the circumstances, the relationship: $\int_c^d f_X(y)dy = F_x(d) - F_x(c)$ also arises from the Fundamental Theorem of Integral Calculus, which states the integral of a function between two limits is the change of its indefinite integrand between these two limits. The upshot is that pdf and CDF are by no means independent descriptors of the behavior of X; when one of them is known, it directly implies the other.

EXAMPLE 1.19 **Our First Continuous Distribution**

Suppose that random variable X has the following CDF:

$$F_X(y) = \begin{cases} 0 & \text{for } y < 0 \\ \dfrac{y}{2} & \text{for } 0 \le y \le 2 \\ 1 & \text{for } y > 2 \end{cases}$$

(i) What is $P(X < 1.5)$?
(ii) What is $P(X > 1.5)$?
(iii) What is $P(0.7 < X < 1.5)$?
(iv) What is $P(4 < X < 5)$?
(v) What is the 70th percentile of X?

Solution

(i) Given that $F_X(1.5) = 1.5/2$, we see from (1-48) that $P(X < 1.5) = \mathbf{.75}$.
(ii) From (1-49), we see that $P(X > 1.5) = 1 - F_x(1.5) = .25$
(iii) From (1-51) and the facts $F_X(1.5) = .75$ and $F_X(0.7) = .35$, we have:

$$P(.7 < X < 1.5) = F_x(1.5) - F_x(.7) = .75 - .35 = \mathbf{.40}$$

(iv) Because both $F_X(4)$ and $F_X(5)$ are both one, (1-51) implies that $P(4 < X < 5)$ must be **zero**. That outcome makes sense: if it is certain that X does not exceed 4, there could not be a positive probability that X falls between 4 and 5. And $F_X(4) = 1$ means that the chance is 100% that X does not exceed 4.
(v) C_{70}, the 70th percentile of X, is the value of y at which $F_X(y) = .7$. Setting $F_X(y) = y/2 = .7$, we get $C_{70} = 1.4$. (Clearly, the 70th percentile of the distribution cannot fall below 0 or above 2, so we ignore those regions.)

EXERCISE 1.20 Your turn. For the CDF in Example 3.1, what is the pdf in the three ranges: below 0, 0–2, and above 2? What is the mean of X? What is the variance?

1.11 The Uniform Distribution

When people are asked about some random quantity X, they sometimes say "I'm pretty sure it's between a and b, but I have no idea where it falls in that interval." In effect, these people are assuming that X is equally likely to take on any value in the range (a,b) (where $b > a$). In that situation, it is said to be **uniformly distributed** over that range. (In notation, X in this case is said to be $\boldsymbol{U(a,b)}$.) To illustrate how to specify a cumulative distribution function $F_X(y)$ (hereafter CDF) in a particular case, we determine it now when X is uniform.

To find $F_x(y)$ when X is uniform on (a,b), it is convenient to break y-values into three distinct ranges: **below a, from a to b, and above b**. If y is below a, then $F_X(y)$, the probability that x *does not exceed* y, must be zero. (See Figure 1.6). By similar reasoning, $F_x(y) = 1$ when $y > b$ (Figure 1.7).

FIGURE 1.6 If X must fall between a and b, then X can't fall below y here so $F_X(y) = P(X \leq y) = 0$.

FIGURE 1.7 If y exceeds b, then it is certain that X falls below y so $F_X(y) = P(X \le y) = 1$.

The more challenging case is that in which $a < y < b$. Here X has a nonzero chance of exceeding y but also a nonzero chance of not doing so (Figure 1.8). To work out $F_X(y)$ precisely, we use a geometric argument: since X is equally likely to be anywhere in (a,b), its chance of falling at or below y is equal to the *fraction* of the line between a and b that falls between a and y (i.e., on the red portion of Figure 1.8). But that fraction is just $(y - a)/(b - a)$. Hence $F_X(y) = (y - a)/(b - a)$ for $a < y < b$.

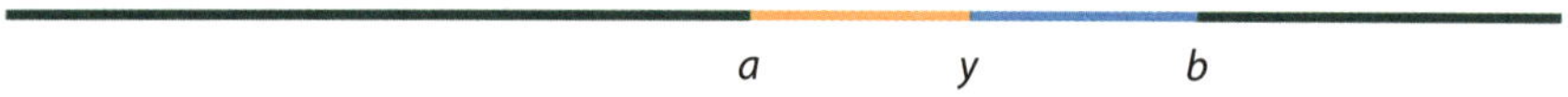

FIGURE 1.8 The chance that X doesn't exceed y here is the chance it falls on the orange part of (a,b) rather than the blue part.

Summarizing, we can write:

$$\text{If } x \text{ is } U(a,b)\text{, then its CDF } F_X(y) = P(x \le y) = \begin{cases} 0 & \text{for } y < a \\ \dfrac{y-a}{b-a} & \text{for } a < y < b \\ 1 & \text{for } y > b \end{cases} \qquad \text{(1-59)}$$

Note, by the way, that random variable X in Example 1.19 was $U(0,2)$.

When X is $U(a,b)$, a graph of $F_X(y)$ vs. y follows:

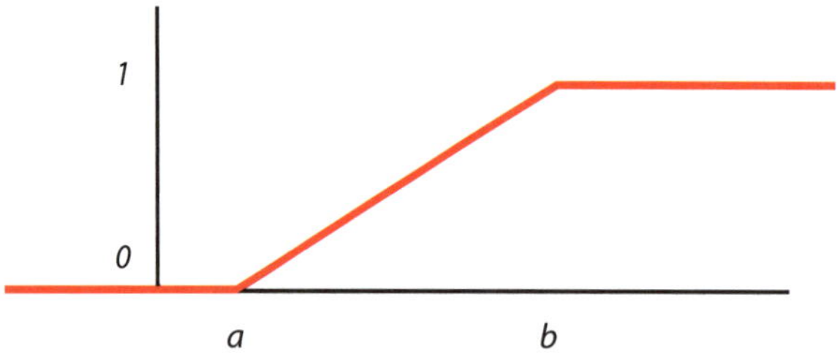

FIGURE 1.9 The CDF of x when x is $U(a,b)$

EXAMPLE 1.20 **A Quickie**

Suppose that X is $U(1,9)$. What are $P(X < 1.2)$ and $P(0.5 < X < 1.2)$?

Solution

If X is $U(1,9)$ (i.e., $a = 1$ and $b = 9$), then $F_X(1.2) = (1.2 - 1)/(9 - 1) = (0.2)/8 = .025$. This outcome means that if X is equally likely to be anywhere from 1 to 9, it has only a 2.5% chance of not exceeding 1.2. As for $P(0.5 < X < 1.2)$, the use of (1-53) and $F_X(0.5) = 0$ again yields the answer .025. That outcome accords with reason: Because X cannot fall below 0.5, the event that $X < 1.2$ is the same as the event that $0.5 < X < 1.2$. If the event is the same, so is its probability.

EXERCISE 1.21 If X is $U(1,9)$, what is the median of X? The 95th percentile? The interquartile range?

To find the probability density function for X when X is $U(a,b)$, we need simply take the derivative of $F_x(y)$ with respect to y. That derivative is zero in the ranges in which $F_X(y)$ is constant, namely, $y < a$ and $y > b$. In the range (a,b), we need simply differentiate $F_x(y) = (y - a)/(b - a)$, a process that yields the derivative $1/(b - a)$. In short:

When x is U(a,b), its pdf fx(y) follows:

$$F_X(y) = \begin{cases} 0 & \text{when } y < a \\ \dfrac{1}{b-a} & \text{when } a \le y \le b \\ 0 & \text{when } y > b \end{cases} \qquad \text{(1-60)}$$

If we use (1-56) to find the mean of X here, we find that $E(X) = (a + b)/2$. That result is not a surprise: we would expect the long-term average of outcomes spread uniformly between a and b to be $(a + b)/2$, the middle of the interval from a to b. Use of (1-57) yields the result $\sigma^2(X) = (b - a)^2/12$, meaning that the standard deviation of X follows: $\sigma(X) = (b - a)/\sqrt{12} = .29(b - a)$. It seems intuitive that the standard deviation of X should be proportional to $b - a$, but it is not obvious that the proportionality constant is 0.29. But that is why we perform calculations rather than rely on our guesses.

If X is $U(a,b)$, then:

$$E(X) = (a + b)/2$$
$$\sigma(X) = (b - a)/\sqrt{12} = .29(b - a) \quad (1\text{-}61)$$

When X is $U(a,b)$, the graph of its pdf follows:

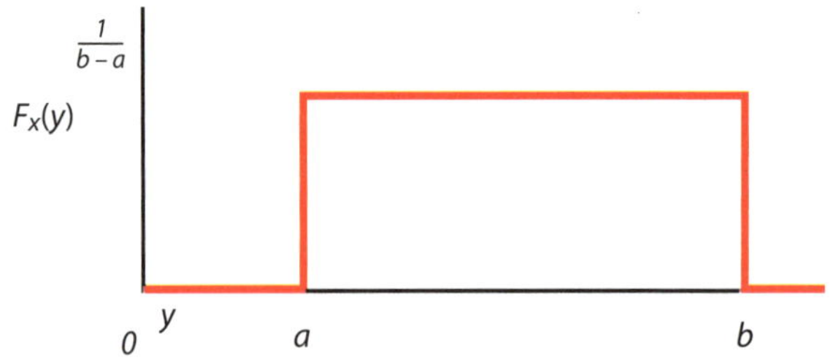

FIGURE 1.10 If X is $U(a,b)$, then its probability density function follows the red line.

EXAMPLE 1.21 **Tunnel Vision**

An electric warning device at a railroad control center reveals that there is rail damage somewhere on a straight four-mile stretch of track. The first three miles are in the open and can be observed from above, but the last mile is in a tunnel. A dispatcher sends a helicopter to look for damage on the outdoor three-mile stretch. However, this search technique is not perfect: even if the helicopter passes over the trouble spot, there is a 20% chance that the crew will not recognize the damage. But if the damage is in the tunnel, it is certain that the helicopter crew will not see it.

A few minutes later, the helicopter reports that it has failed to find the damage. Given this information, what is the probably that the problem occurred in the tunnel?

Solution

Let X be the distance in miles from the start of the four-mile stretch to the point of damage, and assume that the tunnel is in the last mile of the stretch. With the limited information at hand, it is reasonable to assume prior to the search that X is $\boldsymbol{U(0,4)}$. If we define events A and B by:

A: the helicopter crew does not see the damage

B: the damage is in the tunnel,

then what we need here is $\boldsymbol{P(B|A)}$.

We know that $P(B|A) = P(A \cap B)/P(A) = P(B)P(A|B)/P(A)$, and that $P(A)$ follows:

$$P(A) = P(A \cap B) + P(A \cap B^c) = P(B)\,P(A|B) + P(B^c)\,P(A|B^c)$$
(B^c is the complement of B)

We can work out the requisite probabilities with the information at hand. Under the uniformity assumption, $P(B) = .25$, because only the last mile out of four is in the tunnel. (More formally, the tunnel extends from $X = 3$ to $X = 4$, and $P(3 < X \leq 4) = F_X(4) - F_X(3)$ under (1-51). Here $F_X(4) = 1$, and $F_X(3) = (3 - 0)/(4 - 0) = 3/4$, so $P(3 < X < 4) = P(B) = 1/4$.) Similarly, $P(B^c) = 1 - P(B) = .75$. Because the helicopter crew cannot see into the tunnel, $P(A|B) = 1$ under the definition of event A. Because of the helicopter's limitations even when the damage is outside, $P(A|B^c) = 0.2$. Therefore:

$$P(A \cap B) = P(B)P(A|B) = .25 * 1 = .25.$$
$$P(A) = P(A \cap B) + P(A \cap B^c) = .25 * 1 + .75 * (0.2) = .40.$$

Thus: $P(B|A) = P(A \cap B)/P(A) = .25/.40 = 5/8$.

The helicopter report thus *more than doubles* the chance that the problem is in the tunnel (from the original .25 to 5/8 = .625. That may well be bad news, because repairs in the tunnel might be more difficult than repairs out in the open.

EXERCISE 1.22 Suppose that the repair would cost $1000 if outside the tunnel and $3000 if inside.

(i) What was the mean of the projected repair cost when the warning signal in Example 1.21 first reached the tower?

(ii) What was the revised mean of the repair cost after the helicopter reported not seeing the damage?

1.12 The Exponential Distribution

Another continuous distribution of interest to us is the exponential. For an exponential random variable, all non-negative values are possible. If random variable X has an *exponential distribution with parameter* λ *(*$\lambda > 0$*)*, then by definition the probability that it exceeds some value y follows:

$$P(X > y) = e^{-\lambda y} \text{ for } y \geq 0 \qquad (1\text{-}62)$$

Note that, under this formula, $P(X > 0) = 1$ and $P(X > y) \Rightarrow 0$ as $y \Rightarrow \infty$. Given (1-48), the CDF of X is directly available:

When X is **exponential** with parameter λ, its CDF follows:

$$F_X(y) = P(X \leq y) = 1 - P(X > y) = 1 - e^{-\lambda y} \qquad (1\text{-}63)$$

Taking the derivative of (1-63) with respect to y, we find that:

When X is **exponential** with parameter λ, its pdf follows:

$$f_X(y) = \lambda e^{-\lambda y} \tag{1-64}$$

The density function under (1-64) has its peak value at 0 and then declines steadily, in the fashion of exponential decay (Figure 1.11). This pattern means that, as y increases, the chance that x falls in a small Δ starting at y is likewise decreasing. The likeliest "neighborhood" for X of length Δ is $(0, \Delta)$. Large outcomes for X become increasingly rare under this pdf, but they never die out completely.

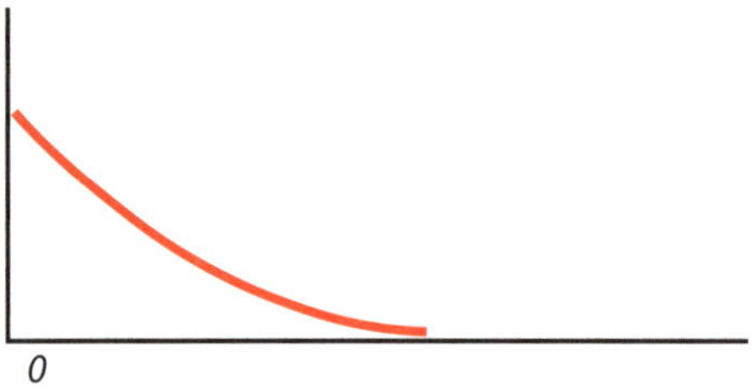

FIGURE 1.11 The probability density function for an exponentially-distributed random variable

Using (1-56) and (1-57) with (1-64) and Calculus establishes that:

If X is exponential with parameter λ, then:

$$E(X) = 1/\lambda$$
$$\sigma(X) = 1/\lambda \tag{1-65}$$

Exponential Distributions are Everywhere

The exponential has often proven to a good approximation to the distribution of time it takes different people to accomplish something. For example, if the times that various people spend at a given ATM (from the time they insert their cards to the time the leave the machine) are plotted, the graph might well look exponential. One of the earliest practical applications of the exponential distribution involved the telephone system of Copenhagen, Denmark. Agner Erlang observed that the durations of different telephone calls approximately followed an exponential distribution: most were very short, some were of moderate length, and a few were interminable.

In Physics, the exponential distribution arises in connection with radioactive decay. In a population of atoms subject all to the same kind of decay (in which the atom emits radiation and is transformed into something else), the time until decay for a given atom follows an exponential distribution. Here the exponential rule is not an approximation: it is a fundamental principle of Chemical Physics.

EXAMPLE 1.22 Memory Loss

Suppose that the length of some interval follows an exponential distribution with parameter λ, and suppose it is known that the interval has just exceeded some number c.

(i) What is the conditional probability that the event occurs in the next q time units ($q > 0$)?

(ii) For example, suppose that the durations of phone calls in minutes are exponentially distributed with parameter λ, and that a given call has just reached the five-minute mark. What is the probability that the call ends in the next ten minutes?

Solution

What we really want here is a conditional probability: $P(c < X \leq c + q \mid X > c)$. (If a call that has lasted c minutes ends within the next q, its total length is in the range $(c, c + q)$.) But we can invoke Bayes' Theorem to write:

$$P(c < X < c + q \mid X > c) = P(c < X \leq c + q \cap X > c)/P(X > c) \qquad \text{(1-66)}$$

From (1-62), we have $P(X > c) = e^{-\lambda c}$. As for $P(c < X \leq c + q \cap X > c)$, we first note that, if $c < X \leq c + q$, then it automatically follows that $X > c$. Therefore, what we really seek in the numerator of (1-67) is $P(c < X \leq c + q)$. To find $P(c < X \leq c + q)$, we can apply our usual rule (1-51) involving the CDF of X, namely:

$$P(c < X \leq c + q) = F_X(c + q) - F_X(c) = (1 - e^{-\lambda(c+q)} - (1 - e^{-\lambda c}).$$
$$= e^{-\lambda c} - e^{-\lambda(c+q)}$$

Fasten your seat belt: a powerful result is coming. When we substitute the expressions just cited into Bayes' Theorem, we get:

$$P(c < X < c + q \mid X > c) = (e^{-\lambda c} - e^{-\lambda(c+q)})/e^{-\lambda c} = 1 - e^{-\lambda q}$$

If X is exponential with parameter λ and $q > 0$, then:

$$P(c < X < c + q \mid X > c) = 1 - e^{-\lambda q} \textit{ regardless of the value of } c \qquad \text{(1-67)}$$

You may be wondering why this result is so powerful. Well, suppose that the exponential process has just started (i.e., $c = 0$) and we want the chance that it ends within q time units. The answer is simply $F_X(q) = 1 - e^{-\lambda q}$. But suppose that the process did not just start; rather, it started c time units ago. The answer is exactly the same: the chance the process ends in the next q time units is $1 - e^{-\lambda q}$ or any value of c. In an exponential process, in other words, the *remaining* time in the process is totally unrelated to the duration of the process thus far.

That this outcome is striking is clearer if we consider a specific case. Suppose that—as in old Copenhagen—telephone calls are exponential in length, and that they have a 60% chance of lasting less than ten minutes. Then a call that has just started has a 60% chance ending within ten minutes. But another call that has already gone one for 20 minutes also has the same 60% chance of ending in the next ten. Same for a call that has gone on so far for 45 minutes. If the question is "how much longer will this process last?" is asked about an exponential process involving time, then the amount of time it has lasted so far is *completely irrelevant* to the answer. (In part (ii) of this problem, the probability of ending in the next ten minutes is $1 - e^{-10\lambda}$, independent of the fact that it has already gone one for five minutes.)

When future developments in a random process do not depend on anything in the past, the process is said to be *memoryless*. We have already described the Poisson process as memoryless; now we know that the exponential process also meets this standard.

EXERCISE 1.23 If the lengths of telephone calls are exponentially distributed in minutes with (positive) parameter λ, then which of the following is more likely (i) a call that has already gone on two minutes will last an additional ten minutes or (ii) a call that has gone on for 45 minutes will last an additional nine minutes? Does the answer depend on the value of λ?

1.13 The Normal Distribution

If we had to identify *the most important probability distribution of all*, we would be wise to go with the normal (or Gaussian). The normal is a continuous distribution that follows a bell-shaped density curve, based on two parameters a and b. Its probability *density* function follows the formula:

If X is **normal** with parameters a and b, then its pdf follows:

$$f_X(y) = \left(\frac{1}{\sqrt{2\pi}}\right)\left(\frac{1}{b}\right)e^{-(y-a)^2/2b^2} \quad \text{where } -\infty < y < \infty \qquad (1\text{-}68)$$

Though it may not immediately be obvious from (1-68), the graph of that function is bell-shaped. Note that the function decays symmetrically as y moves away from a: meaning that (say) $f_X(a + 5) = f_X(a - 5)$.) And one can establish via Calculus that $\int_{-\infty}^{\infty} f_x(y)dy = 1$.

The phrase bell-shaped—which conjures up images of Philadelphia's Liberty Bell—is actually too restrictive: normal density curves can look like rocket ships or turtle shells as well as things that ring (Figure 1.12.). What normal density curves have in common are (i) having only one peak (i.e. being **unimodal**), which occurs at a, and (ii) dropping off the same way in both directions from the peak (i.e., **exhibiting bilateral symmetry**).

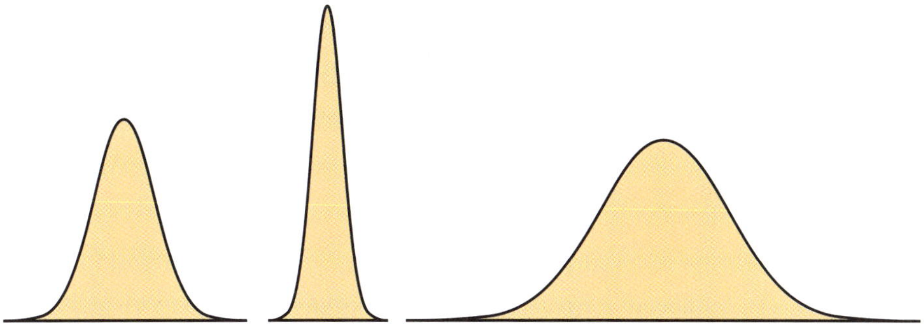

FIGURE 1.12: All in the Family: Three Kinds of Bell Shaped Curves

We might study formula (1-68) a bit to make sure we understand all its implications. Under that formula:

- The peak of the curve is at $y = a$, where $e^{-(y-a)^2/2b^2} = 1$. The density curve exhibits *exponential decay* as y moves away from a, in proportion to the *square* of the distance between y and a.
- As noted, the decay is *symmetric* around a: the density curve drops off just as much when we move to (say) $a + 8$ as it does when we move to $a - 8$.
- The rate of decay depends on b, which enters the exponent via b^2 in the denominator. When b is very small, the term $(y - a)^2$ in the exponent is divided by a very small number, which is the same as being multiplied by a very large number. In consequence, the curve *plunges* as we move away from a, like the middle curve of the three in Figure 3-10. When b is large, however, the decay is far more gradual, as in Figure 3-10's last curve.

How do a and b relate to μ and σ, the mean and standard deviation, respectively, of the normal distribution under study? If we accept that the mean is the long term-average of outcomes if the process is repeated many times, then we see that *it must be true that* $\mu = \boldsymbol{a}$. Over the long run, as many observed x-values will be essentially q units above the peak (i.e., at $a + q$) as q units below (at $a - q$). Thus, the various $a + q$'s and $a - a$'s will average out to a. Because this statement is true for every value of q, the repeated pairing of $a + q$'s and a – q's assures that the long-term average of outcomes must be a.

As for the standard deviation σ of the curve, we sense intuitively that it depends on b given the discussion above. When b is small and thus the density curve drops off quickly, nearly all measurements are very close to the mean and thus the spread around the mean (which is tied to σ) is low. When b is large, by contrast, the density curve drops off slowly, and outcomes far from the mean are fairly common. This higher spread is tied to *higher* values of σ.

But while we see that σ relates to b, who among us would go so far as to assert that $\sigma = b$? Astoundingly, however, *that statement is true*: the parameter b that governs the decay speed of the bell curve as we move away from its peak is *exactly the same* as the square root of the mean-square spread around the peak. One can prove the equality with Calculus using (1-68) and patience.

Given what was just said, *the two parameters* in the normal distribution are really *just its mean and standard deviation*. We can therefore replace a and b with μ and σ, respectively, and rewrite the normal formula:

If random variable X follows a normal distribution with mean μ and standard deviation σ, then its probability density function is given by:

$$f_X(y) = \left(\frac{1}{\sqrt{2\pi}}\right)\left(\frac{1}{\sigma}\right)e^{-(y-\mu)^2/2\sigma^2} \quad \text{where } -\infty < y < \infty \qquad (1\text{-}69)$$

We will say henceforth *that if X is normal with mean μ and standard deviation σ, then* **X is N(μ, σ)**. *Some* people prefer X is $N(\mu, \sigma^2)$, but we focus on σ because it gets greater attention in the discussion of normal random variables.

The normal distribution is unusual in that, in most cases, the parameters of a distribution do not correspond exactly to its mean or standard deviation. When we say X is $B(n,p)$, its mean is neither n nor p (it is np); when X is $U(a,b)$, its mean is neither a nor b but rather $(a + b)/2$. In other words, we lucked out with the normal, especially with respect to σ.

As noted, normal curves are very widely applied. We will soon see that normal curves are relevant even in connection with processes that are *not* originally normal. But what might we want to know about X when X is $N(\mu,\sigma)$? We already know its mean and standard deviation, but what about the probability that an outcome from the distribution falls between two values c and d? Integrating the density function is not easy, because there is no simple function that is the indefinite integrand of the density in (1-69). There are tables to tell us such probabilities based on approximate numerical integration but, before turning to the tables, we should take note of certain well-known properties of **all** normal distributions:

When X is ***N(μ, σ)***, it is *always* true that:

To the nearest percentage point, the probability is 68% that the realization of X (i.e., the observed value) falls within one standard deviation of the mean, between $\mu - \sigma$ *and* $\mu + \sigma$. (1-70)

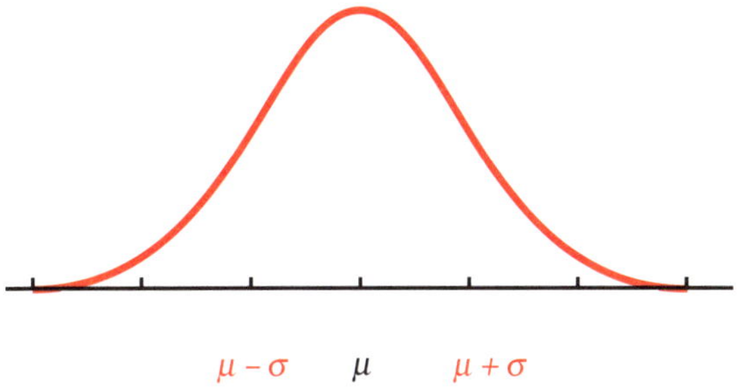

FIGURE 1.13 For *any* normal random variable X, the chance is 68% that the realization of X will fall within one standard deviation of the mean

Furthermore:

When X is $N(\mu, \sigma)$, it is **always** true that :

To the nearest percentage point, the probability is 95% that X falls within two standard deviations of the mean, between $\mu - 2\sigma$ and $\mu + 2\sigma$. (1-71)

When X is $N(\mu, \sigma)$, it is **always** true that :

The probability is 99.7% that X falls within three standard deviations of the mean, between $\mu - 3\sigma$ and $\mu + 3\sigma$. (1-72)

Obviously, not all probabilities associated with a normal curve can be derived from these three statements. We cannot determine, for example, the probability that X falls between $\mu - .6\sigma$ and $\mu + 1.44\sigma$. But given the importance of normal distributions, extensive tables based on numerical approximations have been prepared.

The Normal Table

In the 21st century, a general approach for finding probabilities tied to normal distributions is to use an on-line normal table or a table in a statistical package. The user simply enters the mean μ and standard deviation σ of the normal variable X that is of interest, and the table will typically reveal $P(X < c)$, $P(X > d)$, or $P(c < X < d)$ for any constants c and d that the user specifies. The table, like West Virginia, is almost heaven. At the same time, another application of the table typically reveals various *percentiles* of a given normal distribution. We might want to know for example, the number C_{90} for which $P(X \leq C_{90}) = 90\%$ when X is $N(100,44)$. The table instantly offers the answer.

EXAMPLE 1.23 **California Nightmare?**

Some geological measurements suggest that maximum-strength earthquakes occur on the southern end of California's San Andreas Fault every 160 years on average. But the individual intervals between "Big Ones" vary a bit around this mean. Suppose that an adequate model posits that, given that a huge earthquake has just occurred, the time X until the next one follows a normal distribution with mean 160 and standard deviation 30 years (i.e. that X is $N(160,30)$).

The last huge earthquake on the southern San Andreas Fault—which winds through populous Southern California—occurred in 1857. What is the probability that the next large quake will take place within the next decade? (It is mid-2014 as of this writing.)

Solution

In working through this problem, we take advantage of our knowledge that the first large post-1857 quake had **not** taken place by mid-2014 (i.e. over the first 157 years). Thus, we really face a *conditional* probability problem: given that the fault line was quiet from 1857 to 2014, what is the chance that the next ten years will be shatteringly different?

Let random variable X be the interval between the 1857 earthquake and the next huge one. We do not know what X is, but we do know that it is at least 157 (2014 – 1857). We define two events:

E: $X > 157$, which corresponds to no earthquake from (mid) 1857 to mid-2014.

F: $157 < X < 167$, which entails a big earthquake between mid-2014 and mid-2024.

What we are seeking in this problem is $\boldsymbol{P(F|E)}$.

As usual, we can use Bayes' Theorem to write:

$$P(F|E) = P(E \cap F)/P(E) = P(F)\ P(E|F)/P(E).$$

Given the definitions of E and F (and the observation that $P(E|F)$ must be one), we can write:

$$P(F|E) = P(157 < X < 167)/P(X > 157)$$

From our normal table and the assumption that X is $N(160, 30)$, we learn that $P(X > 157) \approx .540$ and that $P(157 < X < 167) \approx .132$. $P(F|E)$ is therefore estimated as .132/.540, or about .244.

If this geological model is to be believed, therefore, the San Andreas Fault has nearly a 1 in 4 chance of causing a massive earthquake in Southern California within the next ten years. This normal model is not universally accepted, though the mean of 160 years is tied to actual data. And the numerical risk estimate of 24.4% is consistent with other geological predictions.

EXERCISE 1.24 Under the normal-distribution model of Example 1.23, what is the probability of a massive earthquake on the Southern San Andreas Fault over the next quarter-century?

1.14 Combinations of Random Variables

So far, we have been focusing on the behavior of a single random variable. But often it is necessary to consider several random variables at the same time. The financial return on a portfolio of investments, for example, is generally a weighted average of the returns on its various components. If there is uncertainty about how each component will fare, then that uncertainty will carry over to the overall return. In the same way, we might want to know the state of a biological process τ time units from now. That outcome might depend heavily but not completely on the state of the process in $\tau - 1$ time units. We might have a probabilistic model of how the process will evolve until $\tau - 1$, and another model of how it might change between $\tau - 1$ and τ. The issue is how to "synthesize" these two models to describe what could happen atτ.

Joint Distributions of Random Variables

Suppose that X and Z are discrete random variables, and that we want to consider them together rather than separately. Then we might start by examining their joint probability mass function:

Let X and Z be discrete random variables, and let X take on values $(a_1,a_2, \ldots a_m)$ while Z can take values $(b_2,b_2, \ldots b_n)$, Then their **joint probability mass function** p_{ij} is defined by:

$$p_{ij} = P(X = a_i \cap Z = b_j) \quad (1\text{-}73)$$

$$\text{Neccessarily, } \sum_{i=1}^{m}\sum_{j=1}^{n} p_{ij} = 1$$

For continuous random variables X and Z, their **joint density function** $f_{xz}(y,u)$ is defined by:

$$P(y \le X \le y + dy \cap u \le Z \le u + du) \approx f_{XZ}(y,u)dydu \quad (1\text{-}74)$$

In other words, the chance that X and Z fall in the incremental rectangle with lower corner at (y,u) and upper corner at $(y + dy, u + du)$ is essentially $f_{xz}(y,u)dydu$.

Necessarily, integrating $f_{xz}(y,u)$ over two-dimensional space yields:

If X and Z are continuous random variables with joint density function $f_{xz}(y,u)$, then:

$$\iint_{-\infty}^{\infty} f_{XZ}(y,u)\,dy dy = 1 \tag{1-75}$$

We focus in this section on discrete random variables, though an analogous treatment applies to continuous random variables.

EXAMPLE 1.24 **The Die Toss Again**

Suppose yet again that we have two fair dice, one red and one black, and each is tossed. Let X be the outcome on the red die, and Z the outcome on black. What is the joint probability mass function for X and Z?

Solution

X and Z each can assume each value from integer 1 to 6 (i.e., $a_i = i$ for $1 \le i \le 6$ and $b_j = j$ for $1 \le j \le 6$). There are a total of $6 * 6 = 36$ possible (i,j) outcomes and, because they are all equally likely, we have $p_{ij} = 1/36$ for all $1 \le i,j \le 6$.

EXERCISE 1.25 Suppose that the red and both dice are both fair, meaning that all six sides are equally likely to come up. However, the red die has two faces that show two and no faces that show one (i.e., the faces are 2,2,3,4,5,6). What is p_{ij} in this situation for various (i,j) pairs and $1 \le i,j \le 6$?

Sums of Random Variables

Often, we are interested in the sum of two random variables X and Z. For example, if X is the number of men who respond to an advertisement while Z is the number of women who do so, then $S = X + Z$ is the number of people who respond. For simplicity, we will begin with two random variables X and Z, each of which can take on values that are *nonnegative integers*. For the integers i and j, let p_{ij} be $P(X = i \cap Z = j)$. There are $k + 1$ mutually exclusive ways that $X + Z$ can add up to k: X can be 0 while $Z = k$, X can be 1 while $Z = k - 1$, etc.

$P(S = k)$ is given by:

$$P(S = k) = \sum_{i=0}^{k} P(X = i \cap Z = k - i) = \sum_{i=0}^{k} p_{i,k-i} \tag{1-76}$$

Sums of Discrete Random Variables

More generally, suppose that X and Z are discrete random variables that can take on the values $(a_1, a_2, \ldots, a_m)$ and $(b_1, b_2, \ldots, b_n)$,respectively. As in (1-73), let p_{ij} be the *joint probability mass function* of X and Z. When X and Z are discrete, the probability mass function of their sum S follows the probability mass function: $P(S=q) = \Sigma_Q\ p_{ij}$ where Q is the set of (a_i,b_j) combinations for which $a_i + b_j = q$.

Of course, S can take on at most nm values with positive probability, because there are only m possible X-values and n possible Z-values that can be added together.

If X is a discrete random variable that can assume values $(a_1,a_2, \ldots a_m)$ and Z is discrete and can take values $(b_1,b_2, \ldots b_n)$, then their sum $S = X + Z$ is discrete and has a probability mass function given by

$$P(S = q) = \sum\nolimits_Q p_{ij} \tag{1-77}$$

where Q is the set of (a_i,b_j) combinations for which $a_i + b_j = q$, and $p_{ij} = P(X = a_i \cap Z = b_j)$

EXAMPLE 1.25 **The Die Toss Again**

Suppose again we have two fair six-sided dice, one red and one black, and each is tossed. Let X be the outcome on the red die, Z the outcome on black, and S the sum of outcomes (which is the focus of many dice games, like Craps). What is the probability mass function for S?

Solution

As we discussed in Example 1.25, there are 36 possible (i,j) combinations for X and Z with $1 \le i, j \le 6$, and $p_{ij} = 1/36$ for each combination. Then, we can use (1-76) (starting at i =1 rather than i = 0) to write:

$$P(S = 2) = P(X = 1 \cap Z = 1) = p_{11} = 1/36$$

$$P(S = 3) = P(X = 1 \cap Z = 2) + P(X = 2 \cap Z = 1) = p_{12} + p_{21} = 2/36 = 1/18$$

$$\ldots$$

$$P(S = 7) = p_{16} + p_{25} + p_{34} + \ldots p_{61} = 6/36 = 1/6$$

By a symmetry argument, we can establish that $P(S = k) = P(14 - k)$ for each integer k in the range (2,12). For example:

$$P(S = 12) = P(X = 6 \cap Z = 6) = P(S = 2) = P(X = 1 \cap Z = 1) = \frac{1}{36}$$

Therefore:

$$S = \begin{cases} 2 \text{ w.p. } \frac{1}{36} \\ 3 \text{ w.p. } \frac{1}{18} \\ 4 \text{ w.p. } \frac{1}{12} \\ 5 \text{ w.p. } \frac{1}{9} \\ 6 \text{ w.p. } \frac{5}{36} \\ 7 \text{ w.p. } \frac{1}{6} \end{cases}$$

For $8 \le S \le 14$, $P(S = k) = P(14 - k)$

Mean of the Sum

Another derivation at the book website reveals a simple relationship between the mean of sum of random variables and the means of the variables added together:

Whenever $S = X + Z$, we have:

$$\text{If } S = X + Z, \text{ then } E(S) = E(X) + E(Z) \qquad \text{(1-78)}$$

In words, *the mean of a sum of random variables is always the sum of its means.* And—of major importance—the derivation makes no assumptions about whether X and Z are independent (i.e., whether knowing the value of X tells us something about the value of Z). The rule in (1-78) is a universal rule.

Moreover, we have the more general rule:

$$\text{If } S = W_1 + W_2 + W_k, \text{ then } E(S) = E(W_1) + \ldots + E(W_k) \qquad \text{(1-79)}$$

EXAMPLE 1.26 **Die Mean**

Find the mean of the sum when two fair dice are tossed, one black and one red.

Solution

We already know from Example 1.11 that $E(X) = E(Z) = 3.5$ (Here X is the outcome on the red die and Z on the black one.) Thus, if $S = X + Z$, then (1-78) means that

$E(S) = E(X) + E(Z) = 3.5 + 3.5 = 7$. We could have found $E(S)$ directly from the distribution of S in Example 1.25, but isn't it easier using (1-78)?

EXERCISE 1.26 If $U = X^2 + Z^2$, where X and Z are as defined in Example 1.26, what is the mean of U?

EXAMPLE 1.27

Hyper Sum

Find the mean of X, where X is a hypergeometric random variable with parameters (j,m,n).

Solution

Recall that a hypergeometric situation is one in which we select j items at random without replacement from among $m + n$ items, m in group 1 and n in group 2. Random variable X is the number of items selected from group 1. Suppose that, for the ith selection, we define random variable W_i by

$$W_i = \begin{cases} 1 \text{ if the } i^{th} \text{ selection is from Group 1} \\ 0 \text{ if the } i^{th} \text{ selection is from Group 2} \end{cases}$$

Now, the probability that the ith selection is from group 1 is $m/(m + n)$ and from Group 2 is $n/(m + n)$ so we have:

$$W_i = \begin{cases} 1 & \text{w.p. } m/(n+m) \\ 0 & \text{w.p. } n/(m+n) \end{cases}$$

But note that $X = \sum_{i=1}^{j} W_i$ (think about it). Furthermore, note that (1-25) yeilds:

$$E(W_i) = 1 * \left(\frac{m}{m+n}\right) + 0 * \left(\frac{n}{m+n}\right) = m/(m+n)$$

For these reasons and (1-79), we can write:

$$E(X) = \sum_{i=1}^{j} E(W_i) = jm/(n+m)$$

We should acknowledge here that the W_i's are not independent: if $W_1 = 1$, for example, then the probability that $W_2 = 1$ goes down to $(m - 1)/(m + n - 1)$, much as that probability goes up to $m/(m + n - 1)$ if $W_1 = 0$. But the "mean of sums is sum of means rule" does not require that the W_i's be independent: it is true for any sum.

Independent Random Variables

We have been chatting informally about independent random variables, but it is time to be more precise. Two discrete random variables X and Z are said to be *independent*, if that knowing the value of X tells us nothing new about the value of Z (and vice-versa). More formally, can write:

If x and z are **independent** discrete random variables, then:

$$P(X = a_i \cap Z = b_j) = P(X = a_i)P(Z = b_j) \quad (1\text{-}80)$$

Variance of a Sum of Independent Random Variables

We already know that, if $S = X + Z$, then $E(S)$ is always $E(X) + E(Z)$, regardless of whether X and Z are independent. But independence does matter if we want a simple expression for $\sigma^2(S)$.

As we demonstrate at the website, we have a "variance of sum is sum of variances result" when independence prevails:

If X and Z are independent and $S = X + Z$ then:

$$\sigma^2(S) = \sigma^2(X) + \sigma^2(Z) \quad (1\text{-}81)$$

When X and Z are not independent, this "variance of sum is sum of variances" formula is often incorrect.

Note that we do **not** have the rule that the standard deviation of under independence is the sum of the standard deviations of X and Z. Instead, we can use (1-81) to write:

$$\sigma(S) = \sqrt{\sigma^2(X) + \sigma^2(Z)} \quad (1\text{-}82)$$

The result in (1-81) generalizes to several independent random variables:

If $S = W_1 + W_2 + W_k$, and the W_i's are independent, then:

$$\sigma^2(S) = \sigma^2(W_1) + \sigma^2(W_2) + \ldots \sigma^2(W_k) \quad (1\text{-}83)$$

The Difference of Two Independent Random Variables

Suppose that X and Z are random variables and $V = X - Z$. For example, X might be the number of births in a city in the next year, and Z the number of deaths. Neglecting immigration/emigration, V would be the change in the city's population over the year. Because $X - Z = X + (-Z)$ and $E(-Z) = -E(Z)$ (Right?), we can write:

If $V = X - Z$, then $E(V)= E(X - Z) = E(X) + E(-Z) = E(X) - E(Z)$, which shortens to:

$$\text{If } V = X - Z, \text{ then } E(V) = E(X) - E(Z) \quad (1\text{-}84)$$

In other words, the *mean of a difference is always the difference of the means.*

Following up on (1-81), is the variance of a difference of independent random variables the difference of their variances? By no means, as we see if X is the outcome on a fair (six-sided) red die and Z on a fair black one. Here we have $\sigma^2(X) = \sigma^2(Z)$, but could the variance of $V = X - Z$ be $\sigma^2(X) - \sigma^2(Z) = 0$? As we know, the only way to get zero variance is to have $V = X - Z$ take the same value all the time. But $X - Z$ does not always take the same value: it equals zero when $X = Z$ but can be as large as five when $X = 6$ and $Z = 1$. Thus, $\sigma^2(V)$ cannot be zero, and we have a counterexample to the statement "variance of a difference of independent random variables is the difference of their variances."

But is there a simple expression for $\sigma^2(V)$ when X and Z are independent? Happily, there is. If X and Z are independent, then X must also be independent of $-Z$. But then $\sigma^2(V) = \sigma^2(X + (-Z)) = \sigma^2(X) + \sigma^2(-Z)$. Furthermore, it must be true that $\sigma^2(-Z), = \sigma^2(Z)$ (multiplying everything by –1 does not change the squared differences between individual outcomes and the mean outcome). We therefore have $\sigma^2(V) = \sigma^2(X) + \sigma^2(Z)$. The variance of the difference is the **sum** of the individual variances, not their difference.

Summarizing our findings, we have:

If $V = X - Z$ and X and Z are independent, then:

$$E(V) = E(X) - E(Z)$$

But $$\sigma^2(V) = \sigma^2(X) + \sigma^2(Z). \quad (1\text{-}85)$$

EXAMPLE 1.28

Flight Plans

An airline projects that the total cost of operating a particular daily flight during the next quarter will be normally distributed with mean \$630,000 and standard deviation \$120,000, with the uncertainty tied mainly to fluctuations in fuel prices. It projects that the total revenue arising from operating this flight will (independently) be uniformly distributed between \$580,000 and \$700,000. Given these projections, what are the mean and standard deviation of profit for this flight over the next quarter?

Solution

If V is the profit, X the revenue and Z the costs, then:

$$V = X - Z$$

Given that X is $U(580{,}000, 700{,}000)$, we can use formulas about the uniform distribution to conclude that $E(X) = \$640{,}000$ and $\sigma(X) = \$34{,}800$. For normally distributed Z, we are told that $E(Z) = 630{,}000$ and $\sigma(Z) = \$120{,}000$. Because X and Z are independent, we can invoke (1-85) to conclude that:

$$E(V) = E(X) - E(Z) = 640{,}000 - 630{,}000 = \$10{,}000.$$

$$\sigma^2(V) = \sigma^2(X) + \sigma^2(Z) = (34{,}800)^2 + (120{,}000)^2 \Rightarrow \sigma(V) = \$124{,}944$$

EXERCISE 1.27

(i) If the assumption about uniform revenue in Example 1.28 is changed to an assumption of a normal distribution with mean \$640,000 and standard deviation \$34,800, what will happen to the mean and standard deviation of profit?

(ii) Would anything about the distribution of profit change?

Linear Combinations of Random Variables

If random variable Z is related to the n random variables $(X_1, \ldots, X_n)$ by an equation of form

$$Z = c_1X_1 + c_2X_2 \ldots + c_nX_n$$

where the c_i's are known constants and the X_i's are random variables, then Z is said to be a *linear combination* of the n random variables. When the n variables *are independent*, what can be said about the mean and variance of Z?

Well, quite a bit. We recall from (1-33) and (1-34) that, if $U = cX$, then $E(U) = cE(X)$ and $\sigma^2(U) = c^2\sigma^2(X)$. We can therefore move to the expression:

$$E(Z) = E\left(\sum_{i=1}^{n} c_iX_i\right) = \sum_{i=1}^{n} E(c_iX_i) = \sum_{i=1}^{n} c_iE(X_i)$$

(As before, this equation does not require that the X_i's be independent).

We can also write:

$$\sigma^2(Z) = \sigma^2\left(\sum_{i=1}^{n} c_iX_i\right) = \sum_{i=1}^{n} \sigma^2(c_iX_i) = \sum_{i=1}^{n} c_i^2\sigma^2(X_i)$$

(Also as before, this variance expression *does* require independence.)

If $Z = c_1X_1 + c_2X_2 \ldots + c_nX_n$ and the X_i's are independent, then:

$$E(Z) = \sum_{i=1}^{n} c_iE(X_i)$$
$$\sigma^2(Z) = \sum_{i=1}^{n} c_i^2\sigma^2(X_i) \qquad (1\text{-}86)$$

The Average of Random Variables that Are Independent and Identically Distributed

Suppose that random variables ($X_1, X_2, \ldots, X_n$) are independent and all share the same distribution. (They are said to be *iid*: independent and *identically distributed*.) Then they clearly have the same mean and variance (hereafter called μ and $\sigma^2(X)$ respectively). But what can we say about the mean and variance of $\bar{X}$, the arithmetical average of the X_i's, which is given by:

$$\bar{X} = \frac{\sum_{i=1}^{n} X_i}{n}$$

Note that $\bar{X}$ is a linear combination of the X_i's, in which all the c_i's are equal to $1/n$. We can therefore use (1-86 to write:

$$E(\bar{X}) = E\left(\frac{\sum_{i=1}^{n} X_i}{n}\right) = (1/n)\big(E(X) + E(X) + \ldots + E(X)\big) = (1/n)n\mu = \mu$$

$$\sigma^2(\bar{X}) = \sigma^2\left(\frac{\sum_{i=1}^{n} X_i}{n}\right) = (1/n^2)\big(\sigma^2(X) + \sigma^2(X) + \ldots + \sigma^2(X)\big) = (1/n^2)\big(n\sigma^2(X)\big) = \sigma^2(X)/n$$

This last equation implies that the standard deviation $\sigma(\bar{X})$ follows:

$$\sigma(\bar{X}) = \sigma(X)/\sqrt{n}$$

If $\bar{X} = \frac{\sum_{i=1}^{n} X_i}{n}$ and the X_i's are independent and identically distributed (*iid*) with common mean μ and common variance $\sigma^2(X)$, then :

$$E(\bar{X}) = \mu \text{ and } \sigma(\bar{X}) = \sigma(X)/\sqrt{n} \qquad (1\text{-}87)$$

These results about $\bar{X}$ are highly important in Statistics. They say that, when we average n independent random variables with a common distribution, the mean of that average $\bar{X}$ is the same as the mean μ that they all share. That result is unsurprising: if we average together eight numbers that each have mean 17, what value other than 17 would make sense for the mean of that eight-number average?

The result about the standard deviation—namely, $\sigma(\bar{X}) = \sigma(X)/\sqrt{n}$ —says that the spread of $\bar{X}$ around the common mean μ is less than that of the individual random variables that were averaged together. More specifically, the spread drops by a factor of $\sqrt{n}$. Does that outcome make intuitive sense?

EXAMPLE 1.29

Moving In

Discuss why (1-87) makes intuitive sense by considering the average of five tosses of a fair six-sided die.

Solution

We showed in Examples 1.11 and 1.12 that, if we toss the die once, the outcome X_1 has mean 3.5 and standard deviation 1.71. If we tossed the die five times and computed the average $\bar{X}$, then $\bar{X}$ would also have mean 3.5, but its spread around 3.5 would be less than the spread around 3.5 associated with a single toss.

To be explicit, a single toss entails a 1 in 3 chance of getting an outcome of distance 2.5 from the mean (namely, getting $X_1 = 1$ or 6). But when the outcomes of five tosses are averaged together, the only way for $\bar{X}$ to be 2.5 from the mean would to get either all 1's or all 6's, a feat that has a probability of only $2 * (1/6)^5 = 1/3888$. Likewise, outcomes very close to the mean 3.5 are more probable for $\bar{X}$ than for X_1 alone, because high tosses can cancel out other tosses that are low. The upshot is we might expect that the mean-squared spread of $\bar{X}$ around 3.5 will be far less than that for any of the X_i's, which is the outcome implied by (1-87). More specifically, $\bar{X}$ has a variance of $(1.71)^2/5$, and thus a standard deviation of $1.71/\sqrt{5} = 0.76$ (as compared to 1.71 for X_i).

EXERCISE 1.28 If Mendel tosses a fair coin ten times every morning when he gets up, what are the mean and standard deviation of the number of heads on a given Sunday? What are the mean and standard deviation of the average number of heads for the seven days of a particular week?

The Central Limit Theorem

"Look on my works, ye Mighty, and despair!"

—*King Ozymandias of Egypt*

Few would doubt that the Central Limit Theorem is the most powerful result in all Probability and Statistics. It largely explains why the normal distribution is ultimately considered the most important probability distribution (and, indeed, why it is treated as "normal.") Lots of processes that are initially not normal become so if we watch them long enough and average or add the outcomes.

We have already said that, for n independent and identically distributed random variables (iid), their average $\overline{X_n}$ has mean μ and standard deviation $\sigma\sqrt{n}$, where μ and σ are, respectively, the mean and standard deviation of their common distribution. The Central Limit Theorem allows us to go further. It states that, when n is sufficiently large, $\overline{X_n}$ has approximately a *normal* probability distribution, with mean μ and standard deviation $\sigma\sqrt{n}$. We can therefore work out the probability that $\overline{X_n}$ falls in any given numerical range based on μ, σ, and n. For example, the chance that $\overline{X_n}$ falls between $\mu - 2\sigma\sqrt{n}$ and $\mu - 2\sigma\sqrt{n}$ is approximately 95%, while the chance is about 68% that $\overline{X_n}$ falls between $\mu - \sigma\sqrt{n}$ and $\mu - \sigma\sqrt{n}$.

The Central Limit Theorem

We might state the theorem as follows:

Central Limit Theorem:
If the random variables (X_1,X_2, … ,X_n) are independent and identically distributed (iid) with common mean μ and common standard deviation σ then:

For sufficiently large n, their average $\overline{X_n}$ is approximately ***normally*** distributed with mean μ and standard deviation $\sigma\sqrt{n}$. (1-88)

What is extraordinary is that this normality approximation holds *regardless of the original distribution of the variables that are added or averaged*. We might average n outcomes from a uniform distribution, or from a binomial or Poisson distribution. As n goes up and the averaging process continues, the original shape of the common distribution has less and less to do with the behavior of the average $\overline{X_n}$. Instead, a bell-shaped curve takes over. Indeed, for practically *all* distributions with the same mean μ and standard deviation σ, the average $\overline{X_n}$ of n observations is eventually governed by a common bell-shaped curve. To put it another way, the fact that the individual X_i's are not normal eventually becomes *irrelevant* in describing the random variation of their observed average $\overline{X_n}$ around the true mean, μ.

But how does this bell-shaped behavior arise? We get some good intuition on this topic if we consider yet again repeated tosses of a fair die. When $n = 1$, the "average" $\overline{X_1}$ of the outcomes is simply the first outcome, which is equally likely to be any integer from 1 to 6, and is hardly normal in its shape. For $n = 2$, there is greater tendency towards bell-shaped symmetry: $\overline{X_2} = 6$ and $\overline{X_2} = 1$ are now far less likely than $\overline{X_2} = 3$ or $\overline{X_2} = 4$ or $\overline{X_2} = 3.5$. (In this case, $P(\overline{X_2} = 6) = 1/36$, because we need two 6's to get an average of 6; likewise $P(\overline{X_2} = 1) = 1/36$. But $P(\overline{X_2} = 3) = P$(sum of two tosses is 6) = 5/36, as is $P(\overline{X_2} = 4)$. $P(\overline{X_2} = 3.5) = P$(sum is 7) = 1/6.) Table 1.1 reflects the behavior of $\overline{X_n}$ for $n = 1, 2$, and 3, which is becoming increasingly bell-shaped in its distribution as n increases.

Of course, $\overline{X_n}$ based on tosses of a fair die can never literally be normal: it has a discrete distribution, while the normal is continuous. The way to interpret the

TABLE 1.1 Probabilistic Behavior of the *Average* of *n* Tosses of a Fair Die, for $n = 1,2,3$.

	$n = 1$	$n = 2$	$n = 3$
$P(\overline{X}_n < 2)$	0.167	0.083	0.046
$P(2 \le \overline{X}_n < 3)$	0.167	0.194	0.213
$P(3 \le \overline{X}_n \le 4)$	0.333	0.444	0.482
$P(4 < \overline{X}_n \le 5)$	0.167	0.194	0.213
$P(\overline{X}_n > 5)$	0.167	0.083	0.046

(This table arises from direct calculations. For example, $P(\overline{X}_3 = 1) = (1/6)^3 = 1/216$.)

theorem is that the probability that $\overline{X}_n$ falls between two values *c* and *d* gets increasingly close to that probability as computed with the normal table for mean μ and standard deviation $\sigma\sqrt{n}$.

Understandably, you would like a proof (or at least some of you would). A derivation appears at the website.

A broader question about the Central Limit Theorem is: how quickly does $\overline{X}_n$ approach normality? The speed does depend a bit on the original distribution of the X_i's, as well as on our criterion for interpreting the word "approach." As a general rule, *n* does not have to be extremely large for a normality approximation to work well: in general, we are fairly close to normality by the time *n* is a few dozen. We do not require that *n* be several hundred or several thousand.

More specifically, there is something of a consensus that, except for somewhat freakish distributions, the theorem works well for n starting at about 30. (Among other things, we need a finite variance for the X_i's, which is rarely a problem in real-world applications.) But the author would respectfully suggest that the theorem often works well for n as low as 15, as is suggested by calculations at the website.

While the Central Limit Theorem literally applies to an average of *n* independent and identically distributed random variables, their sum is simply the average times *n*. It can be shown that the sum is approximately normally distributed whenever the average itself is. More precisely:

If $S = \sum_{i=1}^{n} X_i$, and *n* is sufficiently large, then:

$$S \approx N(n\mu, \sigma\sqrt{n}) \text{ when the } X_i\text{'s are iid and } \overline{X}_n \approx N\left(\mu, \frac{\sigma}{\sqrt{n}}\right) \quad (1\text{-}89)$$

EXAMPLE 1.30 **Log In**

Suppose that the monthly return (percentage gain) on a particular investment portfolio is estimated to have mean 0.6% and standard deviation 1.0%. The returns in different months are independent.

(i) What can be said about the distribution of the *logarithm* of the five-year return on the portfolio?

(ii) What is the approximate probability that the portfolio gains at least 50% in value over the next five years?

Solution

(i) A portfolio has a return X_i in month i if its value changes by a factor of $(1 + X_i)$ over that month, where X_i is expressed as a decimal. (For example, if the value of the portfolio grows by 2% in a certain month, then its value is multiplied by 1.02 and $X_i = .02$. If it falls by 2%, then $X_i = -.02$). Then if c is the value of the portfolio at the start of the five-year period, its value V_{60} at the end of 5 years = 60 months is:

$$V_{60} = c\prod_{i=1}^{60}(1 + X_i) = cW_{60}$$

where $$W_{60} = \prod_{i=1}^{60}(1 + X_i) \quad \text{(and } \prod \text{ means multiplication)}$$

The quantity W_{60} is the five-year return on the portfolio, as the ratio of V_{60} to c. The *logarithm* of W_{60} follows:

$$\ln W_{60} = \ln\left(\prod_{i=1}^{60}(1 + X_i)\right) = \sum_{i=1}^{60}\ln(1 + X_i) \tag{1-90}$$

(Recall that the logarithm of a product of factors is the sum of their individual logarithms.)

Recall too that, under the Taylor series expansion of $\ln(1 + \varepsilon)$, $\ln(1 + \varepsilon) \approx \varepsilon$ when ε is small. Assuming that is the case here, we can rewrite (1-90) as:

$$\ln W_{60} \approx \sum_{i=1}^{60} X_i$$

Given that each X_i has mean 0.006 and standard deviation .01 and the X_i's are independent, we see under (1-79) and (1-83) that:

$$E(\ln W_{60}) = \sum_{i=1}^{60} E(X_i) = 60 * .006 = 0.36$$

$$\sigma(\ln W_{60}) = \sqrt{\sum_{i=1}^{60}\sigma^2(X_i)} = .01\sqrt{60} = 0.077$$

Moreover, the Central Limit Theorem implies that ln W_{60} has a normal distribution, and therefore that ln $W_{60} \approx N(0.36, .077)$.

(ii) If the portfolio increases in value by exactly 50%, then $W_{60} = 1.5$, and ln $W_{60} = .405$. Because ln $W_{60} \approx N(0.36, .077)$, a visit to the normal table reveals that $P(W_{60} \geq 1.5) \approx .279$.

Given that the logarithm of W_{60} is normal, W_{60} itself is said to have a **lognormal** distribution. This example illustrates why the lognormal distribution is widely used in mathematical finance (where the returns in different periods are often assumed independent under hypotheses about efficient markets).

EXERCISE 1.29 Please return to W_{60} in Example 1.30.

Given that $E(\ln W_{60}) = 0.36$, must the mean of W_{60} be $e^{0.36} = 1.433$? Briefly discuss. (Hint: If some random variable X is equally likely to be e (= 2.718)or e^2, what is the mean of its logarithm?)

While the Central Limit Theorem literally applies to variables that are independent and identically distributed, a normality approximation also works well for several independent variables with distributions that differ to a moderate extent:

> If the X_i's are independent but not identically distributed, their sum and average are approximately normal so long as their means (μ_i's) and their standard deviations (σ_i's) exhibit only modest variation, and the number of X_i's is at least moderately large.

In contrast, the normality approximation would falter if (say) one of the X_i's had a distribution that was decidedly non-normal and its mean and standard deviation were far larger than the mean and standard deviation of the rest of the variables combined. (In that case, the distribution of the sum would be close to that of the dominant X_i.)

Correlation

Taking note of correlations is a fundamental aspect of the way we think. When we meet a new person, we make all kinds of inferences about him based on how he speaks and acts, and most of those inferences prove accurate. Even the adage "where there's smoke there's fire" embodies a legitimate correlation: smoke does not always mean fire, but a fire is more likely to be burning nearby when smoke is present than when it is not.

Correlation is a vital concept in Economics, Medicine, Physics, Sociology, Education, and practically any discipline of interest to the reader. It is therefore important that we understand correlation both intuitively and quantitatively.

Consider two random variables X and Z. In general, the variables are said to be *positively correlated* if, when X is towards the top of its probability distribution, Z tends to be towards the top of its own. Likewise, low X-values tend to correspond to low Z-values. *Negative* correlation typically arises when higher values of X are accompanied by lower values of Z. The two random variables X and Z are *uncorrelated* if knowing that one of the variables is high says essentially nothing about whether the other is high or low. Independent random variables are always uncorrelated.

A meteorological example might be helpful here. Suppose that X is the high temperature on a *randomly-chosen* future date in London, Z the high temperature that day in Paris, U the high in Sao Paulo, and S the high in Singapore. Because London and Paris are close together, the quantities X and Z are positively correlated: a hot day in one of the cities suggests that the randomly-chosen day was in the summer and that it was also hot in the other city. But one would expect both X and Z to be negatively correlated with U, because Sao Paulo is in the Southern Hemisphere while London and Paris are in the Northern. (Summer above the equator corresponds to winter below.) Singapore is practically on the equator and its high temperature scarcely changes over the year: one might anticipate, therefore, that S would be nearly uncorrelated with X, Z, and U.

The "prime-mover" in the study of correlated variables is the *coefficient of correlation,* usually designated ρ. For two random variables X and Z, we designate the correlation ρ_{XZ}:

The **coefficient of correlation** ρ_{xz} for two random variables X and Z is defined by:

$$\rho_{XZ} = E[(X - \mu_X)(Z - \mu_Z)]/\sigma_X\sigma_Z \qquad (1\text{-}91)$$

Although it is far from obvious on inspection, it can be shown that ρ thus defined can *never* fall outside the range –1 to 1. The value 1 signifies a lockstep linear relation under which increases in X are always associated with proportional increases in Z; the value –1 means that, when X goes up, Z always goes down proportionately. A ρ-value of 0 implies no systematic relation between the movements of X and those of Z.

At the website, we prove that ρ_{XZ} cannot exceed one in absolute value.

EXERCISE 1.31 Show using (1-91) and wisdom that, if $Z = aX$ where $a > 0$, then $\rho_{XZ} = 1$.

Another Formula for the Correlation Coefficient

The quantity $E[(X - \mu_X)(Z - \mu_Z)]$ can be written in a simpler form. Please note that:

$$\begin{aligned} E[(X - \mu_X)(Z - \mu_Z)] &= E(XZ) - \mu_X E(Z) - \mu_Z E(X) + \mu_X\mu_Z \\ &= E(XZ) - 2\mu_X\mu_Z + \mu_X\mu_Z = E(XZ) - \mu_X\mu_Z. \end{aligned}$$

Therefore:

The correlation coefficient ρ_{XZ} can also be expressed as:

$$\rho_{XZ} = [E(XZ) - \mu_X\mu_Z]/\sigma_X\sigma_Z \qquad (1\text{-}92)$$

To familiarize ourselves with ρ and its properties, we move at once a simple example.

EXAMPLE 1.31

Three Correlations

Suppose that random variable X is equally likely to take on the values 1 and 11, and that Z has the same distribution. Three possible relationships between X and Z are:

Case I: $Z = X$ all the time

Case II: $Z = 12 - X$ all the time

Case III: Z and X are independent.

What is the correlation coefficient ρ_{xz} in all three cases?

Solution

Note that the first case reflects perfect positive correlation in our intuitive interpretation: when X is high, Z is high; when X is low, Z is low. The second case reflects perfect negative correlation: when X is high, Z is always low, and vice-versa. In the last case, X and Z are uncorrelated.

In calculating ρ_{xz} for the three cases, we note that most of the key quantities in (1-92) are the same in all of them (namely, μ_x, μ_z, σ_x, and σ_z). Given the simplicity of the distributions of X and Z, we can work out with a picture (below) that the means of X and Z are 6, while their variances—their mean squared spreads around the means—must be 25 because X and Z are always five away from their means of 6. The standard deviations of X and Z are therefore $\sqrt{25} = 5$. More formally,

$$\mu_x = \mu_z = 1 * (.5) + 11 * (.5) = 6, \text{ and}$$

$$\sigma^2(X) = \sigma^2(Z) = (1 - 6)^2 * (.5) + (11 - 6)^2 * (.5) = 25.$$

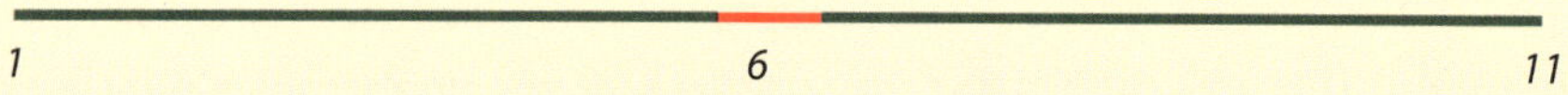

These calculations mean that $\rho_{XZ} = [E(XZ) - \mu_X\mu_Z]/\sigma_x\sigma_z = [E(XZ) - 6^2]/5^2$,or

$$\rho_{XZ} = (1/25)[E(XZ) - 36] \qquad (1\text{-}93)$$

What differs across the three cases is the value of $E(XZ)$. When $X = Z$, the product XZ is equally likely to be $1^2 = 1$ and $11^2 = 121$, so its mean $E(XZ)$ equals $(.5)*(1 + 121) = 61$. In case I, therefore, (1-93) implies that $\rho_{XZ} = (61 - 36)/25 = 1$. In case II ($Z = 12 - X$), the product XZ is always 11 and thus $E(XZ)$ is 11, meaning that ρ_{XZ} = (11 − 36)/25 = −1. In the case of independence, the chance is 25% that XZ will be 121 (i.e, when $X = Z = 11$), 25% that XZ will be 1, and 50% that it will be 11. The mean $E(XZ)$ is therefore 121*(.25) + 11*(.5) + 1*(.25) = 36 in case III, which means that $\rho_{XZ} = 0$. To summarize:

Case I:	$Z = X$	$\rho_{XZ} = \mathbf{1}$
Case II:	$Z = 12 - X$	$\rho_{XZ} = \mathbf{-1}$
Case III:	Z and X independent	$\rho_{XZ} = \mathbf{0}$

This is just what we would have expected based on the intuitive interpretation of correlation. In the derivation of the formula for the sum of two independent random variables (at the website), we establish that $E(XZ) = \mu_X\mu_Z$ when X and Z are independent. But that circumstance leads directly to the conclusion that:

If X and Z are independent, then $E(XZ) = \mu_X\mu_Z$ and therefore $\rho_{XZ} = 0$

There is nothing mysterious about why $E(XZ)$ behaves differently in the three cases. When X and Z get high together (haha), their product XZ gets very high and pulls $E(XZ)$ up. (True, low X times low Z yields low XZ, but the upward force when X and Z are high exceeds the downward force when they are both low.) When X and Z move in opposite ways, either one or the other of them is low at any given time. That low factor drags the product XZ down, and $E(XZ)$ suffers. When X and Z are independent, there is no reason for the $E(XZ)$ to stray from $\mu_x\mu_z$, and it does not.

How do we interpret intermediate correlation coefficients like +0.5 and −0.2? There is no exact answer because many different situations can yield the same correlation coefficient. However, as a crude rule of thumb, a coefficient of correlation is 0.5 suggests that, when one the variables is high in its distribution, the chance is about 75% that the other variable is high in its own. (The same 75% percent agreement applied for low outcomes.) When $\rho = -0.2$, X and Z typically disagree more often than not, but only a bit more often than not.

EXAMPLE 1.32 Of Skill and Luck

A 20-question multiple choice exam on vocabulary words has five choices per question. Random variable X is the number of words that a randomly-chosen test-taker actually knows, and, for words she does not know, the test-taker guesses completely at random among the five choices. Let random variable Z be the number of correct *guesses* that this test-taker makes. Suppose that the probability distribution for X is:

$$X = \begin{cases} 18 \text{ w.p. } .25 \\ 19 \text{ w.p. } .5 \\ 20 \text{ w.p. } .25 \end{cases}$$

(i) Would we expect X and Z to be positively correlated, negatively correlated, or uncorrelated?

(ii) What is the coefficient of correlation of X and Z?

Solution

(i) The higher the value of X, the fewer guesses the test-taker makes and thus the lower the expected *number* of correct guesses. When $X = 20$, for example, we know that $Z = 0$. Thus, we are in the domain of *negative* correlation.

(ii) To find ρ_{XZ}, we use the formula:

$$\rho_{XZ} = [E(XZ) - E(X)E(Z]/\sigma(x)\sigma(z)$$

First we have to work out the various quantities on the right.
From the distribution of X, we can calculate that $E(X) = 19$. Then we can find $\sigma(x)$ by using (1-26) and (1-27):

$$\sigma^2(X) = (18 - 19)^2(0.25) + (19 - 19)^2(0.5) + (20 - 19)^2(0.25) = 0.5$$
$$\text{meaning that } \sigma(X) = \sqrt{0.5} = 0.71.$$

As for Z, it follows the distribution:

$$P(Z = 0) = 0.25 + (0.5) * (0.8) + (0.25) * (0.8)^2 \qquad = .81$$
$$P(Z = 1) = (0.5) * (0.2) + (0.25) * \binom{2}{1}(0.2) * (0.8) \qquad = .18$$
$$P(Z = 2) = (0.25) * (0.2)^2 \qquad = .01$$

Why does Z have this distribution? Well:

- The only way to get $Z = 2$ is to have $X = 18$ *and* both guesses correct. Given a 1-in-5 chance that any given guess is correct, the probability of meeting all requirements for $Z = 2$ is $(0.25) * (0.2)^2$, where $P(X{=}18) = 0.25$.
- As for $Z{=}0$, that outcome can arise for all three values of X, but all guesses that are made (if any) must fail. The probability of k incorrect guesses is $(0.8)^k$.
- For $Z = 1$, we can have $X = 19$ and the single guess correct, or $X{=}18$ and one of the two guesses correct. If two guesses are made, the number correct follows a binomial distribution with two trials and $p = 1/5$; hence the expression $\binom{2}{1}(0.2)(0.8)$.

Given the distribution of Z, we can calculate that $E(Z) = 0.2$ and $\sigma(Z) = 0.42$

Now all that remains is $E(XZ)$. The conditional expectation rule (1-37) tells us that:

$$E(XZ) = \sum_{j=18}^{20} E(XZ|X = j)P(X = j) \tag{1-94}$$

We already have the $P(X = j)$ terms, so we need $E(XZ|X = j)$ for the various values of j. When $X = 18$, there are two guesses, and the number that are correct has the distribution $B(2,0.2)$. (Right?) Thus, the mean number correct is $2 * (0.2) = 0.4$, and the mean value of XZ given that $X = 18$ is $(18) * (0.4) = 7.2$. With similar reasoning, we see that the mean value of XZ given $X = 19$ is $19 * (0.2) = 3.8$. Furthermore, $E(XZ|X = 20) = 0$, because $Z = 0$ in that case. Plugging everything into (1-94), we get:

$$E(XZ) = (7.2) * (0.25) + (3.8) * (0.5) = 3.7$$

Now, returning to the formula for ρ_{XZ}, we find that:

$$\rho_{XZ} = [E(XZ) - E(X)E(Z)]/\sigma(x)\sigma(z) = (3.7 - 19 * (0.2))/(.71 * .42) = -.335$$

As expected, therefore, we obtained a negative correlation coefficient. We might, however, have initially have anticipated a coefficient closer to –1, on the grounds that the *maximum* number of correct guesses—namely, $20 - X$—drops linearly and predictably as X rises. But the negative correlation is reduced because the *actual* number of correct guesses is subject to volatility and thus not closely tied to X: When $X = 18$, for example, the chance is 64%—$(4/5)^2$—that $Z = 0$. Thus, the lowest X is often accompanied by the lowest Z. Because of this volatility, there is only a moderate tendency for lower values of one of the variables to accompany higher values of the other.

EXERCISE 1.30 Make a well … guess about the coefficient of correlation between Z, the number of correct guesses by a student in Example 1.32 and V, the number of incorrect guesses made by that student. (For example, if the student made no guesses, then $Z = V = 0$). Now compute the actual value of ρ_{ZV}.

Two Uses of Correlation

If two quantities are correlated and only one of them is easy to observe, then the visible variable is sometimes treated as proxy for the other. For example, many employers use the Briggs-Meyer Personality Type Inventory to categorize job applicants. Based on how the applicants respond to a questionnaire, they are placed in one of sixteen personality categories. The concept of correlation is also important in creating an investment portfolio. The investor might be urged to buy stock in industries that, historically, have been *negatively* correlated in their fortunes. The idea is that, if one of the industries suffers in the coming period, the other will generally do well, so the overall portfolio will have a "safety net" that makes it more stable. However, the opposite point also applies: if one of the industries does exceptionally well, the other typically acts as a "wet blanket" that dampens the net gain in the portfolio.

Warning: Correlation Is Not Causation

Every student in Probability/Statistics is warned that "correlation is not causation," and now you have received the warning yourself. That fact that X and Z are positively correlated does not mean that increases in X stimulate increases in Z. It could be the other way around: perhaps increases in Z cause increases in X. It is also possible that X and Z simultaneously influence each other, or that X and Z do not affect one another but both are puppets of some third variable U. Positive correlation says that X and Z tend to move the same way over a data set, but does not explain why.

While the warning that correlation is not causation seems eminently logical, it is often ignored in practice. We will see some real-life examples of this problem in Chapter 2.

Sums of Correlated Random Variables

Two Correlated Variables

We know that, when X and Z are independent, the mean and variance of their sum $S = X + Z$ follows:

$$E(S) = E(X) + E(Z)$$

$$\sigma^2(S) = \sigma^2(X) + \sigma^2(Z)$$

But what happens when X and Z are correlated, meaning that ρ_{xz} is nonzero? The "mean of sum is sum of means" rule is unaffected: we noted earlier that its derivation never required that X and Z be independent. But the "variance-of-sum is sum-of-variances" derivation did employ the independence property. Intuitively, a strong *negative* correlation between X and Z would seem to create greater stability for their sum: when X is above $E(X)$, we would expect Z to be below $E(Z)$, meaning that $X + Z$ would vary less around $E(X) + E(Z)$ than it would if X and Z were independent. Similarly, strong positive correlation has the opposite effect: if X and Z get high and low together, then $X + Z$ spends a good deal of time at a considerable distance from $E(X) + E(Z)$, a circumstance conductive to a large average squared distance from the mean. We need to explore the situation.

We show in a derivation at the website that:

If $S = X + Z$, the general formula for $\sigma^2(S)$ is:

$$\sigma^2(S) = \sigma^2(X) + \sigma^2(Z) + 2\rho_{XZ}\sigma(X)\sigma(Z) \qquad (1\text{-}95)$$

where ρ_{XZ} is the correlation coefficient between X and Z.

Thus, the general equation for variance of a sum involves the individual variances, but it also includes the correlation between the two variables that are added together. When that correlation is negative, the variance of the sum falls *below* the variance that would prevail if X and Z were independent; a positive correlation pushes the variance upward. We foreshadowed these outcomes in the previous paragraph. Note that, when X and Z are independent and thus $\rho_{XZ} = 0$, then (1-95) reduces to our earlier formula for $\sigma^2(S)$.

The **covariance** of X and Z is defined by:

$$\text{cov}(X,Z) = E(XZ) - E(X)E(Z) = \rho_{XZ}\sigma(X)\sigma(Z)$$

Thus, (1-95) can be rewritten as:

$$\sigma^2(S) = \sigma^2(X) + \sigma^2(Z) + 2\text{cov}(X,Z) \quad (1\text{-}96)$$

Sometimes, we can compute cov(X,Z) directly and thus can use (1-96) as a shortcut to finding $\sigma^2(S)$.

EXAMPLE 1.31

Will Power

Mendel owns two homes: his primary residence and a vacation home in the desert. In his will, he asks that the two homes be sold upon his death, and the proceeds donated to the fund for a statistical amusement park. Given the uncertainties about his remaining lifespan and the vagaries of the housing market, he estimates that X, the amount that will go to the park from the sale of his primary residence is a random variable with mean \$900K and standard deviation \$300K. The corresponding amount Z from the desert property has mean \$400K and standard deviation \$100K. He estimates the coefficient of correlation of X and Z as 0.65.

What are the mean and standard deviation of the amount of money that will flow to the park fund based on Mendel's will?

Solution

We have $E(X) = 900$, $\sigma_X = 300$, $E(Z) = 400$, $\sigma_Z = 100$, and $\rho_{XZ} = 0.\ 65$. The random quantity $S + X + Z$ is donated to the park. It follows that:

$$E(S) = E(X) + E(Z) = 900\text{K} + 400\text{K} = \$1.3 \text{ million}$$

$$\sigma^2(S) = \sigma^2(X) + \sigma^2(Z) + 2\rho_{XZ}\sigma(X)\sigma(Z) = 300^2 + 100^2 + 2 * (0.65) * 300 * 100$$

Thus, $\sigma^2(S) = 139{,}000$ and $\sigma_S = \sqrt{139{,}000} = \$372.8K$

EXERCISE 1.33 Suppose in Example 1.33 that Mendel's financial advisor tells him that the uncertainty in the combined selling prices of his two homes is 25 percent of their expected total sale price. (By "uncertainty," she means standard deviation.) What is she assuming about the correlation coefficient for the sale prices of the two homes?

The Takeaway Bar

Each chapter of this book will end with a Takeaway Bar, which will summarize the key methodological points that have arisen. It will start with a glossary of terms, presented not in alphabetical order but instead in roughly the order in which the terms arose in the chapter. Then the Bar will list the main formulas from the chapter, and after that go on to offer "some parting thoughts," a brief collection of tips and observations. The hope is that the Bar will be helpful for review, and offer "one-stop shopping" for people who want to locate something important from the chapter without rereading it page-by-page.

Here is the Bar for Chapter 1.

Glossary

Basic Probability:

An **experiment** is a process that yields an outcome/event that is generally not known in advance.

The **sample space** is the set of all possible events/outcomes of an experiment.

An **elementary event** arising from an experiment is one of the "building blocks" of the sample space, and cannot be expressed as the union of two or more other events from that experiment.

An **event** is a subset of the sample space, and may consist of a combination of elementary events.

Kolmogorov's Axioms specify conditions that numerical assignments of probabilities should satisfy.

Subjective probability arises when a person uses a number on a 0–1 scale to express her view about the likelihood of an event.

The **equal probability axiom** states that two events believed equally likely should be assigned the same probability. Two events are considered equally likely if, roughly speaking, neither would be more surprising than the other.

The **relative frequency** concept of probability arises from the idea that the probability of an event is essentially the *fraction* of time it would occur in the long run. if the experiment that could give rise to the event were repeated under independent and identical conditions.

Mutually exclusive events (also known as **disjoint** events) are events that can never occur together in an experiment. Their intersection in the sample space is the empty set.

Exhaustive events have the property that at least one of them will definitely occur.

Complementary events are two mutually exclusive events, one of which will definitely occur. (Complementary events are exhaustive.)

The **conditional probability** of event *B* given event *A* is the probability assigned to *B* when it is known that A has occurred

Bayes' Theorem is a form of the conditional probability law which expresses $P(B|A)$—the conditional probability of event *B* given that event *A* is known to have occurred—under a formula that uses the original probabilities of *A* and *B* ($P(A)$ and $P(B)$, respectively) and also $P(A|B)$, the conditional probability of *A* given *B*.

Independent events have the characteristic that knowing that one of them has occurred does not change the probability that any of the others occurred.

The **joint probability** of several events is the probability that all of them occurred.

Discrete Random Variables:

A **random variable** is a mathematical function that assigns a real number to each elementary event in the sample space.

The **realization** of a random variable is the number it actually assumes, based on which elementary event occurs

A **discrete random variable** can take on a discrete set of numerical values (e.g., integers) with probabilities that are also specified.

The **probability mass function** of a discrete random variable specifies each realization that can arise with nonzero probability, and the probability associated with each realization.

The **mean** of a discrete random variable is a weighted average of its realizations, with the weight of each value equal to the probability assigned to it. Under the Law of Large Numbers, the mean corresponds to the long-term average of outcomes if the experiment that generates the random variable is conducted repeatedly. (The mean is sometimes called the **expected value** of *X*.)

The **variance** of a random variable is a weighted average of squared differences between individual realizations and the variable's mean, with the weight assigned to each squared difference equal to the probability that it arises.

The **standard deviation** of a random variable is the square root of its variance. The standard deviation is expressed in the same units as the mean, and is a widely-used measure of the volatility of the random variable.

A **memoryless** random process has the property that what happened in the past has no relevance to what will happen in the future.

A **binomial process** involves a series of n independent (Bernoulli) trials, each of which can yield one of two complementary outcomes *A* and *B*. $P(A)$, the probability that outcome *A* raises on a particular trial, is the same for all trials. The binomial random variable X is the number of times that A comes up over the n trials. (More precisely, X assigns to each sequence of trial outcomes the number of A's within that sequence.)

A **geometric process** likewise involves a series of independent Bernoulli trials, each of which can yield outcome A or its complement B. $P(A)$, the probability that outcome A raises on a particular trial, is the same for all trials. The geometric random variable X is the number of trials until outcome A arises for the first time.

A **hypergeometric process** is one in which j items are drawn at random without replacement from a finite population that is divided into two distinct groups of known sizes, m and n. The hypergeometric random variable X concerns the number of items selected that come from the first group.

A (time-related) **Poisson process** is one in which events occur randomly over time, at an average rate of λ per unit time. The Poisson random variable X is the number of events that occur over a period of length T.

Continuous Random Variables

A **continuous random variable** is a mapping from the elementary events in the sample space to an interval or intervals on the line of real numbers. In less formal parlance, a continuous random variable can take on any value within some interval(s) of real numbers. One does not speak of the probability that a continuous random variable takes on some *exact* value, but one can quantify the chance that it falls between two values.

The **cumulative distribution function** (CDF) $F_X(y)$ of a continuous random variable X specifies for each number y the probability that X is at or below y.

The **probability density function** (pdf) $f_X(y)$ of a continuous random variable X is the derivative of its CDF, and its graph takes the shape of the histogram of outcomes if the process that generates X is repeated many times. The probability that X falls between y and $y + \Delta$ is $f_X(y)\Delta$ to first order (Δ very small). The height of the density curve at a given value y is proportional to the chance that X falls in the immediate neighborhood of y.

The **Ath percentile** of continuous random variable X is the number C_A for which $P(X \leq C_A) = A\%$.

The **median** of a random variable is its 50th percentile (i.e., its halfway mark). Some people prefer the median to the mean as a measure of the "middle" of the distribution.

The **mode** of a continuous random variable X is the value of y at which $f_X(y)$ is maximized.

A **uniform** random variable is equally likely to take on any value in some range (a,b) but cannot take on values outside of (a,b).

An **exponential** random variable takes on non-negative values, and its pdf has a peak at zero and decays exponentially from there at rate λ. In a Poisson process, the time between consecutive events is exponential.

A **normal** random variable follows a symmetric bell-shaped curve. Its density function drops off exponentially from a peak at the middle of the curve, with the degree of decay tied to the square of the distance from the middle.

Combinations of Random Variables

The **joint probability mass function** of two discrete random variables records the probability that they take a particular combination of numerical values, for all the combinations that have nonzero probability.

The **joint probability density function** of two continuous random variables indicates the probability that these variables fall in various incremental rectangles. The density function must be multiplied by the area of the rectangle to estimate that probability to first order.

Two random variables are **independent** if the probability mass function or probability density function of either of them is the same regardless of the specific value the other variable takes on (i.e., its realization).

The **Central Limit Theorem** states that the *average* of a large number of independent and identically distributed (iid) random variables approximately follows a normal distribution. This normality is achieved even though the common distribution of the variables that are averaged is not normal.

Two random variables X and Z are **positively correlated** if, when one of them is high in its distribution, the other tends likewise to be high. (Also, when one of them is low, the other tends to be low.)

Two random variables x and z are **negatively correlated** if, when one of them is high in its distribution, the other tends to be low in its own.

Two random variables x and z are **uncorrelated** if the fact that one of them is high essentially implies nothing about whether the other one is. Independent random variables are uncorrelated, but lack of correlation does not prove independence.

The **coefficient of correlation** of two random variables is the mean of the product minus the product of their means, divided by the product of their standard deviations. It cannot fall outside the range (–1,1). Values of the coefficient near 1 indicate a strong tendency of the two variables to be high together and low together. Values near –1 indicate that high values of one variable generally are accompanied by low values of the other. Values near zero indicate that having one of the variables high (low) says little about whether the other is high (low).

The **covariance** of two random variables is the mean of their product minus the product of their means. It is the numerator of the coefficient of correlation of the two variables.

Key Formulas

Basic Probability

- If P(A) is the probability of event A, then:

$$0 \leq P(A) \leq 1$$

(A probability of 1 is identified with virtual certainty, while probability 0 is identified with impossibility.)

- If $A_1, A_2, \ldots A_n$ are mutually exclusive events, then:

$$P(A_1 \cup A_2 \ \ldots \ \cup A_n) = \sum_{i=1}^{n} P(A_i)$$

- If $A_1, A_2, \ldots A_n$ are mutually exclusive and exhaustive events, then:

$$\sum_{i=1}^{n} P(A_i) = 1$$

- If B and B^c are complementary events, then:

$$P(B_c) = 1 - P(B)$$

- If B and B^c are complementary events and A some other event, then:

$$P(A) = P(A \cap B) + P(A \cap B^c)$$

- If B_1, B_2, and B_n are mutually exclusive and exhaustive events and A is some other event, then:

$$P(A) = P(A \cap B_1) + P(A \cap B_2) + \ldots + P(A \cap B_n)$$

(This equation is called the Law of Alternatives or the Law of Total Probability.)

- For any two events A and B:

$$P(B|A) = P(A \cap B)/P(A)$$

- Bayes' Theorem:

$$P(B|A) = P(B) * P(A|B) \ / \ P(A)$$

- For *any* events $A_1, A_2, \ldots A_n$:

$$P(A_1 \cap A_2 \ldots \cap A_n) = P(A_1)\ P(A_2|A_1) \ldots\ P(A_3|A_1 \cap A_2) \ldots *\ P(A_n|A_1 \Rightarrow A_{n-1})$$

(This equation is called the Joint Probability Law.)

- If events A and B are independent, then:

$$P(B|A) = P(B);\ P(A|B) = P(A)$$

If $A_1, A_2, \ldots A_n$ are independent events, then:

$$P(A_1 \cap A_2 \ldots \cap A_n) = \pi_{i=1,n} P(A_i)$$

(where π means multiplication)

Discrete Random Variables

- The **mean** μ_x of discrete random variable X (also written as $E(X)$) follows:

$$\mu_X = \sum_{j=1}^{M} a_j p_j$$

where a_j = the jth smallest value that X can assume and $p_j = P(X = a_j)$

- If $Z = cX + b$ where c and b are numerical constants, then:

$$E(Z) = cE(X) + b$$

Conditional Expectation Rule:

- If the A_i's are a set of n mutually exclusive and exhaustive events, then:

$$E(X) = \sum_{i=1}^{n} E\left(X \middle| A_j\right) P\left(A_j\right)$$

where $E(X|A_i)$ is the conditional mean of X given that event A_i occurs

- The **variance** $\sigma^2(X)$of discrete random variable x follows:

$$\sigma^2(X) = \sum_{j=1}^{M} \left(a_j - \mu_X\right)^2 p_j$$

where a_j = the jth smallest value that X can assume and $p_j = P(X = a_j)$

- The standard deviation $\sigma(x)$ of random variable x is the positive square root of its variance:

$$\sigma(x) = +\sqrt{\sigma^2(X)}$$

- If $Z = cX + b$, then:

$$\sigma^2(Z) = c^2\sigma^2(x)$$
$$\sigma(Z) = |c|\,\sigma(x)$$

where $|c|\,\sigma$ is the absolute value of c
Binomial, geometric, hypergeometric, and Poisson random variables can only take on integer nonnegative values.

- If X is *binomial* with parameters n and p, then:

$$P(X = k) = \binom{n}{k} p^k (1-p)^{n-k} \text{ for } 0 \le X \le n$$

where $\binom{n}{k} = n!/\big(k!(n-k)!\big)$

- If X is *geometric* with parameter p, then:

$$P(X = k) = p(1-p)^{k-1} \text{ for } k \ge 1$$

- If X is *hypergeometric* with parameters (j, m, n) then:

$$P(X = k) = \binom{m}{k}\binom{n}{j-k} \Big/ \binom{m+n}{j}$$

where $0 \le k \le \min(m,j)$

- If X is *Poisson* with parameters λ and T, then:

$$P(X = k) = (\lambda T)^k e^{-\lambda t}/k! \quad \text{for } k \ge 0$$

Continuous Random Variables

Some Details about Three Continuous Probability Distributions:

Distribution	Parameters	Notation	Mean	Standard Deviation
Uniform	a,b $(b > a)$	$U(a,b)$	$(a + b)/2$	$.29(b - a)$
Exponential	λ	$e(\lambda)$	$1/\lambda$	$1/\lambda$
Normal	μ,σ	$N(\mu,\sigma)$	μ	σ

- If $F_x(y)$ is the *cumulative distribution function* (CDF) of continuous random variable X, then:

$$P(X \le y) = P(X < y) = F_X(y)$$
$$P(X \ge y) = P(X > y) = 1 - F_X(y)$$
$$P(c \le X \le d) = P(c < X < d) = F_X(d) - F_X(c) \quad (d > c)$$

- The probability density function $f_X(y)$ of a continuous random variable is the derivative of its CDF $F_x(y)$. The probability that X falls in the interval $(y, y + \Delta)$ when Δ is very small is approximately $f_x(y)\,\Delta$.
- If X is *uniform* with parameters a and b, then its CDF $F_x(y)$ and pdf $f_x(y)$ follow:

$$F_x(y) = \begin{cases} 1 & \text{for } y > b \\ \dfrac{y-a}{b-a} & \text{for } a \le y \le b \\ 0 & \text{for } y < a \end{cases}$$

$$f_X(y) = \begin{cases} \dfrac{1}{b-a} & \text{for } a < y < b \\ 0 & \text{for } y < a \text{ and } y > b \end{cases}$$

- If X is exponential with parameter λ, then:

$$F_X(y) = \begin{cases} 1 - e^{-\lambda y} & \text{for } y \ge 0 \\ 0 & \text{for } y < 0 \end{cases}$$

$$f_X(y) = \begin{cases} \lambda e^{-\lambda y} & \text{for } y > 0 \\ 0 & \text{for } y < 0 \end{cases}$$

- If X is normal, $N(\mu, \sigma)$, then:

$$F_x(y) = \left(\frac{1}{\sigma\sqrt{2\pi}}\right)\int_{-\infty}^{y} e^{-\frac{(z-\mu)^2}{2\sigma^2}dz}$$

$$f_X(y) = \frac{1}{\sigma\sqrt{2\pi}}e^{-(y-\mu)^2/2\sigma^2} \quad \text{for any value of } y$$

Combinations of Random Variables

- For any two random variables X and Z:

$$\text{If } S = X + Z, \text{ then } E(S) = E(X) + E(Z)$$

- If X and Z are *independent* random variables, then it is also true that:

$$\sigma^2(S) = \sigma^2(X) + \sigma^2(Z)$$

- If $W = X - Z$ and X and Z are independent, then:

$$E(W) = E(X) - E(Z)$$

$$\sigma^2(W) = \sigma^2(X) + \sigma^2(Z)$$

- For random variables $W_1, W_2, \ldots W_n$, and $\sum_{i=1}^{n} W_i$ it is always true that:

$$E(S) = \sum_{i=1}^{n} E(W_i)$$

If the W_i's are independent random variables, then:

$$\sigma^2(S) = \sum_{i=1}^{n} \sigma^2(W_i)$$

- If $\bar{X}$ is the average of n independent and identically distributed (*iid*) random variables that share mean $E(X)$ and standard deviation $\sigma(X)$, then:

$$E(\bar{X}) = E(X)$$
$$\sigma(\bar{X}) = \sigma(X)/\sqrt{n}$$

- If $\bar{X}$ is the average of a n independent and identically distributed (*iid*) random variables that share mean $E(X)$ and standard deviation $\sigma(X)$ and n is moderately large then:

$$\bar{X} \approx N\left(E(X), \sigma(X)/\sqrt{n}\right)$$

- The coefficient of correlation ρ_{xz} of two random variables X and Z follows:

$$\rho_{XZ} = [E(XZ) - \mu_x\mu_z]/\sigma_x\sigma_z$$

- The quantity $E(XZ) - \mu_x\mu_z$ is called the covariance of X and Z.
- If X and Z are correlated random variables, then:
  If $S = X + Z$:

$$E(S) = E(X) + E(Z)$$
$$\sigma^2(S) = \sigma^2(X) + \sigma^2(X) + 2\rho_{xz}\sigma(X)\sigma(Z)$$

Also:

$$\sigma^2(S) = \sigma^2(X) + \sigma^2(Z) + 2\text{cov}(X,Z)$$

where cov(X,Z) is the covariance of X and Z

Some Parting Thoughts

- While it is rarely possible to eliminate uncertainty, a more precise description of the uncertainties we face might make us better able to cope with them. The study of Probability allows for more precise descriptions.
- Estimating the probability of an event by adding the probabilities for the different ways that it can happen is correct *only* when the ways are mutually exclusive.
- Sometimes, it is easier to find the probability that an event will not occur than the probability that it will. When that happens, it might be wise to find the chance the event won't occur and then, under the rule for complementary events, to subtract that probability from one.
- To find the probability that all events in a series occur (i.e. their *joint probability*), one can multiply their individual probabilities if they are independent. If the events are not independent, one can still find their joint probability through multiplication, provided that one uses appropriate conditional probabilities as factors.
- When it is difficult exactly to determine the probability of some event, it can be useful to approximate that probability under simplifying assumptions.
- When the result of a probability calculation is surprising (and not in error), it is worthwhile trying to understand why the result makes sense. Doing so can improve one's intuition, and might lead to fewer surprises in the future.
- Even the best decision under uncertainty may turn out, when all the facts become known, to be the wrong decision. But the person who takes the trouble to estimate the risks and does the best possible given those risks has nothing to apologize for.
- The basic laws of probability apply just as well in the domain of probability distributions as they did elsewhere. We can use Bayes' Theorem, for example, to find the conditional probability that a random variable falls in a certain range given certain information about that variable.
- *Electronic tables* exist for any well-known probability distributions, and they can all but eliminate tedious calculations. These tables arise in statistical packages, as stand-alone tables on the web, or in even mildly sophisticated hand calculators. One has to get used to the eccentricities of particular tables; many of them, for example, use the word "success" when outcome A arises rather than its complement, even though outcome A might be anything but a success (e.g., a disastrous flood).
- Means and standard deviations offer good first descriptions of how random variables behave. But they are only an "executive summary" of what is going on, and not the whole story.
- Many important discrete random processes involve a series of trials, each of which leads either to event A or its complement B. The binomial, geometric, and hypergeometric processes all fit this framework. In the binomial and geometric processes, $P(A) = p$ on all trials and different trials have independent outcomes; in the hypergeometric setting, different trials are not independent and $P(A)$ varies across trials based on previous outcomes.

- If events occur completely at random over time except that they occur at a known average rate per unit time, then we are in the domain of the *Poisson* distribution for the number of events over a given period.
- If we know the CDF of a continuous random variable, we can use it to work out *any* probability associated with the variable: the chance of falling at or below y, falling strictly below y, exceeding y, or falling between any two values c and d.
- If we think some random quantity X is between two limits a and b ($b > a$) but how no idea where it falls between them, we might well model the situation by assuming that x is *uniform* in the range (a,b), namely, $U(a,b)$. If we have some idea where X falls (e.g., much more likely towards the middle of the range than at its edges), then we should develop a pdf that reflects our beliefs.
- Many powerful results—about the Central Limit theorem and about sums, differences, and products of random variables—depend on assumptions of *independence*. When independence does not prevail, neither in general do these results, so be careful.
- The Central Limit Theorem is remarkable, because it assures the normality of the average (and sum) of a moderate or greater number of independent and identically distributed (iid) random variables, even when the individual variables are not normal.
- In general, the Central Limit Theorem is considered reliable when the number of variables added or averaged is at least 30. However, normality is often approximately present for as few as 15 such variables, unless we are especially interested in probabilities involving the outer tails of the distribution.
- However, the correlation coefficient often allows considerable progress even in the absence of independence.
- Correlation is not causality: the fact that two variables move the same way does not mean that one of them causes the movements of the other. Causality might indeed be present, but the evidence for it must go beyond the correlation itself.

EZ Pass

1. If a fair six-sided die is tossed repeatedly, the probability that the first three arises on the second toss is:
   (A) 1/3
   (B) (5/6) ∗ (1/6)
   (C) (5/6) ∗ (1/5)
   (D) none of these
2. If $P(B|A) = P(A|B)$ and both $P(A)$ and $P(B) > 0$, then the statement $P(A) = P(B)$ is:
   (A) always true
   (B) sometimes true
   (C) never true
   (D) none of these

3. Suppose that $P(B|A) = 2/5$, $P(B) = 3/5$, and $P(A) = 1/2$, and A^c is the complement of event A. Then $P(B|A^c)$ is:
   (A) 1/2
   (B) 3/5
   (C) 2/3
   (D) 4/5
4. If $P(A|B) = 1$, then
   (A) $A \cup B = A$
   (B) $B = A^c$
   (C) $A \cap B = A$
   (D) $A \cap B = B$
5. If three people are in a room, the probability they all have the same birthday (assuming birthdays are uniform over the year is):
   (A) $\left(\frac{1}{365}\right)^3$
   (B) $\left(\frac{1}{365}\right)^2$
   (C) $\frac{3}{365}$
   (D) none of these
6. Discrete random variable X can take on more than one value. Given this fact, the variance of X:
   (A) must be zero
   (B) must be nonzero
   (C) might be either zero or nonzero
   (D) all of these
7. In a discrete random process in which X is a nonnegative integer, $P(X = 1) > P(X = 2)$ is always true if:
   (A) X has a binomial distribution
   (B) X has a Poisson distribution
   (C) X has a geometric distribution
   (D) X has a hypergeometric distribution
8. If a Poisson process with parameter λ is observed over two nonoverlapping periods of equal length, the probability that there were more events in the second period than in the first:
   (A) must exceed 50%
   (B) must fall below 50%
   (C) must equal 50%
   (D) none of these
9. As y goes up, which of the following might decline for continuous random variable X?
   (A) Its probability density function (pdf) but not its cumulative distribution function (CDF)

(B) Its CDF but not its pdf
(C) Both its pdf and its CDF
(D) Neither its CDF nor its pdf

10. Which of the following could not be the probability density function of a continuous random variable X?
(A) $f_X(y) = 1/3$ in $(4,7)$; $f_X(y) = 0$ otherwise
(B) $f_X(y) = (3/127)y^2$ in $(6,7)$; $f_X(y) = 0$ otherwise
(C) $f_X(y) = e^y/\sqrt{\pi}$ in $(0, \sqrt{\pi})$; $f_X(y) = 0$ otherwise
(D) all of these $f_X(y)$'s could be probability density functions

11. If the mean and standard deviation of a continuous random variable are the same, then it could ***not*** be:
(A) uniformly distributed
(B) normally distributed
(C) exponentially distributed
(D) it could be any of these

12. If we make three random selections from the distribution of X when X is $U(a,b)$:
(A) at least one of the selections will be closer to a than b
(B) the second largest of the three selections must be $(a + b)/2$
(C) all three selections must fall in the range (a,b)
(D) all of the above

13. If the lifespan of a battery is exponentially distributed with mean 50 hours, then replacing it when it reaches age 50 hours with another battery with the same lifespan distribution:
(A) is irrational, because both batteries have the same distribution for remaining lifespan
(B) makes sense, because the battery that reaches its mean lifespan is likely to die sooner than the newborn battery
(C) is irrational, because the very fact that the initial battery reached age 50 means that its remaining lifespan will probability exceed the full lifespan of the new battery
(D) all of these

14. In a normal distribution, the mean and median:
(A) are always equal
(B) are sometimes equal but not always
(C) are never equal
(D) always add up to zero

15. If we take the average of six random variables that share mean μ, then the mean of this average:
(A) must always be μ
(B) must be μ only if the variables are independent
(C) must be μ only if the variables are identically distributed
(D) must be μ only if the variables have the same variance

16. The Central Limit Theorem concerns:
(A) the average of any independent random variables

(B) the average of identically distributed random variables, whether or not they are independent
(C) the average of any random variables that have a common mean and variance
(D) none of these

17. If $\sigma(X) > 0$, $E(X + Z) = 0$ and $\sigma(X + Z) = 0$ then:
(A) $E(X) = E(Z) = 0$
(B) $Z = X$
(C) $Z = -X$
(D) Who knows?

18. If X and Z have common variance σ^2 and they have perfect positive correlation, then the variance of their sum is:
(A) 0
(B) $2\sigma^2$
(C) $4\sigma^2$
(D) cannot be determined from the given information

Further Exercises

EXERCISE 1.32 In the "March Madness" basketball tournament in the United States, two pairs of teams meet in the next-to-last of the "brackets." Then the winners of these games face each other in the final bracket. The winner of that game is the victor in the tournament.

Suppose that two teams (East and West) are to compete in the next-to-last bracket, are as North and South. Experts in sports analytics have the following probability estimates about the outcomes for various combinations of teams:

	Probability It Would Win Against:			
Team	East	West	North	South
East	—	2/5	1/2	3/5
West	?	—	3/5	3/5
North	?	?	—	11/20
South	?	?	?	—

(i) Assuming that there are no tie games, there is enough information above to replace all the ?'s with numbers. What are they?
(ii) For each team, find the probability that it will win the tournament. (To win, it must first win its next-to-last bracket and then defeat the winner of the other next-to-last bracket.
(iii) Suppose that it might be possible to change the competitors in the next-to-last bracket (so that (say) East could play South). Could such a rearrangement

change the probability that any given team would win the tournament? Briefly but specifically discuss.

(iv) Would it be possible for South and East to come up with a joint proposal for changing the competitors in the next-to-last bracket under which both of them would have a higher probability of winning the tournament? If so, identify the proposal; if not, make clear why such a proposal is impossible.

EXERCISE 1.33 We will now derive an important probability formula for $P(A \cup B)$ when A and B are not mutually exclusive.

Suppose that A and B are two events, and A^c and B^c their respective complements.

(i) Taking account of the three ways that "A or B" can arise, explain why it is true that:

$$P(A \cup B) = P(A \cap B) + (A \cap B^c) + P(A^c \cap B)$$

(ii) Explain in simple words why it must also be true that:

$$P(A) = P(A \text{ and } B) + (A \text{ and } B^c)$$

(iii) Write an expression for $P(B)$ analogous to that in (ii).

(iv) Now, combining the three equations arising in (i) → (iii), show that:

$$P(A \cup B) = P(A) + P(B) - P(A \cap B).$$

(v) What happens to this last formula when A and B are mutually exclusive? Are you happy?

(vi) Show that when A and B are independent:

$$P(A \cup B) = P(A) + P(B)(1 - P(A)).$$

(vii) Find $P(A \cup B)$ when $P(A) = .4$, $P(B) = .2$, and $P(B|A) = .25$.

EXERCISE 1.34 Suppose that it is known that a given test for lead poisoning has a *zero* false negative rate, and that one per cent of the people who are tested for lead positioning at a given emergency room actually have it. However, because the test has a false-positive rate, the percentage of people for whom the test comes back positive is not one per cent, but instead two per cent. (The false positive rate is the probability that someone who do not suffer lead poisoning yields a positive test result.)

If q is the false positive rate of the test, then what is the value of q? (*Hint:* It is not .01.)

EXERCISE 1.35 Suppose that you had to estimate the temperature based on readings from five thermometers, knowing that each one independently has an 80% chance of showing the correct temperature, a 10% chance of being exactly one degree (Fahrenheit) too high, and a 10% chance of being exactly one degree too low.

(i) If you got the five readings (80°,80°, 81°, 80°, 80°), how would estimate the temperature? (Assume that your estimate must be an integer.)
(ii) What is the probability that your estimate in (i) is correct?
(iii) Suppose you got the five readings (79°,81°, 81°, 81°, 81°), what would your estimate be? (Hint: Don't say 81°. That estimate would surely be wrong.)
(iv) What is the probability that your estimate in (iii) is correct?
(v) Suppose in the general case that the true temperature is T°. What is the probability that your decision rule based on the five readings yield an incorrect answer? (You can assume that T is an integer.)

EXERCISE 1.36 In 2009, the MIT Integrated Global Systems Model estimated that: "there is a 90 percent probability that temperatures will have risen 3.5 to 7.4 degrees Celsius by the end of the century."

Suppose that the increase is assumed to be uniformly distributed between two numbers a and b, and also that the distance of 3.5 above a (i.e., $3.5 - a$) is the same as the distance of b above 7.4 (i.e., $b - 7.4$).

(i) What numerical values of a and b are consistent with the statement above?
(ii) With those values of a and b:
- What is the probability that the rise will be between a and 3.5?
- What is the probability that the rise will exceed 5° Celsius?
- Find an interval centered at $(a + b)/2$ that has a 99% chance of including the actual increase.

EXERCISE 1.37 (Based on a true story) If Travelopedia sells at least \$3 million of tickets on airline X in a given month, the airline will give it a bonus of \$250,000. (The bonus is called an *override commission.*) If the airline fails to reach \$3 million in sales, it gets no bonus at all. Travelopedia projects that it has a 50% chance of meeting the \$3 million target if it does nothing. If Travelopedia itself buys \$60,000 in tickets on airline X—which it is not allowed to sell to anyone—it projects a 95% chance of meeting its target. If it buys \$120,000 in tickets, it projects a 99.9% chance of meeting its target.

(i) Assuming that *average* net gain tied to the bonus is the standard, what should Travelopedia do?

(ii) Suppose that Travelopedia projects that, absent any action on its part, the dollar volume of tickets it can sell in the rest of the month is normally distributed. If it is now short of the \$3 million dollar goal by \$400,000, what is the *mean* of that normal distribution given the statements above?

(iii) Given that Travelopedia will have a 95% chance of meeting its target with a \$60,000 purchase of useless tickets, what is a reasonable estimate of the standard deviation of the normal distribution in (ii)?

(iv) With the mean and standard deviation estimated in (ii) and (iii), is the probability of meeting its with a \$120,000 purchase of useless tickets very close to 99.9%?

EXERCISE 1.38 Derailments on the Vermont Pacific railroad West of the Rocky Mountains occur under a Poisson distribution with parameter 1.2 per year. Derailments East of the Rockies follow a Poisson distribution with parameter 1.3 per year, and derailments on opposite sides of the Rockies are independent events. Each derailment independently has a 40% chance of shutting down the line.

(i) What is the probability that there are at least two derailments on the Vermont Pacific over an 18-month period?

(ii) Given that there were (exactly) two derailments on the Vermont Pacific over 18 months, what is the probability that both were East of the Rockies?

(iii) Given two derailments over 18 months, what is the probability that there was one on each side of the Rockies?

(iv) What is the probability that there are two derailments over 18 months on the railroad and that neither of them shuts down the line?

CHAPTER 2

Statistics: First Stirrings

Strike Four?

It was the end of 2002, and Mendel was worried about his stock market holdings. The Dow Jones Industrial Average had gone down three years in a row, and it seemed that the huge increases of the 1990's were a thing of the past. But then Minerva appeared with what she thought was cheerful news.

"I was just reading *Barron's*," she reported, referring to the esteemed financial magazine. "It said that the Dow Jones has only declined for a fourth year in a row once in its 107-year history. So it estimated the chance of going down again this year as 1 in 104. It said 104 rather than 107 because the Dow Jones couldn't go down a fourth year in a row until the fourth year of its existence."

"And you consider that analysis comforting?" Mendel asked with a certain asperity. "Why not?" Minerva responded, "I know historical data aren't perfectly predictive of the future, but they do say something." "I'm not against history, but I'm against absurd reasoning," Mendel retorted. If this is the way financial experts think, then I'm more worried than ever about the years ahead "

To Be Continued

Statistics is about numbers. It is also about common sense. The numbers (or data) constitute clues that help us answer questions we care about. But, like all good detectives, we have to identify the primary clues, combine them in intelligent ways, and recognize that a particular clue can have more than one interpretation. Good statistical analysis can bring insight and joy; poor statistical analysis can bring suffering and pain.

In this chapter, we get started in Statistics by discussing first how to summarize the pattern in a set of numbers, both through summary statistics and geometric displays. Then we move on to talk about *statistical indicators*, namely, overall metrics like the unemployment rate and temperature/humidity index that most people think of when they hear the word "statistics." From there we proceed to some true stories about statistical mishaps in what seemed like elementary data analysis. The point of these stories is that analyzing data requires clear thinking: our "gut feelings" about how to process numbers can often conflict with what our brains would tell us.

Probability is essentially absent from this chapter. It plays a big role in statistical analysis, because uncertainty of one form or another almost always surrounds the data under study. But those issues can wait until Chapter 3.

2.1 Some Summary Statistics We've Seen Before

Given a set of numbers $(x_1, x_2, \ldots, x_n)$, what can we do to get a better sense what they are saying? We can start out by thinking about how we summarize probability distributions. The two statistics that were preeminent in our earlier discussions were the mean μ and the variance/standard deviation (i.e., σ^2 and σ). We can mimic the calculation of μ and σ using the numbers at hand (the *data*). For the mean, we can find $\overline{X}$, the simple average of the n numbers:

$$\overline{X} = \frac{1}{n}\sum_{i=1}^{n} x_i \qquad (2\text{-}1)$$

For the variance, we can take U^2, the average squared spread of the individual measurements around $\overline{X}$:

$$U^2 = \sum_{i=1}^{n} \frac{(x_i - \overline{X})^2}{n} \qquad (2\text{-}2)$$

The standard deviation U would once again be the square root of the variance:

$$U = \sqrt{U^2} = \sqrt{\sum_{i=1}^{n} \frac{(x_i - \overline{X})^2}{n}} \qquad (2\text{-}3)$$

As before, the mean $\overline{X}$ is a measure of the middle of the distribution, while standard deviation U is a measure of the spread of the measurements around their middle value.

Given a pair of data points (x_i, y_i) from each of n settings, we can compute a *correlation coefficient* to suggest the relationship between x and y. Once again, we mimic the calculation for probability distributions: $\hat{\rho}$, the analog of ρ, is given by:

$$\hat{\rho} = \frac{\left(\frac{1}{n}\right)\left(\sum_{i=1}^{n}(x_i - \overline{X})(y_i - \overline{Y})\right)}{U_X U_Y} \qquad (2\text{-}4)$$

As in probability, any coefficient of correlation calculated under (2-4) must fall in the range (–1,1).

The *coefficient of variation* V is the ratio of U to $\overline{X}$:

$$V = \frac{U}{\overline{X}} \tag{2-5}$$

When we are dealing with nonnegative quantities, V expresses the spread of measurements around the mean as a proportion of the mean. A low value of V suggests high consistency among the measurements, while a high value of V implies that variability among measurements is comparable in magnitude to the measurements themselves.

EXAMPLE 2.1

Call for Drugs?

Tables 2.1 and 2.2 present some recent statistics about revenue and profit at large American telecommunications companies and large American pharmaceutical companies.

(i) What are the mean, standard deviation, and coefficient of variation of profit as percentage of revenue in the two industries?

(ii) What is the coefficient of correlation of percentage profit and revenue among the ten companies in each industry?

TABLE 2.1 Profits and Revenues for the Ten Largest American Telecommunications Companies, 2009

Company	Revenue (Million $)	Profit (Million $)	Profit as % of Revenue
AT&T	124,028	12,867	10.4
Verizon	97,354	6428	6.6
Sprint/Nextel	35,635	–2796	–7.7
Comcast	34,256	2547	7.4
direcTV	19,693	1521	7.7
QWest	13,475	681	5.1
Dish	11,617	902	7.8
Liberty Global	10,561	–789	–7.4
Virgin Media	7440	–1693	–22.8
Cablevision	7230	–228	–3.1

Source: *Fortune* Magazine

TABLE 2.2 Profits and Revenues for the Ten Largest American Pharmaceutical Companies, 2009

Company	Revenue (Million $)	Profit (Million $)	Profit as % of Revenue
Johnson & Johnson	63,747	12,949	20.3
Pfizer	48,296	8104	16.8
Abbott	29,528	4881	16.8
Merck	23,850	7808	32.7
Wyeth	22,834	4418	19.3
Bristol-Myers Squibb	21,366	5247	24.6
Eli Lilly	20,378	−2072	−10.2
Scherling-Plough	18,502	1903	10.3
Amgen	15,003	4196	28.0
Gilead Science	5336	2011	37.7

Source: *Fortune* Magazine

Solution

We can use packages like Excel to do the calculation once we type in the numbers. (In Excel, the commands are AVERAGE for mean and STDEVP for standard deviation.) When we work out the means and standard deviations for *profit percentages* within these data, we get:

Telecommunications: $\bar{X} = 0.4\%$ $U = 10.0\%$

Pharmaceuticals: $\bar{X} = 19.6\%$ $U = 12.6\%$

Obviously, the mean was much higher in pharmaceuticals than telecommunications. The variability around the mean, however, was roughly the same in the two industries.

Do bigger companies (as measured by revenue) tend to achieve higher percentage profits? This sounds like a job for the correlation coefficient between revenue and percentage profit. The coefficients are:

Telecommunications: $\widehat{\rho_{RP}} = .456$

Pharmaceuticals: $\widehat{\rho_{RP}} = -.160$

R = Revenue; P = Percentage Profit

The coefficient for telecommunications suggests a tendency for bigger companies to earn higher profits. The positive sign is no surprise, given that all three of the smallest companies lost money. As for pharmaceuticals, the small negative coefficient indicates that smaller companies tended modestly to outperform their bigger counterparts. A look at the data substantiates that tendency: the average percentage profit was about 18% for the three largest firms listed, and 25% among the three smallest.

EXERCISE 2.1 Now pool the data for the twenty companies into one data set.

(i) Find the mean and standard deviation of the percentage profit for the twenty companies.

(ii) Do you see why the standard deviation of outcomes is higher now than it was when we considered telecommunications and pharmaceutical separately? Briefly explain what happened.

(iii) Find the correlation coefficient of revenue and percentage profit for the twenty companies.

(iv) Suppose we define an indicator variable which is 1 for each pharmaceutical company and 0 for each telecommunications company. What is the correlation coefficient of this variable and percentage profit for the twenty companies? Are you surprised at the outcome?

A Data Set: Life in Europe

To illustrate many of the methods we introduce in this chapter, we work with a data set from Europe, for all of its countries with at least 100,000 citizens in 2010. We do not include countries that are partly in Europe and partly in Asia, meaning Russia and Turkey. For each of the European countries, we list three statistics:

- Population in 2009
- Population Density (measured in people per square kilometer (not per mile: this is Europe)
- Fertility rate (number of children born per woman in her child-bearing years)

We might have many curiosities about Europe that these numbers can satisfy. Which European countries are the most and least populous? Is there much variation in population density across the countries of Europe? Which European countries have such low fertility rates that they seem destined to lose population in the years ahead? Do countries with high population densities have high fertility rates? The European data table below (also available at the website that accompanies this book) allows answers to these questions.

Europe Is Much More than Numbers, but Its Numbers Do Say Something

TABLE 2.3 Europe in 2009

Country	Population (millions)	Population Density (People/Sq. Kilometer)	Fertility (Births/Mother)
Albania	3.6	110	2.06
Austria	8.2	100	1.42
Belarus	9.7	127	1.20
Belgium	10.4	355	1.65
Bosnia	4.6	74	1.23
Bulgaria	7.3	68	1.31
Croatia	4.5	79	1.35
Czech Republic	10.2	132	1.23
Denmark	5.5	128	1.80
Estonia	1.3	30	1.49
Finland	5.2	16	1.83
France	61.0	114	1.89
Germany	82.4	229	1.36

TABLE 2.3 Europe in 2009 *(Continued)*

Country	Population (millions)	Population Density (People/Sq. Kilometer)	Fertility (Births/Mother)
Greece	10.7	86	1.33
Hungary	9.9	108	1.28
Iceland	0.3	3	2.05
Ireland	4.2	63	1.96
Italy	58.1	200	1.38
Latvia	2.2	35	1.29
Lithuania	3.6	51	1.26
Luxembourg	0.5	194	1.66
Macedonia	2.0	82	1.43
Malta	0.4	1318	1.37
Moldova	4.3	144	1.40
Montenegro	0.6	45	1.83
Netherlands	16.7	400	1.72
Norway	4.6	13	1.85
Poland	38.5	122	1.23
Portugal	10.7	115	1.46
Romania	22.2	90	1.30
Serbia	10.2	101	1.79
Slovakia	5.4	111	1.25
Slovenia	2.0	102	1.28
Spain	40.5	91	1.41
Sweden	9.0	21	1.8
Switzerland	7.6	188	1.42
Ukraine	47.4	78	1.22
United Kingdom	60.9	255	1.82

Note: This table only contains countries with at least 100,000 citizens that are wholly contained in Europe.

EXAMPLE 2.2 **Europe at a Glance**

(i) Find the mean, standard deviation, and coefficient of variation of population, population density, and fertility for these countries.

(ii) What is the correlation coefficient for population and population density, population and fertility, and population density and fertility?

(iii) How might one interpret the calculated numbers?

Solution

(i) Applying the above formulas to our European data set, we get the following summary statistics:

Population:	
Mean ($\bar{X}$):	15.4 million inhabitants (per country)
Standard deviation (U):	20.5 million inhabitants
Coefficient of Variation (V):	1.33 (20.5 million/ 15.4 million)
Population Density:	
Mean ($\bar{X}$):	146.8 inhabitants per sq. kilometer
Standard deviation (U):	210.1 inhabitants per sq kilometer
Coefficient of Variation (V):	1.43
Fertility:	
Mean ($\bar{X}$):	1.52 births per mother
Standard deviation (U):	0.26 births per mother
Coefficient of Variation (V):	0.17

(ii) Correlation Coefficients:

Population vs. Population Density:	+.036
Population vs. Fertility:	–.071
Population Density vs. Fertility:	–.087

(iii) What do we make of these numbers? A brief synopsis might be:

- The "average" European country has about 15 million citizens. But there is a good deal of variation around the average, as indicated by both the standard deviation and coefficient of variation.
- The average population density in Europe (under a "one country, one vote" rule) is about 150 per square kilometer. But there is a good deal of variation in population density across European countries, as both the standard deviation and coefficient of variation make clear.
- The average European country experiences 1.52 births per mother over her childbearing years. On this dimension, there is consistency across countries: the coefficient of variation indicates that cross-country variability around this average is only about one-sixth of the average itself.

  These childbirth statistics could be construed as worrisome. For a population to keep steady (absent immigration), there must be at least two births per mother. (Fathers, you will recall, do not bear children.) Not only is Europe on average about 25% below this threshold (1.52 vs. 2), but there is little cross-country variation around this mean. Inspection of the data (which is a healthy supplement to summary statistics) confirms the impression conveyed by $\bar{X}$ and V: of the 38 European countries, only two of them reach the threshold of two (Iceland and Albania), and these two countries just barely do so. Europe, therefore, could well be on a trajectory to a lower population.

- There is scant visible relationship between any pair of the three variables studied. The correlation coefficient was near zero for population and population density. This outcome might seem puzzling, but what it is saying is that the most populous countries tend to be larger in area than the others, meaning that "people per square kilometer" is not higher. The low correlation coefficients for fertility vs. both population and population density suggest that neither bigger nor more densely-populated countries have conspicuously different fertility levels than the rest. (The fertility coefficients add to what we said earlier: we already knew that fertility rates did not vary much across countries, but now we know that the modest variation in fertility that does exist is not appreciably tied to either population or population density.)

EXERCISE 2.2 In the sport cricket, a key metric for measuring the performance of individual players is the batting average, which was described on Wikipedia as follows:

> "In cricket, a player's **batting average** is the total number of runs he has scored divided by the number of times he has been out. Since the number of runs a player scores and how often he gets out are primarily measures of his own playing ability, and largely independent of his teammates, batting average is a good metric for an individual player's skill as a batsman. The number is also simple to interpret intuitively, being approximately the average number of runs the batsman scores per innings. Batting average has been used to gauge cricket players' relative skills since the 18th century."

At the website are the batting averages and some other information for the players on the national team of India as of late 2011. Using those data:

(i) What are the mean, standard deviation, and coefficient of variation of the batting averages of the individual players?

(ii) What is the correlation coefficient between year the player started and his batting average? Do you have any conjecture about why the coefficient takes this value?

(iii) The description of batting average asserts that the statistic—which is NOT the same as runs per innings—is nonetheless a good proxy for the average number of runs per innings played. (The word "innings" is singular in cricket.) Perform a statistical calculation on these data that is relevant to the accuracy of this assertion. What do you conclude?

Weighted Averages

We now return from Europe to discuss general concepts. Sometimes, one wants to average numbers together in a way that gives more emphasis to some of them than others. That desire leads to the concept of weighted averages.

The **weighted average** W_α two numbers x_1 and x_2 follows:

$$W_\alpha = \alpha x_1 + (1 - \alpha)x_2 \tag{2-6}$$

where α is the *weighting factor* and $0 \leq \alpha \leq 1$

Note that when $\alpha = 0$, the weighted average W_0 is just x_1. When $\alpha = 1$, $W_1 = x_2$, and when $\alpha = 1/2$, $W_{1/2} = \bar{X}$. When $\alpha = 3/4$, $W_{3/4}$ is actually the simple average of x_1, x_1, x_1, and x_2: after all, $(x_1 + x_1 + x_1 + x_2)/4 = (3x_1 + x_2)/4 = (3/4)x_1 + (1/4)x_2 = W_3/4$. In that sense, x_1 has three times the weight of x_2 when $\alpha = 3/4$.

When there are n measurements $(x_1, \ldots, x_n)$ their weighted average W follows:

$$W = \sum_{i=1}^{n} \alpha_i x_i \tag{2-7}$$

where $0 \leq \alpha_i \leq 1$ for all $1 \leq j \leq n$ and $\sum_{i=1}^{n} \alpha_i = 1$

EXAMPLE 2.3

More Calls and Drugs

(i) Weighing each telecommunications company in Table 2.1 by its share of revenues for the ten listed companies, find the weighted average percentage profit for the companies.

(ii) Do the same for the ten pharmaceutical companies.

(iii) What is suggested by the difference between these weighted averages and the corresponding simple averages calculated in Example 2.1?

(iv) Suppose we measure industry profitability by summing the dollar profits of the ten Table 1 telecommunications companies and dividing by total revenue for the ten companies. What is the profitability statistic? Is it a coincidence that it is the same as the weighted average calculated in (i)?

Solution

The averages ($\bar{X}$'s) of percentage profitability for telecommunications and pharmaceuticals calculated earlier were based on a "one company, one vote" rule. But one might get a better picture of industry performance if greater weight were given to the bigger companies. For that reason, we might take a weighted average of the individual results, with the weight of each company equal to its share of total revenue for the ten listed companies. For example, AT&T had a 2009 revenue of $124,028 million while the combined revenue for the ten telecommunications companies was $361,289 million, meaning that its weight would be 124028/361289 = 0.343.

(i) Taking the quotient of each telecommunications company's revenue to total revenue for the ten listed companies, we get:

Company	Revenue (million USD)	Revenue Share
AT&T	124028	0.343
Verizon	97354	0.269
Sprint/Nextel	35635	0.099
Comcast	34256	0.095
direcTV	19693	0.055
Qwest	13475	0.037
Dish	11617	0.032
Liberty Global	10561	0.029
Virgin Media	7440	0.021
Cablevision	7230	0.02
TOTAL	**361289**	**1**

Multiplying this share by percentage profit for each company in Table 2.1 and then adding under (2-7), we reach the industry-wide statistic of 5.4%

(ii) Performing the analogous calculation for Pharmaceuticals, we reach the weighted average of 18.4%

(iii) The weighted average for pharmaceuticals is quite similar to the simple average ($\bar{X}$), which was 19.6%. But for telecommunications, the weighted statistic was somewhat higher than the simple one: 5.4% vs. 0.4%. The reason for the difference is that the smallest of the telecommunications firms suffered losses, which pushed the simple average down more than the weighted one.

(iv) The total revenue of the telecommunications companies is $361,289 million and their total profit is $19,440 million. Thus, the profit-to-revenue ratio is 5.4%. That is the same as the weighted average in (i). This outcome is not a coincidence, because each company's contribution to the weighted average was its (revenue/total revenue) $*$ (profit/revenue) = profit/total revenue. When we add the contributions for the ten companies under (2-7), we get total profit for the companies as the numerator and total revenue for the denominator, exactly the profit-to-revenue ratio.

EXERCISE 2.3

(i) Find the weighted batting average for the Indian cricket players, where the weighting factor for each player is the percentage that he played out of total innings played by current players.

(ii) Compare this result to the simple average you found in Exercise 2.2. Was this outcome foreshadowed by the correlation coefficient you calculated in Exercise 2.2?

2.2 Some Other Summary Statistics

We mentioned some other summary statistics in discussing Probability (e.g., the median), but paid less attention to them than to means and standard deviations. Now they get closer to equal billing. Given a set of numbers $(x_1, x_2, \ldots, x_n)$, we can bypass the mean and standard deviation and summarize them by five statistics. Two of these statistics are quite natural:

L: the largest measurement, and

S: the smallest measurement.

Three additional numbers are helpful in breaking the numbers into quartiles (i.e., four parts with equal numbers of measurements). The first of these is the *median M* which, as in Probability, is the 50th percentile of the measurements. Intuitively, the median is the *halfway mark*: as many measurements exceed it as fall below it.

But how do we find the median for the points $(x_1, x_2, \ldots, x_n)$? Suppose we rank the measurements from lowest to highest, and call the ranked set $(a_1, a_2, \ldots a_n)$, where $a_1 = S$ is the smallest, a_2 is the second smallest, etc. (Assume for simplicity that there are no ties. If n is an odd number, then the median M is simply $a_{(n+1)/2}$. If $n = 5$, for example, $M = a_3$, which makes sense because a_3 is the only number for which as many measurements exceed it as fall below it (two above, two below). When n is odd, there are $(n - 1)/2$ measurements above $a_{(n+1)/2}$ and $(n - 1)/2$ below it.

What happens when n is even? When $n = 6$, for example, there is in general no number among the set $(a_1, \ldots, a_6)$ for which as many of the numbers fall below it as above it. But when $a_3 \neq a_4$ every number *between* a_3 and a_4 has three numbers above and three below. In this case, the median is defined as the *middle* of the range from a_3 to a_4, namely, $M = (a_3 + a_4)/2$. More generally, when n is even, the median is the average of $a_{n/2}$ and $a_{n/2+1}$.

To summarize:

The **median** M of $(a_1, a_2, \ldots, a_n)$ follows:

$$M = \begin{cases} a_{(n+1)/2} & \text{when } n \text{ is odd} \\ \dfrac{a_{\left(\frac{n}{2}\right)} + a_{\left(\frac{n}{2}+1\right)}}{2} & \text{when } n \text{ is even} \end{cases} \qquad (2\text{-}8)$$

where a_k = the kth smallest of the n measurements

The median M is a "competitor" for the mean $\bar{X}$ as a measure of the middle of the numbers. And it might seem more appealing, given that it is literally the halfway mark. But consider the case where n is odd, and just over half the measurements are at M or barely below it, while the other half are narrowly bunched *far above M*. Then the median—which is one of the numbers in the set because n is odd—must be in the lower bunch. One might think in this case that the "middle "of the numbers should be about halfway between the two masses. The mean $\bar{X}$ would have that property, but the median M would not. Of course, it need not be an "either/or" proposition for mean and median: it is legal in most countries to calculate both.

The *upper quartile* (*UQ*) of a set of numbers is the *median of those measurements at or above the overall median M*. It is meant to approximate the 75th percentile of the distribution, which is greater than 3/4 of the numbers and smaller than the remaining 1/4. If $n = 5$, for example, *UQ* is the median of (a_3, a_4, a_5), which is a_4. If $n = 10$, *UQ* is the median of $(a_6, a_7, \ldots a_{10})$, which is a_8.

Note that, when $n = 10$, there two measurements above a_8 and seven below it. (We assume here that the different a_i's differ, meaning that there are no ties.) Thus, there are 3.5 times as many measurements below a_8 as above (7/2). But for the 75th percentile, we might expect three times as many below as above, not 3.5 times. Don't sweat it: this is a technicality that withers away as the data set gets larger. As defined, the upper quartile UQ converges on the 75th percentile when n is sizable.

In short:

The **upper quartile** (*UQ*) of a set of numbers is the median of those numbers at or above the overall median M calculated under (2-8).

Similarly:
The **lower quartile** (*LQ*) of a set of numbers is the median of those numbers at or below the overall median M calculated under (2-8).

The five statistics (*S*, *LQ*, *M*, *UQ*, *L*) offers a good quick summary of a set of numbers. Within this framework, the median M is generally considered the middle of the set.

To reflect the spread among the numbers, the **interquartile range** (*IQR*) is often calculated:

$$IQR = UQ - LQ \tag{2-9}$$

Note that, thus defined, the interquartile range only includes about half the measurements. But they are the middle half, suggesting the "central core" of the distribution.

EXAMPLE 2.4

Still More Drugs and Calls

Find the five quartile summary statistics for the telecommunications companies and the drug companies (separately).

Solution

Given Tables 2.1 and 2.2, the five quartile statistics for profit percentage arise as follows:

- The largest and smallest measurements (*L* and *S*, respectively) are obvious from the data
- The median *M* of ten measurements is the average of the fifth and sixth largest profit percentages
- The lower quartile is the third smallest measurement, while the upper quartile is the eighth smallest (or third largest).

Statistic	Telecommunications	Pharmaceuticals
L	10.3%	37.7%
UQ	7.7%	28.0%
M	5.9%	19.8%
LQ	–7.4%	16.5%
S	–22.8%	–10.2%

At this point, some readers will be thinking: do we really need a five-number summary of ten numbers? A fair question, but the quartile summary is more valuable when the number of data points is larger. As with the 38 European countries.

EXERCISE 2.4

Compute the interquartile range for the two sets of numbers in Example 2.4, and compare them with the standard deviations of percentage profit in the two industries. Do you think one of these two measures of spread is more illuminating than the other? Briefly explain.

EXAMPLE 2.5 **Back to Europe**

Find the five-number summary of the numbers about population, population density, and fertility in Europe. Discuss briefly what we learn from each summary.

Solution

In computing the five summary statistics just described for the three variables about European demography (population, population density, and fertility), we note that there are 38 countries in the data set. Thus:

- Given that $n = 38$, the median M is the average of the 19th largest measurement and the 20th largest measurement.
- The upper quartile UQ is the median of the 19 largest measurements, meaning that is it the 10th largest measurement. (It is also the 10th smallest of the 19 measurements at or above the median.)
- The lower quartile LQ is the median of the 19 smallest measurements, meaning that it is the 10th smallest measurement.

The European summary statistics are the following:

	Population	Density	Fertility
S	0.3 (million)	3 (people/sq. km)	1.20 (births/mother)
LQ	3.6	68	1.29
M	7.5	102	1.42
UQ	10.7	128	1.80
L	82.4	1318	2.06

What do these statistics imply? For population, we see that half the European countries have populations at or below 7.5 million, while about half of them (75% – 25%) have populations in the range 3.6–10.7 million. Yet the top quarter of countries have populations as high as 82.4 million (Germany), while the bottom quarter get down to 4% of the median number (0.3 million in Iceland).

It is interesting to contrast these statistics with the previous summary numbers for European population. Earlier, we said that the average population of a European country ($\bar{X}$) was 15.4 million. Now we see that the heavy majority of nations in Europe have far fewer than 15.4 million people. (Based on the UQ, we know that 3/4 of the countries have 10.7 million or fewer people, which is well below the mean of 15.4 million.) Indeed, the median population is less than half the mean (7.5 million vs. 15.4 million). With L at 82.4 million, we see that the outer quartile of measurements from UQ to L spread out much further ($L - UQ = 71.7$ million) than the next quartile between UQ and M (3.2 million). We already knew from the standard deviation U that there was large dispersion among national populations in Europe; now we have a greater sense of what that dispersion looks like. Put briefly, Europe consists of many countries with modest population but a few heavily populous countries that dragged the European mean well above the mainstream of the data. Of course, "heavily populous" is itself a relative term: with 82.4 million people, Germany has only about 1/4 the population of the United States, and 1/12 the population of India or China.

The story is similar with population density, though less extreme. The average population density $\bar{X}$ was 146.2 people/km^2 for Europe. The median population density *M* was only 102, which is only 69% of $\bar{X}$. The upper quartile *UQ* was 128, indicating that, once again, more than 3/4 of European countries had densities below the mean for the continent (146.2). And the largest measurement was more than ten times *UQ* (1318 vs. 128), meaning that, as before, the extreme measurements were really quite extreme.

For fertility, a distinctly different pattern arose. The median and the mean were close together (1.42 and 1.52, respectively). The standard deviation *U* suggested limited dispersion around the mean $\bar{X}$; similarly the interquartile range $IR = UQ - RQ$ (0.51) suggests limited spread around *M*. Importantly, the width of the bottom quartile of the distribution between *S* and *LQ* (i.e., $LQ - S$) is considerably smaller than the width of the next quartile up, which is $M - LQ$ (.09 vs. .13). A similar pattern prevails in the upper half of the distribution: $L - UQ$ is considerably smaller than $UQ - M$ (.26 vs. .38). Thus, instead of riding off towards Kingdom Come, the outermost measurements did not hover far from those in the middle half of the distribution.

It is somewhat a matter of personal taste, but the author thinks that the five measurement summary (*S, LQ, M, UQ, L*)—which divides the data into quartiles—generally gives a clearer sense of what is going on that $\bar{X}$, *U*, and their quotient *V*. These examples from Europe were not chosen with that point in mind, which makes their evidence about the value of working with quartiles all the more impressive.

Before proceeding further, we relax by working an exercise that gives us more of a sense how various quantities are calculated and how they relate to one another.

EXAMPLE 2.6 A Brief Statistical Stroll

For each of the following criteria, pick any five numbers that satisfy it.

(i) The median is the same as the mean.
(ii) The median is twice the mean.
(iii) The mean is twice the median
(iv) The standard deviation of the numbers is zero.
(v) The coefficient of variation of the numbers is one.

Solution

(i) This part isn't too bad. The median of five numbers is the third smallest (or third largest); the mean is the simple average of the numbers. If we choose (1,2,3,4,5) as the numbers, the median and the mean are both 3.

(ii) Suppose the numbers are (A,2,3,4,5), and that A is a constant less than 2. Then the median of the numbers is 3, and their mean is $(A + 14)/5$. If we set $(A + 14)/5 = 1.5$, we can find the value of A that creates a mean half the median. Solving this linear equation, we get $A = -6.5$. (Of course, this is by no means the only five-number set with the median twice the mean.)

(iii) Now suppose the numbers are (1,2,3,4,B), where $B > 4$. The median of the numbers is still 3, but the mean is $(10 + B)/5$. Setting $(10 + B)/5$ equal to 6, we reach $B = 20$, meaning that (1,2,3,4,20), has mean twice the median.

(iv) The standard deviation of the numbers is zero if an only if they are all the same. Thus, (3,3,3,3,3) satisfies the condition $U = 0$.

(v) The coefficient of variation is one if the mean and standard deviation are the same. Suppose that the numbers are $(2 - A, 2, 3, 4, 4 + A)$ for some positive constant A. Then the mean is 3 and the variance is $[(3 - (2 - A))^2 + 1^2 + 0^2 + 1^2 + (4 + A - 3)^2]/5$, which simplifies to $(2(1 + A)^2 + 2)/5$. Equating the variance to 9 (so that the standard deviation equals the mean of 3), we find $A = \sqrt{21.5} - 1$ is a solution, and thus that $A = 3.64$ works. In short, (–1.64, 2, 3, 4, 7.64) has mean 3 and standard deviation 3.

EXERCISE 2.5 Find a set of five numbers for which:

(i) The median is half of the upper quartile
(ii) The interquartile range is four.
(iii) The interquartile range is twice the mean.
(iv) The interquartile ranges equals the standard deviation.
(v) The interquartile range exceeds twice the standard deviation.

2.3 Pictures of Data

It has been said that a picture is worth 1000 words, but how many *equations* is a picture worth? Instead of working with summary statistics (or perhaps in conjunction with such statistics), one can use graphs to convey what is happening in a data set. Now we discuss some of the primary tools in this arsenal, starting with histograms.

Histograms

A **histogram** is a graph that tells us the frequency with which the measurements fall in various ranges. Typically, there are ten-or-so ranges of equal length, and bars are used to reflect either the number or the percentage of measurements in each range. When percentages are used, the histogram is somewhat akin to the probability density curves that we met in Chapter 1.

Histograms are most useful when the observations must fall in a compact and well-defined range. Quiz scores, for example, must fall between 0 and 100%, so one might use the ranges 0–10, 11–20, ... , 91–100. (The first range is slightly longer than the rest, but let it go.) But matters get trickier when the measurements are scattered over a huge range. The population densities in Europe, for example, vary from 3 to 1318 per square kilometer. To create a histogram with ten equal ranges, we would have to use something like 0–150, 150–300, ... , 1350–1500. But then most of the ranges would be nearly or totally empty, while a few would be crowded with lots of measurements that don't really seem that close together (e.g., 152 and 288).

How could we get around this last problem? We could have ranges of unequal length (e.g., 0–50, 50–100, 100–150, ... , 350–400, 400–1400), so long as people read the scale clearly. Or we could work with the *logarithms* of the population densities to base 10. Because $\log_{10}(3) = .477$ and $\log_{10}(1318) = 3.120$, we can comfortably create the ten ranges $0.3 \Rightarrow 0.6$, $0.6 \Rightarrow 0.9$, ... , $3.0 \Rightarrow 3.3$, and get a picture that is easy to look at (Figure 2.1). We need recall, however, that we are using base-10 logarithms, meaning that a 2.2 reflects ten times the population density of a 1.2.

EXAMPLE 2.7

Log On

Create a histogram of the logarithms of population density for the European countries.

Solution

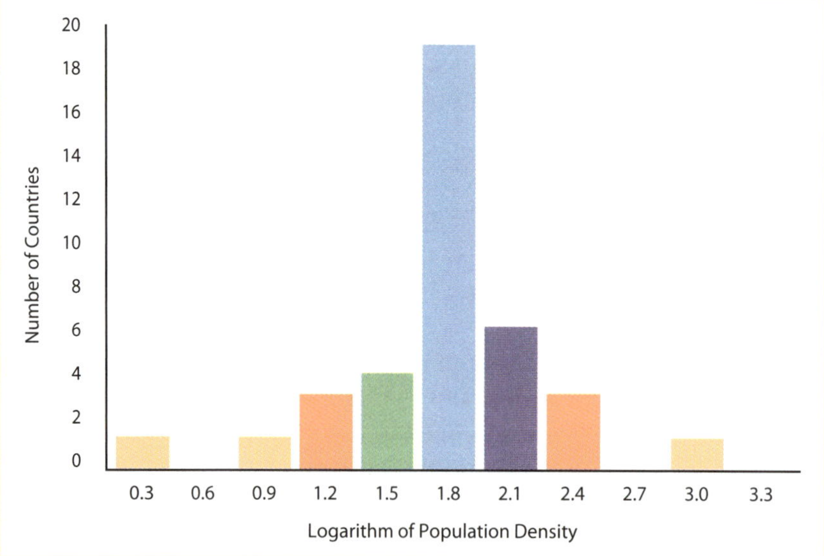

FIGURE 2.1 Histogram of *Logarithms* of European Population Densities

Note that fully half the 38 countries (19 of them) had logarithms of population density within the range 1.8–2.1, which corresponds to densities between $10^{1.8}$ and $10^{2.1}$ (i.e., between 63 and 126).

Stemplots

When data fall in a compact range, sometimes one can record them in a way that effectively creates a histogram while providing additional information. Consider, for example, a stem-and-leaf diagram, often called a *stemplot*. Suppose that a Professor has graded exam papers (by hand!) and wants to list the percentage scores in a way that indicates how well the class did. He might first list the numbers 0 to 10 in a column to the left of a vertical line. Then, as he looks at an individual paper, he can record the last digit of the score right of the vertical line in the row that contains the score's previous digit(s) to the left of the line. With a score of 76, for example, he would list a 6 on the line with a 7 on the left.

EXAMPLE 2.8

Test Message

Suppose that the quiz scores are:

73, 78, 91, 62, 77, 56, 86, 80, 79, 67, 82, 76, 53, 100, 85, 71, 78, 61, 93, 91, 65, 84, 84, 94, 65, 75, 69, 59, 92, 59, 91, 78, 54, 87, 96, 86

Create a stem plot of the results.

Solution

The final stemplot would be:.

0	
1	
2	
3	
4	
5	6 3 9 9 4
6	2 7 1 5 5 9
7	3 8 7 9 6 1 8 5 8
8	6 0 2 5 4 4 7 6
9	1 3 1 4 2 1 6
10	0

This stemplot is a cool way to record numbers. First of all, it saves work: after some minimal preparation left of the vertical, we only have to record one digit from the score rather than two (or even three, if the score is 100%). Thus, for the four scores

73, 78, 77, and 79, we write down five digits instead of eight: the first 7, and then 3, 8, 7, and 9. When we are finished recording, we effectively have a histogram: the length of the line beyond an initial digit reflects how many scores fell in that range. If there were twice as many scores in the 70s as the 80s for example, then the 70's line will be twice as long as the one for the 80's. And we not only know how many scores fell in (say) the 90's, but we also know *where* within that range they fell. Thus, if the 90's column looks like 9|1 2 1 3 2 4 1, we see that scores in that range were all in the low 90's. All in all, a great idea.

Of course, a stemplot suffers the same limitations as a usual histogram: if the measurements are spread out over a wide range, creating ten simple categories of equal size is not easy. A stemplot is great for numbers in the range 0–100; it is not great for numbers between 4.2 and 30 trillion.

EXERCISE 2.6 Round the cricket batting averages for Indian players to the nearest hundredth, and create a stemplot of the results. Express the results as percentages, meaning that 0.26 counts as 26%.

Scatterplots

Given pairs of numbers of the form (x_i, y_i), it can be helpful to plot y vs. x, thereby developing a *scatterplot*. The plot offers a general sense of how y varies with x, and invites us to visualize a trend line that summarizes the general relationship (Figure 2.2). The "scatter" of the points around that trend line is often interpreted as essentially random noise. (In Chapter 8 where we study regression analysis, we will see these ideas formalized.) A scatterplot can amplify on what we learn when we compute a

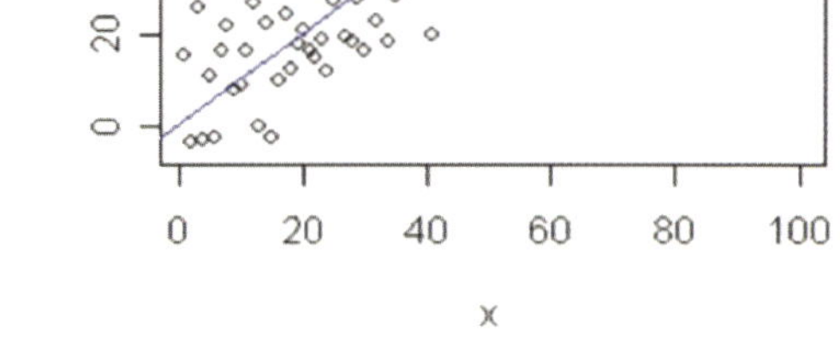

FIGURE 2.2 A Scatterplot, and a Line that Summarizes the y vs. x Trend

correlation coefficient $\widehat{\rho_{xy}}$, although sometimes the coefficient is more informative than our wandering eyes. Scatterplots sometimes make patterns clear that would otherwise not be so. For example, if y and x are related in a way that is decidedly nonlinear (e.g. along a parabola), the scatterplot might make that linkage apparent.

The Box Plot

One pleasant way to summarize a data set geometrically is to create a **box plot**. The box itself is a rectangle which extends from the lower quartile (*LQ*) to the upper quartile of the distribution (*UQ*), with vertical lines at these quartiles and another vertical line at the median (*M*). Then there are horizontal lines outside the rectangle (the "*whiskers*"), one extending down from the lower quartile, and the other extending up from the upper quartile. A simple box plot appears below (Figure 2.3).

FIGURE 2. 3 A Box Plot

How far do the whiskers extend out from the box? There are different conventions on the subject, perhaps the simplest of which is to extend the lower whisker from the smallest measurement to the lower quartile, while the higher whisker goes from the upper quartile to the largest measurement. But this approach can yield unnerving pictures if a few extreme outcomes yield whiskers that are enormously longer than the box itself. (Can you imagine a person with whiskers that go three feet in each direction around his mouth?) So there is an alternate approach that has gained considerable popularity.

This other approach starts with the interquartile range in the data set (*IQR*), which is simply *UQ* – *LQ* (or, geometrically, the length of the rectangle in the box plot). The lower whisker extends a distance of 1.5 *IQR* from *LQ*, *unless* the smallest measurement *S* is closer to *LQ* than that. The upper whisker goes from *UQ* to *UQ* + 1.5 *IQR* or to *L*, whichever is smaller. The measurements beyond the whiskers either to the left or to the right are commonly called **outliers**. Often, the outliers are plotted as *dots beyond the whiskers*, to contribute to the sense that they fall considerably beyond the mainstream of the data (Figure 2.4)

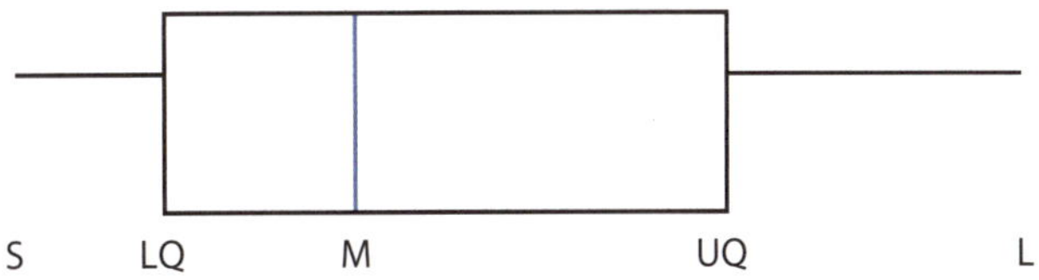

FIGURE 2.4A A Box Plot with No Outliers

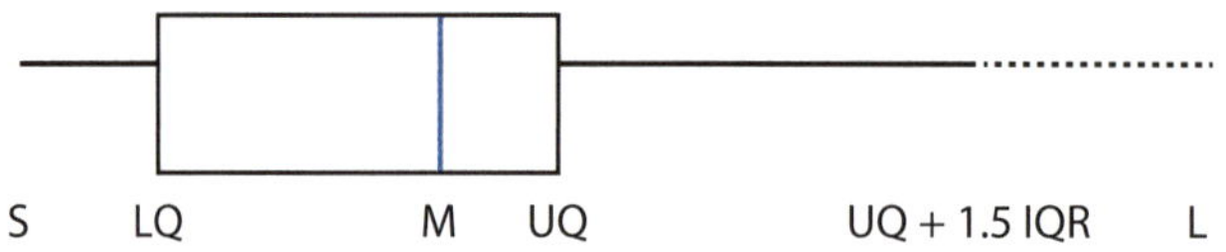

FIGURE 2.4B A Box Plot with Outliers on the Right

In summary:

A box plot commonly consists of:

- A rectangle with vertical sides at *LQ* and *UQ*, with a vertical line at *M*
- A line up to *LQ* from the maximum of *S* and *LQ* – 1.5 *IQR*
- A line up from *UQ* to the minimum of *L* and *UQ* +1.5 *IQR*
- Dots at each point below *LQ* – 1.5 *IQR* down to *S* (if such points exist)
- Dots at each point above *UQ* + 1.5 *IQR* up to *L* (if such points exist)

A Box-Plot Discussion

What are the advantages of box plots? First of all, they offer a visual synopsis of the distribution that reveals many things quickly. Consider the box itself. It tells us at once where the middle half of the data points fall. When the median *M* is closer to (say) *LQ* than to *UQ*, the line at *M* does not bisect the rectangle: the smaller area between *LQ* and *M* than between *M* and *UQ* informs us that the 25% of measurements right below the median are more "bunched together" than the 25% just above. To put it another way, the distribution tends to *skew to the right*: the upward spread from the median within the mainstream of the data is larger than the downward spread.

And the whiskers? If one whisker is longer than the other (because, for example, *L* – *UQ* is smaller than *LQ* – 1.5 IQR), then that circumstance offers additional information about the skew towards the edges of the distribution. And if we see data points beyond the whiskers, both their number and location says something about whether the distribution has a "wild side" far removed from the bulk of the data.

Box plots are especially helpful when comparing two or more distributions. For example, consider the two plots below using the Table 2.1 data about profitability in 2009:

EXAMPLE 2.9

A Tale of Two Boxes

Create box plots for percentage profitability for the Table 1 data about telecommunications and pharmaceuticals.

Solution

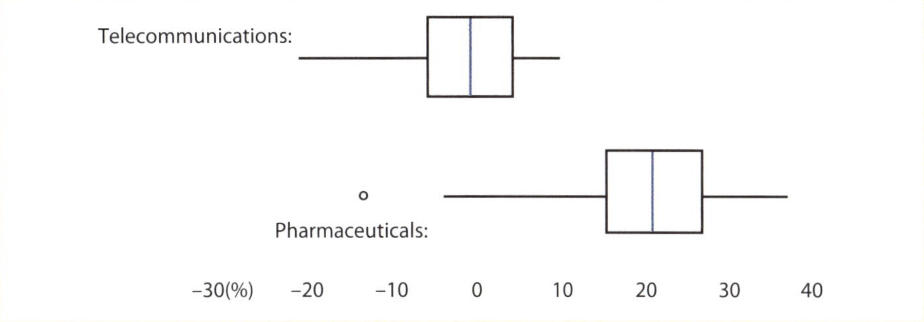

FIGURE 2.5 Box Plots of 2009 Profitability in Telecommunications and Pharmaceuticals (The lone outlier is represented by the dot.)

Recall our calculations in Example 2.4. The interquartile range IQR is 15.1% wide for telecommunications and 11.5% wide for pharmaceuticals. Thus, L is closer to UQ than is UQ + 1.5 IQR in both industries, so the right whisker ends at L. For pharmaceuticals, LQ − 1.5 IQR = −0.8%, meaning that the smallest measurement S = −10.2% is an outlier and is depicted as a dot.

The plots make clear that the profit distribution for pharmaceuticals looks somewhat like the telecommunications distribution after a bodily shift to the right. All five of the quartile statistics—smallest, lower quartile, median, upper quartile, and largest—are larger for pharmaceuticals than telecommunications, and by approximately the same amount.

EXERCISE 2.7 You guessed it: create a box plot for batting averages for the Indian cricket players.

On Outliers

Once we have identified outliers, what should we do about them? The one-word answer is "think." It is always possible that an outlier reflects a simple error, like omitting a decimal point. We should check, therefore, to be sure that outliers are really outliers. But we should not simply hasten to exclude outliers from data analysis: sometimes, they tell us as much or more as the mass of typical data points. (It is no accident that Malcolm Gladwell gave his book about successful individuals the title "Outliers.") If outliers have disproportionate effect on the data analysis we are performing, however, we should wonder whether they warrant such extra weight. Sometimes, they clearly should; other times, though, we do not want the exception to dominate the rule.

But there are outliers and outliers. John Tukey, who originated box plots, described points that are more than 3 IQR above UQ or 3 IQR below LQ as *definite* outliers, and points between 1.5 and 3 IQR away from the nearer of LQ and UQ as *suspected* outliers. There have been suggestions that all suspected or definite outliers should be deleted from the data set in calculating means and standard deviations, to get more "robust" numbers that are not overly influenced by a few extreme measurements. There are other suggestions that only the definite outliers should be deleted. And—importantly—there are also suggestions that all data points, outliers or not, should be treated the same way.

EXAMPLE 2.10 **Outlying European Countries**

Which European countries could be considered outliers in terms of their population densities?

Solution

For European population density, we know from Example 2.5 that:

- $LQ = 68$ and $UQ = 128$
  Thus: $\text{IQR} = UQ - LQ = 60$, meaning that $1.5\ \text{IQR} = 90$ and $3\ \text{IQR} = 180$

From these statistics, we see that $LQ - 1.5\ \text{IQR} < 0$, meaning that there can be no outliers below LQ because population density can never be negative. The range of suspected outliers above UQ begins at $UQ + 1.5\ \text{IQR} = 218$, while the range of definite outliers begins at $UQ + 3\ \text{IQR} = 308$. Reviewing the data, we see that Germany and the United Kingdom are suspected outliers (at 229 and 255, respectively), while Belgium, the Netherlands, and Malta are definite outliers (355, 400, and 1318, respectively).

Let's pursue this answer beyond the requirements of the question. Eliminating all five definite and suspected outliers from the calculation of mean and standard deviation, we get $\bar{X} = 90.9$ and $U = 48.5$. Eliminating only the three definite outliers, we get $\bar{X} = 99.5$ and $U = 58.8$. Compared to the original calculation $\bar{X} = 146.2$ and $U = 200$, these numbers reduce the mean by about 40% but the standard deviation by about 75%. (Because the standard deviation involves squared spread around the mean, deleting a few outliers can drastically affect U.)

Of course, it is unclear whether it is meaningful to announce that the five largest European countries are outliers which should be excluded from continental calculations. A main reason to compute $\bar{X}$ and U without outliers (*in additional to the original calculations that contain them*) is to gauge the extent to which extreme measurements have altered the summary statistics

EXERCISE 2.8 Which European countries are outliers in terms of population? Fertility?

Box Plots vs. Histograms

How do box plots compare to histograms? Histograms do a far better job of conveying the *shape* of the distribution, and making clear where the modes are. Box plots, after all, take the same general form—namely, a box with whiskers—regardless of the shape of the underlying distribution. On the other hand, box plots do better at making clear where some important percentiles of the distribution lie: foremost of these is the median, but the 25th and 75th percentiles are also important benchmarks. And when examining different distributions, their box plots are generally easier to compare than are their histograms, *precisely* because the box plots take the same general shape while the histograms can look quite different. Of course, it is not an either/or question: one can use histograms in some settings and box plots in others, and perhaps occasionally use both.

So far, we have considered how to record numbers and summarize them in illuminating ways. Now we shift gears, and explore some customized statistics that arise in particular contexts and aim to highlight what is happening.

2.4 Statistical Indicators

Many people who hear the word "Statistics" think not about a process for analyzing data but instead about specific numbers that offer important information. Examples of such numbers are the Dow-Jones Industrial Average, the unemployment rate, the wind-chill factor, or the blood cholesterol count. We will refer to such numbers as **statistical indicators**, and will discuss them now because of their obvious importance.

Statistical indicators run a gamut, not just in terms of the topics they address but also in accessibility and value. Some indicators seem impenetrable while others are highly intuitive, much as some are greatly informative while others can be grossly misleading. In thinking about an indicator, we might focus on six basic questions:

- How is the indicator defined?
- What is the purpose of the indicator?
- How transparent is the indicator?
- Does the indicator have a strong logical foundation?
- Is the indicator easy to calculate?
- Does the indicator serve its purpose?

With these questions in mind, we turn to several well-known statistical indicators and assess them. Our broader aim is to develop critical thinking about statistical indicators, so that the reader can go beyond accepting them at face value. We will first introduce the indicators and discuss them a bit; then we end the section by "scoring" the indicators on the criteria listed above.

EXAMPLE 2.11 **The US Department of Transportation's On-Time Statistic**

In late March of 2010, the author went to the webpage of Continental Airlines, and clicked through to get the on-time statistic listed for Flight 1105, the 4:40 PM service from Boston to Newark. The number listed was 60%. Continental Airlines deserves credit because, long before its competitors, it made these statistics readily available on a flight-by-flight basis. But what does this 60% statistic mean, and how useful is it?

Solution

One cannot learn the answers at the Continental website itself: when the author asked "Alex the virtual expert" how on-time statistics were calculated, she responded that "each flight displayed should have an on-time performance associated with it on the flight selection page." But we already knew *that*.

Unless Alex is hiding something, we can surmise that Continental was presenting the on-time statistic for Flight 1105 that the airline reports to the US Department of Transportation (DOT). That number is the percentage of time in a recent period that Flight 1105 reached the gate at Newark no more than 14 minutes after its scheduled arrival time of 6:15 PM. This statistical indicator scores well on two of our criteria: its definition is clear, and it is easy for Continental Airlines to calculate. On the other criteria, matters are less clear.

The apparent purpose of the on-time statistic is to tell travelers about the punctuality of the flight. But to what extent does it do so? The indicator might seem transparent: it tells us that, 60% of the time during the observation period, Flight 1105 reached the gate at Newark before 6:30 PM (6:15 plus 15 minutes). But it leaves open an important question: in the 40% of the time the flight does not arrive before 6:30, when does it get there? Are we talking about 20-minute arrival delays, or two-hour delays? If the passenger hopes to connect at Newark for another flight that leaves at 7:20 PM, what is the chance that Flight 1105 will arrive in time for the connection? (Of course, if Flight 1105 is late, the outbound flight might also be late, but we get no data at all on such a possible correlation.)

Yet we should not be too harsh. Reporting punctuality is a difficult matter, because, over time, a given flight has a *probability distribution* for its arrival time at its destination. Any attempt to summarize such a distribution with a single number is likely to be dissatisfying. Yet every airline schedule in the world entails such a summary: it lists *one* specific time of arrival, not a histogram of possibilities. The DOT statistic gives some indication of how reliably the scheduled time (6:15 PM for Flight 1105) depicts what actually happens. Indeed, the DOT statistic was introduced precisely to encourage airlines to schedule flights realistically rather than optimistically.

EXERCISE 2.9

By late 2011, Continental Airlines presented more information about the punctuality of its flights. For Flight 1611, the 4:45 PM service from Boston to Newark, the September 2011 statistics were:

Percentage On-Time (same definition as above): 44%

Percentage Late (More than 30 minutes) 36%

Percent Cancelled: 0%

(i) Given these data, how many days in September was the flight between 15 and 30 minutes late? (Assume that the flight operates every day.)

(ii) Given this information, what can you say about *S*, *LQ*, *M*, *UQ*, and *L* which we defined in Section 2.2? (You can't determine any of them exactly, but can you specify a range for each? (Assume that flights that are early are assigned a lateness of zero.)

(iii) How much of an improvement is this array of three statistics over the first one alone? Briefly discuss.

EXAMPLE 2.12 **Average Distance Between Repairs**

The Metropolitan Transit Authority in New York uses the *average distance* that New York City subway cars travel between consecutive repairs as a measure of the condition of the fleet. As the average distance grows, it is reasoned, the better the cars are performing their tasks, rather than spending time out of service and spending money to get fixed. To calculate the average, the mileage since the last repair is noted for each car that is repaired over a given period; then the mileages are averaged together.

Are there any problems with this metric that might reduce its reliability?

Solution

The metric "average distance between repairs" is attractive in some ways: it is easy to understand, and easy to calculate. And we apply the underlying idea of the metric in our own lives: if Aunt Zelda is going to the doctor much less often now than before (to get "repaired"), we may well say that she is feeling better lately. But this is a situation where "the devil is in the details," because a problem comes up if we consider the specific method by which the statistic is calculated. As noted, distances since the last repair are calculated when cars go into the maintenance shop for new repairs.

What's wrong with that? The problem is that a subway car that does not need repair in a given year does not go into the shop, and thus no record is made of the fact that it performed flawlessly throughout the year. That circumstance can lead to surprising patterns in the data. It is possible, for example, that the average distance between repairs goes from 28,000 miles in one year to 23,000 miles in the next, which looks bad. But suppose 400 cars needed repair the first year and only 200 the second. This 50% drop between the two years in the number of cars repaired is totally missing from the average distance calculation. Thus, the drop from 28,000 to 23,000 miles—which suggests a worsening condition of the fleet—may give a highly misleading impression.

What might be done to reduce the difficulty? At a minimum, "number repaired" could be listed alongside average distance between repairs to give a fuller picture of what is happening. But perhaps a modification of the metric could deal with the issue. Suppose, for example, that "average distance between repairs" is changed to *"miles travelled per repair performed,"* with "miles travelled" based on *all* subway cars. Thus, a car that went 30,000 miles over the year with no repairs would be counted in the numerator of the ratio, while another car that went 28,000 miles with two repairs would enter both the numerator and the denominator. In the example above, the drop in the denominator from 400 repairs the first year to 200 repairs the second would enter the revised statistic.

Of course, "miles travelled per repair performed" seems harder to calculate than "average distance between repairs." It seems necessary to figure out for every single car how far it travelled over the year. But there might be ways to simplify the calculation. For example, suppose that 200 trains with six cars apiece operate daily in each direction on a particular line that is nine miles long. Then the cars collectively travel $200 * 2 * 9 = 3600$ car miles per day, which could be multiplied by 365 to get an annual figure. (Perhaps fewer trains operate on weekends, so one would adjust the calculation.) What is important is that this method—based on the overall schedule rather than the travels of individual cars—makes the "miles travelled" calculation tractable, and thus makes the metric "miles travelled per repair performed" feasible to compute.

EXERCISE 2.10 Suppose that a 12-mile long subway line has 100 cars, and that each car travels the line an average of 16 times per day. Over the course of ten years, the 99 cars that are well-built need repairs every 50,000 miles, while the last car is highly trouble-prone and needs repairs every 500 miles.

(i) What is the mean-distance-between-repairs statistic for the 100 cars, assuming that a distance is recorded only when a car reaches the repair shop?
(ii) What is the mean number of miles travelled per repair performed?
(iii) Would these two statistics be the same if the trouble-prone car were retired once and for all?

EXAMPLE 2.13

The Dow Jones Industrial Average

One could easily argue that the Dow Jones Industrial Average (DJIA)—which we have already mentioned in this chapter—is the most famous financial statistic in the United States. During the American financial crisis in late 2008, newspaper headlines routinely rushed to the DJIA to describe the turmoil on the New York Stock Exchange. But does the DJIA really deserve its status as a barometer of stock-market movements?

Solution

On the surface, the popularity of the DJIA is incomprehensible. The index is based on a highly nonrandom subset of 30 large companies among those listed on the New York Stock Exchange (or on NASDAQ, which contributes two companies to the DJIA). The indicator basically asks the question "if someone wanted to purchase one share of stock in each of the 30 companies, what would be the total cost?" (In effect, the DJIA is a sum rather than an average.) There is a "divisor" in the DJIA to deal with some technical issues: if a stock splits two-for-one, for example, then one share at $30 becomes two shares at $15 apiece. Once the divisor is at hand, the DJIA is elementary to calculate: one need merely add 30 numbers together.

One thing the DJIA has in its favor is tradition. It has been around for 114 years, and one should not be too quick to dismiss the advantages of familiarity. When people want to compare the economic crisis of 2008–09 with that of 1929–32, it is nice to measure stock-price movements with a common yardstick.

But given the way the DJIA operates, how it can possibly depict overall trends on the New York Stock Exchange or NASDAQ? There are only 30 companies in the average, all are large, and all are classified as industrial (though some make pharmaceutical products or computer chips). Some of the 30 companies have far more shares of stock outstanding than others, but that circumstance is ignored when the focus is on buying a *single* share. Moreover, DJIA treats a $1 increase in share price the same way regardless of which company sustained it. If the share price rises from $10 to $11 at company A and from $50 to $51 at company B, the percentage increase is five times as high at A than B (10% vs. 2%). But the DJIA treats this difference as irrelevant.

Yet the drawbacks of a statistical indicator in theory may greatly exceed its disadvantages in practice. One all-inclusive measure of trends on the New York Stock Exchange is the Composite Index, which essentially answers the question: how much would it cost to buy the Exchange? (For each firm, the share price is multiplied by the number of shares outstanding to find the firm's *market capitalization;* then the market capitalizations of all companies are added together.) The Composite Index avoids several of the problems we cited for the DJIA: it includes essentially all stocks, and takes account of how many shares are being traded for each. However, it is obviously less easy to calculate than the DJIA.

How does what the DJIA reveals about price movements for single shares in 30 companies compare with what the Composite Index tells us about all shares in all 4,000-or-so companies on the New York Stock Exchange? Table 2.4 presents the closing values of the DJIA and the Composite Index at six-month intervals starting from December 2003 to December 2009. (Conventions for computing the Composite Index were changed in 2003, so these dates all involve the new procedures.)

TABLE 2.4 Dow-Jones Industrial Average and Composite Index Values at Six-Month Intervals Starting on December 31, 2003

Month	DJIA Close	Composite Index Close
December 2003	10453.92	6464.00
June 2004	10435.48	6602.99
December 2004	10783.01	7250.06
June 2005	10274.97	7217.78
December 2005	10717.50	7753.95
June 2006	11150.22	8169.07
December 2006	12463.15	9139.02
June 2007	13408.62	9873.02
December 2007	13264.82	9740.32
June 2008	11350.01	8660.48
December 2008	8776.39	5757.05
June 2009	8447.00	5905.15
December 2009	10428.05	7184.96

For these data, the coefficient of correlation $\hat{\rho}$ of the DJIA and the Composite Index is a formidable .958. On inspection of the various numbers, we see strong similarities in recent trends. Between June 2009 and December 2009, for example, both indicators increased by about 20%. Both reached their high points over the twelve periods at the same time, namely, June 2007. True: the movements are not perfectly synchronized: between December 2003 and December 2009, the DJIA was almost unchanged while the Composite Index increased by about 10%. On balance, however, the two indicators tell *essentially the same story* about New York Stock Exchange movements over the observation period. And the similarity in this period is not atypical.

In short, the DJIA has three things going for it: it is simple, it is traditional, and it does what it purports to do. The author is among its fans.

EXERCISE 2.11

The "Standard & Poors 500" is an alternative to the Dow Jones Industrial Average as a barometer of US stock-market movements. The closing values of these indicators over six-month periods over 2003–09 were as follows:

Month	DJIA Close	S & P 500 Close
December 2003	10453.92	1111.92
June 2004	10435.48	1140.84
December 2004	10783.01	1211.92
June 2005	10274.97	1191.33
December 2005	10717.50	1248.29
June 2006	11150.22	1270.20
December 2006	12463.15	1418.30
June 2007	13408.62	1503.35
December 2007	13264.82	1468.36
June 2008	11350.01	1280.00
December 2008	8776.39	903.25
June 2009	8447.00	919.32
December 2009	10428.05	1114.11

(i) Using these data, compute the correlation coefficient of the two indices at the thirteen times listed from 2003 to 2009.
(ii) Beyond the correlation coefficient, what do you see as the similarities and differences in movement over time in the two indices?
(iii) Go to an appropriate web site and learn how the Standard and Poor (S & P) 500 index is calculated. In simple words, how would you explain how the S & P 500 differs from the Dow Jones Industrial Average?

EXAMPLE 2.14

The Auto Fatality Rate

Among the statistical indicators about the hazards of driving on US roads is "deaths per 100 million miles driven." The numbers are calculated annually on a state-by-state basis. Is New Jersey, which had the lowest such statistic in a 2009 comparison of American states, really the safest state in which to drive?

Solution

As Table 2.5 suggests, some US states are considerably safer than others under the risk stated risk criterion:

TABLE 2.5 Deaths per 100 Million Miles Driven in Eleven American States in 2009

State	Auto Fatality Rate	Rate as Multiple of New Jersey Rate
Alabama	1.63	2.04
Colorado	1.15	1.44
Georgia	1.37	1.71
Iowa	1.34	1.68
Maryland	1.07	1.34
Missouri	1.41	1.76
New Jersey	0.80	1
Ohio	1.10	1.38
South Carolina	1.85	2.31
Vermont	1.00	1.25
Wyoming	1.68	2.10

Notes: These illustrative states are every fifth one (starting from the first) in an alphabetical listing of the 51 American states including Washington, D.C. New Jersey is the baseline in the last column because it had the lowest auto fatality rate among the states shown.

The indicator "deaths per 100 million miles driven" has some intuitive appeal: it compares the cost of driving in terms of lives lost with the benefit in terms of transportation provided. It is easy to understand, and presumably quite easy to calculate. (Estimating "miles driven" in a given state might be challenging, but procedures have evidently been worked out to do so.) And it does seem to answer a simple question: if I must drive one "random" mile in some state, in which state might I be least likely to perish?

One can, however, raise some objections to the calculation. It pays no attention to *injuries* in auto accidents, although they are far more numerous than deaths and are sometimes very serious. And it misses the point that people might have to do more driving in some states to go about their daily lives than in other states. Furthermore, some special issues arise because this statistic pertains an "average" mile driven. We know that averages can reflect conditions that cannot actually arise: if a fair die is tossed repeatedly, the average outcome of 3.5 can never come up on a single toss. That concern can be raised for many averages, but extra problems can arise when averages are used to make comparisons. New Jersey emerged as the safest state in Table 2.5, but consider Table 2.6 about what happened in 2004 when the auto mortality rate is broken into two components: that on rural roads and that on non-rural roads (namely, urban roads and highways). Once again, New Jersey had the overall lowest fatality rate. But it had the *highest* fatality rate among the five states listed in rural driving and was only third lowest in non-rural driving.

TABLE 2.6 Rural and Non-rural Auto Fatality Rates in Five States in 2004

State	Rural Fatality Rate	Non-rural Fatality Rate	% of Driving That is Rural	Overall Fatality Rate
Colorado	2.94	1.01	23	1.45
Georgia	2.06	1.14	35	1.46
Iowa	2.03	0.89	46	1.42
Maryland	2.56	0.87	19	1.19
New Jersey	3.12	0.90	8	1.08

Note: Tables 2.5 and 2.6 concern different years, so their overall fatality rates can differ.

How then did New Jersey emerge with the lowest overall mortality rate? As we see, the answer is that only 8% of driving in New Jersey takes place on rural roads, which are systematically less safe than non-rural roads. The corresponding percentage is at least twice as high in the other states. Thus, the New Jersey statistic is more heavily weighted than the others toward the safer non-rural driving. Heavily urbanized New Jersey obviously benefits from these differences in weighting. It is not demonstrably safer than Iowa—which outperforms it on both kinds of roads—but it gets to deemphasize its rural record to a much greater extent.

This situation reflects a general hazard in comparing averages that is known as **Simpson's Paradox**. It is possible that organization A outperforms organization B on every dimension on which comparisons are made, and yet the overall average score for B is higher than that for A. That can happen because the *weighting* of the different dimensions might be different for the two organizations. For example, firm A might pay its white-collar workers an average of $60,000 apiece per year and its blue-collar workers $40,000. For firm B, the corresponding statistics might be $57,000 and $38,000. But suppose half the workers at A are white collar, as are 2/3 of those at B. Then the statistic "average salary per worker" will be $50,000 at A and $50,667 at B (namely, (2/3) ∗ 57,000 + (1/3) ∗ 38,000). B might look like the higher paying firm based on the overall average, but neither white-collar nor blue-collar workers there are better compensated than their counterparts at A.

The moral of this story is *not* that we should avoid indicators like the auto fatality rate, which are averages. Nor should we refuse to make comparisons among averages. But we should think carefully when we work with averages, and perhaps something delve more deeply into the data before reaching final conclusions.

EXERCISE 2.12 US Department of Transportation (DOT) statistics showed the following on-time records for Alaska Airlines and American West Airlines in June 1991 for five cities:

	Alaska Airlines		America West Airlines	
City	Total Number of Arrivals	Percent On Time	Total Number of Arrivals	Percent On Time
Los Angeles	559	88.9	811	85.6
Phoenix	233	94.8	5255	92.1
San Diego	232	91.4	448	85.8
San Francisco	605	83.1	449	71.3
Seattle	2146	85.8	262	76.7
TOTAL	**3775**		**7225**	

(These five cities were the only ones served by Alaska Airlines among the 30 cities used in computing DOT on-time ratings.)

(i) Using the data above, work out what percentage of Alaska Airlines' 3775 flights into the five cities arrived on-time. Do the same for America West's 7225 flights.

(ii) What percentage of America West's 7225 flights landed in each city among the five cities listed? What are the corresponding percentages for Alaska Airlines?

(iii) Referring to (ii), explain why America West's five-city on-time record was better than Alaska's, even though Alaska *had a better on-time record in every one of the five cities*.

(iv) Do these data illustrate Simpson's Paradox?

Source: J. Caulkins, A. Barnett, P. Larkey, Y. Yuan, and J. Goranson. (1993). "The On-Time Machines: Some Analyses of Airline Punctuality." *Operations Research* 41(4), pp. 710–720.

EXAMPLE 2.15 **The Windchill Index**

The Windchill Temperature Index (WC) of the US National Weather Service reflects the point that the combination of low temperatures and high winds creates greater discomfort and danger to health than either of those phenomena on its own.

It follows the formula:

$$WC = 35.74 + 0.6215\,T - 35.75\,S^{0.16} + 0.4275\,T(S^{0.16})$$

where T = temperature in degrees Fahrenheit, S = windspeed in miles per hour

Is the index worth using?

Solution

The webpage at the US National Weather Service explains that WC is "based on a human face model," and "incorporates heat transfer theory" to model heat loss from the body to its surroundings. The lower the value of WC, the more rapidly will frostbite arise for facial skin exposed to that combination of temperature and wind. Different combinations of T and S that yield that same WC are treated as equivalent. The wind chill factor "assumes no impact from the sun."

While the wind-chill formula is not exactly intuitive (who would have imagined that it is essentially the *sixth* root of wind speed ($S^{0.16}$) that matters?), the formula is easy to use if a computer is nearby. It takes account what many of us know from experience: cold weather is especially horrid when the wind is blowing briskly. And it goes beyond impressions to devise an *objective* scientific measure of the hazards posed by cold weather, tied to the danger of frostbite. The author is by no means a meteorologist, but the WCT strikes him as an especially helpful indicator.

EXERCISE 2.13 Given the formula for the wind-chill factor:

(i) What is the wind chill factor when the temperature is 20°F and the wind is blowing at 10 miles per hour?
(ii) What temperature has the same wind-chill factor as in (i) when there is no wind?
(iii) What wind speed accompanied by a temperature of 15°F yields the same wind-chill factor as in (i)?
(iv) Suppose that the wind-chill factor an hour later is three degrees higher than in (i). If the increase was caused solely be a drop in wind speed, what was the wind speed an hour later? (Do not hesitate to use approximation methods based on a spread sheet.)
(v) Identify three pairs of temperature and wind-speed for which the wind-chill factor is zero. (Any three would be fine.)

EXAMPLE 2.16 **A Baseball Performance Measure**

At one time, a baseball player was evaluated largely based on his batting average, which was the percentage of time that he achieved a hit (whether a single, double, triple, or home run) when he was a bat. (As we have seen, a batting average also arises in connection with Cricket.) But in the modern era of "sabermetrics," more attention is placed on a players" *on-base plus slugging* (OPS) statistic. This statistic is the sum of two others:

- The player's *on-base percentage* (OBP), which is essentially the percentage of time at bat that he reaches first base or further. (It differs a bit from the traditional batting average, because it takes account of occasions when the player gets on-base not because he got a hit but instead because he was "walked" by the pitcher or was hit by a pitched ball.)
- The player's *slugging percentage* (SLG), which award the player four points for a home run, three points for a triple, two points for a double, one point for a single, and zero points for any other outcome. The slugging percentage is the *average* of the player's scores over all his times at bat.

As noted, OPS = OBP + SLG

What can we say about the viability of this indicator?

Solution

Taken literally, this statistic is rather confusing. The OPB statistic falls in the range (0,100) as a percentage, or (0,1) as a fraction. But the SLG can vary from zero to four. Thus, the OPS has the range (0,5), with the bulk of its determination tied to the SLG component. And while the on-base percentage is a percentage, the slugging percentage is not: it is an *average* of outcomes, not a percentage of anything. And what are the units of measure of OPS? If a player reaches base 30% of the time and on average gets 3/4 of a single per time at bat (i.e. SLG = 3/4), what does the sum of 1.05 mean? And if he gets a single on a turn at bat, it affects both OPB and SLG, so are we effectively counting the single twice?

The reader might recall that we talked this way at the start of the discussion about the Dow-Jones Industrial Average (DJIA). Yet eventually, we concluded that our initial fears were overwrought because, in the real world, the DJIA works well. The author is no authority on baseball, but he is not aware of clear corresponding evidence that OPS works, meaning that those players who rank highest on OPS are unquestionably the best. To be sure, OPS has some good points: It takes take into account that a home run is worth more than a single, and that getting on base without a hit is better than striking out. But one wonders whether these point might be might be made in a simpler index with a stronger foundation in logic. For example, might a modification of (say) SLG that credited walks and perhaps "sacrifice flies" work as well as OPS, without causing mathematicians to tear their hair out?

EXERCISE 2.14 In baseball, suppose that Mendel never gets more than a single when he is at bat.

(i) What is the maximum possible value of his OSP given that circumstance?
(ii) Ezekiel always gets a double when he gets a hit. If he makes hits the same percentage of the time as Mendel (and both of them either get hits or strike

out: they never reach first base for other reasons), could his OSP be twice that of Mendel? (Ignore the case when neither of them ever gets a hit.)

(iii) A particular player always gets either singles or home runs when he gets a hit. Also:

- He gets equal numbers of singles and home runs
- He either gets a hit or strikes out (meaning that is never "walked" to first base)

If his OSP is 1.05, what percentage of the time does he get a hit?

EXAMPLE 2.17 **Relative Risk Factors**

What is a relative risk factor, and what are its strengths and weaknesses?

Solution

Widely used in medicine, a relative risk factor compares the risk among people who engage in a certain activity with that among people who do not. The factor is the *ratio* of the two risks.

For example, suppose that 40% of the population smokes cigarettes, and 85% of lung-cancer sufferers are cigarette smokers. Then, assuming that Q is the number of cases of lung cancer and N the overall population, the relative risk factor R can be estimated as follows:

Per capita lung-cancer risk among smokers $\approx (.85Q)/(.4N) = 2.125\ (Q/N)$

Per capita lung-cancer risk among non-smokers $\approx (.15Q)/(.6N) = .25(Q/N)$

$$\boldsymbol{R} = \text{risk among smokers}/(\text{risk among nonsmokers}) = 2.125/(.25) = 8.5$$

Note that this outcome is quite different from the ratio 85/40, a quotient which suggests that smokers have roughly "twice their share" of lung-cancer cases. This last assessment is correct, but it doesn't mean that smokers have twice the risk. The risk-multiplier is far higher: smokers suffer nearly six times as much lung cancer as non-smokers (85% vs. 15%), even though non-smokers are considerably more numerous than smokers. To draw sound conclusions, work out the ratio carefully based on its definition, instead of performing quick maneuvers on the first numbers you see.

A possible problem with R is that, as its name makes clear, it is a *relative* risk factor. When $R = 3$, a risk has tripled, but what was the original risk level? A tripling from 5% to 15% is of obvious importance, an increase from 1 in a trillion to 1 in 333 million seems rather less so. In medical literature, an emphasis on risk factor R is not always accompanied by estimates of the baseline level of risk. It is not R's fault, but we might be better informed if the original risk level were reported at the same time as R.

EXERCISE 2.15 In a research study, it was estimated that 90% of US men drink coffee, as do 96% of male sufferers of cancer of the pancreas. Based on these two numbers, what is the relative risk of pancreatic cancer for those who drink coffee relative to those who do not? (Relax: subsequent studies suggest that coffee does *not* raise pancreatic cancer risk.)

EXAMPLE 2.18

The Gini Coefficient

The aim of the ***Gini coefficient*** in Economics is to depict inequality (or dispersion) in a community's distribution of wealth.

(i) How is the Gini coefficient defined?
(ii) What is the range for a Gini coefficient?
(iii) How are Gini coefficients of zero and one interpreted?

Solution

The calculation of a Gini coefficient starts with the determination of the total income of all community members over a given period. Then the members of the community are (figuratively) lined up from richest to poorest, with the highest-income person at the front of the line and the lowest-income person at the end. The analyst then goes down the line, noting after each person both the fraction of community members observed thus far and the fraction of total community income already accounted for. For example, the first 10% of community members ranked according to income might account for 32% of total community income.

The analysis then graphs "fraction of population" versus "fraction of total income." (See Figure 2.6.) The curve must start out at (0,0): just before the process has begun, the analyst has considered 0% of the population and seen 0% of the

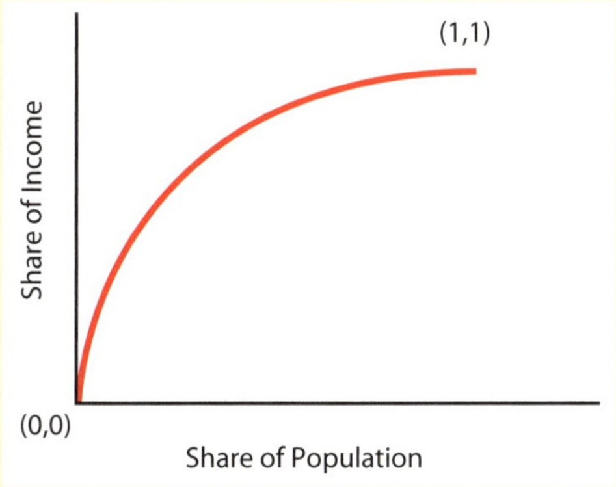

FIGURE 2.6 A Gini Curve

income. It must end at (1,1): when the analyst has observed 100% of the population, he has necessarily seen 100% of the income. The question is how the curve has moved from (0,0) to (1,1). We would expect the curve to grow fastest at the beginning, because the wealthy citizens who are first to be counted have a disproportionate share of community income. Towards the end of the process, the rate of growth might drop considerably, as the last citizens to be counted have the lower per capita income.

In the Gini calculation, the next step is to consider the diagonal line $y = x$. If all citizens had exactly the same income, then the curve "fraction of population" versus "fraction of income" would actually be this diagonal line. (In this case, the citizens would presumably line up in random order given that there is no basis of ranking by income.) The Gini coefficient then asks us to find the area A between the actual curve and the "equal income" diagonal, the idea being that, the greater the area, the greater the lopsidedness in the distribution of income (Figure 2.7).

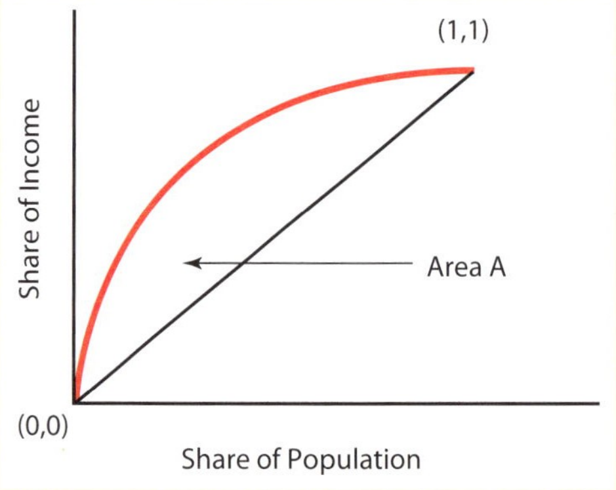

FIGURE 2.7 The area *A* between the Gini curve and the Diagonal $y = x$ Is Key to the Gini Coefficient

Wait: we're not home yet. The Gini cofficient *G* is not the area *A* but rather *twice* the area:

$$G = 2A \tag{2-10}$$

Why twice the area? Well, the most extreme inequality in income arises when an extremely small fraction of the population gains all of the income while the rest the population gets none. In that case, the "share of income/share of population" curve shoots up from (0,0) to (0^+,1), where 0^+ is the minuscule fraction of the population that gets all the income. The curve then traces the horizontal path from (0^+,1) to (1,1). Thus, the area between the curve and the diagonal $y = x$ is essentially a right triangle with legs both equal to one (Figure 2.8), and its area A equals 1/2 based on the usual formula from geometry. Because the area A in an actual income distribution must fall in the range 0 to 1/2, twice that area has the range (0,1). It's considered nice to have (0,1) as the range of a statistical indicator, which is why G is defined as it is.

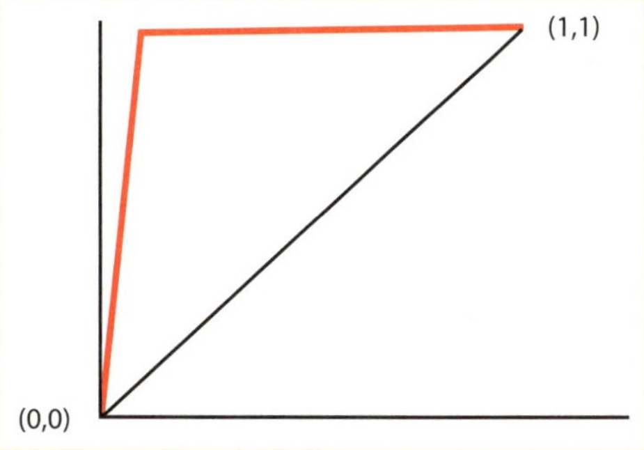

FIGURE 2.8 When a Few People Have Essentially All the Income, the Area Between the Gini curve (Red) and the Diagonal is Essentially 1/2.

In short, a Gini coefficient near zero means almost no divergence in income (which some people consider the ideal situation but others do not), while a Gini coefficient near one means an almost total concentration of income in a few hands. Gini coefficients between zero and one signify that the distribution falls between these two extremes.

EXAMPLE 2.16

Half to Fifth

Suppose that, in a particular community of population N, 80% of the members all have income I while the remaining 20% all have income 4 I. What Gini coefficient would correspond to this pattern?

Solution

Under the procedure for lining up, the high-income fifth of the population would occupy the first 20% of the places, while the remaining 4/5 of citizens would line up behind. If I is per capita income in the poorer part of the population, the total community income T would follow:

$$T = N\left(\frac{4}{5}\right)I + N\left(\frac{1}{5}\right)4I = 1.6NI$$

The total income among the "wealthy fifth" is $N\left(\frac{1}{5}\right)4I = .8$ NI, as is the total income for the remaining 4/5 of the population. Thus, the wealthy fifth of the population possesses half the community's total income, and the Gini curve passes through (.2,.5).

The graph of income share vs. population share grows linearly from (0,0) to (.2,.5). As we enter the low-income group, graph of "share of income vs. share of population" grows linearly from (.2,.5) to (1,1). Thus, the Gini curve consists of two connected lines as shown in red below.

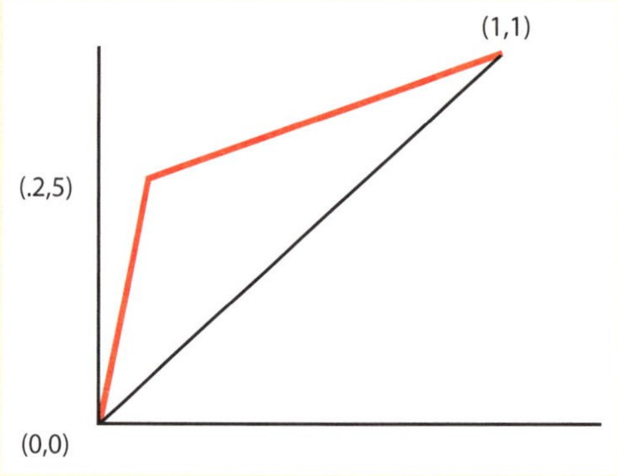

FIGURE 2.9 A Gini graph for this example

To find the area A in this situation, we can proceed in two steps:

- We first find the total area below the Gini "income share/population share" red curve and above the line $y = 0$.
- Then we subtract the area *below* the equal-income line $y = x$, and above the line $y = 0$.

Actually, the second quantity is just 1/2, because the region bounded by (0,0), (1,0), and (1,1) is a *right triangle* with legs of one.

To find the area under the curve of "income share/population share," we return to high-school geometry. We can break the region of interest into three parts (Figure 2.10):

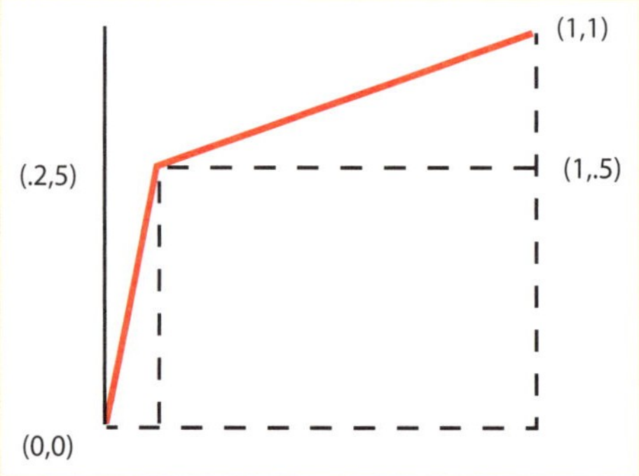

FIGURE 2.10 The Area Below the Gini curve

R_1:The triangle bounded by (0,0), (.2,.5) and (.2,0)

R_2:The rectangle bounded by (.2,0), (1,0), (.2,.5) and (1,.5)

R_3:The triangle bounded by (.2,.5), (1,.5) and (1,1)

We then use familiar formulas to establish that the area of R_1 is .05, the area of R_2 is .40, and the area of R_3 is .20. Thus, the area below the curve is .05 + .40 + .20 = .65, and the area A that is below the curve *and* above the diagonal follows: A = .65 – .50 = .15. Thus, the Gini coefficient in this situation follows:

$$G = 2A = 2(.15) = .30$$

Some Discussion of the Gini Coefficient

The Gini Coefficient reflects a sophisticated attempt to measure the discrepancy between the actual income distribution and the distribution that would arise if everyone had the same income. It works directly with the *entire* distribution, and is not content to present a "snapshot" like: the 10% of the population with the highest incomes gets 29% of total income. G is restricted to in the range (0,1), which is obviously a more accessible range than 22 to 833 or the entire set of non-negative numbers. It is not all that easy to calculate by hand, but this is the 21st century. Not only is it easy to write a program in a spreadsheet to find the Gini coefficient, but there are packages on the web that will accept the data (ranked richest to poorest) and instantly compute the coefficient.

If there is one dimension on which the Gini might not excel, it is transparency. What exactly does it mean to say that the Gini coefficient in a given setting is (say) .43? It doesn't "exactly" mean anything, because many different income distributions all share that Gini coefficient. (In a future exercise, we will encounter three very different income distributions that share a common Gini coefficient.) But do we have even a good intuitive sense?

Perhaps not. Given that a Gini coefficient of zero represents pure equality and of one reflects almost total inequality, a coefficient of 0.3 might seem to indicate a modest degree of dispersion. In our example, however, 1/5 of the population had four times the per capita income as the remaining 4/5, and the Gini coefficient was 0.3. Not everyone will view that degree of inequality as modest. And, if we are unsure what to make of a Gini coefficient of (say) 0.36, what does it mean when one community has a coefficient of 0.36 and another of 0.45? A difference of 0.09 does not seem large on a scale from 0 to 1, but the actual distributions that generate these two coefficients might not seem similar at all. If we worked with Gini coefficients extensively, we would get a better feel for what they mean, but we would never totally dispel an element of mystery.

In some ways, the Gini coefficient is the opposite of the DOT on-time statistic. While Gini works painstakingly with the entire distribution of income, the DOT score breaks the lateness distribution into two large categories "less than 15 minutes late/ 15 or more minutes late." Yet, while we are unsure what a particular Gini coefficient means, we know exactly what the DOT score is saying. A DOT score of 74%, for example, means that 74% of the flights considered reached the arrival gate no more than 14 minutes beyond their scheduled arrival times.

What this comparison suggests is that, in devising statistical indicators, transparency and intellectual depth need not go hand in hand. Indeed, the more deeply we think about any issue, the more difficult it might seem to summarize it with a single statistic. But we live in a world in which people do not have endless attention spans or times to spare; often, we have to get to the point quickly. In the tradeoff between simplicity and analytical rigor, we should not automatically prefer the latter to the former.

EXERCISE 2.16 The following data reflect the income distribution for US families in 2007:

Range	Number of Families (Millions)
$25,000–$29,999	3.8
$30,000–$34,999	3.7
$35,000–$39,999	3.7
$40,000–$44,999	3.7
$45,000–$49,999	3.3
$50,000–$59,999	6.5
$60,000–$74,999	8.8
$75,000–$84,999	5.1
$85,000–$99,999	6.1
$100,00–$149,999	12.0
$150,000–$199,999	4.4
$200,000 and up	3.8

Source: US Census Bureau

For simplicity, assume that all families in each range except the highest have incomes at the *middle* of the range. For the last range, assume that the mean income is $250,000.

(i) Based on these numbers, what is the average income of US in 2007? Do the calculation by working out total income divided by total number of residents.

(ii) Now take a weighted average of 2007 US family incomes, where the weight for each range is the percentage of families with income in that range.

(iii) Do the two calculations yield the same answer? If not, go back and see what you did wrong.

(iv) What is the Gini coefficient for US family incomes in 2007 given these data? (Hint: A trapezoid of width c with height a on one side and b on the other has an area of $c(a + b)/2$:

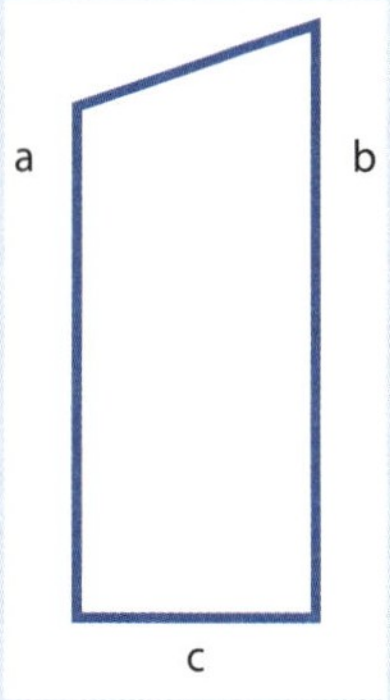

Rating the Indicators

Perhaps we should summarize these discussions by rating each indicator on four dimensions:

- *Transparency*, which is essentially the answer to the question: how easy is it to describe the indicator to a nonmathematical friend?
- *Underlying logic*, which is tied to the question: does the indicator reflect clear reasoning that suggests it will accomplish its purpose?
- *Ease of calculation*
- *Utility*, which means how successfully does the indicator accomplish its goal?

Here are the author's (admittedly subjective) assessments about the eight indicators discussed:

Indicator	Transparency	Logic	Ease of Calculation	Utility
Dow-Jones Industrial Average	High	Low	High	High
Mean Distance Between Repairs	High	Low	High	Low
DOT On-Time Score	High	Medium	High	Medium
Auto Fatality Rate	High	Medium	Medium	Medium
Relative Risk Factor	High	Medium	High	Medium
Gini Coefficient	Low	High	Medium	Medium
Windchill Index	Medium	High	High	High
On-base plus Slugging (OPS)	Low	Low	High	??

The "winners" among these indicators are Windchill Index and the Dow Jones Industrial Average. Note that the Dow Jones scores low for logic: when we hear about how it is calculated, we might think it outlandish to use DJIA as a stock-market barometer. But DJIA's empirical record of success speaks for itself, so the Dow can be forgiven its theoretical shortcomings. The relative risk factor and auto fatality rate are simple, but they leave unclear how dangerous a particular disease is or which localities are really safest to drive in. The Gini coefficient is elegant, but it isn't obvious what to make of a calculated value.

More important than the assessment of a particular indicator is a broader point: Before we take an indicator seriously, we should think hard about its strengths and weaknesses. That is especially true, of course, for any indicators that we ourselves devise.

EXERCISE 2.17 The *unemployment rate* is officially defined as the ratio of the number of individuals who are not working but looked for work in the last four weeks to the number of individuals in the labor force (working plus looking for work). How would you rate the unemployment rate on the dimensions of transparency, logic, ease of calculation, and utility?

2.5 Some Statistical Mishaps

> "Statistics. That's something that is wrong."
>
> —*a juror, quoted in The New Yorker, January 24–31, 2005*

For the last part of this introduction to statistical thinking, we discuss several true-life data analyses that contain errors. The review is not meant to demean the study of Statistics: The truism that statistics can be misleading has no more content than the statement that paragraphs can be misleading. But we should recognize that misleading statistical "analyses" do exist, creating misconceptions on a whole host of important matters. Here we consider several true-life examples of fallacious statistical reasoning, loosely dividing the examples into categories based on the nature of the error., The aim is to stress that even simple data analyses can be can be worse than useless if they are not conducted with care. And—we should admit—it's simply fun to watch other people's mistakes as a means of avoiding our own.

Spurious Correlation

The point cannot be emphasized enough: the fact that two quantities are correlated in a data set does not mean that one of them is governing the behavior of the other. If X and Y move together it could be because X and Y are simultaneously affected by a third variable Z which governs them both. Or it might be mostly a coincidence. We consider three examples of intellectual confusion in this area.

Illustration A: Wake Up Call

The *Boston Globe* reported a positive correlation between the number of hours a positive sleeps per night and the risk of a heart attack. The apparent implication was that long sleepers might do well to purchase an alarm clock.

That advice would make sense if the extra sleeping somehow weakened the heart, and thereby increased the risk of coronary problems. But one can think of other reasons for the correlation. It is possible that some people have weak hearts, which make the exertions of day-to-day living especially tiring for them. These people might sleep more than average, and yet still have higher heart attack rates. However, if they did *not* get extra sleep, then the strain on the hearts might be greater, and their risk of heart attack might grow from (say) twice the population average to four times the average. In that case, buying an alarm clock would be *exactly the wrong thing* to do.

Illustration B: The Sound of Music?

On December 28, 2003, *Parade* magazine informed us that "Kids who study music get higher scores on their SATs. ... Music helps to improve overall academic performance." Thus, it argued, eliminating after-school programs in music would have far-reaching consequences.

It is of course possible that music soothes some savage beasts, and makes them calm enough to master right triangles. But it is also conceivable that the very characteristics that lead people to pursue music predispose them to other cerebral activities, which is the real reason for their higher test scores. Indeed, fewer music programs could actually raise SAT scores: Would-be participants might spend more time reading. (No, that is *not* an argument for ending the programs.)

EXERCISE 2.18 *The Wall Street Journal* of October 23, 2010 quoted the provost of the University of Houston as saying "I know from experience that everyone who taught Statistics got a lower (student) evaluation than those who taught courses that are a little less challenging." Might this assessment involve spurious correlation? Briefly discuss.

Nonrandom Sampling

The issue here is that findings about a sample of a population are extrapolated to the whole population, despite ample reason to fear that the sample is not representative of the broader group.

Illustration C: Thin Air

Several years ago, ads in The New York Times and The Wall Street Journal proclaimed that "84% of Frequent Business Travelers (from New York to) Chicago Prefer Midway Airlines to American, United, and TWA." The figure was surprising because, at the time, Midway was only carrying 8% of the air traffic between New York and Chicago. Why was there such a big discrepancy between what people prefer and what they actually do?

In small letters at the bottom of the ad, it was reported that the survey yielding the 84% figure was "conducted among Midway passengers between New York and Chicago." In other words, the only travelers eligible for inclusion in the study were the small group of eccentrics who were already flying Midway! If there was any surprise in the results, it was that one in six Midway passengers would themselves prefer to be elsewhere.

The poll was almost a parody of nonrandom sampling. Not all members of the relevant population were equally likely to enter the sample: a Midway ticket was required for admission. To treat the sample as representative of all frequent flyers between New York and Chicago, one would have to make the startling assumption that Midway's popularity among those who fly it is the same as that among those who don't.

EXERCISE 2.19 The person who books a hotel room on *Priceline* is often asked afterwards to rate the hotel on a 1–10 scale. Might the average score among those who responded be misleading as a guide to the average score among all *Priceline* users who stayed at that hotel? Briefly discuss.

Compared to What?

Sometimes it is suggested that a number is unusually high or low, and that one can draw some conclusion from this deviation from the expected level. However, no attempt is made to estimate what the expected level actually is.

Illustration D: Award-Winning Titles

In the excitement before the 2000 Academy Awards show, *Time* magazine offered us the news that "Oscar loves films with concise titles." In the previous twenty years, the average title length of the film voted Best Picture of the Year was only 2.7 words. The median length of the winning title was only 2.5 words.

The unstated premise in *Time*'s analysis is that films that fail to win Oscars have titles that drag on. But it is easy enough to investigate that possibility. This author picked up a nearby copy of the *Boston Globe*, and prepared a histogram of the title lengths for all 39 American films that were then showing in greater Boston. (Foreign films are not eligible for Best Picture of the Year.) It turned out that the 39 titles averaged 2.6 words long, and had a median length of 2.0 (vs. 2.3 and 2.5, respectively, for the winners). In other words, Oscar hardly gravitates to the shortest titles. Conceivably, the Academy's voters care more about the caliber of a movie than the length of its name.

Illustration E: Missing in Action?

In 1999, the *National Journal* reported that, of the 234 sons of members of Congress who reached military age during the Vietnam War, only 28 (12%) served in Vietnam. The clear implication was that Congressional sons were largely absent from a war to which Congress was sending other people's sons.

But what proportion of Congressional sons would have served in Vietnam had they participated at the same rate as other young Americans? In pondering this question, three points are essential:

- 2 million American men turned 18 (draft age) every year during the Vietnam War.
- Over the ten-year conflict from 1963–72, the average US troop strength in Vietnam was about 250,000.
- A serviceman's usual tour of duty was one year.

In a typical year of the war, therefore, 2 million men reached military age, while about 250,000 men went to Vietnam for one-year tours of duty. The ratio of 250,000 to 2,000,000 is *one in eight* (12.5%). That figure, of course, is very close to the corresponding number for the Congressional sons. It is unfortunate, therefore, that the *National Journal* advanced the 28-in-234 statistic without making any effort to provide a baseline for comparison. Simple arithmetic makes clear that the statistic is not in itself sinister.

EXERCISE 2.21 It was noted that 80% of babies who have breathing problems at night sleep on pillows. Therefore, it is suggested that, for safety reasons, no babies should sleep with pillows. Why might that argument be misleading?

Aggregated Assault

We might be asked to draw a conclusion from a general pattern in a data set, even though putting the data under a "statistical microscope" would produce evidence that works against the conclusion.

Illustration F: Safe Passages

Airlines with home countries in the economically-advanced First World have traditionally had better safety records than airlines from the Developing World. Over 1987–96, for example, the death risk per scheduled international jet flight was 7 times higher for Developing World airlines than for First-World ones (1 in 400,000 vs. 1 in 3 million). Such a difference has led some people to conclude that, given a choice between a First and a Developing World airline on a given route, it is safer to go with the former.

Such reasoning might suffer from what is called the *ecological fallacy*. Consider the attempted syllogism:

1. John is better in Math than Bill
2. I have a problem in Trigonometry
3. I should therefore go to John rather than Bill

The problem with this reasoning is that Trigonometry might Bill's strong point and John's weak point. Thus, whatever the overall differences in their Math skills, Bill might be better in this particular instance.

The analogy with the airline example arises when we recognize that *First and Developing World airlines essentially compete only on "hybrid" routes* between the two "worlds" (e.g. Miami-Buenos Aires, Paris-Karachi, Tokyo-Jakarta). The germane question for the traveler, therefore, is whether First and Developing World airlines differ in safety on hybrid routes. (If not, then overall differences are not relevant in cases where passengers have a choice.)

To investigate that issue, one has to gather safety data for the routes of interest. The result for 1987–96 is the following:

TABLE 2.6 Death Risk per Scheduled International Jet Flight Between the First World and the Developing World, 1987–96

Type of Airline	Death Risk per Flight
First-World	1 in 600,000
Developing World	1 in 600,000

Source: A. Barnett and A. Wang. (2000). "Passenger Mortality Risk Estimates Provide Perspectives about Airline Safety." *Flight Safety Digest* (Flight Safety Foundation), April 2000, pp. 1–11.

As we see, the factor-of-seven difference *withers away* when First and Developing World airlines fly in comparable conditions. The common risk level of 1 in 600,000 is closer to the overall Developing World record than to the First World record. What the data might be saying, therefore, is that certain hazards in Developing World aviation pose perils to all carriers that fly there, and not just local ones. Nor was the pattern limited to 1987–96: this author found the same outcome using more recent data about flights between the First and Developing Worlds.* Over 2000–05, the death risk per flight was approximately one in 1.5 million on both First World and Developing World carriers.

* A. Barnett, "Global Passenger-mortality Risk Decreased Substantially from 2000 to 2005," *Flight Safety Digest*, Jan–Feb 2006, pp. 14–18.

EXERCISE 2.22 Consider the argument:

- Affluent people in America generally favor Republican candidates for President.
- Connecticut has the highest per capita income in the United States.

Therefore:

- Connecticut will generally favor the Republican candidate for President.

Why might this reasoning illustrate the ecological fallacy? (Incidentally, Connecticut has voted for the Democrat in the last six presidential elections.)

Confused Contrasts

In some instances, a comparison between two numbers tempts us to draw quick conclusions that could falter under further scrutiny.

Illustration G: Doctor Glut?

US News and World Report reported in 1983 that US physicians were "growing in number but not in pay." Between 1970 and 1982, the number of US doctors increased from 330,000 to 480,000, but the average salary per doctor (in 1982 dollars) fell from $104,000 to $100,000. The magazine seemed to discern in these statistics yet another application of the law of supply and demand: more doctors, less pay for each.

Simple calculations, however, raise questions about whether the market's "invisible hand" explains the drop in average pay. It seems plausible that perhaps 25% of the doctors practicing in 1970 (i.e., about 80,000 of them) had retired by 1982, leaving about 330,000 – 80,000 = 250,000 of them at work. Thus would mean that roughly half of the 480,000 doctors (480,000 – 250,000 = 230,000) working in 1982 had begun work during the previous twelve years. Because of this large influx, the typical physician in 1982 could well have been *younger* than his 1970 counterpart. And, because younger doctors have lower salaries than older ones, the overall decline in mean salary might have reflected a downward shift in the *age distribution*, rather than reduced compensation at any given age.

The point here is that shifts in complicated averages can occur for several reasons. (We discussed this issue earlier in connection with Simpson's Paradox.) We should hesitate to seize upon one possibility without identifying and considering the others.

Illustration H: Unemployment and Poverty

In November 2013, The *Boston Globe* reported that "20% of workers from Massachusetts households with annual incomes of less than $20,000 were unemployed compared with 3.3% of workers from households with annual incomes of $150,000 or more." The goal was to illuminate the point that poor families lost jobs in the recession to a greater extent than more affluent ones.

But there is an issue with direction of causality here. If the prime bread-winner in a family loses her job, then *as a consequence* one would expect its annual income to fall. The fact that income was below $20,000 after the job loss does not imply that the income fell in that dollar range prior to the loss. Even if all families had the same income before unemployment took its toll, their incomes presumably would differ sharply after some lost their jobs. In short, the *Globe*'s statistic might make little more than the obvious point that unemployment costs money.

Illustration I: Fare Fight

In 1995, New York's Metropolitan Transit Authority (MTA) raised fares by 20% on the New York City subway and by 9% on the commuter-rail lines serving the city. The MTA was harshly criticized (and, of course, a lawsuit was filed) for raising fares by a greater percentage for often-impoverished city residents than for rich suburbanites. The Authority said, however, that its critics had gotten things backwards: suburbanites typically take commuter trains into the city and then take the subway the rest of the way to work. Thus, suburban commuters were getting hit twice, and they were the ones bearing the heaviest brunt of the fare increase.

Was the MTA correct? If we assume that the increase should be measured as a *percentage* of the original fare, the following table for a typical commuter tells the story:

	Old Fare	New Fare
Subway	$1.25	$1.50 (+20%)
Commuter Train	$3.00	$3.27 (+9%)
Total	**$4.25**	**$4.77 (+12%)**

In a way, the table belabors the obvious. If part of the suburbanite's fare increases by 20% and the remainder by 9%, then her overall increase must be *between* 9% and 20%. In this case, 9% was closer to the final figure because the commuter-train fare is the greater proportion of the travel expense.

Perhaps unintentionally, the MTA was hinting that we should *add* the two fare increases together, and conclude that suburban commuters were socked with a 29% rise in fare (20% + 9%). The absurdity of that calculation is clear if we consider an inflation-induced 4% increase in the price of every food item. A person who bought 25 items would not pay 100% more, but only 4%. Because adding may be more intuitive than averaging, confusing the two concepts is more common than we might realize.

Illustration J: Polls Apart

On December 19, 2002, The *Wall Street Journal* ran a story titled "Year after Year, the Big Divide in Politics is Race." To buttress its point, it presented some recent polling results, including some on the topic of national security:

Question: Will the US do too little to identify potential terrorists, or might it violate citizens' rights by going too far?

The responses by race (black and white) were depicted graphically as follow

The chart noted that "figures don't include unsure/other responses; margin of error (of poll) is +/– 3.2 per cent"

This graphical display seems most unusual. It encourages readers to compare the percentage of whites who answered one way with the percentage of blacks who answered *the other way (!!)*. Especially strange is that the appearance of a racial difference seems depends crucially on this backwards comparison. If, after all, 45% of blacks think that the US will go too far, then 55% chose the other alternative "won't go far enough." (Note that "unsure/other" responses have been deleted from the results.) And that 55% figure is very close to the white statistic of 57%: indeed, the difference falls within the poll's 3.2% margin-of-error. The very chart that supposedly illustrates "racial differences" indicates the *absence* of such differences on the question at hand. And the chart appeared on page one of the *Wall Street Journal*!

Illustration K: Follow the Crowd?

On the smash TV hit *Who Wants To Be A Millionaire?*, a contestant could seek help in two ways on a multiple-choice question with four possible answers. She could poll the studio audience, or she could phone someone she considered an expert on the topic. Tabulations indicated that, while the experts who were contacted knew the right answer 65% of the time, the corresponding figure for the audience was 91%. This difference was cited in a respected book about "the wisdom of crowds," which argued that collective assessments tend to outperform even the judgments of individual experts.

Such a comparison is only reliable, however, if the experts and the audience confronted questions of equal difficulty. If—as this author, a *Millionaire* fan, suspects—contestants generally polled the audience on easier questions and postponed contacting the experts until harder ones, then we have no idea what would have happened in a "head to head" competition. In theory, a fair comparison could have been arranged: The difficulty of a question was tied to its dollar value, so the audience success rate at every cash value could have been matched with the expert success rate at that value. But no such matching was performed.

Furthermore, the audience and the expert were held to different standards of correctness. The audience got full credit whenever the correct choice received more votes than any of the other three choices, while the expert actually had to be right. If, for example, the correct answer was choice (*A*), then the audience would succeed if the voting pattern was 36% for *A*, 24% for *B*, 20% for *C*, and 20% for *D*. This despite the fact that 64% of the audience was wrong. Indeed, if the correct choice was A and the audience voted 36% for *A* and 64% for *B*, then the audience would have failed, even though the proportion who knew the right answer was the same in the earlier case. It seems odd that a statistic about the wisdom of crowds depends strongly on how errors were distributed among those in the crowd who showed no wisdom.

Illustration L: Get the Picture?

On September 1, 2003, *The New York Times* listed the "top summer movies" of that year. The five biggest hits were listed in order as follows:

Movie	Release Date
Finding Nemo	May 30
The Matrix Reloaded	May 15
Pirates of the Caribbean	July 9
Bruce Almighty	May 23
X2: X-Men United	May 2

The ranking was based on "box office receipts, May 2-August 24."

Note that four of the five top movies were released in May. Perhaps that circumstance is a coincidence, or perhaps the distributors deliberately sent their most promising pictures out early. But there might be another explanation for the pattern. A movie released in early May had something like fifteen weeks to amass box office receipts by August 24. A picture released in mid-August, by contrast, had something like one week. Treating receipts through August 24 as a proxy for popularity, therefore, hardly seems fair to the late arrivals.

Adjusting for differing "exposure times" through late August might be difficult: many movies get the bulk of their receipts in the first few weeks, while others gradually gain momentum as word of their virtues gets around. But perhaps the worst strategy to deal with differing times since release is to ignore the differences altogether, and that—weirdly—is what the *Times* decided to do.

EXERCISE 2.23 Some of the illustrations in this section (Confused Contrasts) could reflect Simpson's Paradox if full data were available. Which illustrations are they, and what general pattern in the data would reflect Simpson's paradox (if it arose)?

Strange Statistics

Some statistical analyses are simply odd. They exude illogic, though that circumstance is not always easy to recognize.

Illustration M: Home Pregnancy Testing

An advertisement for a home pregnancy test boasted that, under laboratory conditions, the test was 99.5 % accurate.

This statistic is a bit vague because there are two kinds of mistakes an early pregnancy test can make: telling a woman who is pregnant that she isn't, or telling a woman who is not pregnant that she is. Evidently, 99.5% is some weighted average of the two accuracy rates, though what weighting was used is impossible to determine.

A brochure put out by the manufacturer of the test kit, however, raised a more disturbing issue. The 99.5 % figure was based on laboratory testing of 200 women, for whom the result was accurate in 199 cases. But the distribution of cases was most surprising:

		Actual Status:	
		Pregnant	*Not Pregnant*
Test Result:	*Pregnant*	197	0
	Not Pregnant	1	2

Two questions arise at once. Why were 99% of the women in the test (198 out of 200) actually pregnant? And, even more strangely, why was the accuracy for non-pregnant women estimated from a sample of size *two*?

Things got worse as the brochure went on. The 2-for-2 statistic about non-pregnant women was based on an analysis of the test results by laboratory technicians. But the whole point of a home pregnancy test is that women can use it themselves. The brochure took account of this issue by reporting what happened when women interpreted test results on their own: of 101 such women who were not pregnant, 8 mistakenly concluded that they were.

In other words, the manufacturer had two accuracy results about non-pregnant women who took the test. One, based on a (presumably) random sample of the product's users, showed an error rate of 8% in 101 trials. The other, based on a "sample" of laboratory technicians, obtained a 0% error rate over 2 trials. In its advertising, the manufacturer worked with the 0% rate in the small expert sample and ignored the 8% rate in the large, more typical one.

Illustration N: Did Obamacare Flop?

A February 2014 article in *The Wall Street Journal* suggested that Americans who lacked health insurance were not exactly rushing to sign up for care under the new Affordable Care Act ("ObamaCare"). It noted that "A McKinsey & Co. survey of those thought to be eligible for ObamaCare health-care exchanges found that only 11% of those who bought new coverage between November 2013 and January 2014 were previously uninsured."

However, this 11% statistic is of scant relevance to the point it supposedly supported. What one would want to know is how the number of uninsured who signed up compares to the number of uninsured who did *not* sign up. Yet the 11% instead compares uninsured signups to signups among those already insured. Thus, the McKinsey statistic has the wrong denominator to address the issue that interested the *Journal*. The *Journal*'s columnist missed the point that, if relatively few Americans are uninsured, then they might be a small minority of all signups even if 100% of the uninsured signed up.

Illustration P: Innocent Victims?

In the bitter debate about capital punishment in America, a major concern is about whether innocent people are being executed. Perhaps the most prominent statistic in the controversy around 2000 was a 1-in-7 ratio: as explained by *Newsweek*, the ratio means that "for every seven executions nationwide since the death penalty was reinstated in 1976, one death-row inmate has been set free." (The person was set free because of major errors in his trial.) William F. Buckley Jr. probably reflected the common understanding of this statistic when he wrote that "if the figures work out retroactively, then one out of seven (of the 640) executed Americans was, in fact, innocent."

Greatly upset by the ratio, the *Economist* noted that "if an airline crashed once for every seven times it reached its destination, it would surely be suspended immediately." A bit of probing makes clear, however, that the 1-in-7 ratio makes no sense.

We start with two important statistics. There is an obvious interest in the *error rate E for capital-sentencing*, which follows:

$$E = \frac{\text{Number of Innocents Sentenced to Death}}{\text{Total Number of People Sentenced to Death}}$$

Also of importance is the *error rate X for actual executions*:

$$X = \frac{\text{Number of Innocents Executed}}{\text{Total Number of People Executed}}$$

In an ideal world, both these rates would be zero.

The 1-to-7 ratio, however, represents neither of these two rates but rather a confused amalgam of their components. It divides the number of known innocents sentenced to death by the number of executions. In other words, it divides the numerator of E by the denominator of X. Such a calculation is of no value: It is akin to computing an earnings-per-share statistic by dividing the earnings of one company by the number of shares of a completely different one.

Suppose that there are 10,000 people on death row and that, over a given period, one of them is found innocent and freed while one is executed. The only reliable inference from these statistics is the obvious point that, during this period, both executions and known sentencing errors were extremely rare. To take the quotient "one divided by one" while ignoring the number 10,000 altogether does not demonstrate that executions are fraught with errors; it is a meaningless act that yields no insight.

Illustration Q: Strike Four?

As reported at the start of this chapter, the Dow Jones Industrial Average (DJIA) declined in 2002 for the third year in a row. Trying to reassure investors, *Barron's* noted on 1/3/03 that:

Only once in its 107-year history has the Dow Jones Industrial Average suffered a lower close four years in a row. So … the odds against it happening this year are something like 104-to-1."

(The reporter used 104 rather than 107 because that the DJIA could not have dropped a fourth time during the first three years of its existence.)

Alas, this reassurance had very little logical basis. For the index to have suffered its fourth consecutive drop in (say) 1958, it would have to have dropped in the three prior years 1955–57 (which it did not). In other words, there were not 104 past opportunities to fall the fourth time in a row, but some smaller number.

How much smaller? If we exclude the years when the DJIA was in no position to fall for a fourth time in a row, we are left with exactly *three* possibilities: 1904, 1932, and 1942. The index did fall in 1932, but it rose in 1904 and 1942.

In other words, a historical estimate of the chance of further fall given exactly three previous falls is not 1 in 104, but rather 1 in 3. And given that this estimate is based on a sample size of three, it is highly unstable. Moreover, the very idea of using a historical estimate is questionable, given that the forces that influence investors in 2003 might bear scant relation to those at work more than half a century earlier. A sounder forecast was perhaps offered by the analyst who, when asked whether the market would go up or down in the next year, simply answered "yes."

Illustration R: Middle Aged

"All four were elderly. By his very presence, Jeremy was lowering the *median age* considerably." Thus wrote Jonathan Kellerman in his mystery novel *The Conspiracy Club* (2003, page 91). How Jeremy could lower the median age is itself something of a mystery. The four people were around eighty years old and he around thirty. Before he entered the room, the median age was roughly 80, and after he entered the median age—the age of the third oldest person present—was still roughly 80.

To be sure, Jeremy lowered the *mean* age by about ten years. What we have is an example of how "median" and "mean" are sometimes treated as interchangeable parts. We observe that phenomenon when we are told the median age of the US population, yet we see the lifespan of that population summarized by life expectancy: mean age at death. The conflation of the two words is unfortunate because, normal curves aside, the 50th percentile of a distribution can differ appreciably from its average.

Illustration S: Personality Disorder?

In her 2005 book Bait and Switch, Barbara Ehrenreich criticized major US corporations for using personality tests in their hiring decisions. Discussing the Briggs-Meyer test that classifies people into sixteen personality categories, she stated that:

> "The Briggs-Meyer Type Indicator has *zero predictive value* even on its own terms. Only 47 per cent of people tested fell into the same category on a second administration of the test." (Italicizing added.)

It is unclear why she spoke of "zero" predictive value. Suppose that the sixteen personality types arise with about equal frequency in the test results. Then, if the test really had no predictive power, the probability that a person's second test result would be the same as the first would be about 1/16 (6%). At 47%, the actual degree of replication is about eight times as high. There is a middle ground between perfect predictive power and none whatsoever, and Briggs-Meyer might fall there.

Also, there is the issue of how greatly a second test result differs from the first. Moving from one category to another one that is quite similar does less to damage the credibility of the test than a huge change in the assessment (e.g., from extreme extrovert to extreme introvert). The 47% statistic does not address this issue, which further weakens its ability to demonstrate that Briggs-Meyer is worthless.

EXERCISE 2.24 In 2008, the *Boston Globe* reported that the huge rise in gas prices had increased ridership on mass transit. As evidence, it noted that there were 10% more riders in March 2008 than in February 2007. Do you see a problem with this comparison? If not, look again.

Non-Statistics

Sometimes a "data analysis" is worthless because it is not based on accurate data. The "analyst" thought it so obvious that the data were reliable that, unbeknownst to his readers, he never bothered checking the numbers.

Example T: Left Out

A statistical expert testified for Vice President Gore in the legal proceedings in Florida that followed the 2000 US presidential election. (In that election, the two candidates were separated by 537 votes out of several million cast, and there were suspicions that technical problems made the vote count unreliable.) The expert contended that, long before the 2000 election, the defects of Florida punch-card ballots were evident.

In the 1998 election in Palm Beach County, for example, about 3% fewer votes were cast in the Senate race—in which voters had to punch through the *left* side of the ballot—than in the Governor's race on the *right* side. The most likely explanation for the difference, the expert suggested, was that it required more strength to punch through the left side of the ballot than the right, and elderly voters could not always do so successfully. Hence, contests that appeared on the left side (like the 2000 Presidential race) involved a sizeable proportion of lost votes. For that reason, the expert argued, the loss of votes because of flawed ballots left unclear which candidate had actually carried the state.

The opposing lawyer asked whether the expert had ever looked at the 1998 Palm Beach ballot. When the statistician answered that he hadn't, the lawyer provided the ballot and asked the expert to inspect it. In 1998, both the Governor's race and the Senate race had *both appeared on the left side* (!!). Comparing the two tallies, therefore, could not possibly demonstrate some difference between left and right. The expert's analysis was therefore preposterous. He looked pale.

What happened in the courtroom could have happened to any of us, so we would do well to learn from the expert's mistake. Often, we are provided data by others and asked to perform a specific analysis. Before we begin, however, we should make sure we understand what the data actually consist of, and how they were collected. We should ask to see a few individual data points, and then probe carefully if anything seems the least bit puzzling. Rest assured that, if we prepare an analysis based on tainted data, we might not be viewed as innocent victims of someone else's mistake. We might instead be thought of as incompetent, irresponsible, or both.

EXERCISE 2.25 Can you think of any data-based conclusion you have heard that was based on incorrect data? If not, go on the web and search for an example of erroneous data (e.g. in medical records). In the situation that you (briefly) describe, what steps might have prevented the erroneous analysis that occurred?

A Comment about the Mishaps

Obviously, not all data analyses are flawed, and the ones just cited are a very non-random sample. Nor is the aim to ridicule the organizations cited above, which are among the most careful and distinguished in the United States. The point is that enough analyses do have difficulties that the reader should be alert and sometimes skeptical. And the analyses subject to potential trouble certainly include those that we ourselves perform.

None of the analyses just cited was terribly complicated: the math rarely consisted of more than simple arithmetic. But if even simple analyses are subject to pitfalls that can invalidate them, the danger is even greater in more complex analyses when it is harder to see what is going on. We should *not* be discouraged: there are ways to reduce the risk of error, and one of them is trying to learn from the mistakes of others. Hence this discussion.

Still, we should remember the words of Ringo Starr: it don't come easy. Data analysis presents a continuing challenge, and one that we have no choice but to accept if numbers are to help us rather than confuse us. Welcome to Statistics.

The Takeaway Bar

Glossary

The **mean** or **average** of a set of numbers is their sum divided by the number of numbers.

A **weighted average** of a set of numbers assigns a weighting factor to each, where the factors are all nonnegative and add up to one. The greater the weighting factor, the greater the influence of that number on the final weighted average.

The **standard deviation** of a set of numbers, defined in analogy to the standard deviation in probability, is the square root of the average squared spread of the individual numbers around their mean.

The **coefficient of variation** of a set of numbers is the ratio of their standard deviation to their mean.

The **median** of a set of numbers has the property that the number of numbers that fall below that value is the same as the number that fall above. With an odd number of numbers, the median is one of them. With an even number of numbers, the median is halfway between the two adjacent numbers nearest the "halfway mark" of the ranked numbers.

The **lower quartile** of a set of numbers is the median of those numbers at or below the median for the full set. Essentially, the lower quartile is at the 25th percentile of the ranked numbers.

The **upper quartile** of a set of numbers is the median of those numbers at or above the median for the full set. The upper quartile is essentially at the 75th percentile of the ranked numbers.

The **interquartile range** of a set of numbers is the distance between its lower quartile and its upper quartile.

An **outlier** among a set of numbers is a number well beyond the interquartile range, whether above or below. If IQR is the interquartile range, UQ the upper quartile, and LQ the lower quartile, then measurements either above $UQ + 1.5IQR$ or below $LQ - 1.5IQR$ are often classified at outliers.

A **histogram** of a set of numbers is a graph that tells how many (or what proportion) of the numbers fall in each of a series of adjacent ranges. The histogram indicates the shape of the distribution of numbers.

A **box-plot** is a geometric representation of a set of numbers, which contains a rectangle with the lower quartile and the upper quartile as vertical sides. Within the rectangle is a vertical line at the median, while prongs (or "whiskers') extending from the rectangles and (sometimes) dots beyond the prongs indicate how far the distribution extends beyond the interquartile range.

A **statistical indicator** is a number that aims to summarize what is happening in some process of interest.

Simpson's Paradox arises when A outperforms B on every one of a series of dimensions, and yet B outperforms A when weighted averages across the dimensions are taken. This reversal occurs because the weighting factors in the B average differ from those in the A average, in ways that benefit B.

The **ecological fallacy** is present when one assumes that some pattern that holds in general must also hold true in every specific instance.

A **relative risk factor** expresses the risk for entities with a certain characteristic as a multiple of the risk for those entities that do not have the characteristic.

Formulas

Mean:

$$\bar{X} = \frac{1}{n}\sum_{i=1}^{n} x_i$$

Standard Deviation:

$$U = \sqrt{U^2} = \sqrt{\sum_{i=1}^{n} \frac{(x_i - \bar{X})^2}{n}}$$

Coefficient of Variation:

$$V = \frac{U}{\bar{X}}$$

Coefficient of Correlation:

$$\hat{\rho} = \frac{\left(\frac{1}{n}\right)\left(\sum_{i=1}^{n}(x_i - \bar{X})(y_i - \bar{Y})\right)}{U_X U_Y}$$

Weighted Average:

$$W = \sum_{i=1}^{n} \alpha_i x_i$$

where $0 \le \alpha_i \le 1$ for all $1 \le j \le n$ and $\sum_{i=1}^{n} \alpha_i = 1$

Median of n measurements:

$$M = \begin{cases} a_{(n+1)/2} & \text{when } n \text{ is odd} \\ \dfrac{a_{\left(\frac{n}{2}\right)} + a_{\left(\frac{n}{2}+1\right)}}{2} & \text{when } n \text{ is even} \end{cases}$$

where a_k = the kth smallest of the n measurements

Upper Quartile:

Median of those numbers at or above the overall median M

Lower Quartile:

Median of those numbers at or above the overall median M

Interquartile Range:

Distance from upper quartile to lower quartile

Some Parting Thoughts

- A quartile-based summary of a distribution based on (S, LQ, M, UQ, L) generally seems more informative than the combination of mean and standard deviation. Then again, one would hope a five-parameter summary says more than a summary with two parameters.
- Histograms are much better than box-plots at conveying the shape of a distribution. But the very fact that box plots have the same general look for all distributions—a rectangle with prongs and perhaps a few dots—makes it easier to compare distributions using box-plots.
- Outliers are measurements a good distance away from the central "core" of a distribution of numbers. Except when they are recorded in error, there is rarely a clear case for excluding them for data analysis. One might, however, perform certain calculations both with and without the outliers, to see how sensitive the results are to the outlying measurements.
- It cannot be said too often, so we will say it yet again: correlation is not the same as causality. The fact that two variable move the same way in a data set says nothing in itself about which variable influences the other or even whether either variable influences the other.
- The ideal statistical indicator is easy to understand, easy to calculate, based on logical assumptions, and a source of reliable answers to a sensible question. In practice, many well-known statistical indicators fall short or some of these dimensions. Yet one should not be a perfectionist: some shortcomings can reasonably be overlooked. Think of the Dow Jones Industrial Average.
- In creating a statistical indicator, considerations of reliability and simplicity can sometimes pull in opposite directions.
- Some statistical indicators seem unhelpful because they distract our attention for what is perhaps the most important question. A relative risk factor, for example, can divert attention away from measuring absolute levels of risk. Measuring average time or distance since last repair can obscure the question of how many items needed repair.

- Because of Simpson's paradox, the entity with the best overall statistic might be outperformed by another entity on every dimension considered. This reversal might not be noticed because the overall statistic is calculated in ways that obscure the dimension-by-dimension calculations.
- Unfair comparisons of performance go beyond the domain of Simpson's paradox. Among other things, the lengths of observation periods may differ, as can the conditions under which performances are measured. Moreover, the very statistics used to measure performance might be inappropriate.
- Even the most excellent newspapers and magazines—like *The New York Times*, *The Wall Street Journal*, *Time*, and *The Economist*—sometimes present flawed statistical analyses. That circumstance does not mean that these publications are unreliable, but it does mean that we should think carefully about what we read. It also implies that data analysis sometimes looks easier than it actually is.

EZ Pass

1. In a particular set with n numbers, the distance between S and L was twice the interquartile range. The number of outliers in this data set was:
   (A) $n/2$
   (B) 0
   (C) 2
   (D) cannot be determined from the given information
2. In two classes with 59 students apiece, the median scores on the Statistics exam were, respectively, 68% and 62%. If the scores for the two classes are combined, then the median of the 118 scores:
   (A) could exceed 68%
   (B) must fall below 68% but above 62%
   (C) must be 65%
   (D) none of these
3. When temperatures were taken by two different thermometers, the two readings (t_1,t_2) were recorded and a scatter plot was made. If the points tended closely to adhere to the 45° line through the origin, then one can conclude that:
   (A) the readings on the two thermometers were highly consistent with one another
   (B) the readings on one thermometer were uncorrelated with those on the other
   (C) the readings on both thermometers were highly accurate
   (D) when the temperature increased (in real life) by five degrees, the estimated temperature increased by 45 degrees.
4. If 77% of those suffering severe headaches who take aspirin get relief within a half hour, then the effectiveness of the aspirin:
   (A) is approximately 77%
   (B) is unknown, because we don't know how many people with severe headaches get relief within 30 minutes without any aspirin

(C) exceeds 77%, because many people who benefit from aspirin do so more than thirty minutes after they take it
(D) would be higher than 77% if people took more aspirin tablets for their headaches

5. Which of the following would be most likely to illustrate Simpson's Paradox?
(A) People who eat yogurt do better in Math, so yogurt is said to increase Math skills.
(B) To assess whether a newspaper's abrasive style is turning away readers, its subscribers are polled.
(C) The Gini coefficient of a nation's income has grown over time.
(D) The salaries of lawyers at every age are higher in 2010 than in 2000, yet the average salary for all lawyers is lower in 2010 than 2000.

6. When the temperature is 25° Fahrenheit and there is no wind but 50% humidity, the windchill factor:
(A) must be 25° Fahrenheit
(B) must be 12.5° Fahrenheit
(C) must be zero
(D) cannot be determined from the given information

Further Exercises

EXERCISE 2.26 The *breaking point* for a given wire is the level of strain at which it becomes too deformed to function. Twenty experiments on wires of this kind of that yield the following numbers for its breaking point (in pounds):

22.6	18.9	22.1	20.4	23.1	21.5	21.6	19.8	21.7	20.5
21.5	20.6	22.8	22.5	20.8	20.4	18.8	20.8	22.4	21.5

(i) Find (S, LQ, M, UQ, L) for this data set.
(ii) Create a stemplot of the results, which the numbers on the left of the line are the integers from 18 to 23.
(iii) Create a histogram of the results, using ten "bins" of equal size.
(iv) Are there any outliers among the results?
Suppose it is asserted based on theory that the median breaking point for these wires is 22 pounds. Do the data above seem reasonably consistent with that possibility? (If 22 is in fact the median, how improbable is it to get the observed number of outcomes above the median in a sample of 20 wires?)

EXERCISE 2.27 At the website are data about the number of times at bat for each New York Yankees player in 2009 with at least 50 at-bats, as well as his number of home runs.

(i) What are the mean, standard deviation, and coefficient of variation of the *numbers* of home runs for the various players?
(ii) What are the quartiles of the distribution of home runs?
(iii) Any outliers?
(iv) Calculate home runs as a *percentage* of times at bat for each player.
(v) Now find the mean, standard deviation, and coefficient of variation of the home-run *percentages* for the players.
(vi) Is the variability across players in home-run percentages appreciably less than that for the number of home runs? (Hint: Is it more appropriate for this purpose to compare the standard deviations for the two quantities, or the coefficients of variation?)
(vii) What is the correlation coefficient of times at bat and home-run percentage?
(viii) Does the correlation suggest a reason why some players go to bat more than others? Briefly discuss.

EXERCISE 2.28 Please pick a set of 17 distinct numbers.

(i) For these numbers, list the five quartile statistics (*S*, *LQ*, *M*, *UQ*, *L*).
(ii) Find the mean $\bar{X}$ of these numbers.
(iii) Find a set of 17 numbers that have the same quartile statistics as in (i) but are not all the same as the original numbers.
(iv) Find a set of 17 numbers with the same quartile statistics and the same mean as in (i), but are not all the same as the original numbers. (Hint: You need only fiddle with two of the 20 numbers to achieve these goals.)
(v) Find a set of 17 numbers with the same quartile statistics as in (i) but for which the mean $\bar{X}$ is maximized.
(vi) Find a set of 17 numbers with the same quartile statistics as in (i) but for which the standard deviation is minimized.

EXERCISE 2.29 Suppose that we have a large number of data points from a normal distribution with mean μ and standard deviation σ. If we calculate the five quartile statistics (*S*, *LQ*, *M*, *UQ*, *L*) then:

(i) Approximately where would we expect *LQ*, *M*, and *UQ* to fall, in terms of μ and σ? (Hint: Return to the normal probability table, which tells us where say)the 75th percentile falls in terms of μ and σ. More specifically, if A_{75} the 75th percentile for $N(0,1)$, then the 75th percentile for $N(\mu, \sigma)$ is $\mu + A_{75}\,\sigma$.)
(ii) What would be interquartile range be?
(iii) Approximately what percentage of the data points would be classified as outliers?

(iv) For $n = 1000$, make a reasonable estimate of S and L in terms of μ and σ. (Hint: What is the 99.9 percentile of the normal distribution with parameters μ and σ?)

EXERCISE 2.30 Suppose that we want to come up with an overall fertility statistic for Europe based on the 38 national statistics in Table 5.3.

(i) What is the simple average of the 38 national fertility statistics?
(ii) Now take a weighted average, where the weight for a given country is its proportion of total population for the 38 countries. (For example, if 1% of citizens of these countries live in country Z, then Z gets a weight of 0.01)
(iii) Which of these two statistics do you think better reflects overall fertility in Europe? Briefly explain.

EXERCISE 2.31 One of the bestselling books about business is *In Search of Excellence* by Tom Peters and Robert Waterman, which explored the management strategies of 43 companies that appeared to be highly successful. For example, these companies tend to "stick to their knitting," meaning that they focus on the businesses they know. If the book is viewed as a guide to how to create a successful company, is there a statistical reason to view its conclusions as tentative? Briefly discuss.

EXERCISE 2.32 An airline sends an e-mail message to each passenger on selected flights after they have completed their journeys. Suppose that, on a particular flight:

- 30% of passengers considered the service excellent
- 30% considered the service good
- 30% considered the service fair
- 10% considered the service poor

However, not all passengers complete the surveys, which are detailed and tedious. Suppose that:

- 20% of passengers who thought the service excellent completed the form, and gave the airline 4 out of 4 as an overall rating
- 10% of passengers who thought the service good completed the form, and gave the airline 3 out of 4.
- 10% of passengers who thought the service fair completed the form, and gave the airline 2 out of 4.

- 40% of passengers who thought the service poor completed the form, and gave the airline 1 out of 4.

Also, assume that there were 100 passengers in total on the flight.

(i) If all 100 passengers had completed the survey, what would the flight's average rating have been on the 1–4 scale?
(ii) How many of the flight's passengers completed the survey form?
(iii) What was the flight's average rating among passengers who completed the form?
(iv) In words, how would you explain why the answers to (i) and (iii) differed?

EXERCISE 2.33 Suppose that men who watch ESPN are three times as likely to be crazy as men who do not.

(i) What is the relative risk factor for craziness for men who watch ESPN compared to men who do not?
(ii) If 40% of crazy men are ESPN watchers, what proportion of the male population watches ESPN? (Hint: Work backwards from the definition of relative risk factor, letting N be the total number of men and Q the total number of crazy men.)

EXERCISE 2.34 Suppose that half the residents of some community have an annual income of \$B and the other half have an income of \$B + 10,000.

(i) What fraction of the community's total income is gained by the half of its residents who have incomes of \$$B$?
(ii) Write an expression for the community's Gini coefficient in terms of B. (Assume that the line that plots percentage of income vs. percentage of residents grows linearly with percentage of residents in the range 0–50%, and linearly between 50% and 100% though with a different slope.
(iii) For what value of B is the Gini coefficient maximized here? What is the maximum value of the Gini coefficient?
(iv) For what value of B is the Gini coefficient equal to 0.10?
(v) If everyone's income dropped by 15%, would the Gini coefficient change?
(vi) Suppose that residents with income of \$(B + 10,000) suffer a 15% drop in income, while other residents do not. For what value of B would this change push the Gini coefficient to zero? (Hint: This is not hard if you think about it.)

EXERCISE 2.35 Find the Gini coefficient in each of the following three situations:

(i) 80% of the population has income X and 20% has income $4X$ (Yes: You have seen this before).
(ii) 50% of the population has income X and 50% has income $4X$
(iii) 1/3 of the population has income \$10,000, 1/3 has income \$90,000, and the remaining third has income \$100,000
(iv) Do these results tell us something about the variety of situations that lead to the same Gini coefficient?

EXERCISE 2.36 Suppose that it is known that people who do not exercise have twice the risk of heart problems as those who do, and that obese individuals have twice the risk of heart problems of those who are not obese.

(i) Does it follow that a person who is obese and does not exercise has four times the risk of heart problems as those who exercise and are not obese?
(ii) "Even if the relative risk factor for heart problems is two for those who don't exercise compared to those who do, it does not follow that lack of exercise in itself doubles the risk." Do you agree with this statement? Explain.

EXERCISE 2.37 On April 3, 2010 *The New York Times* performed an analysis about "the best places to catch a cab" in Manhattan. The places it listed were those places where the largest number of passengers were actually picked up in 2009. Might there be a logical flaw in this analysis?

EXERCISE 2.38 In the 1995 trial in which O.J. Simpson was accused of killing his wife, evidence was introduced that he had previously subjected her to domestic violence. But his lawyers argued that this point was of minimal importance. Professor Alan Dershowitz (a member of the Simpson defense team) later wrote:

> "We knew we could prove, if we had to, that an infinitesimal percentage—certainly fewer than 1 of 2,500—of men who slap or beat their domestic partners go on to murder them."

But some statisticians like I.J. Good and J. Caulkins were uncomfortable with this application of statistics.

(i) The number 1 in 2500 is a conditional probability of the form $P(B|A)$. What are events A and B?

(ii) Suppose that C is the event that a woman is killed, and B the event that her partner was the killer. Is $P(B|C)$ the same as $P(B|A)$ above?

(iii) Of every 100,000 women with abusive partners, approximately how many would be killed based on the 1 in 2500 statistic?

(iv) Suppose that 100 of these 100,000 women would be killed by people other than their partners. Given the answer to (iii), what is an estimate of $P(B|C)$?

(v) Do you see why the statisticians were uncomfortable?

EXERCISE 2.39 Some years ago, *National Review* magazine announced that it had "good news for egalitarians and redistributionists." The US tax structure was becoming more progressive: over a ten-year period, the share of all taxes paid by the top 10% of wage earners had grown from 48% to 54%. Thus, the magazine argued, US tax policies were soaking the rich all the more drenchingly.

Could this argument be misleading? (Hint: Is it possible that the top 10% are paying a larger share of all taxes while the *percentage* of their incomes they pay in taxes is going down? Perhaps a numerical example would help clarify things.)

EXERCISE 2.40 Explain why each of the following procedures might yield biased results:

(i) The percentage of individual investors who lost money today on the New York Stock Exchange is approximated by the percentage of stocks that went down.

(ii) To gauge support for a change in Social Security policy, a pollster calls landline telephone numbers chosen by a random-number generator. He estimates the percentage of adults who support the change by the fraction of people he reached who support it.

(iii) To assess whether a company's hiring policies are gender-neutral, a committee examines the credentials of all current employees at the time they were hired.

EXERCISE 2.41 *CBS News* noted that the life expectancy (essentially, average lifespan) in a certain country was fifty years. Thus, it explained, a person approaching fifty could not expect to live much longer. Why might this analysis be problematic?

EXERCISE 2.42 In the United States in 2008, 118 firefighters lost their lives at work, out of a total of 1,148,850 firefighters. That works out to one death at work per 9736 firefighters.

Among all Americans, the death rate in 2008 was 1 in 119. Therefore, firefighting is one of the safest professions in America.

Do you see a possible flaw in that reasoning? What is it?

EXERCISE 2.43 The 8/22/11 issue of *Time* magazine had an item titled "Early Morning Smoking Risk." It noted that cigarette smokers who lit up a cigarette within 30 minutes of waking up in the morning had 79% higher risk of lung cancer and 59% higher risk of cancer of the head and neck than those who waited at least an hour until their first cigarette. Might it be misleading to suggest that the *timing* of the first cigarette in itself affects cancer risk? (Hint: With what factor that might plausibly affect cancer risk could time until the first cigarette be correlated?)

Introductory Inference and Statistical Sampling

Scallop Treatment (Based on a true story)

In 1990, a fishing vessel docked at New Bedford, Massachusetts with a harvest of 11,000 bags of scallops. The Fisheries and Wildlife Service requires that, in any such harvest, the average meat per scallop weigh at least 1/36 of a pound. The requirement is aimed at protecting baby scallops, though less to guarantee them happy childhoods than to preserve enough adult scallops so that the species does not disappear.

To assess whether the ship had complied with the weight requirement, the harbormaster randomly selected 18 bags for weighing. His agents then computed the meat-per-scallop ratio for each of the bags. Based on the 18 numbers thus generated, the harbormaster estimated that each of the ship's scallops possessed on average 1/39 of a pound of meat (i.e. that they were about 7 percent lighter than the minimum requirement). Viewing this outcome as clear evidence that the weight standard had been violated, federal authorities confiscated 95% of the scallops and sold them at an auction. The fishing voyage was thus transformed into a financial catastrophe for its participants.

The ship's owner was as displeased with the US government as Captain Ahab had been with Moby Dick. He declared that the vessel had fully satisfied the weight standard, and saw lunacy in the assertion that sampling 18 bags out of 11,000 could yield a reliable estimate of the mean weight of all the vessel's scallops. He filed a lawsuit against the government and arranged for a Boston law firm to represent him. The firm, in turn, asked the statistician for advice on whether the authorities had acted improperly.

How much accuracy could possibly arise, the lawyers inquired, from a sample of only 18 bags? Clearly, they were hoping the statistician would say "practically none at all." They were thus disappointed when he offered the equivocal response "it depends."

(To be continued)

Statistics is about numbers, but numbers come in two broad categories. There are some numbers that are known exactly, like the Dow Jones Industrial Average once the stock market has closed or the average exam score once the papers have been graded. But there are other numbers that, as a practical matter, can only be approximated. When we estimate a number, we are naturally concerned with how far off our estimate might be. Yet precisely because we don't know the actual number, we can't know exactly what our estimation error was. Realistically, we should be satisfied with a probability distribution for the estimation error we incurred.

This chapter concerns itself with ***parameter*** estimates, where by "parameter" we mean some number of interest. We will discuss criteria by which we might make and evaluate ***point estimates*** (specific numerical estimates for the unknown parameter) as well as ***margins of error*** and ***interval estimates*** (numerical indications about the uncertainty that surrounds our point estimates). Because much parameter estimation is based on random samples of data, we will move quickly to the basic theory of statistical sampling. And because we are inferring from data where the true parameter might lie, the analyses that we are undertaking are examples of ***statistical inference***. Let us infer.

3.1 Uncertainty Associated With Data

When we estimate, the available data do not exactly reveal what we want to know. That can happen for two distinct reasons. The first is that the data points themselves are inexact, meaning that the observed value differs from the true value by an uncertain amount. You need go no further than your cell phone to see such imprecision in action. If your phone is like the author's, it tells time by the minute. It will read 3:23 PM throughout the 60 seconds between 3:23:00 and 3:24:00. Thus, when we open the phone and see time $\ddot{t}$, the actual time t *to the nearest minute* is given by:

$$t = \ddot{t} + \varepsilon$$

where ε is a random variable with distribution $\varepsilon = \begin{cases} 1 \text{ w.p. } \frac{1}{2} \\ 0 \text{ w.p. } \frac{1}{2} \end{cases}$

This distribution for ε reflects the facts that (i) the number of seconds since the display changed to $\ddot{t}$ is uniform $U(0,60)$, and (ii) $\ddot{t} + 1$ is the time to the nearest minute when the number of seconds since $\ddot{t}$ exceeds 30.

The second reason for uncertainty is that, while the individual data points are exact, they constitute only a *sample* from a broader population whose behavior interests us. We have seen numerous examples of such sampling; when a pollster seeks to forecast an election result, for example, he selects a random sample of likely voters and asks them their preferences. Even if those canvassed know how they will vote and tell the truth, the precision with which their preferences reflect those of *all* voters depends on the luck of the draw. Thus, there is an element of *sampling error* associated with the available data. As we will see, we can make probabilistic statements about the extent of such sampling error, but we can never eliminate it.

We will deal with both kinds of uncertainty in data in this chapter, but mainly the issues associated with sampling error. We will say much more about uncertainty in the individual data points when we study regression analysis.

3.2 Maximum Likelihood Estimation

The general question we tackle in this chapter is: How might we estimate numbers (parameters) in a sensible way? We describe a few important approaches here, but we postpone applying the methods to actual data until we understand what the methods are doing. Please be patient.

To start the treatment of parameter estimation, we turn to the well-known method of maximum likelihood. We begin with an example.

EXAMPLE 3.1

Tax Shelter?

In a random poll taken in June 2013, 119 of 220 adults from the US Midwest expressed their opposition to charging sales tax on items purchased over the Internet.

Suppose that p is the fraction of all Midwestern adults who oppose an Internet sales tax.

(i) Given the poll outcome, what estimate for p arises intuitively?

(ii) Let $\breve{p}$ be the value of p at which the observed results would have the highest probability of arising. Is $\breve{p}$ the same as the answer to (i)?

Solution

(i) A natural estimate of p given the data is 119/220, which is .541. Perhaps there are other outcomes with intuitive appeal, but the simple rule "what you see is what you get" seems dominant to this author.

(ii) To find out whether $\breve{p}$ as defined above equals .541, we should simply calculate $\breve{p}$. We can interpret the poll as a binomial process, in which each person canvassed independently has probability p of opposing the tax. We can therefore write:

$$P(119 \text{ tax opponents out of 220 questioned} | p = \alpha) = \binom{220}{119} \alpha^{119}(1-\alpha)^{220-119}$$

We call $L(\alpha)$ =(P(119 tax opponents out of 220 questioned$|p = \alpha$) the ***likelihood*** of the observed outcome as a function of α. Because $L(\alpha) = \binom{220}{119}\alpha^{119}(1-\alpha)^{220-119} = \binom{220}{119}\alpha^{119}(1-\alpha)^{101}$, the value of α at which $L(\alpha)$ is maximized (namely, $\breve{p}$) is obtained by setting $dL(\alpha)/d\alpha$ equal to zero. If we do so by using the rule for differentiating a product of functions, we reach: $\breve{p}$ = 119/220. Thus, the intuitive estimate of p is the same as the estimate $\breve{p}$ under which the observed result is most likely.

EXERCISE 3.1 Suppose that a six-sided die (which is not necessarily fair) is to be tossed 12 times. If p_1 is the true probability of getting a one on this die and three ones arose among the 12 tosses, which estimate $\breve{p}_1$ of p_1 yields the highest probability for getting that outcome?

Example 3.1 has quietly introduced an important approach to making point estimates of parameters: the maximum-likelihood method. The maximum likelihood estimate of a parameter is defined by:

> The **maximum likelihood estimate** of a parameter is the numerical value of that parameter under which the observed data outcome has the highest probability of arising.

The maximum likelihood estimate depends on which probability distribution the data are assumed to follow. A few examples illustrate how maximum likelihood estimates are calculated.

EXAMPLE 3.2

The Natural

Suppose that a binomial process has been observed for n trials, and event A has arisen k times, with $0 < k < n$. What is the maximum likelihood estimate of p?

Solution

This example is just the generalization of Example 3.1. If α is the chance of event A in the binomial process, then the probability of the observed outcome (i.e., k A's and $n - k$ B's) follows:

$$L(\alpha) = P(k\ A\text{'s over } n \text{ trials} | p = \alpha) = \binom{n}{k}\alpha^k(1-\alpha)^{n-k}$$

The derivative $\frac{dL(\alpha)}{d\alpha}$ is given by

$$\begin{aligned}\frac{dL(\alpha)}{d\alpha} &= \binom{n}{k}\left(k\alpha^{k-1}(1-\alpha)^{n-k} - (n-k)\alpha^k(1-\alpha)^{n-k-1}\right)\\ &= \binom{n}{k}\left(\alpha^{k-1}(1-\alpha)^{n-k-1}\left(k(1-\alpha)-(n-k)\alpha\right)\right)\end{aligned}$$

This derivative is zero at $\alpha = 0$, $\alpha = 1$, and $\alpha = \frac{k}{n}$. In the range (0,1), $\alpha = 0$ and $\alpha = 1$ are minima of the likelihood function $L(\alpha)$ given that $0 < k < n$, while $\alpha = \frac{k}{n}$

is a maximum. Therefore, $\breve{p}$, the maximum likelihood estimate (***hereafter MLE***) of p, is given by:

$$\breve{p} = k/n.$$

While this formula for $\breve{p}$ was obtained assuming $0 < k < n$, it also works for $k = 0$ and $k = n$. Thus, we have:

In a binomial process with parameter p in which A comes up k times over n trials, the maximum likelihood estimate of p is k/n. (3-1)

EXERCISE 3.2 Suppose that a binomial process is observed for n trials, and A arises on the first k trials while B arises on all the remaining $n - k$ trials. Given this information:

(i) Do you still think that the MLE of the binomial parameter p is k/n? (No calculation; just go with your instincts.)

(ii) Please maximize the likelihood function, and find out whether your intuition in (i) was correct.

EXAMPLE 3.3 **First Time**

Suppose that Mendel wants to estimate p, the probability that he will encounter a green light at a given traffic intersection. To do so, he will count the number of times he arrives there until the first time he finds the light green. If that number turns out to be j, he will estimate p as $1/j$. Is that the maximum likelihood estimate of p?

(Assume that the probability the light is green is p whenever he arrives at it, and that different arrivals are independent.)

Solution

Under the key assumption (probability p of green on each arrival, with different arrivals independent), the number of arrivals until the first green light is geometrically-distributed with parameter p. Thus, we can write:

$$L(\alpha) = P(\text{first green on } j\text{th arrival} \mid p = \alpha) = \alpha(1 - \alpha)^{j-1}$$

Here $dL(\alpha)/d\alpha = (1 - \alpha)^{j-1} - (j - 1)\alpha(1 - \alpha)^{j-2} = (1 - \alpha)^{j-2}((1 - \alpha) - (j - 1)\alpha)$

Thus, $dL(\alpha)/d\alpha = 0$ when $\alpha = 0$ and when $(1 - \alpha) - (j - 1)\alpha = 0$. The value of α that satisfies this last equation is $\alpha = 1/j$. (The value $\alpha = 0$ minimizes $L(\alpha)$ in the range (0,1), while $\alpha = 1/j$ maximizes it. If you're skeptical of this last statement, check out the second derivative of $L(\alpha)$ at $\alpha = 1/j$.) Thus, $\breve{p}$, the MLE of p, follows: $\breve{p} = 1/j$. Mendel's estimate of p is in fact the maximum likelihood estimate.

In a geometric process with parameter p, the maximum likelihood estimate of p is $1/j$, where j is the number of trials until the first A comes up. (3-2)

EXERCISE 3.3 Suppose that Mendel instead counted arrivals until the second time he encounters a green light. Because that happened on his j^{th} arrival, he estimated p by $2/j$.

(i) In terms of p, what is the probability that the second green light is on Mendel's j^{th} arrival? (Hint: If the second green is on the jth arrival, how many greens were there on the first $j-1$ arrivals?)
(ii) Is k/j the maximum likelihood estimate of p?
(iii) Suppose that the second green was on the jth arrival but we also learn that the first green was on his first arrival. Would this new information change the MLE of p?

EXAMPLE 3.4 **At This Rate**

Suppose that a time-related Poisson process with parameter λ is observed for T time units, and exactly k events are observed. Show that the maximum likelihood estimate of λ is k/T.

Solution

Under the Poisson formula, we can write:

$$P(k \text{ events in period } T|\lambda = \alpha) = (\alpha T)^k e^{-\alpha T}/k!$$

If $L(\alpha) = (\alpha T)^k e^{-\alpha T}/k!$, then $dL(\alpha)/d\alpha$ is given by:

$$dL(\alpha)/d\alpha = kT\frac{(\alpha T)^{k-1}e^{-\alpha T}}{k!} - T\frac{(\alpha T)^k e^{-\alpha T}}{k!}$$

If we set this derivative equal to zero and cancel out the common factor $\frac{(\alpha T)^{k-1}e^{-\alpha T}}{k!}$, we reach $k - \alpha T = 0$, in which case $\alpha = \frac{k}{T}$.

Thus $\breve{\lambda}$, the MLE of λ, follows: $\breve{\lambda} = k/T$.

In a time-related Poisson process with parameter λ, the maximum likelihood estimate of λ if exactly k events occur over period T is k/T. (3-3)

EXERCISE 3.4 The table below presents the numbers of hurricanes that struck the United States during the years 1981 to 2009.

Year	Number of US Hurricanes	Year	Number of US Hurricanes
1981	0	1996	2
1982	0	1997	1
1983	1	1998	3
1984	1	1999	3
1985	5	2000	0
1986	2	2001	0
1987	1	2002	1
1988	1	2003	2
1989	3	2004	6
1990	0	2005	6
1991	1	2006	0
1992	1	2007	1
1993	1	2008	3
1994	0	2009	0
1995	1		

Suppose that it is posited that hurricanes in the United States arise under a Poisson process with rate λ per year, with λ constant over the period 1981–2009.

(i) What is the maximum likelihood estimate of λ for the period shown? (Hint: Use (3-3) for the entire period.)

(ii) What is the maximum likelihood estimate of λ based only on the data for the 21st century?

There are cases when maximum likelihood estimates of parameters are not the ones that first come to mind. We consider one such case now.

EXAMPLE 3.5

Narrow Minded

It is known that X is uniform on the *integers* from a to b (i.e., it has a discrete distribution), and three consecutive observations of X are 8, 2, and 9. Given these data, what are the maximum likelihood estimates of a and b? Does that outcome accord with intuition?

Solution

If the integers range from a to b, it is obvious that $a \leq 2$ and $b \geq 9$. (Otherwise, the probability of getting the observed pattern is zero.) If $b - a = q$, then each integer from a to b would be assigned probability $1/(1 + q)$ under uniformity. Then the probability of getting (8, 2, 9) would be $1/(1 + q)^3$. That quantity is maximized when q is minimized, which happens when q =7 and a = 2 and b = 9. Thus, the maximum likelihood estimates of a and b are, respectively, 2 and 9, corresponding to the narrowest possible interval consistent with the data.

This result seems unintuitive because we would not expect the smallest of three observations from a uniform distribution to be at the *very minimum* of that distribution and the largest to be at the very maximum. But any other choices of a and b imply a lower probability of the observed outcome. A definition is a definition: 2 and 9 are the MLE's.

Actually, this example makes an important point. In the abstract, maximum likelihood estimation is very attractive: why not estimate a parameter by the value under which the observed result seems most "natural?" But not only is maximum likelihood not guaranteed to yield a correct estimate; in some instances, it produces parameter estimates that seem neither natural nor believable. As happened here.

EXERCISE 3.5 Suppose that discrete random variable X can take on integer values from 1 through n, and that $P(X = k)$ is given by:

$$P(X = k) = \frac{k}{\binom{n+1}{2}} \quad \text{for } 1 \le k \le n$$

However, the value of n is not known.

(i) Verify that the $P(X = k)$'s for permissible values of k add up to one.
(ii) Five independent selections from the distribution are obtained, and they are 5, 3, 6, 8, and 6. What is the maximum likelihood estimate of n?

EXERCISE 3.6 Suppose that it is known that X is $N(0, \sigma)$, and we get the two observations 0.4 and 1.8.

(i) What is the maximum likelihood estimate of σ given these data? Treat 1.4 as meaning in the infinitesimal interval from 0.4 to 0.4 + Δ, and 1.8 in the interval (1.8, 1.8 + Δ). (Hint: Recall that σ appears in two separate places in the normal density function.)
(ii) Does the estimate seem intuitively attractive to you?

3.3 Estimators and Bias

A quantity like the maximum likelihood estimate can only be calculated when data are at hand. However, the general formula under which the calculation occurs can be specified in advance. For a binomial process, for example, the MLE of parameter p is X/n, where n is the number of trials and random variable X is the number of A's that arise over those trials. It follows that the estimator $\breve{p} = X/n$ is itself a random variable prior to the inspection of data, and that the calculated MLE is the realization of $\breve{p}$.

The last paragraph makes a distinction that we will follow in this chapter.

An **estimator** is a random variable that embodies the rule for approximating the quantity of interest.

The **estimate** is the number that arises when the estimation rule is applied to data.

In assessing the accuracy of an estimator, it is often worthwhile to take account of its first two moments, its mean and variance. Perhaps the most fundamental question about an estimator is: Is it biased? We discuss that issue now.

Unbiased Estimators

If θ is the true value of the quantity of interest and $\ddot{\theta}$ is an estimator of θ, then $\ddot{\theta}$ is said to be **unbiased** if $E(\ddot{\theta}) = \theta$. An unbiased estimator is therefore correct on average although, except in the rare situation when $\ddot{\theta}$ has zero variance, it is not guaranteed to be correct in a given case. Unbiasedness means that, if the estimator is used repeatedly in similar circumstances, there is no net estimation error: the individual overestimates and underestimates cancel out over the long run.

An estimator $\ddot{\theta}$ of θ is ***unbiased*** if $E(\ddot{\theta}) = \theta$.

Again, we are treating $\ddot{\theta}$ as a random variable prior to the estimation. $\ddot{\theta}$ therefore has a probability distribution, and unbiasedness concerns whether the mean of that distribution equals the quantity that is being estimated.

EXAMPLE 3.7

Bias in the Trials?

Suppose that, in a binomial process with parameter p, p will be estimated as $\breve{p} = X/n$, where random variable X is the number of A's over n forthcoming trials.

(i) Is $\breve{p}$ an unbiased estimate of p?

(ii) Suppose that p is instead estimated by $\ddot{p} = \frac{Z}{n/2}$, where Z is the number of A's over the first $n/2$ trials. Is $\ddot{p}$ also an unbiased estimate of p?

(iii) Given that (i) and (ii) have the same answer, do you have a preference for one of the estimators? If so, which one and why; if not, why not?

Solution

(i) Because $\breve{p} = X/n,\ E(\breve{p}) = E(X)/n$

But because X is $B(n,p)$, we know that $E(X) = np$.

Therefore, $E(\breve{p}) = E(X)/n = np/n = p$, and $\breve{p}$ is unbiased.

(ii) Because Z is $B(n/2, p)$, $E(Z) = (n/2)$p. Thus:

$$E(\ddot{p}) = E\left(\frac{z}{n/2}\right) = \frac{\left(\frac{n}{2}\right)p}{\left(\frac{n}{2}\right)} = p, \text{ and } \ddot{p} \text{ is also an unbiased estimator of } p.$$

(iii) Both estimators are "on average" correct, but that is quite different from saying that either is sure to be correct in a particular case. Intuitively, we might expect that, most of the time, the estimate $\breve{p}$ will be closer to p than $\ddot{p}$ because it is based on twice as many trials. We will formalize this line of reasoning shortly when we consider the variance of an estimator as well as its mean.

EXERCISE 3.7 In a Poisson process, the rate parameter λ will be estimated by $\ddot{\lambda} = X/\tau$, where X is the number of events in the process over a forthcoming period of length τ. Show that $\ddot{\lambda}$ is an unbiased estimator of λ.

From Example 3.7 and Exercise 3.7, we have:

- In a binomial process with n trials of which X resulted in event A, X/n is an unbiased estimator of p.
- In a Poisson process in which X events arise over a fixed period of length τ, X/τ is an unbiased estimator of λ (3-4)

Are Maximum Likelihood Estimates Unbiased?

MLE's are not necessarily unbiased, for the same general reason that the mode of a probability distribution need not be at its mean. One has to perform analysis on a case-by-case basis. Example 3.7 and Exercise 3.7 imply that the MLEs that arise under binomial and Poisson processes are unbiased. (Note that $\breve{p}$ and $\ddot{\lambda}$ are in fact the MLE's we obtained earlier.) However, the situation is different for the (continuous) uniform distribution, as we will see now.

EXAMPLE 3.8 **A Bias Emerges**

Suppose that it is known that X is $U(0,b)$, and six observations of X are 2.5, 6.88, 2.71, 4.39, 1.22, and 2.99 (to the nearest trillionth).

(i) What is the maximum likelihood estimate of b?
(ii) Can we show that the maximum likelihood estimate of b is biased?

Solution

(i) Under the reasoning of Example 3.5, the MLE of b is the highest observed value, which is 6.88.
(ii) We have six independent observations from $U(0,b)$.
If $Z_6 = \max(X_1, X_2, \ldots X_6)$, Z_6 is the MLE of b. $F_6(y)$, the DFF of Z_6, follows:

$$F_6(y) = P(Z_6 \le y) = \begin{cases} 1 & \text{for } y > b \\ \left(\frac{y}{b}\right)^6 & \text{for } 0 \le y \le b \\ 0 & \text{for } y < 0 \end{cases}$$

This expression is correct because the only way to have the maximum below y for $0 \le y \le b$ is to have all six observations below y. Each observation independently has a chance of y/b of falling below y (and a zero chance of being exactly at y).
Differentiating the formula for the CDF, we find that the pdf $f_6(y)$ follows: $f_6(y) = 6y^5/b^6$ in $(0,b)$ and 0 elsewhere. Therefore:

$$E(Z_6) = \int_0^b y f_6(y)\,dy = \int_0^b 6y(y^5/b^6)\,dy = \left(\frac{6}{b^6}\right)\left(\frac{b^7}{7}\right) = \left(\frac{6}{7}\right)b$$

It follows that the MLE is *not* unbiased: it biased downward by on average 14% of the value of b (i.e., by $b/7$). We had hinted at such a downward bias in Example 3.5, and now we have quantified it.

EXERCISE 3.8 Suppose that it is proposed that b be estimated in Example 3.8 by twice the average of the six random observations from $U(0,b)$. Would that estimator be unbiased?

3.4 Variance and Mean Squared Errors of Estimators

Even if an estimator is on average correct (i.e., is unbiased), there is the question of how far off it might be. *Mean squared error* is a common and intuitive indicator about the spread of the estimator around the true parameter value.

If θ is the parameter of interest and $\ddot{\theta}$ is an estimator of θ (i.e., θ is the true value of the parameter), then its **mean squared error** (MSE) is given by:

$$MSE(\ddot{\theta}) = E\left((\ddot{\theta} - \theta)^2\right) \tag{3-5}$$

When $\ddot{\theta}$ is unbiased, $MSE(\ddot{\theta})$ is the squared spread of $\ddot{\theta}$ around its mean θ. In other words:

When $\ddot{\theta}$ is an unbiased estimator of θ, its mean-squared error is the same as its variance:

$$MSE(\ddot{\theta}) = \sigma^2(\ddot{\theta}) \tag{3-6}$$

What if $\ddot{\theta}$ is not an unbiased estimator of θ? Then we have a different relationship between MSE and variance:

$$\text{If } E(\ddot{\theta}) = \gamma \text{ and } \gamma \neq \theta, \text{ then } \text{MSE}(\ddot{\theta}) = \sigma^2(\ddot{\theta}) + (\gamma - \theta)^2 \tag{3-7}$$

The derivation of this last formula is brief. We note that:

$$\begin{aligned}(\gamma - \theta)^2 &= \left((\ddot{\theta} - \gamma) + (\gamma - \theta)\right)^2 \\ &= (\ddot{\theta} - \gamma)^2 + 2(\ddot{\theta} - \gamma)(\gamma - \theta) + (\gamma - \theta)^2\end{aligned}$$

Taking expectations of both sides of this equation, we reach:

$$E\left((\gamma - \theta)^2\right) = \sigma^2(\ddot{\theta}) + 2(\gamma - \theta)E\left((\ddot{\theta} - \gamma)\right) + (\gamma - \theta)^2 \tag{3-8}$$

Here we are using the definition of variance, and the fact that γ and θ are constants. We complete the derivation by noting that $E(\ddot{\theta}) = \gamma$ and that therefore $E\left((\ddot{\theta} - \gamma)\right) = 0$. In consequence, (3-8) becomes (3-7).

EXAMPLE 3.10

Bias Is Better?

Suppose that $\theta = 4$ and that $\ddot{\theta}$ and $\ddot{\theta}$ are two estimators of θ based on data subject to uncertainty. Suppose further that $\ddot{\theta}$ is $N(4.2)$ while $\ddot{\theta}$ is $N(5,1)$.

(i) Are either of these estimators unbiased?
(ii) Which has lower mean squared error?
(iii) Which has a higher probability of being correct to within 1/2?
(iv) Is one of these estimators better than the other?

Solution

(i) Because $E(\ddot{\theta}) = \theta = 4$, $\ddot{\theta}$ is unbiased. Because $E(\dddot{\theta}) = 5$, $\dddot{\theta}$ suffers an upward bias.
(ii) Because $\ddot{\theta}$ is unbiased, its mean squared error is simply its variance, which equals $2^2 = 4$. For $\dddot{\theta}$, we use (3-7) to write $MSE(\dddot{\theta}) = 1^2 + (5-4)^2 = 2$. Thus $\dddot{\theta}$ has the lower mean squared error.
(iii) Here we note that "correct to within 1/2" means falling between 3.5 and 4.5. We turn to the normal table and find:

$$P\left(3.5 \le \ddot{\theta} \le 4.5 \middle| \ddot{\theta} \text{ is } N(4,2)\right) = .198$$
$$P\left(3.5 \le \dddot{\theta} \le 4.5 \middle| \dddot{\theta} \text{ is } N(5,1)\right) = .241$$

Thus, the biased estimator once again has the edge.
(iv) This question is deliberately irritating, for the answer depends on how we define "better." If we put higher priority on avoiding bias than on anything else, then $\ddot{\theta}$ is a "better" estimator than $\dddot{\theta}$. If mean-square error is the standard, then $\dddot{\theta}$ emerges as better. Should we have another standard (e.g., correct to within δ), then the status of "better" could depend on δ. The question is meant to remind us that, unless we think about what we want in an estimator, we can't reliably assess whether it performs satisfactorily.

EXERCISE 3.10
(i) In Example 3.10, is there a numerical value of δ for which $\ddot{\theta}$ is more likely than $\dddot{\theta}$ to be correct to within δ?
(ii) Which of the two estimators is more likely to overestimate θ by at least one? (Hint: You don't need a normal table to answer this question.)

Minimum Variance Linear Unbiased Estimates

In many instances, people combine their preferences about the first two moments of an estimator, desiring that it be unbiased *and* that it achieve the minimum possible variance among unbiased estimators. Finding that estimator is tractable under the assumption that the estimator can be expressed as a linear combination of random variables, each of which is itself unbiased.

If random variables $(X_1, \ldots, X_n)$ are independent and are all unbiased estimators of θ, linear combinations of these estimators take the form:

$$g(X_1, X_2, \ldots, X_n) = \sum_{i=1}^{n} c_i X_i$$

where the c_i's are constants.

Because $E(X_i) = \theta$ for all i, we can write:

$$E(g(X_1, X_2, \ldots, X_n)) = \sum_{i=1}^{n} c_i E(X_i) = \theta \sum_{i=1}^{n} c_i$$

It follows that $g(X_1, X_2, \ldots, X_n)$ is an unbiased estimator of θ if and only if $\sum_{i=1}^{n} c_i = 1$. All c_i-combinations that satisfy this constraint—and there are an infinite number of them—produce an unbiased linear combination.

But which of these unbiased combinations yields minimum variance? For $n = 2$, we derive the answer to that question at the website using Calculus:

When $E(X_1) = \theta$ and $\sigma^2(X_1) = \sigma_1{}^2$, while $E(X_2) = \theta$ and $\sigma^2(X_2) = \sigma_2{}^2$ and X_1 and X_2 are independent, then the **minimum variance linear unbiased estimator** (MVLU) of θ based on X_1 and X_2 follows:

$$\text{MVLU}(\theta | X_1, X_2) = \frac{\sigma_2^2}{\sigma_1^2 + \sigma_2^2} X_1 + \frac{\sigma_1^2}{\sigma_1^2 + \sigma_2^2} X_2 \qquad (3\text{-}9)$$

When $\sigma_1^2 = \sigma_2^2$, then $\frac{\sigma_2^2}{\sigma_1^2 + \sigma_2^2} = \frac{1}{2}$, and (3-9) implies that the MVLU is a simple average of X_1 and X_2.

EXAMPLE 3.11

Middle Way

Suppose that X_1 and X_2 are independent unbiased estimators of μ, and that $\sigma^2(X_1) = 4$ while $\sigma^2(X_2) = 9$.

(i) What linear combination of X_1 and X_2 has the lowest variance?
(ii) What is the variance for this combination?
(iii) Why isn't variance minimized when we assign all the weight to X_1?

Solution

(i) Using (3-9), we would assign a weight of $\frac{\sigma_2^2}{\sigma_1^2 + \sigma_2^2} = \frac{9}{13}$ to X_1 and a weight of $\frac{4}{13}$ to X_2, yielding the linear combination:

$$g(X_1, X_2) = \left(\frac{9}{13}\right) X_1 + \left(\frac{4}{13}\right) X_2$$

(ii) The variance of this linear combination is $4\left(\frac{9}{13}\right)^2 + 9\left(\frac{4}{13}\right)^2 = 2.77$.

(iii) Suppose that α is the weight placed on X_1, and $(1 - \alpha)$ the weight placed on X_2. When $\alpha = 1$, then the linear combination has a variance of $\sigma^2(X_1) = 4$. A graph of the variance of the linear combination as a function of α (i.e., a graph of $4\alpha^2 + 9(1 - \alpha)^2$) looks like Figure 3.1

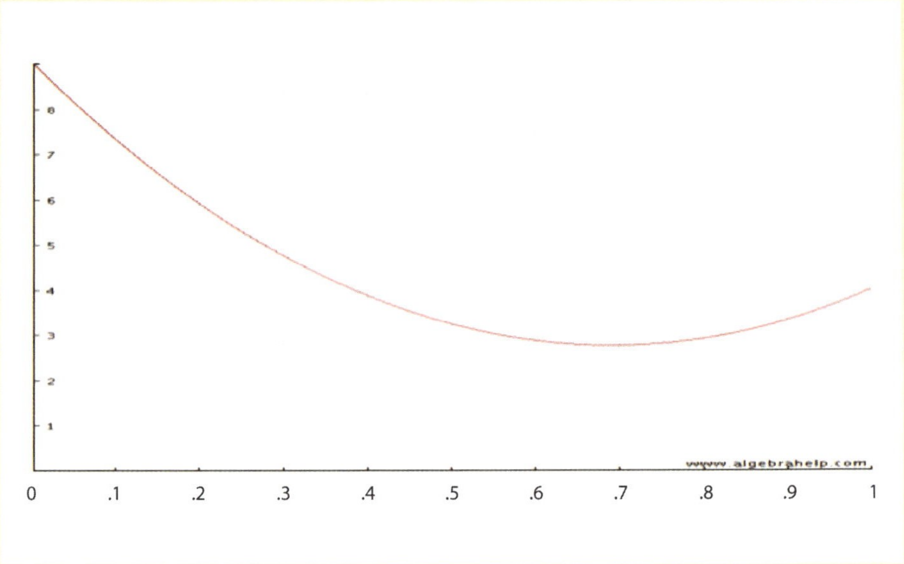

FIGURE 3-1 Graph of $4\alpha^2 + 9(1 - \alpha)^2$ versus α for $0 < \alpha < 1$

The outcome makes intuitive sense if we recall why the average of two consecutive die tosses has lesser spread around its mean of 3.5 than does a single toss. In the present setting, getting a very high value of the weighted average of X_1 and X_2 requires that they both be very high. That joint outcome is less probable than getting a very high value for X_1 alone. The same is true of a very low value. In consequence, values of $\left(\frac{9}{13}\right)X_1 + \left(\frac{4}{13}\right)X_2$ tend to stray less from μ than do values of X_1. For that reason, the variance of $\left(\frac{9}{13}\right)X_1 + \left(\frac{4}{13}\right)X_2$ is less than that of X_1.

EXERCISE 3.11 Suppose that X_1 in Example 3.11 is $N(1,2)$ while X_2 is $N(1,3)$. (Note that these standard deviations correspond to the variances in the Example.)

(i) Assume that $\left(\frac{9}{13}\right)X_1 + \left(\frac{4}{13}\right)X_2$ is normally distributed. (It is.) Compare the probability that $\left(\frac{9}{13}\right)X_1 + \left(\frac{4}{13}\right)X_2$ exceeds its mean by at least two with the corresponding probability for X_1 alone.

(ii) Do you see a relation between your answer to (ii) and the fact that $\left(\frac{9}{13}\right)X_1 + \left(\frac{4}{13}\right)X_2$ has a lower variance that X_1? Briefly discuss.

If we study (3-9), we see something interesting. In (3-9), the multiplier of X_1 has σ_2^2 and not σ_1^2 as its numerator, while X_2 has σ_1^2 as numerator. That outcome means that, when $\sigma_2^2 > \sigma_1^2$, the weighting factor for X_1, namely, $\frac{\sigma_2^2}{\sigma_1^2 + \sigma_2^2}$ is larger than that for X_2. Put another way, X_1 gets more weight than X_2 when X_1 has lower variance than X_2. But that pattern makes intuitive sense: If minimizing variance is the goal, we would expect that the estimator with lower variance should get greater priority than the higher-variance estimator. Indeed, giving greater weight to the more volatile of the two unbiased estimators would make no sense.

Equation (3-9) can be generalized to the case of n independent unbiased estimators and, like (3-9), the broader result would give equal weight to all estimators if they all have the same variance. That outcome yields a result of great importance to us:

Given n independent unbiased estimators with a common variance, the linear combination of the estimators with the lowest variance is their **simple average**. More mathematically:
If $X_1, X_2, \ldots, X_n$ are independent and all unbiased estimators of θ and $\sigma^2(X_i) = \sigma^2$ for all i, then

$$\bar{X} = \frac{1}{n}\sum_{i=1}^{n} X_i$$

is the minimum-variance linear unbiased estimator of θ based on the X_i's. (3-10)

As we noted in Chapter 1, the variance of $\bar{X}$ follows:

$$\sigma^2(\bar{X}) = \frac{\sigma^2}{n} \tag{3-11}$$

As we will see later in this chapter, (3-10) and (3-11) are central to statistical random sampling.

EXAMPLE 3.12 **Economies of Scale?**

Suppose that ten reputable economists make forecasts of GDP growth (in %) in the Eurozone for next year, and it is known that each one of them independently suffers a forecasting error that has mean zero and standard deviation 0.5%. Then another 15 reputable economists also make independent forecasts, with the same distribution of forecasting error.

(i) Is the forecast by an individual reputable economist unbiased?
(ii) What is the minimum-variance linear unbiased estimator of next year's GDP growth based on the first ten forecasts?
(iii) Suppose the minimum-variance forecast based on the first ten economists and—separately—the minimum-variance forecast based on the other 15 economists is to be taken. What linear combination of these two group forecasts achieves minimum variance?
(iv) Is the answer to (iii) the same as a simple average of the 25 forecasts?
(v) Suppose that it is found that two of the original ten reputable forecasters do not make independent forecasts: they analyze the data the same way and always reach exactly the same estimate of GDP growth. There are two proposals for how to treat this circumstance: (i) ignore it, and compute the group forecast the usual way or (ii) treat the two forecasts as one and thus reduce the original sample from ten to nine. Which of these approaches will yield a group forecast with lesser variance?

Solution

(i) If $\ddot{\theta}_\iota$ is the ith economist's forecast and θ is the actual GDP growth that occurs next year in the Eurozone, then we are told that $E(\ddot{\theta}_\iota - \theta) = 0$. But $E(\ddot{\theta}_\iota - \theta) = 0 \rightarrow E(\ddot{\theta}_\iota) = \theta$. Therefore, $\ddot{\theta}_\iota$ is an unbiased estimator of θ.
(ii) Based on (3-10), minimum variance is achieved by the simple average of the ten individual forecasts. (The variance associated with that average is $(.5)^2/10 = .025$ under (3-11).
(iii) For the other 15 economists, minimum variance is achieved by the average of their forecasts, and that variance equals $(.5)^2/15 = .017$. Under (3-9), the weighted average of the two group forecasts that achieves minimum variance assigns a weight of $.017/(.025 + .017) = .40$ to the ten-economist average (which has the higher variance) and .60 to the 15-economist average.
(iv) Yes, the result in (iii) is equivalent to averaging the 25 forecasts. For an economist in the first group, the weighted average multiplies her individual forecast by $.4(1/10) = 1/25$. An economist in the second group has his forecast multiplied by $.6(1/15) = 1/25$. Thus, all 25 forecasts are weighted equally, which makes sense if all the reputable economists have the same distribution of errors and make independent forecasts.

(v) Suppose that the two forecasters who proceed identically are numbered 1 and 2 in the first group of ten. Because $\ddot{\theta}_1 = \ddot{\theta}_2$, the original average is effectively transformed to $(2\ddot{\theta}_1 + \ddot{\theta}_3 + \ldots \ddot{\theta}_{10})/10$. The variance of this sum equals $(4(.5)^2 + (.5)^2 + (.5)^2)/100 = 12(.5)^2/100 = .03$. This outcome is higher than the variance .025 in (i) because X_1 and X_2 move in lockstep here: when one of them is high, for example, so is the other. Thus, $X_1 + X_2$ has higher variance than it would if X_1 and X_2 were independent.

If we delete $\ddot{\theta}_2$ because of its redundancy rather than ignore the fact that $\ddot{\theta}_1 = \ddot{\theta}_2$, then the average of the nine remaining forecasts has variance of $9(.5)^2/81 = .028$. So it is better to delete $\ddot{\theta}_2$ than to proceed as if the deviation from independence had not occurred.

EXERCISE 3.12 Suppose in Example 3.12 that two of the forecasters among the initial ten reputable economists are known to make correlated forecasts. Each makes a forecast that has mean zero and standard deviation 0.5%, but the coefficient of correlation of their forecasting errors is known to be 1/2.

(i) What is the variance of the simple average of the two forecasts?
(ii) Without mathematics, argue that the simple average of these two forecasts yields lesser variance than other linear combinations of the two forecasts.
(iii) What weighted average of the simple average in (i) and the minimum variance forecast for the other eight forecasters achieves the lowest overall variance?
(iv) What is the lowest overall variance?

The minimum-variance linear unbiased estimator of a quantity is often described as the ***best linear unbiased estimator*** (BLUE). Using variance to reflect estimation error effectively applies a penalty of b^2 when the estimator errs by b. That could be a reasonable way to proceed, though it is a matter of judgment whether it deserves the description "best."

The various concepts that we have just discussed (e.g., maximum likelihood, unbiasedness, error variance) will be important throughout the rest of our treatment of Statistics. We will encounter them shortly in the context of a statistical enterprise of huge importance: using data from a sample to make inferences about a broad population.

3.5 Random Statistical Sampling to Estimate a Population Mean

We have been discussing certain characteristics of "good" parameter estimators, but have not discussed how to collect data in ways that facilitate the calculation of (say) unbiased estimates. Here we consider the approach to data gathering that is as potent as it is pervasive: simple random sampling.

We estimate parameters when determining them exactly would be prohibitively expensive and/or difficult. A simple strategy for approximating the parameter is to take a **random sample** from the population of interest, under which each member has the same chance of being selected. Then the parameter for the full population is estimated from the data from the sample. Going further, we can use information about the sample to accompany the parameter estimate with an assessment about its accuracy.

Estimating a Population Mean

In a particularly important specific case, *every member of the population has a relevant number associated with him, her, or it*. Every citizen of Akron, Ohio, for example, has a height. The average of these numbers over the entire population is called the ***population mean***. One can take a random sample, find the **sample mean**, and use it as a point estimate for the mean in the population. Fortunately, there is a well-developed theory that indicates how closely the sample mean approximates the population mean. For that reason, one can offer ***interval estimates*** for the parameter, based on the sample mean and the ***margin of error*** associated with the approximation.

Some notation:

N = size of the full population

μ = average of individual numbers over the full population

σ^2 = variance of individual numbers over the full population (defined as the average squared difference from μ over the full population)

n = size of the random sample (In this chapter, we assume $n \ll N$)

X_i = number for the ith member selected into the random sample (which is a random variable prior to the random selection)

To estimate μ once $(X_1, X_2, \ldots X_n)$ are at hand, it is common to use the sample mean $\bar{X}$ defined by:

$$\bar{X} = \sum_{i=1}^{n} X_i.$$

But are we sure that $\bar{X}$ is an unbiased estimator of the population mean μ? Let's explore the situation.

EXAMPLE 3.13 **No Bias**

(i) Why is X_1 an unbiased estimator of μ?
(ii) If X_1 is unbiased, does that mean that $\bar{X}$ is unbiased?

Solution

(i) Suppose we number the individual members the population from 1 to N, and let the number associated with member j be called b_j. Before we make the first random selection from the population, each member has a $1/N$ chance of being selected. (Right?) Therefore, X_1 has the discrete distribution:

$$X_1 = \begin{cases} b_1 \text{ w.p. } 1/N \\ b_2 \text{ w.p. } 1/N \\ \quad \dots \\ b_N \text{ w.p. } 1/N \end{cases}$$

(It is not necessary that the b_j's all be different: for instance, b_2 and b_5 could be the same.)
The mean of X_1 then follows:

$$E(X_1) = \sum_{j=1}^{n} b_j \left(\frac{1}{N}\right) \tag{3-12}$$

But the right hand side of (3-12) equals μ by definition. Therefore, X_1 is an unbiased estimator of μ.

(ii) The same argument implies that all the other X_i's are unbiased estimators of μ. It follows that:

$$E(\bar{X}) = \left(\frac{1}{n}\right)\sum_{i=1}^{n} E(X_i) = \left(\frac{1}{n}\right)(n\mu) = \mu$$

Thus, $\bar{X}$ is itself an unbiased estimator of μ.

EXERCISE 3.13

(i) Explain why X_1 and X_2 in the random selection in Example 3.13 are not fully independent, using a specific numerical example. (We are assuming that we are sampling *without replacement*, meaning that the population member chosen first cannot also be chosen second.)

(ii) Suppose that $X_1 = 5$, and the population mean in a population of size one million is μ. What is the conditional mean of X_2? (Hint: What is the average among the 999,999 population members who were not chosen first?)

(iii) Does the answer to (ii) suggest that X_1 and X_2 are so weakly correlated that we can as a practical matter think of them as independent?

Because of Example 3.13 and the "*de facto* independence" assumption for the X_i's justified in Exercise 3.13, we can invoke (3-10) to say that:

When $N >> n$, the sample mean $\bar{X}$ is essentially the minimum-variance linear unbiased estimator of μ based on a random sample of size n. (3-13)

The qualifier "essentially" would disappear if N were infinite and we could think of the sample points as independent. In the remainder of this chapter, we will assume N is so large relative to n that we can assume independence.

But even though $\bar{X}$ is on average correct and it constitutes a reasonable point estimator of μ, how much error might we encounter in making the approximation? Here we do well to consider the random variable W defined by:

$$W = \bar{X} - \mu \qquad (3\text{-}14)$$

W is the **sampling error** in using the sample mean as an estimate of the population mean. Its value depends on the luck of the draw: if we are lucky in choosing the random sample, then W will be at or near zero. If we are unlucky and choose disproportionately-many sample points at the high end (or low end) of the population's distribution of numbers, then W could be large in absolute value.

Now an example to increase our familiarity with W.

EXAMPLE 3.14

Car Pool

In Upper Bourgeoisia, 60% of the families have four cars apiece and 40% have five. Minerva, who does not know these facts, seeks to estimate the mean number of cars per family in Upper Bourgeoisia. She will choose *five* families at random from the town's huge population of families, and will use the sample mean of their numbers of cars to estimate the mean for the town.

(i) What is the true mean for the number of cars per family in Upper Bourgeoisia?

(ii) What is the maximum sampling error (corresponding to what we've been calling W) that Minerva can get in her exercise? What is the probability she will wind up getting this maximum error?

(iii) What is the minimum sampling error she can get? What is the probability that she will achieve this minimum error?

Solution

(i) If N is the number of families in Upper Bourgeoisia (hereafter UB), then $.6N$ of them have four cars and $.4N$ have five. Therefore, the average number of cars per family follows: $\mu = (.6N * 4 + .4N * 5)/N = \mathbf{4.4}$.

(ii) In this situation, the value of $W = \bar{X} - \mu$ is maximized if all five UB families selected each have five cars. In that case, $W = 5 - 4.4 = 0.6$, which is the maximum possible sampling error. As for $P(W = 0.6)$, we note that each family selected has a 40% chance of having five cars; assuming (the essential) independence of the five outcomes, we get:

$$P(W = 0.6) = (.4)^5 = .0102$$

If Minerva gets an $\bar{X}$ of 5, she will not know that she was too high by 0.6. (If she knew μ, why would she undertake sampling to estimate it?) Thus, she would make the estimate of 5 as her best guess given the data.

(iii) The minimum sampling error here is zero, which will arise if Minerva selects three UB families with four cars and two families with five. (In that case, $\bar{X}$ will be 4.4, and thus $W = \bar{X} - \mu = 4.4 - 4.4 = 0$.) Under random sampling from a large population, we assume that each of Minerva's selections independently has a 60% chance of being four and 40% of being five. It follows that V, the number of families she selects that have four cars, is binomially distributed; more specifically, V is $B(5, .6)$. Minerva's estimate will be exactly correct if $V = 3$; from an electronic binomial table, we learn that $P(V=3) = .3456$.

EXERCISE 3.14 Work out the full probability mass function for W under Minerva's five random selections in Example 3.14. Then show that $E(W) = 0$.

Now, back to more general discussion about W. Because $E(\bar{X}) = \mu$, we can write:

$$E(W) = E(\bar{X} - \mu) = E(\bar{X}) - \mu = \mu - \mu = 0$$

What else can we say about W? In fact, we can write expressions for $E(W^2)$ and $\sigma^2(W)$ based on what we did earlier in this chapter. Before doing so, however, we benefit from discussing qualitatively what factors should influence $E(W^2)$ and $\sigma^2(W)$. We should note at the outset that:

$$\text{Because } E(W) = 0,\ \sigma^2(W) = E(W - 0)^2 = E(W^2)$$

Hence the variance of sampling error and mean-square sampling error are *one and the same.*

What factors about the sampling situation should influence the size of $\sigma^2(W) = E(W^2)$) in a given case? One obvious answer is n, the number of data points used to compute $\bar{X}$. All other factors being equal, we would expect $E(W^2)$ to go down as n goes up: the more data points used in computing the sample mean, the less vulnerable $\bar{X}$ would seem to taking an extreme value far from μ (and hence the less likely that $W^2 = (\bar{X} - \mu)^2$ is large).

Less immediately obvious is the fact that $E(W^2)$ depends on the **spread** in the distribution across the population. But that dependence becomes quite plausible if we consider Figures 3-2 and 3-3 below, which are histograms or probability density functions for two possible distributions over the population (i.e., the X-distribution). Suppose that you had to estimate μ by $\bar{X}$ for a fixed sample size n, and would be "penalized" in proportion to the square of your estimation error (i.e., of your W-value). Would you have any preference between the two x-distributions shown below?

In the first case above, virtually all x-values are close to μ; thus, regardless of which members of the population enter the random sample of size n, $\bar{X}$ cannot stray very far from μ. But, in the second case, it is not outlandish to fear that a freakish sample could produce an $\bar{X}$ enormously distant from μ. Disciplined by a penalty function that grows with the square of the error, therefore, we would find the Figure 3-3 situation much less appealing than that of Figure 3-2. These comments suggest that the **variance** σ^2 of numbers over the population being sampled affects the accurately of estimating its mean μ from a random data sample. *The larger the variance of the population distribution, the lesser the likelihood that $\bar{X}$ will be a highly accurate approximation for μ.*

Of course, we know where this journey ends, because we know the formula for $E(W^2) = \sigma^2(W)$. From (3-11), we know that $\sigma^2(\bar{X}) = \frac{\sigma^2}{n}$. And because $W = \bar{X} - \mu$ where μ is a constant, $\sigma^2(W) = \sigma^2(\bar{X})$.

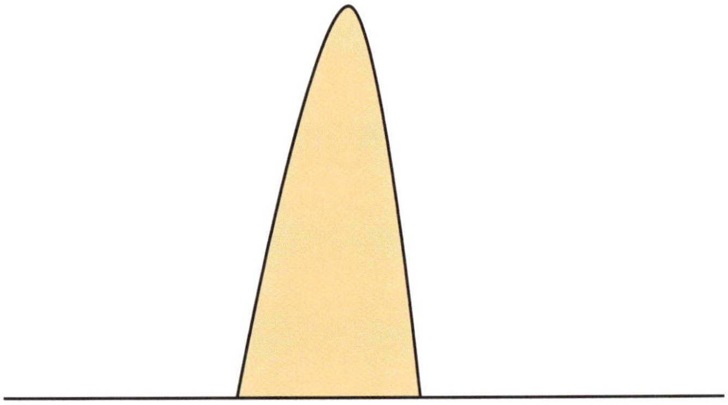

FIGURE 3-2 An X-distribution with Little Spread

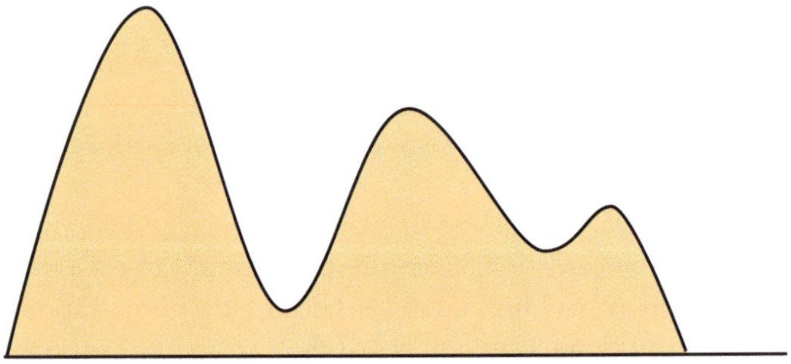

FIGURE 3-3 An X-distribution with Much Spread

If $W = \bar{X} - \mu$ where $\bar{X}$ is the sample mean in a random sample of size n from a large population with variance σ^2, then:

$$E(W^2) = \sigma^2(W) = \sigma^2/n \qquad (3\text{-}15)$$

Note that the features of (3-15) were foreshadowed by the discussion that preceded it. As sample size n increases when sampling from a given population, the mean-squared sampling error goes down because n is in the denominator of (3-15). And at a fixed value of n, mean-squared sampling error goes up with population variance because σ^2 is in the numerator.

Returning to the situation that started this chapter, (3-15) explains why the author responded "it depends" when asked about the accuracy that can be achieved with a sample of 18 bags of scallops. The fact that $n = 18$ certainly matters, but so does the spread in the population being sampled. If σ^2 were very low, a sample of 18 might achieve high accuracy (indeed, if σ^2 were very low, even a sample of size three might do very well). Because the author knew nothing about the variance of meat-per-scallop ratios across the 11,000 bags of scallops (the population), he could not address the accuracy of the sampling results. (At that point, anyway.)

Readers may recall that we went further with $\bar{X}$ in Chapter 1 than determining its mean and variance. We showed that the Central Limit Theorem implied that, for moderately-large n, $\bar{X}$ was approximately *normally* distributed, with a mean of μ and a standard deviation of $\sigma/\sqrt{n}$. Because $W = \bar{X} - \mu$, the Central Limit Theorem leaves no doubt about the approximate distribution of W:

If $W = \bar{X} - \mu$ and $\bar{X}$ is based on even a moderately large sample from a population with variance σ^2, then:

$$W \approx N\left(0, \sigma/\sqrt{n}\right) \qquad (3\text{-}16)$$

As we discussed in Chapter 1, "moderately large" is widely taken to mean at least 30, although we argued that the normal approximation is nearly as good in several common situations for n as low as 15.

We now know a good deal about the sampling error in estimating μ by $\bar{X}$. For example, it follows from (3-16) that:

- There is approximately a 95% chance that the sampling error W falls between $-1.96\sigma/\sqrt{n}$ and $1.96\sigma/\sqrt{n}$ or, equivalently, that the absolute sampling error does not exceed $-1.96\sigma/\sqrt{n}$. As we know, the 95% range in a normal distribution literally extends 1.96 standard deviations in each direction from the mean. (People often use 2 instead of 1.96 because the normal approximation arising from the Central Limit Theorem is just that: an approximation. However, we will use 1.96 in our own approximations related to 95% probability.)

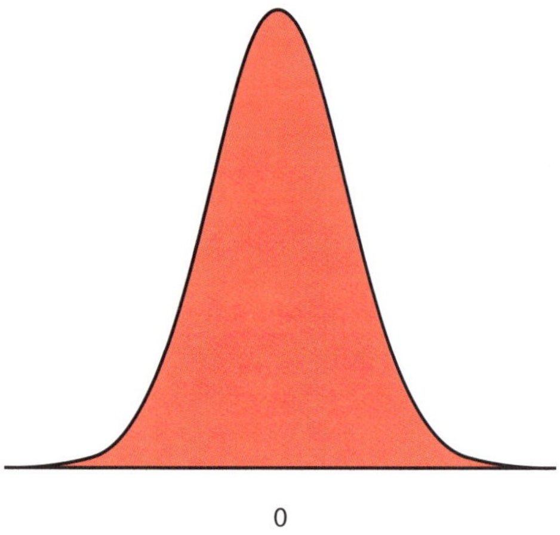

FIGURE 3.4 Even if an original distribution has a weird shape (e.g., as in Figure 3.3), the sampling error in estimating the *mean* of that distribution by a sample mean is roughly zero-mean bell-shaped, for moderate or larger sample sizes.

- There is approximately a 99% chance that W falls between $-2.58\sigma/\sqrt{n}$ and $2.58\sigma/\sqrt{n}$.
- Using (3-16) and a normal table, we can work out the probability that W falls in any numerical range of interest. For example, we could find $P(-5 < W < 5)$, which is the same as $P(|W| < 5)$.

We should ponder once again how amazing the Central Limit Theorem is. Statement (3-16) about sampling error does not assume that the distribution of numbers over the full population was normal. Regardless of the shape of the original distribution, (3-16) says that the *error in estimating the mean* of that distribution by a sample mean has a bell-shaped normal distribution, assuming a moderate sample size. That bell-shaped curve is centered at zero (Figure 3.3a).

EXAMPLE 3.15

Cereal Number

Suppose it is believed that the weight of the dry cereal that a machine puts in various boxes varies from box to box, with a standard deviation of 1/32 lb. However, there is some uncertainty about the average weight of the cereal across the boxes. It is decided to take a random sample of 100 filled boxes from the many thousand produced each day, weigh the cereal in each, and use the sample mean as an estimate of the true mean weight over all boxes filled. How confident can the company be that the sampling error in this procedure is less than 0.01 lbs.?

Solution

By "less than 0.01," the company presumably means that the *absolute* sampling error is less than 0.01. If the sampling error were less than .01 because it was –0.54, it seems unlikely that the company would celebrate the exactitude of the result.

Using (3-16) with $\sigma = 1/32 = .031$ and $n = 100$, we find that $W \approx N\ (0, .0031)$. We can therefore use a normal table to determine that $P(-0.01 < W < 0.01) \approx .999$. In short, it is exceedingly likely that the (absolute) sampling error in estimating μ by $\bar{X}$ is less than 0.01 lb.

EXERCISE 3.15 Suppose that the company in Example 3.15 decides to save money in future sampling by using a sample size of 50. Now what is the probability that the sampling error is less than 0.01? Would you have expected cutting the sample size in half to have a greater effect on this probability? Briefly discuss.

Trouble Develops

Before we croon about our successes, however, we should recognize a huge obstacle to using (3-15) and (3-16). The whole premise of the sampling exercise is that we do not know μ; that being the case, why would we know σ^2, the average squared spread around μ over the population? At the moment, therefore, (3-15) confers a dubious benefit: we can express one unknown quantity in terms of another. Can we find a way out of this mess? Fortunately, yes.

Estimating the Variance σ^2 of the Population Distribution

A moment's thought makes clear that we are not paralyzed by the absence of σ^2, for we have a basis for estimating it from the data at hand. We are already making a data-based estimate of μ, namely, $\bar{X}$. Can't we also make a data-based estimate of σ^2? More specifically, why not estimate σ^2 under the approximation that the *sample squared-spread around the sample mean* $\bar{X}$ is a reasonable proxy for the true squared-spread σ^2 around the true mean μ?

The equation corresponding to this last sentence would have us estimate σ^2 by U^2 which follows:

$$U^2 = \sum_{i=1}^{n} \frac{\left(X_i - \bar{X}\right)^2}{n} \tag{3-17}$$

where X_i = the ith observation in the random sample and $\bar{X} = \dfrac{\sum_{i=1}^{n} X_i}{n}$.

The game plan could be: Use (3-17) to estimate σ^2 by U^2, and then return to (3-15) and estimate $\sigma^2(W)$ by $U^2/n \approx \sigma^2/n$.

But while equation (3-17) seems to put us back on course, we should hold the champagne. For, although the reason is not obvious, (3-17) yields an estimator of σ^2 that tends on average to be *too low*.

What Was Wrong with U^2?

Why is U^2 biased downward? The problem is suggested by Figure 3.5 below, which arises from a case in which $n = 4$. All four of the data points in the random sample fall above the population mean, an outcome that is unusual but not terribly rare. (If the mean is also the median, the probability of the Figure 3.5 pattern is $(.5)^4 = 1/16$.) The problem is that, should the chosen points fall by chance towards the upper end of the population-distribution, then their sample mean $\bar{X}$ has little choice but to go along for the ride. Thus, as the figure implies, the squared spread of the four observations around their sample mean $\bar{X}$ can fall well below their squared spread around μ. In other words, $U^2 < \frac{\sum_{i=1}^{n}(x_i - \mu)^2}{n}$ here.

Nor can Figure 3.5 be dismissed as a peculiar worst-case outcome. If there is about a 1 in 16 chance of getting four points all greater than the mean, there is about an equal chance of getting all four below the mean, in which case U^2 computed under (3-21) is once again too low. Admittedly, there are many cases in which the spread around $\bar{X}$ will approximate that around μ (e.g. two points above μ and two below). But such outcomes do not contradict the assertion that there is a *net* downward bias in estimating σ^2 by U^2. To be too low on average, after all, does not require that one be too low 100% of the time.

But how serious is this issue of downward bias, and how do we respond to it? To answer the first question, we can work out $E(U^2)$, the expected value of U^2, and can then compare it to σ^2. The derivation appears at the website, and it leads to the outcome:

If we estimate the population variance σ^2 by U^2 which follows:

$$U^2 = \sum_{i=1}^{n} \frac{(X_i - \bar{X})^2}{n}$$

our estimate is on average too low because $E(U^2) = \frac{(n-1)\sigma^2}{n} < \sigma^2$ (3-18)

FIGURE 3.5 A Random Sample Of Four Points That All Exceed The Population Mean

Equation (3-18) is saying that the estimate we get when we use U^2 to approximate σ^2 is on average too low by a factor of $(n - 1)/n$. When $n = 10$, for example, the estimate is on average 9/10 as large as the population variance, which corresponds to a shortfall of 10%.

Because $(n - 1)/n < 1$ for all n, the downward bias is present at every sample size. But the effect is much larger at small n than large n. When $n = 4$, for example, $(n - 1)/n = 3/4$. But when $n = 100$, $(n - 1)/n = .99$. That pattern makes sense: the horror story we depicted in Figure 3.5 is far more likely to arise at small n than at larger n, when we would expect all parts of the distribution to be represented in the sample.

Yet if U^2 is too low on average by a factor of $(n - 1)/n$, there is an easy adjustment that eliminates the bias (i.e., yields an estimate that is on average correct). If we define S^2 by:

$$S^2 = \left(\frac{n}{n-1}\right)U^2$$

Then

$$E(S^2) = \left(\frac{n}{(n-1)}\right)E(U^2) = \left(\frac{n}{(n-1)}\right)\left(\frac{n-1}{n}\right)\sigma^2 = \sigma^2$$

Also:

$$S^2 = \left(\frac{n}{n-1}\right)U^2 = \left(\frac{n}{n-1}\right)\sum_{i=1}^{n}\frac{(X_i - \bar{X})^2}{n} = \sum_{i=1}^{n}\frac{(X_i - \bar{X})^2}{n-1}$$

In short:

If a random sample of size n is taken from a large population, an unbiased estimator S^2 of the population variance σ^2 follows:

$$S^2 = \sum_{i=1}^{n}\frac{(X_i - \bar{X})^2}{n-1} \qquad (3\text{-}19)$$

To correct for the downward bias in U^2, (3-19) says that all we need do is revise the denominator from n to $n - 1$. In Excel, the command STDEV uses (3-19) to yield an unbiased estimate of a population variance given data from a random sample.

The quantity S^2 in (3-19) is generally called the **sample variance** and its square root S is called the **sample standard deviation.** Much as the sample mean estimates the population mean μ, these two quantities estimate the variance σ^2 and standard deviation σ, respectively, in the broader population.

Back to Sampling Error

Now that we have an unbiased estimate of the population variance, we can use it to obtain an unbiased estimate of the sampling-error variance $\sigma^2(W)$. If $S^2(W)$ is defined by: $S^2(W) = \frac{S^2}{n}$, then:

$$E\left(S^2(W)\right) = E\left(\frac{S^2}{n}\right) = \frac{E\left(S^2\right)}{n} = \frac{\sigma^2}{n} = \sigma^2(W)$$

If $W = \bar{X} - \mu$ for a random sample of size n, then an unbiased estimator $S^2(W)$ of $\sigma^2(W)$ follows:

$$S^2(W) = \frac{S^2}{n} = \sum_{i=1}^{n} \frac{\left(X_i - \bar{X}\right)^2}{n(n-1)} \qquad (3\text{-}20)$$

Notice that $S^2(W)$ has $n(n-1)$ as its denominator. That happens because:

- the $n - 1$ reflects the averaging of squared deviations around the mean to get an estimate of σ^2
- the n reflects the additional fact that larger samples are more accurate than smaller ones.

The next question we might have is: Can we use $S^2(W)$ as an estimator of $\sigma^2(W)$ and retain the normal approximation for W in (3-16)? In other words, can we assume that, for sufficiently large n, we can make the approximation that:

$$W = \bar{X} - \mu \approx N(0,\sigma(W) \approx N(0,S(W))$$

The answer is not obvious because, even though S^2 is on average equal to the population variance σ^2, S^2 exhibits some random spread around σ^2. Hence, like the sample mean, the sample variance is itself subject to some sampling error (Say that quickly as a sobriety test). As n gets large, however, S^2 approaches σ^2 and thus $S^2(W)$ approaches $\sigma^2(W)$, and it can be shown that:

For sufficiently large n:

$$W = \bar{X} - \mu \approx N(0,S(W)) \qquad (3\text{-}21)$$

where

$$S(W) = \frac{S}{\sqrt{n}} = \sqrt{\sum_{i=1}^{n} \frac{\left(X_i - \bar{X}\right)^2}{n(n-1)}}$$

What do we mean by "sufficiently large n"? Fortunately, the sample size needed to use (3-21) is not appreciably larger than the one required by (3-16). Extensive simulations have shown that the approximation generally works well for n at least 30 and often works well for n as low as 15. In other words, even though we have something of a double approximation—a normal approximation with a standard deviation that it itself is an approximation—we have a formula we can fruitfully use.

EXAMPLE 3.16 **Double or Nothing?**

We are about to begin random sampling from a population, and plan to compute S^2 (the estimate of the population variance) when $n = 50$ and again when n increases to 100. Would we expect the earlier estimate of S^2 to be larger than the later one?

Solution

The word "expect" is a little vague, but the most reasonable interpretation of the question is: Would the expected value of S^2_{50} exceed that of S^2_{100}, where $S_h{}^2$ is the sample variance based on a random sample of size h? Equation (3-19) implies that both S^2_{50} and S^2_{100} have the same mean, namely, σ^2. But if they are on average the same, their mean difference is zero. (Granted, S^2_{50} and S^2_{100} are not independent, but the rule "mean of a difference is the difference of means" is always true.) Under the circumstances, we do not "expect" S^2_{50} to exceed S^2_{100}.

EXERCISE 3.16 Do you understand why we said that S^2_{50} and S^2_{100} are not independent? Briefly explain why they aren't.

Where do things stand? As of now, we have a point estimate of μ (namely, $\bar{X}$), and a probability distribution for sampling error W associated with the estimate. Our achievements, however, are but means to a mean: the real task is not to highlight $\bar{X}$ or W but to be *as specific as possible about where μ is.* We are almost in the Promised Land, but have a few more things to cover before entering it.

3.6 Interval Estimates for μ

Probabilistic Statements about Where μ Is

Suppose that we encounter three assertions pertaining to an imperfect thermometer:

(I) 95% of the time, the thermometer is correct to within one degree Fahrenheit.

(II) The thermometer's present reading is 83°F.

(III) There is a 95% chance that the true temperature now is between 82° and 84°F.

Question: Do statements (I) and (II) imply (III)?

The most careful way to answer would be "not necessarily." Should the accuracy of the thermometer, for example, vary with the actual temperature, then (II) might reduce the relevance of (I). Perhaps some additional facts beyond (I) and (II) would make it appropriate to modify (III). But when caveats of this kind need not be taken seriously, treating (III) as the consequence of (I) and (II) seems wholly consonant with common sense. If we adopt the viewpoint of classical Statistics, we effectively accept the "(I) and (II) imply (III)" logic in the context of random sampling.

Pursuing this viewpoint, we might think about assigning a probability to the event that μ falls in a particular range. Suppose that, in a given setting, $\bar{X} = 12$ and $P(-4 \le W \le 4) \approx .95$. Then:

- Our point estimate of μ is 12.
- Sampling theory tells us that there is a 95% chance that the estimate is correct to within 4 one way or the other (i.e., $P(-4 \le W \le 4) \approx .95$.)

Taken together, what do these two statements tell us about the whereabouts of μ? The question comes down to another: Under what circumstances is an estimate of 12 correct to within 4? This happens when and only when the true quantity is between 8 and 16. Thus, with $\bar{X} = 12$, the events "$\bar{X}$ is correct to within 4" and "μ is between 8 and 16" are *one and the same*. It follows that if $P(-4 \le W \le 4) \approx .95$ here, then so is $P(8 \le \mu \le 16)$. (We cannot change an event's probability by moving from one description of it to an equivalent one.) To put it directly, we can write that $P(8 \le \mu \le 16) \approx .95$.

Wait. Given that μ is a numerical constant, in what sense does it have a probability distribution? More than a few people object to statements like "there is a 95% chance that μ falls between 8 and 16." They say that either μ is between 8 and 16—in which case it falls there with probability one—or it does not, in which case it falls there with probability zero. It is one thing to say that X, the next outcome on the toss of a fair die, has a probability distribution, but what about a quantity that is already fixed?

While respectful of that concern, the author would make an important distinction. Using a probability distribution does not deny that a numerical constant has an unchanging value; rather, it reflects uncertainty about what that value is. Going back to the fair die, suppose that it was tossed ten minutes ago in another room, and we have not yet learned the outcome. Even though the outcome is now a constant, we can still meaningfully say that the probability is 1/6 that a five came up.

Or, to go to a different setting, the author of this book is nearsighted, meaning that he needs glasses to see into the distance but must take them off to read and write. He tends to misplace them with monotonous regularity and then, precisely because he has taken them off, to have considerable trouble finding them. Based on plenty of past experiences when his glasses were missing, he might say that there is a 30% chance they are in the living room, 25% in the bedroom, 15% in the kitchen, etc. But that statement does not mean that the glasses are sneaking from room to room: they are in one place, but probability arises because the author is ignorant of that place. He uses probability to reflect the relative likelihood of the different possible locations.

In the same way, we fully acknowledge that μ is a constant, but believe that the results of random sampling tell us something about where μ is likely to be and where it almost surely cannot be. In that sense, μ can be treated as a random variable, and the probability distribution for μ is a succinct summary of what we have learned. In this book, we will speak of a probability distribution for μ but, in deference to the sensitivity of the topic, we will introduce some alternate vocabulary that makes some people more comfortable.

A Probability Distribution for μ

The sampling error $W = \bar{X} - \mu$ is a random variable prior to the selection of the random sample. A key assumption in classical Statistics is that, *once we have completed the sampling, W has the same distribution as before*. True, we now know what $\bar{X}$ is, but that information does not change our assessment about the accuracy of $\bar{X}$ as an estimate of μ. If, for example, we originally believed based on sampling theory that W would have a 95% chance of being correct to within 3, we would continue to hold that belief after learning that $\bar{X} = 25$.

The definition $W = \bar{X} - \mu$ can be rewritten as $\mu = \bar{X} - W$. After sampling determines $\bar{X}$, we can treat μ as a random variable equal to the sample mean minus the unknown sampling error. It follows that:

Once random sampling has been completed and $\bar{X}$ is a known constant:

- $E(\mu) = E(\bar{X} - W) = \bar{X} - E(W) = \bar{X} - 0 = \bar{X}$
- $\sigma^2(\mu) = \sigma^2(\bar{X} - W) = \sigma^2(-W) = \sigma^2(W) = \sigma^2 / n$ (3-22)

When the sample size is sufficiently large, we can approximate σ^2 in the original X-distribution by S^2 and also assume that μ has approximately a normal distribution:

Once random sampling is completed for sufficiently large n, we know $\bar{X}$ and we continue to treat W as approximately $N(0, S/\sqrt{n})$. Therefore, we treat the unknown location of $\mu = \bar{X} - W$ as $N(\bar{X}, S/\sqrt{n})$. (3-23)

Sufficiently large n

What do we mean by "sufficiently large n" in (3-23) in the context of random sampling? As noted, it is widely believed that the normal approximation works when $n \geq 30$ but the author believes that the approximation is generally adequate for n as low as 15. Therefore, we will use the rule of thumb $n \geq 15$ here; if your instructor prefers a higher threshold, however, always listen to your instructor.

We are almost ready to use (3-23) in real examples, but we should stop a moment to marvel at it. We could never expect that random sampling would reveal the exact value of μ; the most we could reasonably hope for is a point estimate for μ, some information about the sampling error that attends the estimate, and a way of estimating the probability that μ falls in any given numerical range. Given (3-23), these wishes have come true when a moderately-large data sample is in hand. In both conceptual and practical terms, sampling theory is uncannily friendly.

EXAMPLE 3.17 **Grand Central**

When $n = 49$, $\bar{X} = 80$, and $S = 14$, approximately what is the probability that $76 \leq \mu \leq 84$? Assume that the data constitute a random sample from the population.

Solution

Under (3-23), we have $\mu \approx N\left(\bar{X}, \frac{S}{\sqrt{n}}\right) = N\left(80, \frac{14}{\sqrt{49}}\right) = N(80,2)$. Therefore $P(76 \leq \mu \leq 84) \approx .95$ based on the two-standard deviations rule for normal distributions.

EXERCISE 3.17 Suppose that $n = 49$, $\bar{X} = 80$, and $P(76 \leq \mu \leq 84) \approx .68$. Given that information (and assuming a random sample), what was the sample standard deviation?

95% Confidence Intervals and Their Cousins

Often it is considered desirable not simply to offer a point estimate of a quantity of interest, but also to indicate a numerical *interval* that with high likelihood contains the quantity's true value. The most common such ranges are those known as *95% confidence intervals*. Given a probability distribution for a parameter's true value, the 95% confidence interval is a range of numbers that (not surprisingly) has a .95 chance of including the true value. Generally, the range chosen goes from the 2.5%ile of the parameter's distribution to the 97.5%ile.

If $\mu \approx N(\bar{X}, S/\sqrt{n})$, then the 95% Confidence Interval for μ as defined above would extend about two standard deviations in each direction from the mean of the distribution. Using the more precise 1.96 rather than 2, we would write:

For sufficiently large n, the **95% confidence interval** for μ is approximated by:

$$\left(\bar{X} - 1.96S/\sqrt{n}, \bar{X} + 1.96S/\sqrt{n}\right) \quad (3\text{-}24)$$

Do not be shocked or dismayed if you see an analysis that uses two standard deviations in (3-24); it's a defensible approximation in an expression that is itself an approximation.

People who wish to avoid saying that there is a 95% probability that μ falls in the 95% confidence interval sometimes say that they are "95% confident" that μ falls there. They make that statement because, 19 times out of 20, the interval so constructed will include μ. In practical terms, the two perspectives lead to the same place.

In the day-to-day use of Statistics, people often construe the 95% confidence interval (*hereafter 95% CI*) as the *range of plausible values* for the parameter of interest. By this they mean that, barring unusually bad luck in drawing the random sample, the confidence interval includes the true value of the parameter. Of course, the parameter's actual value need not fall in the 95% CI: the very construction of the interval assumes a .05 chance that the true value falls outside it. Dismissing values outside the interval, therefore, runs a 5% risk of dismissing the true value of the parameter.

Moreover, at the level of instinct, many people not only assume that the parameter's true value is inside the 95% CI but that it is *equally likely* to fall anywhere within that interval. Such a "uniformity" assumption, however, is usually unjustified. For example, if sampling leads to the approximation that μ is $N(50,10)$, then the 95% CI for μ is (30.4, 69.6). But while the intervals (32, 35) and (46, 49) are both contained in the 95% CI and are both equally long, the normality of μ centered at 50 means that the true mean is far more likely to fall in (46,49) than in (32,35). The true value of μ more often shows up *towards the center of the 95% CI* than towards its edges.

The 95% CI is not the only range estimate used for an unknown parameter's value. Sometimes, the 90% CI—which ranges from the 5th percentile of the distribution to the 95th—is reported instead. Sometimes the 99% CI—from the 0.5th percentile to the 99.5th—is computed. How do these intervals compare with the 95% CI in a particular case? The 95% CI is *wider* than the 90% CI and fully contains it. (The 2.5%ile of the distribution, after all, is lower than the 5th, much as the 97.5th is higher than the 95th.) Similarly, the 95% CI is *narrower* than (and fully contained in) the 99% CI. (See Figure 3.6.)

But while the 95% CI does not enjoy monopoly status, it is considered the standard confidence interval and refusals to use it do not pass unnoticed. That point was made vividly in 1993, when the U. S. Environmental Protection Agency (EPA) released a 90% CI for the carcinogenic risk posed by passive cigarette smoke (i.e., smoke that nonsmokers inhale from nearby cigarettes). The EPA was criticized by some parties for not using the traditional (and wider) 95% CI. Irritation with the EPA was especially

FIGURE 3.6 Three Familiar Confidence Intervals of Varying Width

intense because the particular parameter value that indicated no extra risk fell within the 95% CI but not the narrower 90% CI. It was suggested that a desire to reach a certain conclusion influenced the somewhat eccentric choice of the 90% CI.

How wide are the 90% and 99% confidence intervals? We noted in Chapter 1 that, for all normal distributions, a particular percentile of the distribution falls the same number of standard deviations away from its mean. Use of an electronic table makes clear that:

- The 0.5%ile of a normal distribution is 2.58 standard deviations below the mean, while the 99.5%ile is 2.58 standard deviations above the mean
- The 5%ile of a normal distribution is 1.64 standard deviations below the mean, while the 95%ile is 1.64 standard deviations above the mean

From these statements, we can write:

For sufficiently large n:

- the **99% confidence interval** for $\mu \approx \left(\bar{X} - 2.58S/\sqrt{n}, \bar{X} + 2.58S/\sqrt{n}\right)$
- the **90% confidence interval** for $\mu \approx \left(\bar{X} - 1.64S/\sqrt{n}, \bar{X} + 1.64S/\sqrt{n}\right)$ (3-25)

EXERCISE 3.18

(i) Between what two percentiles of the distribution of μ would the 80% confidence interval for the population mean fall?

(ii) Find a general expression for the 80% confidence interval analogous to those in (3-25). (To find how by many standard deviations the 90th percentile of a normal distribution exceeds the mean, find the 90th percentile of $N(0,1)$ from a table. That number is 1.28, so the 90th percentile of $N(\mu,\sigma)$ is $\mu + 1.28\sigma$.)

3.7 Sampling Theory Applied to Actual Data

Now we reinforce what we discussed in the last section with several examples, starting with the one about scallops at the beginning of this chapter.

EXAMPLE 3.18 **Scallop Treatment**

What does statistical sampling suggest about whether the scallop ship that started this chapter was in violation of the rules against harvesting baby scallops?

(continued from start of chapter)

"Depends on what?" the lawyers pressed. "Isn't it absurd to think that 18 bags can adequately represent 11,000?"

Well, answered the author (who was the statistician in the case), suppose that we worked out the meat-per-scallop ratio for each bag on the ship, and then plotted the histogram of the 11,000 results. If the histogram was highly peaked (as in Figure 3.2), then, regardless of which 18 bags were selected in the random sample, the average of their readings could not stray far from the average for the whole catch. If, on the other hand, there was great dispersion across bags in the meat-to-scallop ratio (Figure 3.3), then, with bad luck, the sample mean for 18 bags could diverge sharply from the population mean.

But how would we know what this histogram would look like when the 11,000 bags were never examined? We'd never know for sure, the author explained, but we could make an educated guess based on the extent of spread among the 18 sample measurements. In the low-variance case (Figure 3.2), all the measurements would be close together; in the other (Figure 3.3), the measurements would probably (though not definitely) range from the lower end to the upper end of the widely varying outcomes. Notice that the author was discussing S, the sample standard deviation among the 18 measurements, but he was avoiding technical language and equations in favor of an intuitive explanation with pictures. Such an approach can make general concepts accessible to those who have a phobia for Mathematics (a very large group in the United States).

It was time to turn to the actual numbers for the 18 bags sampled and to start using sampling theory. To make things easier, we expressed each of the 18 meat-per-scallop ratios as a multiple of 1/36 of a pound, the minimum legal ratio for the entire catch. Thus, a meat-per-scallop ratio for 1/36 of a pound would be 1 in the new system, while 1/18 of a pound would be expressed as a 2 (i.e., (1/18)/(1/36). Assume for simplicity that the key question was whether the average ratio across all the 11,000 bags was at least one (i.e., assume that the rule applies at the level of bags rather than the level of individual scallops). The actual meat-per-scallop statistics for the 18 sampled bags were as follows:

0.93	0.88	0.85	0.91	0.91	0.84	0.89	0.98	0.87
0.91	0.92	0.99	0.90	1.14	0.98	1.06	0.88	0.93

Before any formal calculations, we can already see that these numbers do not look good for the captain. Fully 89% of the bags—16 out of 18—fell below the legal standard. And because there was little spread among the 18 ratios, we are close to the Figure 3.2 situation in which the sample mean is a reliable guide to the population mean. While the numbers suggest that the population mean was only slightly below one, the Harbormaster took a very dim view of even a small shortfall relative to the weight standard.

We can move beyond initial impressions to actual numbers. Given the data, we can turn directly to (3-23) for probabilistic statements about the location of μ, the average meat-per-scallop ratio for all 11,000 bags. For these 18 sample measurements, the sample mean $\bar{X}$ is 0.932 (7% below the legal standard of 1), and the sample standard deviation S of the distribution for μ (i.e., $\sqrt{S^2}$ in (3-19)) is .075. (Why not work these numbers out on your own?) With $n = 18$, we can treat the normal approximation as reasonably satisfactory. Therefore, noting that $S/\sqrt{n} = .075/\sqrt{18} = .018$, we can make the approximation that $\mu \approx N(.932, .018)$. Then we can make two important calculations.

The first calculation, using (3-24), tells us that the 95% CI for μ is (.896, .968). That outcome is bad news for the ship owner: the standard portrayal of the range of plausible values for μ—namely, the 95% CI—does not include the critical value of 1, the number suggesting compliance with the law. However, the 95% CI is not the only way to reflect uncertainty in using the sample mean to estimate the population mean. We could also ask: What is the probability that the ship owner was innocent despite the results of the sampling?

The last question can be translated into "given the observed results, approximately what is the probability that $\mu \geq 1$?" Answering that question using a normal table yields an outcome that is not encouraging: If $\mu \approx N(.932,.018)$, then $P(\mu \geq 1)$ is less than 1 in 10,000. Needless to say, this statistic does not advance the claim that the sample mean fell short because of bad luck in the sampling.

For the reasons, the author concluded that a sample size of 18 was not really too small to answer the question at issue. So long as the sampling procedure was random (as opposed to (say) a deliberate attempt to include bags with small scallops), then the sample if 18 bags told us a great deal about whether the ship violated the law. Direct statistical analysis could not possibly get the client off the hook (as it were).

There was, however, a point that could be made on the client's behalf. While sampling error cannot fully explain why the observed mean fell below 1, it might explain *some* of the deficit. If we view the 95% CI—from .896 to .968—as the range of plausible values for the ship's overall meat-per-scallop ratio, then .968 is among those values. Under an innocent-until-proven-guilty logic, one could argue the captain deserved the "benefit of the doubt" in the statistical calculation, and thus that .968 rather than .932 should have been used in setting penalties. Had that been done, only about half the ship's scallops would have been confiscated rather than the actual 95%.

It isn't clear whether that argument would have been persuasive in court. About two weeks before the trial, an out-of-court settlement was reached. The settlement was unfavorable to the ship owner but, given our calculations about μ, he may have been wise not to fight to the end.

EXERCISE 3.19

(i) Given the data in Example 3.18, what is the 99% confidence interval for the average meat-per-scallop ratio μ over the 11,000 bags?

(ii) *True or False*: In the distribution for μ, the chance is 4% that the population mean falls in the 99% confidence interval just calculated but not in the 95% confidence interval calculated in Example 3.18.

(iii) *True or False*: If the sample size had been 36 but $\bar{X}$ and S had been the same, the 99% confidence interval would have been half as wide (from top to bottom) as the interval calculated in (i).

EXAMPLE 3.19

A North Poll (Based on a True Story)

Fifteen Boeing 747 pilots who performed simulated landings on the North Airfield of Los Angeles International Airport (LAX) were asked to compare the safety of that airfield with that at other US airports at which these pilots routinely landed. More specifically, they were asked to rate the safety of LAX-North on a 1–7 scale, in which a 1 meant that LAX was far safer than the other airports while a 7 meant that the other airports were far safer than LAX. Their individual responses were as follows:

4	4	4
4	3	4
4	2	6
3	6	6
4	4	3

Assuming that these fifteen pilots were a random sample of 747 pilots that fly in the US, what can we infer about μ, the mean rating of the LAX North Airfield that the full population of 747 pilots would have given had they all answered the question?

Solution

Using the commands AVERAGE and STDEV in Excel, the author found that $\bar{X}$ = 4.07 and $S = 1.16$ for these pilots. Accepting $n = 15$ as the threshold for the normal approximation, the 95% confidence interval for m would be approximately $(4.07 - 1.96 * 1.16/\sqrt{15},\ 4.07 + 1.96 * 1.16/\sqrt{15}\ ) =$ **(3.49, 4.65)**. If we wanted $P(\mu \le 4)$—which is the chance the full population of 747 pilots would have deemed LAX North was at least as safe as other US airports—we could make the approximation that $\mu \approx N(4.07, 1.16/\sqrt{15}\ = N(4.07,.30)$, and then use a table to learn that $P(\mu \le 4) \approx .41$. What these results suggest is that the pilots viewed LAX North as very close to average in safety among its peer US airports (where 4 on the scale means average). Moreover, the modest standard deviation of .30 suggest that the conclusion would be unlikely to change sharply if a far larger of number of 747 pilots were questioned.

EXERCISE 3.20

For the 15 pilots in Example 3.19:

(i) What distribution of the 15 individual outcomes would have yielded a sample mean $\bar{X}$ of 4.07 and the widest possible 95% confidence interval for μ? (Hint: Start out by trying to get 1's and 5's that add up to $4.07 * 15 = 61$.)

(ii) Approximately what would the 95% confidence interval be under those circumstances?

EXAMPLE 3.21

Fission Expedition

A random sample of 30 fuel rods is taken at a nuclear reactor, and the percentage of uranium enrichment is determined for each. (There are about 30,000 fuel rods in the reactor.) The measurements are as follows:

2.98	2.79	3.01	2.96	2.88	2.81	2.94	2.88	2.82	2.89
2.99	2.80	2.87	2.85	2.87	2.96	2.79	2.86	2.95	2.89
2.93	2.85	2.80	2.98	3.00	2.84	2.90	2.88	2.86	2.83

(i) What is an unbiased estimate of μ, the mean percentage of uranium enrichment for all fuel rods in the reactor?
(ii) What is a 90% confidence interval for μ given the data?
(iii) What can we say about the probability that μ is greater than every single one of the 30 measurements listed above?

Solution

(i) A familiar unbiased estimate for μ is of course $\bar{X}$, the sample mean of the 30 observations. Here $\bar{X} = 2.89$ (to the nearest hundredth).
(ii) To find a 90% confidence interval based on $\bar{X}$, we need S, the standard deviation estimate for the population. From (3-19), we find that $S = .067$. Therefore, (3-25) implies that the 90% confidence interval extends from $2.89 - 1.64 * .067/\sqrt{30}$ to $2.89 + 1.64 * .067/\sqrt{30}$, meaning that the interval is (2.87, 2.91).
(iii) We can certainly say that the probability is infinitesimal. If the mean of the distribution is the same as the median (as would happen, for example, if the enrichment percentages were normally distributed across the rods), then the chance that all 30 observations would fall below the true mean would be $(1/2)^{30}$. Whatever the exact distribution of enrichment percentages, the approximate normal distribution for μ has mean 2.89 and standard deviation .012. The largest observation is 3.00, so the chance that μ exceeds all 30 measurements is $P(\mu > 3.00 | \mu \text{ is } N(2.89,0.12))$. Because $(3 - 2.89)/.012) = 9.2$, 3.00 is 9.2 standard deviations above the mean of μ's distribution. The probability that μ exceeds 3.00 is so close to zero that normal tables just offer a string of zeros.

EXERCISE 3.21

Given the data in Example 3.20, approximate the probability that μ is greater than all but seven of the observed measurements (i.e., that exactly seven of the measurements exceed the true mean).

3.7 What if the Sample Size is Small?

We have a wonderful theory for estimating μ from random sampling when n is at least about 15. But meeting that sample size requirement is not always easy, particularly in fields like medicine and engineering. Can we say anything about how reliably $\bar{X}$ represents μ when n falls short of 15?

Well, we can continue to say that the sampling error $W = \bar{X} - \mu$ has mean zero and standard deviation $S/\sqrt{n}$. But we can't readily move to a probability distribution for μ, unless we are willing to make an additional assumption. One assumption that leads to highly tractable results is that the *original distribution* over the population follows a bell-shaped normal curve.

Let's make sure that we understand what we are saying here. The beauty of the Central Limit Theorem is that, regardless of the shape of the original distribution across the population, the probability distribution for μ is approximately bell-shaped for even moderately large n. Here we are imposing the *new* requirement that the original curve is itself bell-shaped. Doing so allows us to set a distribution for μ even for sample sizes in the single digits. Ironically, though, the distribution of μ when n is small is *not* normal even when we assume that the original distribution is normal. Let's discuss what happens.

Suppose that we have n data values X_1, X_2, ... X_n, that represent independent picks from a *normal* distribution with mean μ. Then we can compute $\bar{X}$ and S as usual. Mr. William Sealy Gosset of the Guinness Brewery in Dublin determined a probability distribution *not for* μ but instead for the quantity t that is given by:

The t-ratio is defined by:

$$t = \left(\bar{X} - \mu\right)/\left(S/\sqrt{n}\right) \qquad (3\text{-}26)$$

Mr. Gosset (working under the pseudonym "student") showed that t follows what is usually called the *student's t-distribution with n – 1 degrees of freedom*. We need to discuss now what a t-distribution is.

There is one key parameter in a t-distribution, called the *number of degrees of freedom*. Regardless of this number, every t-distribution is continuous, has a mean of zero, and its histogram/density curve is symmetrically distributed about zero (Figure 3.7). From these properties, it follows that:

$$P(t_k > c) = P(t_k < -c) \text{ for any constant } c$$

where t_k has a t-distribution with k degrees of freedom

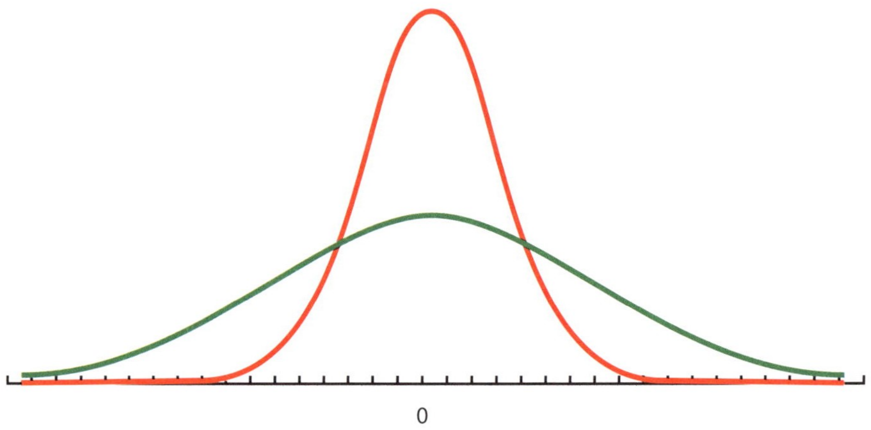

FIGURE 3.7 Every t-distribution is zero-mean symmetric but the density curve varies with the number of degrees of freedom

But how do we find (say) $P(t > 1.5)$ for a given t-distribution? We make note of its number of degrees of freedom and turn to an electronic table and face a display something like:

Degrees of Freedom = ______

Area right of ______ = ______

Compute

If we have eight degrees of freedom, we type 8 on the first line. If we want $P(t > 1.5)$, we type 1.5 after "area right of" and before the equal sign. Then we press "Compute!" and then $P(t > 1.5)$ appears to the right of the second equal sign. With eight degrees of freedom, we learn that $P(t > 1.5) = .0860$.

What if we want a particular percentile of a *t*-distribution, like the 97.5th percentile? Then we list degrees of freedom after the top equal sign and the percentile expressed as a decimal to the right of the bottom equal sign. When we enter .975 for seven degrees of freedom, we learn that the answer is 2.37.

But how does learning about t help us to learn about μ? Be patient. Recall that, for n data points, the quantity $t = (\bar{X} - \mu)/(S/\sqrt{n})$ has a t-distribution with $n - 1$ degrees of freedom. (Again, we are assuming that the distribution over the full population follows a normal curve.) It is easiest to proceed from here with a specific example. Suppose that we have eight data points, meaning that we have seven degrees of freedom (eight minus one). Thus, the 97.5th percentile of the distribution is 2.37, and another use of the table reveals that the 2.5th percentile is –2.37. (Given the zero-mean symmetry of the distribution, it makes perfect sense that the 2.5th percentile is the negative of the 97.5th percentile.)

This exercise is especially useful because, if the 2.5th percentile of the t-distribution is –2.37 and the 97.5th percentile is 2.37, then there is a 95% chance that t_7 will fall in the range (–2.37, 2.37). (The subscript seven refers to the number of degrees of freedom.) *But this probabilistic statement about t can be converted into a probabilistic statement about* μ. If:

$$P(-2.37 \le t_7 \le 2.37) = .95$$

Then

$$P\left(-2.37 \le (\bar{X} - \mu)/(S/\sqrt{n}) \le 2.37\right) = .95$$

Multiplying through by $S/\sqrt{n}$, we get:

$$P\left(-2.37(S/\sqrt{n}) \le \bar{X} - \mu \le 2.37(S/\sqrt{n})\right) = .95$$

Now, multiplying by –1 and then adding $\bar{X}$ throughout, we get:

$$P\left(\bar{X} - 2.37(S/\sqrt{n}) \le \mu \le \bar{X} + 2.37(S/\sqrt{n})\right) = .95 \qquad \text{(3-27)}$$

when $\bar{X}$ is based on a random sample of size eight from a normal distribution

With (3-27), we now have a 95% confidence interval for μ, based on $\bar{X}$ and S as before. But instead of moving by 1.96 $(S/\sqrt{n})$) in each direction around $\bar{X}$ (as in (3-28)), we now go 2.37 $(S/\sqrt{n})$. In a sense, multiplying by 2.37 rather than 1.96 is part of the price we pay for having only eight data points. (The other part of the price is that the divisor $\sqrt{n}$ is smaller when n is smaller.)

We can generalize this last calculation for a useful formula. Suppose that A_k is the 97.5th percentile of the t-distribution with k degrees of freedom, meaning that $-A_k$ is the 2.5th percentile (by zero-mean symmetry). Then:

If we have n independent data points sampled from a normal distribution, a 95% confidence interval for its mean μ extends from:

$$\bar{X} - A_{n-1}(S/\sqrt{n}) \text{ to } \bar{X} + A_{n-1}(S/\sqrt{n}) \qquad \text{(3-28)}$$

where A_{n-1} = 97.5%ile of t-distribution with $n - 1$ degrees of freedom

And that is not all, as the following example suggests.

EXAMPLE 3.21 ***t*-Party**

Suppose that $n = 7$, $\bar{X} = 6$, and $S = 2$. If the data come from a normal distribution, approximate the probability that $\mu \geq 5$.

Solution

We can proceed by converting the probabilistic question about μ into an equivalent probabilistic question about the t-distribution with $7 - 1 = 6$ degrees of freedom. First we write:

$$t_6 = \left(\bar{X} - \mu\right)/\left(S/\sqrt{n}\right) = \left(6 - \mu\right)/\left(2/\sqrt{7}\right) = 1.32\left(6 - \mu\right) \qquad (3\text{-}29)$$

If we want $P(\mu \geq 5)$, we note from the (3-29) that, when $\mu = 5$, $t_6 = 1.32$. Furthermore, when $\mu > 5$, t_6 must be *below* 1.32. Therefore

$$P(\mu \geq 5) = P(t_6 \leq 1.32).$$

A visit to the t-table reveals that $P(t_6 \leq 1.32) = P(\mu \geq 5) = .883$. Given that $\bar{X}$ was 6 here, it is not surprising that the chance exceeds 50% that μ is at least five. It is noteworthy, though, that the probability is as high as 88.3% with only seven data points (and an assumption of that the original distribution is normal).

In summary:

If we assume sampling from a normal distribution, then, even with a small sample size:

- The appropriate t-distribution yields a 95% confidence interval for μ
- The appropriate t-distribution yields the probability that μ falls in a particular range

While we have illustrated how to work with a t-distribution to approximate a confidence interval for μ, we have not indicated why doing so works (even assuming that the data come from an underlying normal distribution). Unfortunately, the proof involves ideas that go well beyond what can readily be covered in an introduction to Applied Statistics. Readers who are curious can find derivations at the web (e.g., by Googling "derivation of t-distribution), but they are likely to find the reading tough going.

EXERCISE 3.22 Suppose that nine data points from a normal distribution with unknown mean and variance are:

6.4 7.8 7.1 7.2 7.5 6.7 7.3 6.8 7.0

(i) What is a 95% confidence interval for μ?

(ii) What is $P(\mu \leq 7)$?

The use of the t-distribution to make inferences about μ raises a question: How would we know whether the data at hand arise from a normal distribution? If we have only a few data points, we can hardly expect that their plot will be bell-shaped even if the population is normal. In some instances, though, the data will make clear that the normal assumption is silly. For example, suppose that the eight data points are 1, 2, 3, 4, 5, 6, 7, and 8772. Then it is hard to imagine that the distribution is normal with a mean near $\bar{X} = 1100$. In many cases, however, the evidence will be inconclusive, and we can assume at least approximate normality to apply the t-distribution.

Extensive simulations have shown that moderate deviations from normality do not greatly compromise calculations based on t-distributions. (Of course, once n is about 15, we have a normal approximation for μ from the Central Limit Theorem.)

EXAMPLE 3.22 Hot Times

Mendel was irked in early 2010 by *The Boston Globe*'s listing of the "average high temperature" on particular dates based on long-term historical data. Because he believes in global warming, he thinks that high temperatures several decades ago will tend to understate what we might expect now or in the future. Thus, to estimate an average high in Boston for April 15 of 2010, he gathered the actual high temperatures over 2001–2009:

Year	High Temperature on April 15 in Boston
2001	60° (Fahrenheit)
2002	58°
2003	81°
2004	53°
2005	48°
2006	77°
2007	44°
2008	53°
2009	50°
(Yes; these are actual data.)	

(i) Suppose that Mendel believes that the high temperature in Boston on April 15 is normally distributed with a mean μ *that was the same over 2001–2009 and which will prevail thereafter.* Using these data, what estimate might Mendel make of the average high in Boston on April 15?

(ii) Under his assumption, what is a 95% confidence interval for μ?

(iii) Under his assumption, what is $P(\mu > 56°)$?

Solutions

(i) The sample mean of the nine observations is $\bar{X} = 58.22°$. That might be the best statistic for average high, although there is a case for rounding off to 58° given that the original measurements were themselves rounded off to the nearest degree.

(ii) For these nine measurements, the sample standard deviation S is 12.76° under (3-19). Thus, a 95% confidence interval for μ under (3-28) would follow:

$$95\% \text{ CI for } \mu: \quad \bar{X} - A_{n-1}(S/\sqrt{n}) \text{ to } \bar{X} + A_{n-1}(S/\sqrt{n})$$

$$= (58.22 - A_8(12.76/3), 58.22 + A_8(12.76/3))$$

where A_8 = 97.5%ile of t-distribution with eight degrees of freedom

From a t-table, we find that $A_8 = 2.306$. Thus, the 95% CI for μ extends from 48.41°F to 68.03°F. A fairly wide interval, but the high variability among the nine observations leaves considerable uncertainty about where μ is.

(iii) To find $P(\mu > 56°)$, we convert the question about μ into an equivalent probabilistic question about t_8. We note that:

$$t_8 = (\bar{X} - \mu)/(S/\sqrt{n}) = (58.22 - \mu)(12.73/3) = .236(58.22 - \mu).$$

When $\mu = 56$, then $t = .524$ (i.e., .236 * 2.22) under this last formula. And when $\mu > 56$, $t_8 < .524$. Therefore, $P(\mu > 56°) = P(t_8 < .524)$. From the t-table for eight degrees of freedom, we find that $P(t_8 < .524)$ is .693. Thus, $P(\mu > 56°) \approx .693$.

The particular reason we found $P(\mu > 56°)$ is that 56°F is the average high for April 15 that was listed in the *Boston Globe* around 2010. The solution to (iii) means that the 2001–09 data suggest that it is more likely than not (a 69% chance) that the April 15 mean had increased from 56°F. But when there is a 31% chance that the mean had fallen below 56°F, this evidence for a warming trend is far from definitive.

What the 95% CI means here is that, if we recorded the April 15 high temperatures for several decades starting in 2010 and averaged them together, then Mendel's assumptions imply that the chance is about 95% that the average would fall between about 48° and 68°. Of course, you might ask whether Mendel is being inconsistent: if he believes that a continuing warming trend makes data from 1971–79 unreliable now, why should data from 2001–2009 be reliable several decades hence? (You're getting good at this!).

EXERCISE 3.23 The high temperatures in Boston over 2010–13 were:

2010: 60°

2011: 47°

2012: 77°

2013: 49°

(i) Assuming that these outcomes arise from a normal distribution with mean μ, what is the probability that $\mu > 58.22$?

(ii) Does the answer to (i) suggest that the mean high temperature in Boston based on 2010–13 data is higher than the mean suggested by 2000–09 data. Discuss briefly.

3.8 Polling and the Estimation of Population Proportions

That polls have become central in the political life of so many democracies testifies to the widespread confidence in their accuracy. Yet at first glance, the claims of the pollsters seem outrageous: how can Rasmussen interview a random sample of (say) 1500 American voters and announce that he knows the will of a nation with 200 million voters to within three percentage points? Actually, we have already answered that question through our study of estimating μ by $\bar{X}$.

Consider a yes/no question, and let p be the proportion of members of the population for whom the answer is "yes." (The answer is "no" for the remaining fraction $1 - p$ of the population; there are no members for whom the answer is unknown.) Suppose that a random sample of size n is chosen and the yes/no status of each of its members is determined. If the answer is "yes" for k members and "no" for the remaining $n - k$ say no, how might we estimate p?

It would seem natural to estimate p by $\hat{p}$ that follows:

$$\hat{p} = k/n. \tag{3-30}$$

$\hat{p}$ is simply the *fraction* who said yes in the random sample. It turns out that estimating p by $\hat{p}$ is a special case of estimating μ by $\bar{X}$. Thus, the detailed sampling theory we've been discussing is directly relevant to the accuracy of working with $\hat{p}$.

To see this, we need to convert the sampling error W, which is now given by $W = \hat{p} - p$, into the general form $W = \bar{X} - \mu$. We can do so if we introduce a clever indicator variable.

Suppose that a "yes/no" question is asked for every member in a very large population of size N is asked a "yes/no" question. We define the indicator variable $Z_j (j = 1, \ldots N)$ by:

$$Z_j = \begin{cases} 1 & \text{if the answer is yes for member } j \\ 0 & \text{if the answer is no for member } j \end{cases}$$

The average μ of these indicator variables over the entire population is simply the population's fraction of "yes members." (Think about it: if the answer is "yes" for 3/5 of the population of size N (i.e., $p = 3/5$), then the population mean would be $((3/5)\,N * 1 + (2/5)N * 0)/N = 3/5$.) In other words, $\mu = p$. And within a random sample of size n, the same reasoning shows that the average $\bar{X}$ for the zeroes and ones among the recorded answers (i.e., for the indicator variables) is simply the fraction of yeses, namely $\hat{p}$ under (3-30). Thus, as stated above, estimating p by $\hat{p}$ is a special case of estimating μ by $\bar{X}$ when the underlying distribution takes a binary (0 or 1) form.

As an estimator, $\hat{p}$ has some attractive properties. Because estimating p by $\hat{p}$ is a form of estimating μ by $\bar{X}$, $\hat{p}$ is an unbiased estimator of p. (Indeed, $\hat{p}$ is the minimum variance linear unbiased estimator of p.) Furthermore, polling is a binomial process in which the two outcomes are "yes" and "no" and each person questioned independently says "yes" with probability p. Therefore, (3-1) means that $\hat{p}$ is the maximum likelihood estimate of p.

But in estimating p by $\hat{p}$, what happens to the sample standard deviation S in the probability distribution for μ ? We work through the algebra at the website, and reach:

The sample standard deviation S when a proportion p is estimated by $\hat{p}$ in a random sample of size n is given by:

$$S = \sqrt{\frac{n\hat{p}(1-\hat{p})}{n-1}} \tag{3-31}$$

We can, moreover, apply the Central Limit Theorem in this setting. The general rule $\mu \approx N(\bar{X}, S/\sqrt{n})$ is customized to:

$$\mu \approx N\left(\bar{X}, S/\sqrt{n}\right) \Rightarrow p \approx N\left(\hat{p}, \sqrt{\frac{\hat{p}(1-\hat{p})}{n-1}}\right) \tag{3-32}$$

Simulations suggest that we should strive for $n \geq 20$, $n\hat{p} \geq 5$, and $n(1-\hat{p}) \geq 5$ to use (3-36) comfortably.

In summary:

If $\hat{p}$ is the fraction of "yes" outcomes in a random sample of size n in a large population, then an approximate probability distribution for the population fraction p follows:

$$p \approx N\left(\hat{p}, \sqrt{\frac{\hat{p}(1-\hat{p})}{n-1}}\right) \tag{3-33}$$

assuming that $n \geq 20$, $n\hat{p} \geq 5$, and $n(1-\hat{p}) \geq 5$

EXAMPLE 3.23

Approval of President Obama

In a Rasmussen poll based on "1500 likely voters" reported on May 27, 2010, US President Barack Obama's job performance was approved by 45% of the voters and disapproved by 53%, with the remaining 2% expressing no opinion. If we set aside the 2% of 1500 = 30 votes who did not approve or disapprove, we have 1470 voters who took a stand. If we assume that these 1470 voters were a random sample of all American voters who have an opinion about the president's performance, what is the probability that a majority of such voters approved of that performance?

Solution

Here we have 1500 * .45 = 675 people who expressed approval of President Obama, and 1500 * .53 = 795 who disapproved. Therefore, $\hat{p} = 675/(675 + 795) = .459$. Applying (3-33) here, we get:

$$p \approx N\left(\hat{p}, \sqrt{\frac{\hat{p}(1-\hat{p})}{n-1}}\right) = N(.459, .013)$$

Turning to the normal table, we find the $P(p > 1/2) \approx .0008$.

That outcome means that the chance is only about 1 in 1200 that the majority of American voters who had opinions about President Obama's performance around 5/27/10 had positive opinions. That seems a very strong conclusion given that the President had 46% support in a modest sample. But the standard deviation of .013 (1.3%) means that the 46% estimate is considered highly accurate. In fact, 50% is more than three standard deviations higher than the 45.9% estimate, and we know that outcomes that far from the mean are very rare in a normal distribution.

To be sure, there can be questions of whether the sample was really random (meaning that all American voters were equally likely to enter the sample), and whether the respondents answered honestly. But Rasmussen has a reputation for reliability, so raising these questions does not imply that they cannot be answered satisfactorily.

EXERCISE 3.24 In the Rasmussen poll about President Obama reported on 11/25/11, 44% of American "likely voters" approved of the President's job performance, while 53% disapproved. The sample size was described as approximately 3000.

(i) Among those respondents who expressed an opinion (i.e., excluding the 3% who were undecided), what percentage approved of President Obama's performance around 11/25/11?

(ii) What is the approximate probability that his true approval rating was at least 45.9% in the population of likely voters with a clear opinion (i.e., at least the level that was observed in the May 2010 poll in Example 3.23)?

Margins of Error in Polling Results

An especially important consequence of (3-33) concerns the 95% confidence interval:

In a random poll with n respondents to estimate a proportion p in a population, the **95% confidence interval for p** extends from approximately

$$\hat{p} - 1.96\sqrt{\frac{\hat{p}(1-\hat{p})}{n-1}} \text{ to } \hat{p} + 1.96\sqrt{\frac{\hat{p}(1-\hat{p})}{n-1}} \quad \text{(3-34)}$$

so long as $n \geq 20$ and both $n\hat{p}$ and $n(1-\hat{p})$ are at least five.

Equation (3-34) is generally the basis for a pollster's summary of her results (e.g., 64% plus or minus three percentage points). Note that the pollster is not literally claiming to be right to within three points; she is making the less-sweeping claim that, 19 times out of 20, the poll will be accurate to within that margin. The quantity $1.96\sqrt{\frac{\hat{p}(1-\hat{p})}{n-1}}$ —which is the distance from the middle of the 95% confidence interval at $\hat{p}$ to either end of the interval—is generally referred to as the ***margin of error*** of the poll. (In calculating margins of error with (3-34), it is assumed that the number n in the poll is a small fraction of the population size N.)

In a poll, the **margin of error** is generally defined as the distance between the center of the 95% confidence interval for the key statistic and the upper end (or lower end) of the confidence interval. That margin equals $1.96\sqrt{\frac{\hat{p}(1-\hat{p})}{n-1}}$.

In the poll above about President Obama on 5/27/10 in which his approval p was approximated as N(.459,.013), the margin of error of the poll was 1.96 ∗ .013 = .025. That outcome means that the chance is about 95% that, had all American likely voters been polled, their approval of President Obama would have been within 2.5 percentage points of the poll outcome of 45.9%.

A Conservative Margin of Error

Pollsters usually like to set the sample size n in advance. But how can they do so when, as (3-34) makes clear, the accuracy of estimating p by $\hat{p}$ depends on the (yet unknown) value of $\hat{p}$? The mathematical form of (3-34) offers an escape from the difficulty. For *any* $\hat{p}$ between 0 and 1, the product $\hat{p}(1-\hat{p}) = \hat{p} - \hat{p}^2$ *never exceeds 1/4*. That maximum product arises when $\hat{p} = 1/2$. (Figure 3.8). Verifying this last statement is an elementary exercise in Calculus.

Nice to know, but how does this help either us or the pollster? In the formula for margin of error—namely, $1.96\sqrt{\frac{\hat{p}(1-\hat{p})}{n-1}}$ —the pollster can make *the worst-case assumption* about $\hat{p}(1-\hat{p})$, namely, that it will equal ¼. That is a worst case assumption because setting $\hat{p}(1-\hat{p})$ at 1/4 yields the highest possible value of $1.96\sqrt{\frac{\hat{p}(1-\hat{p})}{n-1}}$ at sample size n, and thus the widest possible 95% confidence interval for p under (3-34). (Right?). She can therefore write:

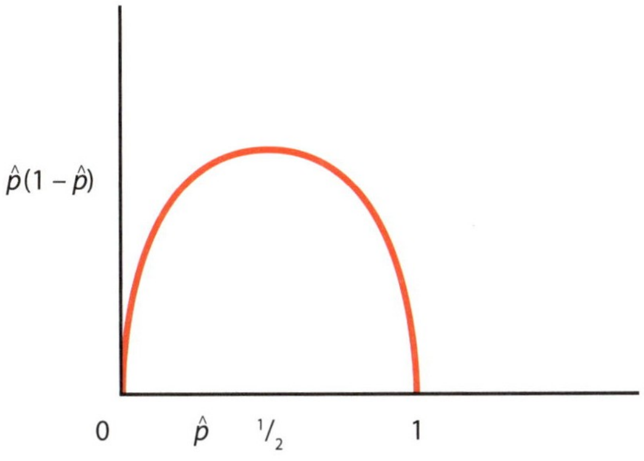

FIGURE 3.8 In the range (0,1), the function $\hat{p}(1-\hat{p})$ has a maximum value of 1/4 at $\hat{p} = 1/2$.

For $n \geq 20$ and $n\hat{p}$ and $n(1-\hat{p}) \geq 5$, the 95% CI for p cannot be wider than:

$$\hat{p} - 1.96\sqrt{\frac{1}{4(n-1)}} \quad \text{to} \quad \hat{p} + 1.96\sqrt{\frac{1}{4(n-1)}}$$

Because $\sqrt{1/4} = \frac{1}{2}$, we can simplify this expression to:

For $n \geq 20$, $n\hat{p} \geq 5$ and $n(1-\hat{p}) \geq 5$, the 95% CI for p cannot be wider than:

$$\hat{p} - .98\sqrt{\frac{1}{(n-1)}} \quad \text{to} \quad \hat{p} + .98\sqrt{\frac{1}{(n-1)}} \qquad (3\text{-}35)$$

Rounding off .98 to 1 in (3-35) yields the statement that the margin of error in the poll ***cannot exceed* $1/\sqrt{(n-1)}$**.

Formula (3-35) is such a strong statement that we repeat it: so long as $n \geq 20$, $n\hat{p} \geq 5$ and $n(1-\hat{p}) \geq 5$, the margin of error (hereafter MOE) of a poll cannot exceed $1/\sqrt{(n-1)}$ regardless of the value of $\hat{p}$. That circumstance suggests an "insurance policy" for the pollster in choosing the sample size n in advance of the polling. First, decide what margin-of-error is the largest tolerable (e.g., .03). Then find the value of n for which $1/\sqrt{(n-1)}$ is equal to that MOE. The pollster can be certain that, whatever actual result $\hat{p}$ arises in the poll, the calculated MOE will be $1/\sqrt{(n-1)}$ or smaller (assuming, of course, that n is large enough that the Central Limit Theorem applies). Thus, the accuracy requirement will be satisfied. (This discussion assumes that all n responses are usable: if some must be discarded, then the effective sample size falls below n. Pollsters typically assume a "wastage" percentage, and try to arrange for n usable responses after losses to wastage.)

EXAMPLE 3.24 **How Big a Sample?**

How large must the sample size be in a random poll so that the margin of error (MOE) in the 95% confidence interval for p cannot extend more than *three percentage points* around the sample proportion $\hat{p}$?

Solution

Applying (3-35) and treating $1/\sqrt{(n-1)}$ as the MOE, we can write:

$$.03 = 1/\sqrt{(n-1)} \Rightarrow (.03)^2 = 1/(n-1) \Rightarrow n = 1112$$

Because political pollsters typically strive for a MOE of three percentage points, they often start with a random sample of size 1500. That number seems simple and round, is a bit higher than 1112, and allows for some wastage.

EXERCISE 3.25 Suppose that the pollster uses a 99% confidence interval, and wants its width around $\hat{p}$ to be no more than three percentage points. Then what sample size would be required?

Mixed Blessings: Square Root Laws, Heartbreak Hills, Diminishing Returns, and Silver Linings

What if a pollster wanted to keep the MOE down to *two* percentage points? Then he would solve the equation: $1/\sqrt{(n-1)} = .02$, which has the solution $n = 2501$.

Wait: that seems unfair. 2501 is more than twice 1112; yet, when we more than double the sampling effort, the MOE does not drop by a factor of two. Rather, it drops by only one-third (from .03 to .02).

We have just arrived at one of the "heartbreak hills" of Statistics: the accuracy of a sampling result—as indicated by (say) the width of its 95% confidence interval—varies not with the sample size but instead with the *square root* of the sample size. To cut the MOE in half, we cannot simply double the sample size but instead have to *quadruple* it. Additional sampling follows a "law of diminishing returns," under which increasing the sample size n by a fixed amount (e.g., by 100) is progressively less beneficial as n goes up. Bummer.

There is, however, a silver lining to the square-root law. If increasing the sampling size gains us less than we might hope, reducing the sample size hurts us less than we might fear. We have said, for example, that $n = 1112$ yields an MOE of three percentage points in polling. Suppose that we took a sample only half as large, meaning that $n = 556$. Then the quantity $1/\sqrt{(n-1)}$ does not double from .03 to .06; instead, it only increases to .042. Indeed, we need a sample of size 1112 to achieve an MOE of .03, but only a sample of 401 to achieve an MOE of .05.

The upshot of this discussion is that, before we begin sampling, we need to think hard about how much accuracy we really need. In the next example, we temporarily digress from polling to discuss setting a sample size when we lack a simplifying formula like (3-35).

EXAMPLE 3.25

Weight Watchers

Every few years, the US Federal Aviation Administration (FAA) requires that each American airline use random sampling to estimate μ, the average weight of its passengers. But the FAA's request has been highly specific: the airline is to create a 95% confidence interval for μ, and the width of the interval around its center $\bar{X}$ is not to exceed 1% of that number (i.e., the 95% CI cannot go beyond $(.99\bar{X}, 1.01\bar{X})$). It is up to the airline to choose a sample size n that will meet this requirement. Approximately what value of n will suffice?

Solution

The 95% CI for μ is given by:

$$(\bar{X} - 1.96S/\sqrt{n}, \bar{X} + 1.96S/\sqrt{n})$$

The requirement that this interval correspond to $(.99\bar{X}, 1.01\bar{X})$ translates to the following condition on the interval's upper end:

$$\bar{X} + 1.96S/\sqrt{n} = 1.01\bar{X} \quad \Rightarrow \quad 1.96S/\sqrt{n} = .01\bar{X} \qquad (3\text{-}36)$$

But $\bar{X}$ and S are not known before sampling takes place. So how can an airline use (3-36) to figure out how to keep the FAA happy?

What the airline has to do is make preliminary estimates of $\bar{X}$ and S to get started. There are various ways it could proceed. There are national tabulations of the weight distribution for US citizens. Moreover, the airline may have results from similar sampling a few years ago, which might furnish a good approximation of what can be expected now. However it gets the numbers, suppose that it starts out with the estimates that $\bar{X} \approx 150$ and S is somewhere between 40 and 50. Substituting $\bar{X} \approx 150$ and S = 50 into (3-36) yields:

$$98/\sqrt{n} \approx 1.5 \Rightarrow \sqrt{n} \approx 65 \Rightarrow n \approx 4200$$

This outcome will presumably cause shock and dismay at the airline.

Things are better for $\bar{X} \approx 150$ and the lesser estimate S $\approx$ 40, in which case (3-36) implies that:

$$78/\sqrt{n} \approx 1.5 \Rightarrow \sqrt{n} \approx 52 \Rightarrow n \approx 2700$$

It is interesting that reducing the S-estimate from 50 to 40 (i.e., by 20%) cuts the sample-size requirement by 36% (from 4200 to 2700). Here as elsewhere, the square-root law for sample sizes is at work.

The initial outlook, therefore, is that the sample size to meet the FAA requirement will be in the range 2500–4500. But what if our estimates of $\bar{X}$ and S are not accurate? If so, that point should be apparent as we start collecting data and observing $\bar{X}$ and S. It might be prudent, therefore, *to specify in advance* that, once we reach a sample size of (say) 1000, we will substitute the *observed* $\bar{X}$ and S into (3-36) and thereby make a "mid-course correction" for the estimated final sample size (n). We would not have to start afresh: the 1000 data points already collected could be part of the final sample.

Once we reach the revised value of n, we would look at the 95% CI If it meets the FAA rule, we would stop; if the spread of the CI were still too great, we would use the latest $\bar{X}$ and S to estimate how much more sampling is needed. Using this multi-stage (iterative) procedure, we're likely to complete the sampling at an n-value pretty close to the lowest sampling size that would meet the FAA's standard.

A key point here is that the familiar question "how big a sample do we need?" cannot usually be answered at the outset. Sampling accuracy depends on the *spread* in the original population (σ, as estimated by S). Furthermore—as happened here—the degree of accuracy that is sought sometimes depends on the sample mean (and thus indirectly on the true mean). Given that the parameters μ and σ are not yet known, a reasonable strategy is first to estimate the key parameters to get a "ballpark" estimate of n, and then to modify the estimate once enough data have been collected.

Readers might not wind up weighing passengers for the FAA, but they might find themselves asked to perform sampling under similar conditions. For example, a magazine publisher might want to offer potential advertisers a very precise estimate of the mean family income among its upscale readers. He might call for random sampling and specify, as FAA did, that the margin-of-error in the 95% CI should not exceed 1% of the middle value. Thus, if the sample mean is \$186,000, the 95% CI should be roughly (184,000 to 188,000).

An opportunity presents itself. Would it be tolerable, you might ask the publisher, to have a 2% margin of error, in which case the 95% CI in the case above would be about (182,000 to 190,000)? It's hard to imagine, after all, that an advertiser who would purchase space under the original accuracy standard would decline to do so under the slightly-looser one. The reason you bring the matter up, you might quietly say, is that, under the square-root law about sampling accuracy, achieving a 2% margin-of-error requires only about *one-quarter* as much sampling as a 1% margin. One could cut sampling costs by 75% with virtually no loss in advertising revenue. If the publisher sees your point, you might find that some of the cost savings finds its way into your annual bonus.

EXERCISE 3.26 Suppose that the FAA rule had instead been: the 95% confidence interval for average weight cannot extend more than three pounds around its center. What sample size would meet that requirement under the initial guesses that the mean is approximately 150 and the standard deviation is approximately 50?

EXAMPLE 3.26

Singapore Swing?

Prior to the parliamentary election in May 2011, a telephone poll of 611 Singapore citizens aged 18 or higher answered the question "Do you think Singapore needs a stronger opposition in Parliament?" The results were reported as 66% yes, 20% no, and 14% undecided. There were also other questions in the poll. The report about the poll stated that its estimated margin of error was 3.96%.

(i) Does the margin of error of 3.96% for results correspond to a 95% confidence interval? Under what circumstances?

(ii) If all Singaporean adults had been questioned rather than a random sample of 611, approximately what is the probability that at least 3/4 would have answered yes?

Solution

(i) Given that there were several different questions in the poll that yielded different values, it seems reasonable to assume that the pollsters used the worst-case approximation $\hat{p}(1-\hat{p}) = \frac{1}{4}$ in their calculations. In that case, the standard deviation of the distribution of p would be $\sqrt{\frac{1}{4(n-1)}} = \sqrt{\frac{1}{4(*610)}}$ = .0202. When .0202 is multiplied by 1.96, the result is .0396, exactly what was reported as the margin of error. In short, the pollsters were apparently working with a worst-case 95% confidence interval, and using 1.96 standard deviations in computing the margin of error. What they did, in other words, was traditional and sensible.

(ii) The polling result $\hat{p}$ = .66 for "yes" responses was rounded off to the nearest percentage point. Multiplying 611 by .66, we get 403.26, which suggests that 403 respondents out of 611 answered yes (yielding a $\hat{p}$ of .659 which got rounded to .66). (We are treating "no" and "undecided" as both indicating an unwillingness to say "yes.") We should use $\hat{p}$ = 403/611= 0.659 rather than the worst-case $p = 1/2$ in setting a standard deviation for p, in which case (3-33) implies that $p \approx N(.659, .019)$. The normal table reveals that $P(p \geq 0.75) \approx .000001$. What this outcome is saying is that, given the sample size and the value of $\hat{p}$, a random sampling error of nine percentage points is all but out of the question.

Of course, all these calculations assume that a random telephone phone among Singaporean adults yields a random sample of such adults. If (say) younger adults are less likely than older ones to answer their phones because they are addicted to text messaging, then the sample might not be completely random.

EXERCISE 3.27 Exclude those polled who answered "yes" in example 3.26, and estimate the probability that more Singaporean adults oppose a stronger opposition in Parliament than are undecided on the issue.

Polls Are Everywhere

Some of the most familiar polls concern politics, but the idea of estimating a population fraction p based on the fraction $\hat{p}$ in a random sample arises in many diverse settings. For example:

- In ***marketing***, a random sample of potential users of a new product might be asked whether they would purchase it, and the fraction who say "yes" is extrapolated to the full population.
- In ***quality control***, a sample of items produced might be "polled" to see whether they are defective. The fraction found defective is assumed to be a point estimate of the defect rate in the full population of items.
- In ***medicine***, the fraction of randomly-chosen patients with some disease who benefit from a given treatment is used to estimate the effectiveness of the treatment in the full population of disease sufferers.

In a more specific setting, one issue in the lawsuits following the 2008 financial crisis was what proportion of subprime mortgages were offered to borrowers who did not meet eligibility requirements. Because reviewing the files for a particular mortgage can be very time-consuming, a random sample of the population of files was selected and scrutinized. The proportion $\hat{p}$ of ineligible borrowers in the sample was used to approximate the proportion among all borrowers in the population.

A Pause in Sampling

We have already accomplished quite a bit in understanding and applying the concepts of statistical sampling. When estimating a population mean μ or a population proportion p, we have:

- Identified point estimates for the two quantities
- Discussed how sample size and the variability in the full population affect the accuracy of approximating μ or p by their point estimates
- Found approximate probability distributions for μ and p assuming the sample size is sufficiently large (or is small but taken from a normal distribution)
- Used these probability distributions to work out confidence intervals and margins of error in estimating μ and p
- While recognizing that sample size is not the only determinant of sampling accuracy, explored the "square root law" that suggests how changing the sample size affects the precision of sampling results

But there are other questions that we have not addressed, including:

- What happens when the sample size is not a minuscule fraction of the population size, meaning that a sizable fraction of the population is included in the random sample?
- How do we actually achieve a random sample from a population? Are there potential biases in the sampling method that mean that some members of the population are more likely to make it into the sample than others?
- What if we are particularly interested in a small subgroup of the population, which would only contribute a few data points in a simple random sample from the population?
- What if we are interested not in estimating μ or p, but instead some other statistic about the population (e.g., the median)?
- What exactly is *bootstrapping*, a modern sophisticated method tied to statistical sampling that offers confidence intervals and margins of error in situations in which existing sampling theory does not apply?

We will consider all these questions in Chapter 6, which is titled "More Statistical Sampling." In Chapter 4, however, we turn to another key component of statistical inference, which is the use of data subject to randomness to test hypotheses about how the world works.

The Takeaway Bar

Glossary

Estimation is a process by which a parameter of interest is approximated from data that are subject to randomness.

A **maximum likelihood estimate** of a parameter is the parameter value under which the observed results have the highest probability of arising (under an assumed distribution).

An **estimator** is a random variable that embodies the rule under which a parameter will be estimated, using data subject to randomness.

An **unbiased** estimator of a parameter is one that on average is equal to that parameter (i.e., its mean equals the parameter).

The **mean squared error** of an estimator is the average squared difference between the estimator and the true value of the parameter.

A **minimum-variance unbiased estimator** of a parameter is the unbiased estimate that has the lowest mean-squared estimation error among all unbiased estimates.

A **minimum-variance linear unbiased estimator** of a parameter is a linear combination of unbiased estimators that has the lowest variance among all such linear combinations.

A **random sample** from a population is a sample of its members, in which every member has the same chance of being selected and different selections are independent.

The **sampling error** is a random variable equal to the difference between the sample-based estimate of a parameter and the true parameter value.

The **sample mean** (*often denoted* $\bar{X}$) is the arithmetic average of the outcomes among members of a data sample.

Assuming a random sample, the **sample variance** (often denoted S^2) is an unbiased estimator of the population variance σ^2. It is based on a slight adjustment of the mean squared spread within the random sample around the sample mean.

The **sample standard deviation** (S, the square root of S^2) is an estimator of the standard deviation σ of a population.

The **95% confidence interval** for a parameter is an interval that has approximately a 95% chance of including the true value of the parameter. The parameter generally has a 2.5% chance of falling above its 95% confidence interval, and a 2.5% chance of falling below the interval. The 95% confidence interval is often treated as the *range of plausible values* for the parameter given the available data.

The **A% confidence interval** for a parameter has an A% chance of including its true value. Generally, the chance is (100-A%)/2 that the parameter falls above the interval, and (100-A%)/2 that the parameter falls below it.

A **sample proportion** (often denoted $\hat{p}$) is the proportion of members of a sample that have a certain characteristic. $\hat{p}$ is frequently used to estimate p, the fraction of population members with that characteristic.

The **margin of error** in a poll generally half the width of the 95% confidence interval for p based on the data. Equivalently, it is the distance between the middle of the confidence interval (usually $\hat{p}$ under random sampling) and the upper end (or lower end) of the interval.

Some Key Formulas

Some Noteworthy Estimators

(Only those discussed in the text are reported below)

Binomial process with n trials and parameter p:

MLE: $\breve{p} = k/n$	k = #A's over n trials
Unbiased: $\hat{p} = k/n$	

MLE = Maximum Likelihood Estimate

Geometric with parameter p:

MLE $\breve{p} = 1/j$	j = # trials until first A

*Poisson with parameter **λ**:*

MLE: $\breve{\lambda} = k/\tau$	k = # events over period τ
Unbiased: $\hat{\lambda} = k/\tau$	

*Uniform with parameters **a** and **b**:*

MLE of a: $\breve{a}$ =the smallest among the observations from the distribution

MLE of b: $\breve{b}$ =the smallest among the observations from the distribution

Sample Mean:

$$\bar{X} = \sum_{i=1}^{n} X_i / n$$

where n = sample size and X_i= i^{th} data point in sample
$\bar{X}$ is a minimum variance linear unbiased estimator of population mean μ

Sample Variance:

$$S^2 = \frac{\sum_{i=1}^{n}(X_i - \bar{X})^2}{n-1}$$

S^2 is an unbiased estimator of the population variance of σ^2

$S = \sqrt{S^2}$ is an estimator of σ, the population's standard deviation

Sampling error W when estimating μ by $\bar{X}$: $W = \bar{X} - \mu$

When $\bar{X}$ is based on n data points in a random sample then:

- $E(W) = 0$ and $\sigma(W) = \frac{\sigma}{\sqrt{n}} \approx S\sqrt{n}$
- If $n \geq 30$ or so, then $W \approx N(0, S\sqrt{n})$

Approximate Probability Distribution for μ:

$$\text{If } n \geq 30, \text{ then } \mu \approx N(\bar{X}, S\sqrt{n})$$

(As discussed, the normal approximation generally works well for $n \geq 15$.)

Confidence intervals for μ:

If $n \geq 30$, then:

$$\begin{aligned}
&\text{95\% confidence interval (CI)} \approx (\bar{X} - 1.96S/\sqrt{n}, \bar{X} + 1.96S/\sqrt{n}) \\
&\text{90\% confidence interval} \approx (\bar{X} - 1.64S/\sqrt{n}, \bar{X} + 1.64S/\sqrt{n}) \\
&\text{99\% confidence interval} \approx (\bar{X} - 2.58S/\sqrt{n}, \bar{X} + 2.58S/\sqrt{n})
\end{aligned}$$

Sample proportion $\hat{p}$ = fraction of members in a random sample who have a certain characteristic

If p is fraction with the characteristic in a population and $\hat{p}$ is the fraction with the characteristic in a random sample of size n, then the sampling error W *is* $\hat{p} - p$.

If $W = \hat{p} - p$, then:

- $E(W) = 0$, and $\sigma(W) \approx \sqrt{\frac{\hat{p}(1-\hat{p})}{n-1}}$
- For $n \geq 20$ and both $n\hat{p}$ and $n(1-\hat{p}) \geq 5$, then $W \approx N\left(0, \sqrt{\frac{\hat{p}(1-\hat{p})}{n-1}}\right)$

Approximate Probability Distribution for p:

If $n \geq 20$ and both $n\hat{p}$ and $n(1-\hat{p}) \geq 5$, then $p \approx N\left(\hat{p}, \sqrt{\frac{\hat{p}(1-\hat{p})}{n-1}}\right)$

95% confidence interval for p:

If $n \geq 20$ and $n\hat{p}$ and $n(1-\hat{p}) \geq 5$, the 95% confidence interval (CI) for p is:

$$\left(\hat{p} - 1.96\sqrt{\frac{\hat{p}(1-\hat{p})}{n-1}}, \hat{p} + 1.96\sqrt{\frac{\hat{p}(1-\hat{p})}{n-1}}\right)$$

Regardless of $\hat{p}$, the 95% CI for p under these conditions cannot be wider than:

$$\left(\hat{p} - .98/\sqrt{n-1}, \hat{p} + .98/\sqrt{n-1}\right)$$

Some Parting Thoughts

- Maximum Likelihood Estimation and Unbiased Estimation constitute two distinct approaches to approximating a parameter. Maximum Likelihood Estimates are (typically) unique, while there are generally an infinite number of unbiased estimators based on available data. Some maximum likelihood estimators are unbiased, while others are not.
- A biased estimate of a parameter can sometimes achieve lower mean-squared estimation error than an unbiased one. That circumstance reminds us that deciding which estimator is "best" depends on the criterion for comparing estimators.
- Statistical sampling is a powerful process. Through sampling only a small percentage of a population, one can often learn about its key characteristics to within a very limited margin of sampling error.
- The accuracy of extrapolating from a random mean to a population mean depends to an important extent on the sample size, as is obvious to many people. But it also depends on the degree of variability within the original population, which can be estimated by the sample standard deviation. That point is *not* obvious to many people. Because one rarely knows in advance what sample standard deviation S will arise in the data, one cannot usually specify at the outset what sample size will guarantee a particular level of precision.
- If a particular value like zero falls in the 95% confidence interval for a parameter, that circumstance does not prove that the parameter is zero. But it does mean that zero fell within the "range of plausible values" for the parameter. Values outside the 95% confidence interval may be less "plausible" as estimates for the parameter, but they are not altogether excluded. After all, the chance is 5% that the 95% confidence interval for a parameter does not include its true value.

- While the 95% confidence interval is considered the range of plausible values for a parameter, the probability distribution for the parameter generally assigns greater likelihood to values in the *interior* of the confidence interval than to those at its edges.
- Sampling formulas for confidence intervals typically use $\sqrt{n}$ (or, in the case of proportions, $\sqrt{n-1}$) in the denominator of the expression for the width of the intervals. Because of that *square root law,* one typically has to quadruple the sample size to shrink the confidence interval by a factor of two. More broadly, the benefits of greater sampling do not grow linearly with sample size. At the same time, the drawbacks of reducing the sample size are less than linear: cutting the sample size by half does not double the width of the confidence interval but instead raises it by a factor of 1.41 (i.e., $\sqrt{2}$).
- The Central Limit Theorem continues to be remarkable. In the context of random sampling to estimate a mean, it implies that the probability distribution for the mean is approximately normal given a random sample of at least moderate size. There is no requirement that the original distribution itself be normal. Thus, we can often make probabilistic statements about where the mean is located based on the sample mean, the sample standard deviation, and the sample size, and *nothing else.*

EZ Pass

1. Suppose that $\bar{X}$ is an unbiased estimate of μ_x and $\bar{Z}$ is an unbiased estimate of μ_z. Then $\bar{X} + \bar{Z}$ is:
   (A) always an unbiased estimate of $\mu_x + \mu_z$
   (B) an unbiased estimate of $\mu_x + \mu_z$ only if X and Z are independent
   (C) an unbiased estimate of $\mu_x + \mu_z$ only if $\mu_x = \mu_z$
   (D) never an unbiased estimate of $\mu_x + \mu_z$
2. Two estimators of μ are both unbiased, and both have the same nonzero variance σ^2. The weighted average of the estimators that has the lowest variance:
   (A) weights the two estimators equally
   (B) assigns all weight to one of the estimators
   (C) is not unique, because all weighted averages of the two estimators have the same variance
   (D) none of these
3. Among thousands of financial experts, 50 are sampled at random. The average of their forecasts for American Airlines' profits in 2015 is $600 million. If all the experts had provided their forecasts, the probability that the average would be at least $650 million:
   (A) is less than .01
   (B) is between .01 and .20
   (C) is between .20 and .50
   (D) cannot be determined from the given information

4. In a random sample of 400 Belgian people, 49% favor privatizing the national railroad system. Under classical sampling theory, the *probability* that, over the entire Belgian population, a majority favor privatizing the rail system:
   (A) must be below .5
   (B) must be above .5
   (C) can be either above or below .5
   (D) must be exactly .5

(Read (EZ-4) carefully.)

5. In a random sampling exercise based on 144 data points, the 95% confidence interval for the mean waiting time for service at a telephone hot line extends from 2.2 minutes to 2.6 minutes. The 68% confidence interval for mean wait—which is centered at the sample mean and has a 68% chance of including the true mean—extends from:
   (A) 2.26 to 2.54 minutes
   (B) 2.3 to 2.5 minutes
   (C) 2.34 to 2.46 minutes
   (D) cannot be determined from the given information
6. A paper in a marketing journal reports that, in a random sample of Toyota owners in the US, 56% consider Toyotas the safest cars on the road. The article notes that the 95% confidence interval for *p*—the fraction of all Toyota owners who consider Toyota the safest cars—does not include the value 50%. But the paper does not present the actual confidence interval; nor does it report the sample size.

   Based on the information above, we can conclude that the sample size:
   (A) must exceed 250 but need not exceed 500
   (B) must exceed 500
   (C) cannot exceed 500 but can exceed 250
   (D) cannot exceed 250

Further Exercises

EXERCISE 3.27

(i) Suppose that X is $U(0,b)$, and we have five random observations ($X_1, X_2, \ldots, X_5$) from the distribution of X. Devise an unbiased estimator of b based on the observations. Show that it is unbiased.

(ii) Now suppose that X is $U(a,b)$ and we have five random observations ($X_1, X_2, \ldots, X_5$) from the distribution of X. Devise unbiased estimators of a and b based on the observations, and show that they are unbiased.

EXERCISE 3.28 We return to the scallops case in Example 3.18

Suppose that the government's intended schedule of fines violations of the weight-standard for scallops was the following:

True Value of μ	Percentage of Catch Confiscated
1 or greater	None
.97 to .999	35%
.94 to .969	65%
.90 to .939	95%
Below .90	100%

(i) Given the information about μ arising from the 18 data points, estimate for each of these four possible penalties the probability that the ship deserved it.

(ii) Do you think that the actual penalty (95% confiscated) was too harsh given your calculations above?

(iii) Suppose that the harbormaster chose his sample of bags by taking the first 18 he saw stacked together. Are there reasons such a sample might not be random? Explain.

EXERCISE 3.29 Suppose that, within a random data sample of size n ($n \geq 30$), the sample mean was $\bar{X}$ and the sample standard deviation was S. Given this information:

(i) What is the 2.5th percentile of the distribution for the population mean μ?

(ii) What is the 5th percentile?

(iii) What is the 10th percentile?

(iv) What is the 25th percentile?

(v) What is the median?

Assume that an A% confidence interval for μ extends from its $(100 - A)/2$ percentile to its $50 + A/2$ percentile.

(vi) What is the 95% confidence interval for μ?

(vii) What is the 90% confidence interval?

(viii) What is the 80% confidence interval?

(ix) What is the 68% confidence interval?

(x) What is the 99% confidence interval?

EXERCISE 3.30 Over 1980–2009, the high and low temperatures on March 15 in Boston were:

Year	High	Low	Year	High	Low
1980	36°(F)	25°	1995	44°	37°
1981	53°	25°	1996	51°	41°
1982	48°	34°	1997	39°	26°
1983	46°	39°	1998	39°	30°
1984	46°	26°	1999	34°	32°
1985	44°	30°	2000	46°	31°
1986	57°	37°	2001	50°	36°
1987	37°	28°	2002	49°	39°
1988	39°	32°	2003	37°	28°
1989	57°	37°	2004	52°	37°
1990	72°	41°	2005	46°	31°
1991	37°	30°	2006	41°	32°
1992	32°	21°	2007	57°	38°
1993	53°	25°	2008	40°	34°
1994	43°	34°	2009	52°	34°

Assume that these 30 years are a random sample of years for Boston, and let random variable V be the difference between the high and low temperatures in Boston on March 15 in a given year. Let μ be the long-term average value of V.

(i) Without any formal analysis, why is it obvious from the data above that $\mu > 0$? (Hint: Can V be negative?)
(ii) What is the sample mean $\bar{V}$ for the data above?
(iii) What is a 95% confidence interval for μ?
(iv) What is a 90% confidence interval for μ?
(v) Approximately what is $P(\mu > 10)$ given these data?
(vi) Let p be the fraction of years in which V exceeds 10°F in Boston on March 15. What is a 95% confidence interval for p given these data?

EXERCISE 3.31 A set of n individuals picked at random are asked their views about permanent tax cuts. The usual 95% confidence interval for the proportion who favor the cuts extends from 0.48 to 0.58.

(i) What were $\hat{p}$ and n? Assume that none of those polled refused to answer or was undecided, and that the normality approximation was appropriate.
(ii) Suppose that it is desired to do more sampling until the 95% confidence interval for p is entirely above 0.50. Given the results so far, roughly how much more sampling would be needed to achieve that outcome? What

approximations are you making in your calculation? (For example, are you assuming that $\hat{p}$ doesn't change as the sample size grows?)

(iii) If the amount of additional sampling indicated by (ii) is actually performed, approximately what is the probability that the new 95% confidence interval is all above 0.50? (Hint: You don't need a table to answer this.)

EXERCISE 3.32 (Based on a true story) In a random sample of 400 full-time workers, the sample mean of earnings was \$ 10,442 over the first quarter of 2006 (i.e. over the first three months of the year). It is reported that width of the 95% confidence interval for the mean quarterly earnings among all full-time workers—measured from its top to its bottom—"does not exceed" \$100.

(i) Given this information, what is the *maximum* possible standard deviation among the 400 data points in the sample?

(ii) The information above in insufficient to allow us to work out $P(\mu > c)$ for particular values of c like \$10,000. But are there *any* numerical values of c for which we can estimate the probability that the population mean μ exceeds c, even with the limited data provided?

(iii) Suppose that it is decided to sample another 500 full-time workers, and then to combine the results with those for the first 400 sampled. Will it be possible to report that the width of the 95% confidence interval for all workers sampled "does not exceed" \$50? Please discuss briefly.

EXERCISE 3.33 Mendel knows that the true standard deviation of a certain distribution is half the true mean, but he does not know the numerical value of either quantity. He will take a random sample of data and obtain a 95% confidence interval for the mean, and he wants the width of the interval—measured from the middle to the upper end—to be no more than 5% of the middle value itself. Given the information above, roughly what sample size will he need to meet his requirement? (Hint: Approximately how will S relate to $\bar{X}$?)

EXERCISE 3.34 A random sample of 100 people who have tried the two search engines Google and Bing is selected, and each person is asked about his or her preference between them. Suppose that random variable X for a randomly-selected individual is defined by:

$$X = \begin{cases} 1 & \text{if the person prefers Google to Bing} \\ -1 & \text{if the person prefers Bing to Google} \\ 0 & \text{if the person considers them equally good} \end{cases}$$

The Author Googled to Find This Logo

Let μ be the mean X-score over the full population of individuals who have tried both search engines.

Suppose that, over the 100 individuals questioned, 22 preferred Bing to Google, 39 preferred Google to Bing, and the rest considered the two search engines equal.

(i) Find the sample mean of $(X_1, \dots, X_{100})$ given these data.
(ii) Find a 95% confidence interval for μ given these data.
(iii) Mendel hypothesizes that the two search engines are equally popular among people who have tried both. Are the sample data reasonably consistent with this hypothesis? Discuss briefly.
(iv) Is the probability that μ exceeds .25 at least 2%? (You need not calculate the actual probability.)

EXERCISE 3.35 In an electronic poll within an MBA class of 90 US students, 36 people said "yes" for nationalizing a certain oil company and 50 said "no." But because of a technical glitch, four of the votes were lost.

(i) Let $\hat{p}$ be the fraction of the 90 respondents who voted yes. Given the data above, what are the maximum and minimum values that $\hat{p}$ could take on?

Suppose that the students can be treated as a random sample of size 90 from the population of all US MBA students, and let p be the true fraction of all such students who favored yes. Let γ be the number of lost votes that were "yes" votes

(ii) For what value of γ would the 95% confidence interval for p be the widest?
(iii) For what value of γ would the center of the 95% confidence interval be .422?
(iv) For what value of γ would $P(p > .5)$ be the highest?
(v) Would this highest value of $P(p > .5)$ exceed 20%? (You need not calculate the probability; just explain your answer.)

EXERCISE 3.36 A manufacturer needs a precise estimate of p, the fraction of items it produces that are defective. It insists that the 95% confidence interval for p not extend more than

.001 in either direction from the middle of the interval. Thus, if the middle of the interval is .084, the 95% CI cannot extend beyond (.083, .085)

(i) After sampling 500 items, the company finds 40 that are defective. What is the 95% CI for p based on these data?
(ii) Based on the existing sample, what is approximate probability distribution for p? Be as specific as possible.
(iii) Based on the results at hand, approximately how much additional sampling should the company do to achieve a confidence interval that meets the company's standards related to width? (Hint: What is $\sqrt{\hat{p}(1-\hat{p})}/\sqrt{n-1}$ here?

EXERCISE 3.37 In a CNN/Opinion Research poll conducted in December 2009 with 523 US adult respondents, 45% said that global warming exists and is caused mostly by humans. (The remaining 55% said that global warming exists but has mostly natural causes, or that global warming is "an unproven theory.")

(i) Given these data, estimate the probability that the majority of American adults believed at that time that global warming exists and is caused mostly by humans.
(ii) CNN said that the margin of error of the poll was 4.5 percentage points. Does that number correspond to your own calculated margin of error? (Yes: you should calculate something to compare with CNN.)

EXERCISE 3.38 Suppose that R—the radius of a circular computer chip produced on a given machine—is subject to random variability, and that μ is the average radius for all the chips produced. A 95% confidence interval for μ is created based on a random sample of size 144, and it extends from 1.04 to 1.16. Suppose that it is decided to double the number of chips sampled, with the existing sample counted as the first half of the full sample. When the final results are calculated, the upper end of the new 95% CI might *most reasonably* be estimated as:

(A) 1.22
(B) 1.16
(C) 1.14
(D) 1.13

EXERCISE 3.39 Suppose that μ is the average change in the condition of a patient with a particular disease who is treated with a new drug. The drug is helpful so long as $\mu > 0$.

(i) Suppose that the 95% confidence interval for μ based on random sampling contains the value zero. Then:
(A) the drug is equally likely to help a patient as to hurt him
(B) the drug is definitely worthless
(C) we have no idea whatever about whether the drug confers any benefit
(D) we have no strong statistical evidence at present that the drug is beneficial

(ii) If the confidence interval in (i) extends from –2 to +5 based on a sample of 86, then the probability that μ exceeds zero:
(A) exceeds 50%
(B) is exactly 50%
(C) falls below 50%
(D) all of these

EXERCISE 3.47 A poll on the website *Fanpop.com* asked viewers of the TV show "Gossip Girl" which boy would be the best mate for Serena: Dan, Nate, or Chuck. The results were 51% Nate, 41% Dan, and 8% Chuck. Assume that there were 500 respondents to the on-line poll, and that they constitute a random sample of viewers of "Gossip Girl."

(i) Given the results, what is the probability that more viewers of the show favor Nate as Serena's mate than favor Dan and Chuck combined?

(ii) It is desired to assess whether viewers of the show favor Nate as the mate over Dan. Suppose that those who favored Chuck are deleted from the poll, and the remaining voters are treating as a random sample of voters with a Nate/Dan preference. Under this assumption, what is the probability that Nate is ahead among all viewers with a Nate/Dan preference?

(iii) Why might it be a mistake to delete Chuck's supporters in estimating the overall preference for Nate over Dan? (Hint: If Chuck were not a candidate, do we know how his supporters would have split between Nate and Dan?)

CHAPTER 4

Judgment Calls:

Formulating, Evaluating, and Testing Theories

In Vitro Fertilization

The couple was concerned that *in vitro* fertilization would not work for them. Especially because their insurance company wouldn't cover the procedure, they wanted to know whether they should stop if they didn't succeed in (say) the first two attempts. They went to speak with a doctor at the hospital where they would undergo the procedure.

"There's no reason at all to be discouraged with two failures," the doctor responded. "All couples have the same chance of success, and that chance is the same on every trial." Turning to the man, he said "You're an engineer, and you know that getting three heads in a row when you toss a coin doesn't change the chance of tails on the next toss. Same thing here. If you keep at it, sooner or later you're bound to succeed."

The engineer was skeptical, and wondered whether early failures hospital's *in vitro* fertilization really said nothing about the chances of later ones. So he decided to review the research literature about success patterns with the procedure.

To Be Continued

Testing hypotheses against data is a fundamental component of the scientific method and a central concern of Statistics. Over many years, statisticians have codified testing procedures that reflect the elegant use of common sense as well as profound awareness of the frailties of human nature. We scrutinize some of these procedures in this chapter, distinguishing tasks that the methods perform well from others they cannot be expected to perform. But rather than overwhelm the reader with a barrage of statistical tests, we concentrate on a few of them. Once we understand the logic and procedures of those tests, we will be in a position to generalize quickly to other tests.

This chapter is focused on a *classical* treatment of hypothesis testing. Roughly speaking, that tradition emphasizes the question: if the hypothesis H_0 being tested is true, what is the conditional probability of getting an outcome like the one that actually arose in relevant data? That conditional probability is *not* the same as the probability that H_0 is actually true (i.e., it is of the form $P(B|A)$ rather than $P(A)$).

4.1 The General Test Regimen

Classical hypothesis testing starts with a specific conjecture such as "birthdays are uniformly distributed throughout the year." This conjecture is often called the *null hypothesis* and designated H_0. In the author's' view, empirical hypothesis testing typically has four primary steps:

(i) State H_0 as specifically as possible.
(ii) Gather data that directly pertain to whether H_0 is true.
(iii) Consider all possible patterns that might appear in the data and, for each, decide whether to treat it as consistent with H_0 (i.e., in the **acceptance** region for the hypothesis) or as incompatible with H_0 (in the **rejection** region).
(iv) Examine the data and then, in accordance with the rules developed in step (iii), either accept or reject H_0.

Let's consider how this procedure would work for the "uniform birthdays" hypothesis:

- To fulfill Step (i), we might express the hypothesis of uniform birthdays in a probabilistic manner: For a person chosen at random, the chance that she was born on any given date (month and day) is 1/365. (As before, we are neglecting February 29.)
- To get some relevant data, we might then ask some number of randomly selected people (perhaps 240) to write their birthdays on sheets of paper and put the sheets in some box (Step (ii)).
- Then, without knowledge about what was written, we would specify how we would react to each conceivable outcome (Step (iii)). (For example, we might well decide that if all of the 240 respondents were born in January, we would cease to believe H_0. We would thus place the outcome "all in January" in the rejection region.)
- Finally, we would open the box, determine the outcome, and make the accept/reject decision based on our prior decision about how to handle that particular case (Step (iv)).

Note that the procedure suggests that the criteria for accepting or rejecting H_0 be set *before* the test data are inspected. The reason for this convention is that, in many settings, the person performing the test is not neutral about the conclusion: he may be hoping, for instance, to be able to accept H_0. If this person saw the data before setting the test rules then, consciously or unconsciously, he might specify the rules to achieve the desired outcome. To avoid even the appearance of partiality, the test rules should be set before the data are examined.

Hypothesis testing is necessarily an imperfect business. However sensible our test rules, the data can lead us to reject H_0 even though it is true, or to accept H_0 even though it is false. For example, no rational individual would accept the uniform-birthdays hypothesis if all 240 people questioned were born in January. Yet, *even under that hypothesis*, the probability of such a bizarre coincidence is greater than zero. (It would be something like $(1/12)^{240}$.) Similarly, a true tendency of birthdays to concentrate in April, May, and June might not be apparent in the data arising from random sampling. More generally, accepting a hypothesis is not tantamount to swearing in blood that it is true, and rejecting a hypothesis does not prove its falsity beyond all doubt. We can aspire to make reasonable decisions, not choices that are invariably correct.

Here we need to repeat an important point. *When we say we "accept" H_0, we are **not** saying that we are sure that H_0 is correct.* We are making the weaker statement that it is reasonably compatible with available data. To stress the point that accepting a hypothesis H_0 does not prove its correctness, some people prefer to say *"fail to reject H_0"* rather than "accept H_0." While we will not use that language, we respect the concerns that led others to adopt it. Indeed, we will see instances in which the data would allow us to "accept" several different hypotheses. By that we mean that all these hypotheses are reasonably compatible with the data at hand, not that we believe all of them are true.

4.2 Error Probabilities in Hypothesis Testing

It would be helpful, of course, to go beyond conceding that hypothesis testing is risky and to quantify the chance of falling victim to a mistake. A worthwhile quantity to know would be α, the conditional probability that, if H_0 is true, the data and test at hand will lead us falsely to reject it (i.e. to commit what's called a **Type I error**). Correspondingly, we would want the probability β of Type II error (false acceptance). To repeat:

$$\boldsymbol{\alpha = P(\textbf{Type I error})} = P(\text{reject } H_0 | H_0 \text{ is true}) \tag{4-1}$$

$$\boldsymbol{\beta = P(\textbf{Type II error})} = P(\text{accept } H_0 | H_0 \text{ is false}) \tag{4-2}$$

Obviously, a test of H_0 that yields both lower α and lower β than another test seems the more desirable of the two, unless the cost of achieving the greater accuracy is unacceptably high.

Warning: Don't Add the Error Probabilities!

Given the definitions of α and β, there can be a "gut feeling" that $\alpha + \beta$ is the overall chance that the hypothesis test will lead to an incorrect conclusion about H_0. This perception is simply incorrect. First of all, α and β are *conditional* probabilities. And they are contingent not on the same event but rather on two complementary events (one being the truth of H_0 and the other, its falsity). None of the laws of probability specify the sum of two such probabilities: What can we make of $P(A|D) + P(B|D^c)$?

Moreover, a simple example makes clear that the chance of reaching the wrong decision about H_0 is **not** $\alpha + \beta$. Suppose that $\alpha = .1$ and $\beta = .1$. Then, if H_0 is true, there is a 10% chance of falsely rejecting it and, if H_0 is false, there is a 10% chance of false acceptance. In other words, the chance of making an error is 10% regardless of the truth or falsity of H_0. The probability of a wrong decision is thus 10%, not $\alpha + \beta = 10\% + 10\% = 20\%$.

EXAMPLE 4.1

Alpha Male

Mendel has taken possession of a six-sided die that is asserted (hypothesized) to be fair. To test this hypothesis (H_0), Mendel plans to toss it twice. Then he will:

- Accept H_0 if the two tosses yield different numbers
- Reject H_0 if the same number comes up both times

His reasoning is that getting the same number twice suggests a bias towards that number, which contradicts the notion that all outcomes are equally likely.

What can we say about the error probability α in this situation?

Solution

What we can say about α is its precise numerical value. Under (4-1), we can write:

$$\alpha = P(\text{reject } H_0 | H_0 \text{ true}) = P(\text{same number twice} | \text{die fair})$$

No matter which number came up first, the chance that it comes up again on the next toss is 1/6 if the die is actually fair. Thus: $\alpha = 1/6$.

What about β, the chance of accidental acceptance (Type II error) under Mendel's tests? We will turn to it momentarily.

EXERCISE 4.1

Suppose that Mendel will toss the die three times, and reject the hypothesis that the die is fair if (and only if) the same number comes up all three times. What is the probability of Type I error for this test?

A Beta Blocker

Suppose that a die is tossed, and we are asked "what is the probability of getting a one?" Told that the die is fair, we would instantly answer 1/6. But if simply told that the die is *unfair,* we would be at a loss for words. Not only could the probability of a one be as low as zero and as high as one, but we couldn't even rule out the value 1/6. (Whatever makes the die unfair, after all, need not affect the chance of a one.) More generally, computing a probability under the condition that X is true is usually far easier than doing so given that X is false. If X is false, then many other alternatives could be true, and the probability in question could vary from one alternative to another.

It follows from this last paragraph that a vague statement like "H_0 is false" will not lend itself easily to probability calculations that are conditioned on that statement. This shortcoming has implications for the calculation of error rate β, and they are not pleasant ones. We can see the issue readily if we return to Mendel's test in Example 4.1.

EXAMPLE 4.2

Beta Male

For the test in Example 7.1 based on two tosses of a die, find β, the probability of erroneously accepting H_0 (Die Is Fair), assuming that H_0 is false because:

(i) All six sides of the die have 1's on them
(ii) Three sides of the die have 1's, and the other three have 5's
(iii) Two sides have 1's, and the other sides are 3,4,5, and 6.

Solution

We can work out $\beta = P(\text{accept } H_0 | H_0 \text{ false})$ in all three cases, but we will not reach the same answer. Recall that Mendel accepts H_0 unless he gets the same number on both tosses.

- In case (i), it is immediate that $\beta = 0$: Mendel only accepts if he gets two different outcomes, which is impossible if all sides have 1's.
- In case (ii), we can deduce that $\beta = 1/2$: whether a one or a five comes up on the first toss, the chance is 50% (3/6) that the other number will come up on the second toss (and thus that H_0 will be accepted under Mendel's criterion)
- In case (iii), the calculation is a little harder: we have to flex our probability muscles:

$$\beta = P(\text{two different outcomes} | \text{two sides 1's, and the others 3,4,5, and 6})$$

There are two mutually exclusive ways of getting different outcomes in this situation:

A_1: one on the first toss, and then something else on the second

A_2: something other than one on first toss, and something different on the second

Given what we know about the die, we can write:

$P(A_1) = P(\text{one on first toss}) * P(\text{something else on second}) = (1/3) * (2/3) = 2/9$

(Note that different tosses yield independent outcomes even though the die is not fair)

$P(A_2) = P(\text{other than one on first toss}) * P(\text{something else on second})$
$= (2/3) * (5/6) = 5/9$

Because $\beta = P(A_1) + P(A_2)$, we have b = 2/9 + 5/9 = 7/9 .

In summary:

Sides of Die	Probability of Falsely Accepting "Die is Fair" Hypothesis
All 1's	0
Half 1's, Half 5's	50% (1/2)
Two 1's, and 3,4,5,6	78% (7/9)

(Based on two tosses, and the rule "accept if the tosses have different outcomes"

These calculations illustrate the point that, most of the time, *there is no unique value of* β when a given hypothesis is tested. The statement that "H_0 is false" typically allows many alternatives to H_0 that might be true, and the chance of "fooling" the test into mistakenly accepting H_0 can vary dramatically from one alternative to another.

EXERCISE 4.2 As in exercise 4.1, assume that Mendel will toss the die three times to assess whether it is fair (H_0), and will accept H_0 *unless* the three tosses all yield the same outcome. Find the risk of accidental acceptance (β) if (i) all sides of the die are 1's, (ii) three sides are 1's and the other three 5's, and (iii) the six sides are (1,1,3,4,5,6).

An Asymmetry in Hypothesis Testing

We have been evading a troubling point with these various β-calculations. Because the statement "H_0 is true" is very specific, we can determine a single value of α when we test a hypothesis H_0 in a particular way. But the vagueness of the statement "H_0 is false" usually means that we cannot come up with a single value of β. We can work out β if we know precisely why H_0 might be false (as we have seen), but why would we generally expect to know that? In an odd asymmetry, therefore, we know the probability of accidentally rejecting a hypothesis in a statistical test, but rarely know the chance of accidentally accepting it. α is at hand, but β is not. This asymmetry has major implications about how statistical tests are conducted, as we will soon see.

Now we turn from these general considerations to the specifics of some well-known hypothesis tests.

4.2 The One Sample *t*-Test

Many statistical tests have the following structure:

- A **test statistic** is identified, which relates to the degree of compatibility between hypothesis H_0 and the test data.
- The probability distribution of the test statistic *assuming that* H_0 *is true* is identified.
- The test statistic is then calculated with the data, and is then expressed as a percentile of its hypothesized distribution.

The hypothesis is typically rejected if the test statistic is at an extreme of its distribution under H_0. The rejection region might extend upwards from (say) the 95th percentile of the distribution (i.e., outcomes far out on the right tail lead to rejection). Alternately, the rejection region might extend downwards from the 5th percentile (i.e., might be on the left tail). Sometimes, outcomes either below the 2.5%ile or above the 97.5%ile lead to rejection (i.e., an extreme outcome at either tail is treated as unfavorable to H_0).

To illustrate what happens in practice, we first consider the one-sample *t*-test. In that test, it is assumed that we have n independent selections from a *normal* distribution with an unknown mean μ. The null hypothesis H_0 being tested is that $\mu = d$, where d is some specified value (e.g., zero). How might we use the observed values $(X_1, X_2, \dots, X_n)$ to assess whether it is plausible that $\mu = d$?

Intuitively, we would find the sample mean $\bar{X} = \sum_{i=1}^{n} X_i / n$, and formulate a test rule of the general form:

To test H_0: $\mu = d$

- Accept H_0 if $|\bar{X} - d| \le c$
- Reject H_0 if $|\bar{X} - d| > c$ (4-3)

where c is some nonnegative constant

In words, we would accept H_0 so long as the sample mean is within c of the hypothesized mean of d. But if the sample mean is further than c away from d, we would reject H_0.

But how would we choose c? We gain some perspective if we consider two extreme cases: $c = \infty$, and $c = 0$. If $c = \infty$, the test rule really amounts to "always accept." That is certainly an easy rule to follow: we wouldn't even need to compute $\bar{X}$. But what would be the error probabilities associated with that test? Here we can calculate both α and β.

$$\alpha = P(\text{reject } H_0 | H_0 \text{ true}) = 0$$

$$\beta = P(\text{accept } H_0 | H_0 \text{ false}) = 1$$

After all, if we always accept, the danger of erroneous rejection has disappeared. But if H_0 is false, we will definitely make the mistake of erroneously accepting H_0.

What about $c = 0$? Here we are saying that we will accept the hypothesis that $\mu = d$ only if the sample mean is exactly equal to d. But in a continuous distribution—which $\bar{X}$ must surely follow if the X_i's are themselves normal—the chance that $\bar{X}$ will be exactly equal to d is zero even if $\mu = d$. If $\mu \neq d$, the probability that $\bar{X} = d$ is again zero because $\bar{X}$ is continuous. We therefore conclude that:

$$\alpha = P(\text{reject } H_0 | H_0 \text{ true}) = 1$$

$$\beta = P(\text{accept } H_0 | H_0 \text{ false}) = 0$$

These statements reflect the fact that, when a test is impossible to pass, the test rule becomes "always reject." That's a splendid rule when H_0 is false but a monstrous one when H_0 is true.

To summarize results for the two extreme cases $c = \infty$ and $c = 0$:

Case	α	β
$c = \infty$	0	1
$c = 0$	1	0

Presumably, we would choose some test rule between "always accept" and "always reject," which would call for finite value of c greater than zero. But what value might we choose?

To answer that question, it would be helpful to work out graphs of α vs. c and β vs. c. Then we would choose some value of c that gives us error probabilities that we could live with. We sense that α vs. c would follow a graph with a general shape something like that of Figure 4.1. The figure arises because, as c increases, the test is becoming more "lenient." Thus, the chance of accidentally rejecting H_0 goes down.

If $\mu = d$ and the X_i's come from a normal distribution, we can work out the distribution of $\bar{X} - d$, and thus $P(|\bar{X} - d| \geq c)$ for any value of c. (As we saw in Chapter 3, this is true whether we know σ for the underlying normal distribution or must estimate it by S and use the t-distribution) Therefore, we could work out the specific curve in Figure 4.1 for the situation at hand.

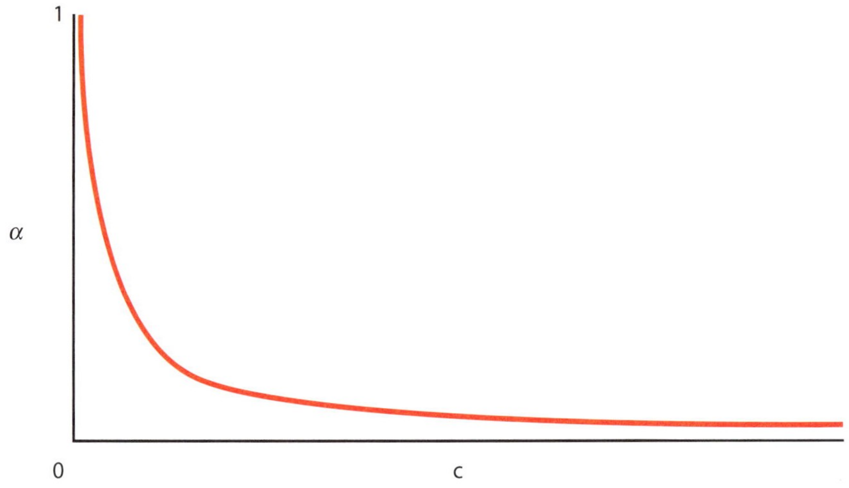

FIGURE 4.1 As *c* goes up, Type I error risk (α) goes down

On the other hand, there is no way we could plot a curve of β vs. c. If H_0 is false, all we know is that $\mu \neq d$. Yet $P(|\bar{X} - d| \leq c)$ depends on *why* $\mu \neq d$: the probability will not be the same if $\mu = d + 1$ as it would be if $\mu = d - 700$. We are in the situation we discussed earlier in this chapter: we can quantify α but not β.

Rather than think anxiously about how we might choose c, we turn now to what people usually do. The most common convention is to choose the numerical value of c at which $\alpha = .05$. This custom, which is tied to such celebrated statistical phrases as "the 5% significance level," warrants some explanation.

In many languages, there is a proverb of the general form: *better the devil you know than the devil you don't know.* In the present context, Type I error could be viewed as the devil we know, because we can quantify the risk (i.e., α) that we suffer that error. Given that α can be calculated but that β cannot be, it seems reasonable to arrange to keep α low while trying to prevent β (namely, Type II error) from running wild. As we have seen, choosing c so high that the α-value is essentially 0 also allows β to approach 1, so that option is excluded. But there is a general consensus that, in most applications, we can tolerate a risk of Type I error as high as 5% (i.e., 1 in 20). The hope is that the value of c for which $\alpha = .05$ is sufficiently low that the test of H_0 is quite strict, meaning that there is little likelihood of erroneously passing when μ is far from d. If that happens, we are reducing the risk of accidental acceptance even if we cannot quantify the achievement.

In testing a hypothesis H_0, it is common to set the test rule so that:

$$\alpha = P(\text{reject } H_0 | H_0 \text{ true}) = .05$$

Though α = .05 is an especially common test standard, sometimes tests are conducted with alternate α-values like .01 or .10. We will have more to say about setting test rules later in this chapter; for now, we return to the question of how we could choose c in test (4-3).

From Chapter 3, we know that when the n X_i's come from a normal distribution with unknown standard deviation σ, the random variable t_{n-1} defined by:

$$t_{n-1} = \left(\bar{X} - \mu\right) / \left(S/\sqrt{n}\right)$$

has a t-distribution with $n - 1$ degrees of freedom, regardless of whether or not $\mu = d$. We know further from (3-28) that a 95% confidence interval for μ based on the data is:

$$\left(\bar{X} - A_{n-1}\left(S/\sqrt{n}\right)\right) \text{ to } \left(\bar{X} + A_{n-1}\left(S/\sqrt{n}\right)\right)$$

where A_{n-1} = 97.5%ile of t-distribution with $n - 1$ degrees of freedom

Now consider the following test of H_0:

- Find the 95% confidence interval (CI) for μ given the X_i's.
- If the interval includes the value d, then accept H_0: $\mu = d$ (4-4)
- If the interval does not include the value d, then reject H_0: $\mu = d$

Recall that the 95% confidence interval for μ is constructed to have a 95% chance of including the true mean. It follows that, if $\mu = d$, the probability is 95% that the interval will include d and 5% that it will fail to do so. If we base the test of H_0 on the 95% CI as in (4-4), therefore, the chance of accidental rejection will be 5%. But that is precisely the value α = .05 that we are seeking.

It is a matter of algebra (and common sense) that the 95% confidence interval $\bar{X} - A_{n-1}\left(S/\sqrt{n}\right)$ to $\bar{X} + A_{n-1}\left(S/\sqrt{n}\right)$ includes the value d if and only if:

$$\left|\bar{X} - d\right| \le A_{n-1}\left(S/\sqrt{n}\right)$$

After all, only if d is within $A_{n-1}\left(S/\sqrt{n}\right)$ of $\bar{X}$ will it fall with the calculated 95% CI for μ.

In summary, the choice $c = A_{n-1}\left(S/\sqrt{n}\right)$ in (4-3) yields a test of H_0 with α = .05.

Putting it all together, we have the one-sample t-test about the mean of a normal distribution:

The one-sample t-test of H_0: $\mu = d$ with $\alpha = P(\text{reject } H_0 | H_0 \text{ true})$ = .05 is:

- Accept H_0 if $\left|\bar{X} - d\right| \le A_{n-1}\left(S/\sqrt{n}\right)$
- Reject H_0 if $\left|\bar{X} - d\right| > A_{n-1}\left(S/\sqrt{n}\right)$ (4-5)

where A_{n-1} = 97.5% ile of t-distribution with $n - 1$ degrees of freedom
n = sample size
S = sample standard deviation obtained via (3-19)

The test assumes that the underlying X-distribution is normal.

EXAMPLE 4.3 **The Scallops Are Back**

Suppose that the harbormaster in the scallops problem 3.18 had decided to use the data about the 18 randomly sampled bags to test the hypothesis H_0: $\mu = 1$ for the *SS Massachusetts*, via a one-sample t-test. Here weights would be measured in units of 1/36 lb., the legal standard.

We saw in Chapter 3 that the 18 data points were:

0.93	0.88	0.85	0.91	0.91	0.84	0.89	0.98	0.87
0.91	0.92	0.99	0.90	1.14	0.98	1.06	0.88	0.93

(i) Why might the harbormaster be interested in H_0: $\mu = 1$?
(ii) Set up the standard t-test with $\alpha = .05$ for this situation.
(iii) Perform the test.
(iv) What assumption about the population of scallop bags underlies the t-test?
(v) Does that assumption seem reasonable here?
(vi) Is there a disadvantage of this test, given what the harbormaster was investigating?
(vii) In Chapter 3, we analyzed the data using the Central Limit Theorem and not the t-distribution. Was that a viable approach?

Solution

(i) Because $\mu = 1$ is the legal standard, the harbormaster would be testing the hypothesis that, among all 11,000 bags of scallops aboard the *SS Massachusetts*, the average meat-per-scallop ratio just barely complies with the law. Only if the data argue for rejecting that hypothesis of innocence would he decide that the ship deserved a fine.
(ii) The general t-test here would take the form:
- Accept H_0 if $|\bar{X} - 1| \le A_{17}(S/\sqrt{n})$
- Reject H_0 if $|\bar{X} - 1| > A_{17}(S/\sqrt{n})$

where A_{17} = 97.5% ile of t-distribution with 17 degrees of freedom

Here $n = 18$, $A_{17} = 2.11$ (from table), and $S = .075$ (from example 3.18), meaning that the test is:
- Accept H_0 if $|\bar{X} - 1| \le 0.37$
- Reject H_0 if $|\bar{X} - 1| > 0.37$

(iii) To perform the test, $\bar{X}$ has to be calculated. Because $\bar{X}$ = .932 for these data, $|\bar{X} - 1|$ = 0.68. Because .068 > .037, the test calls for rejection of H_0.

(iv) The *t*-test assumes that, if meat-per-scallop ratios were calculated for all 11,000 bags, the histogram of results would be bell-shaped. Of course, no one knows whether that would actually happen.

(v) A simple stem-and-leaf plot for the 18 data points is:

0.84–0.85	■ ■ ■
0.86–0.87	■
0.88–0.89	■ ■ ■
0.90–0.91	■ ■ ■ ■
0.92–0.93	■ ■ ■
0.94–0.95	
0.96–0.97	
0.98–0.99	■ ■ ■
1.00–1.02	
1.02–1.04	
1.05–1.06	■
1.07–1.08	
1.09–1.10	
1.11–1.12	
1.13–1.14	■

That histogram is not literally bell-shaped, but a random sample of size 18 cannot be expected to reflect the overall pattern perfectly. The data do not make normality seem absurd, and extensive simulations have shown that moderate deviations from normality do not seriously harm the *t*-test. (More precisely, the accept/reject threshold *c* based on assuming normality is close to the exact threshold for many other distributions.)

(vi) If H_0 is rejected, that could be because (i) $\bar{X} < .963$ or (ii) $\bar{X} > 1.037$. But only the first reason for failure suggests noncompliance with the minimum-weight standard. Later in the chapter, we will discuss how to construct tests so that rejection is clearly related to the reason that the test was performed.

(vii) Our use of the Central Limit Theorem (CLT) was based on the assumption that, with $n = 18$, μ was approximately normally distributed with mean $\bar{X}$ and standard deviation $\frac{S}{\sqrt{n}}$. Simulations have shown that this assumption is quite reliable, perhaps even more so than the normality assumption about the overall population of bags. The alternative to the *t*-test would involve calculating the 95% confidence interval for μ based on the CLT, and rejecting H_0 if the interval does not include the value one. (In practice, the test based on the CLT here is very similar to that based on the *t*-ratio.)

EXERCISE 4.3 Using the Central Limit Theorem, formulate a test of H_0 in Example 4.3 of the form:

- Accept H_0 if $|\bar{X} - 1| \le c$
- Reject H_0 if $|\bar{X} - 1| > c$

Choose c so that the test has $\alpha = .05$. (*Hint:* if H_0 is true and $\mu = 1$, what is the approximate distribution of $\bar{X} - 1$ given S and $\sqrt{n}$?)

4.3 A Two-Sample Difference of Means Test

Professor Richard Lowry of Vassar College has described the "t-Test for the Significance of the Difference between the Means of Two Independent Samples" as "probably the most widely used statistical test of all time, and certainly the most widely known." He further opined that "no statistical toolbox should ever be without it." Under the circumstances, we do well to discuss the test.

Suppose that there are two populations (or two distributions) and one wants to test the hypothesis H_0 that their means are the same. For example, is the average income the same among those purchase Toyota Corollas as among those who buy Honda Civics? The t-test of the equal-means hypothesis assumes that *both distributions are normal* (or—equivalently—that the quantity of interest has a normal distribution over the population).

Under a t-test, one takes random samples from each distribution. Suppose that X is the first random variable and V the second, and that the sample size is n_X from the first distribution and n_V from the second. The two samples are assumed independent. A "high level" overview of the test is:

- Find the sample mean $\bar{X}$ for data points from the first distribution and the sample mean $\bar{V}$ for the points from the second
- Conduct a test of H_0: $\mu_X = \mu_V$ of the general form:

$$\text{Accept } H_0 \text{ if } |\bar{X} - \bar{V}| \le c$$

$$\text{Reject } H_0 \text{ if } |\bar{X} - \bar{V}| > c \tag{4-6}$$

Not surprisingly, the choice of c depends on the sample sizes n_X and n_V, and on the estimated standard deviations in each population. It also depends on the risk of accidental rejection (α) that the test is designed to achieve.

Now we get more specific. If H_0 is true and $\mu_X = \mu_V$ while $W = \bar{X} - \bar{V}$, then:

$$E(W) = E(\bar{X}) - E(\bar{V}) = \mu_X - \mu_V = 0$$

In one important form of the test, the equal variance assumption $\sigma_X^2 = \sigma_V^2$ is made. In that case:

$$\sigma^2(W) = \sigma^2(\bar{X}) + \sigma^2(\bar{V}) = \frac{\sigma^2}{n_X} + \frac{\sigma^2}{n_V} = \sigma^2\left(\frac{1}{n_X} + \frac{1}{n_v}\right)$$

Under these circumstances, the ratio t_{XV} defined by:

$$t_{XV} = \frac{\bar{X} - \bar{V}}{S\sqrt{\left(\frac{1}{n_X} + \frac{1}{n_v}\right)}} \quad (4\text{-}7)$$

where S is an estimate of the standard deviation σ that is assumed equal in both distributions

can be shown under H_0 to follow a t-distribution with $(n_X - 1) + (n_v - 1) = n_X + n_v - 2$ degrees of freedom. Thus, we can develop a t-test of H_0 similar to the one sample t-test we discussed in the last section.

One important issue is how we calculate S. Unless the two sample sizes are equal, we wouldn't think that the individual variance estimates (namely, S_X^2 and S_V^2) deserve equal weight in estimating the common variance σ^2. It can be shown that S^2, the minimum-variance estimate of σ^2 follows:

$$S^2 = \gamma S_X^2 + (1 - \gamma)S_V^2 \quad (4\text{-}8)$$

where $\gamma = (n_X - 1)/(n_x + n_V - 2)$ and thus $1 - \gamma = (n_V - 1)/(n_x + n_V - 2)$

Note that this expression for γ gives greater weight to the sample variance based on the larger sample size (i.e., if $n_x > n_v$, then $\gamma > 1 - \gamma$). That makes intuitive sense.

Putting it all together, we can create a t-test with $\alpha = 0.5$ analogous to the one in Section 4.2.

To test H_0: $\mu_X = \mu_V$ when X and V are both normal with common variance σ^2:

- Compute $\bar{X}$, $\bar{V}$, and S given the sample data
- Find the 95% confidence interval for $\mu_X - \mu_V$, which extends from

$$\left(\bar{X} - \bar{V} - \left(S\sqrt{\left(\frac{1}{n_X} + \frac{1}{n_v}\right)}\right)A_{n_X+n_v-2} \text{ to } \bar{X} - \bar{V} + \left(S\sqrt{\left(\frac{1}{n_X} + \frac{1}{n_v}\right)}\right)A_{n_X+n_v-2}\right)$$

where $A_{n_X+n_v-2}$ is the 97.5th percentile of the t-distribution with $n_X + n_v + 2$ degrees of freedom and n_X and n_v are, respectively, the sample sizes from the X and V distributions

- $$\begin{cases} \text{Accept } H_0 \text{ if 0 is in the 95\% CI} \\ \text{Reject } H_0 \text{ if 0 is not in the 95\% CI} \end{cases} \tag{4-9}$$

This t-test has a 5% risk of accidentally rejecting H_0.

This test is algebraically equivalent to the test rule:

Two Sample t-Test of H_0: $\mu_X = \mu_V$:

$$\begin{cases} \text{Accept } H_0 \text{ if } \left|\bar{X} - \bar{V}\right| \le \left(S\sqrt{\left(\dfrac{1}{n_X} + \dfrac{1}{n_v}\right)}\right) A_{n_x+n_v+2} \\ \text{Reject } H_0 \text{ otherwise} \end{cases} \tag{4-10}$$

where $A_{n_X+n_v+2}$ is the 97.5th percentile of the t-distribution with $n_x + n_V + 2$ degrees of freedom and n_x and n_v are, respectively, the sample sizes from the X and V distributions

Unequal Variances

In some instances, it is implausible to believe that the X and V distributions share a common variance σ^2. There is a modification of the t-test above that allows testing for equal means when the sample variances should be treated separately. The more general t-statistic follows:

$$t_{XV} = \frac{\bar{X} - \bar{V}}{\sqrt{\left(\dfrac{S_X^2}{n_X} + \dfrac{S_V^2}{n_V}\right)}} \tag{4-11}$$

where

$$S_X^2 = \sum_{i=1}^{n_X} \frac{\left(X_i - \bar{X}\right)^2}{n_X - 1} \quad \text{and} \quad S_V^2 = \sum_{i=1}^{n_V} \frac{\left(V_i - \bar{V}\right)^2}{(n_V - 1)}$$

Random variable t_{XV} approximately follows a t-distribution under the normality assumption for X and V, but the number of degrees of freedom ψ follows a relatively complicated formula:

$$\psi = \frac{(U_X + U_V)^2}{\frac{U_X^2}{n_X - 1} + \frac{U_V^2}{n_V - 1}} \tag{4-12}$$

where $U_X = \frac{S_X^2}{n_X}$ and $U_V = \frac{S_V^2}{n_V}$

When the calculated degrees of freedom is not an integer, one should round up to the nearest integer (e.g., a 21.3 would become a 22).

The test of H_0: $\mu_X = \mu_V$ corresponding to (4-10) would be:

$$\begin{cases} \text{Accept } H_0 \text{ if } |\bar{X} - \bar{V}| \le \left(\sqrt{\frac{S_X^2}{n_X} + \frac{S_V^2}{n_V}}\right) A_\psi \\ \text{Reject } H_0 \text{ otherwise} \end{cases} \tag{4-13}$$

where A_ψ is the 97.5th percentile of the t-distribution with ψ degrees of freedom, where ψ is given by (4-12).

EXAMPLE 4.4

No Lectures?

The biology majors at UCZ are required to take Physics, and an experiment has been conducted about whether an online version of the course would be as effective as a traditional lecture format. The 24 majors in the class of 2017 are divided at random into two groups of size 12, the first of which takes the online course while the second takes the lecture-based version. All students will take the same final exam, and the scores will be used to test the hypothesis that mean score under the two course formats is the same for Biology majors. It is assumed that the variances of scores will be the same for both groups.

(i) What other assumptions must be made if a two-sample t-test is used to compare the results?

(ii) Suppose that the final exam scores are the following

On-Line	Lecture
86	77
99	88
79	80
85	93
82	85

On-Line	Lecture
83	89
76	85
86	78
92	90
81	94
87	87
82	84

Perform the t-test on the hypothesis that mean exam score is the same under the two teaching methods.

(iii) If we accept H_0 under the t-test, does that mean that the two methods are equally effective with the Biology majors?

Solution

(i) Several assumptions would underlie the t-test. The first is that the Biology majors in the class of 2017 can be treated as a random sample from the UCZ Biology majors over a large number of future years. The score distributions for the online and lecture students are both assumed to be normal. The exam is assumed to be a sensible measure of achievement in the course, and the class average can be considered the appropriate statistic to measure a group's mastery of Physics. Moreover, it is assumed that the students in the online group do not discuss the course with students in the lecture-based version. (If they did talk, then a given student might be influenced by both teaching methods, which would muddle the intended comparison.)
None of these assumptions is likely to be met exactly. But the t-test is only an exercise in approximation, asking whether the difference in average exam scores is so large than it cannot plausibly be attributed to "sampling error."

(ii) With the exam data:

$$\bar{X} = 84.83,\ \bar{V} = 85.83,\ S_x^2 = 36.88,\ S_v^2 = 29.97,\ n_x = 12,\ n_z = 12$$

(Here X refers to on-line teaching and V the lecture method)
Furthermore, $S = \sqrt{(S_x^2 + S_v^2)/2} = 5.78$, and $A_{x+v-2} = A_{22} = 2.07$.

Thus, $S\sqrt{\left(\frac{1}{n_x} + \frac{1}{n_v}\right)} = 2.36$ and $A_{22}S\sqrt{\left(\frac{1}{n_x} + \frac{1}{n_v}\right)} = 4.89$

The two-sample t-test thus becomes:

$$\begin{cases} \text{Accept } H_0 \text{ if } |\bar{X} - \bar{V}| \le \mathbf{4.89} \\ \text{Reject } H_0 \text{ otherwise} \end{cases}$$

Because $|\bar{X} - \bar{V}| = 1$ here, we would accept H_0.

(iii) Accepting H_0 here does not mean that the long-term means under the two formats must be exactly the same. While zero is in the confidence interval for the difference in means, so are many other values (including one, the actual difference between the sample means). What the *t*-test does mean is that someone who asserted in advance that the two teaching methods are equally effective need not abandon that view in light of the exam scores. But the statement that the methods *might* be equally effective is not the same as the far stronger statement that they *must* be equally effective. On the other hand, the data work against speculation that one of the methods is far more effective than the other.

EXERCISE 4.4 Test the equality-of-means hypothesis in Example 4.4 without assuming that the two populations have a common variance.

4.4 A Large-Sample Test Difference of Means

If n_X and n_V are sufficiently large, we can invoke the Central Limit Theorem to investigate whether the hypothesis H_0: $\mu_X = \mu_V$ is reasonably consistent with available data. Suppose that $\bar{X}$ and S_X are, respectively, the sample mean and sample standard deviation for the n_X selections from the *X*-distribution, while $\bar{V}$ and S_V are the corresponding quantities for the n_v *V*-selections. Then, with sufficient sample sizes, we can proceed as in Chapter 3 to write:

$$\mu_X \approx N\left(\bar{X}, S_X/\sqrt{n_X}\right)$$
$$\mu_V \approx N\left(\bar{V}, S_V/\sqrt{n_V}\right)$$

We can treat the random variables μ_X and μ_V as independent: there is no reason to believe that the sampling error that distances $\bar{X}$ from μ_X has anything to do with sampling error in estimating μ_V by $\bar{V}$. Because the difference between two independent normal random variables is itself normal (Chapter 1), we can write:

$$\mu_X - \mu_V \approx N\left(\bar{X} - \bar{V}, \sqrt{\frac{S_X^2}{n_X} + \frac{S_V^2}{n_V}}\right) \quad (4\text{-}14)$$

With this probability distribution for $\mu_X - \mu_V$, we can conduct the following test of H_0: $\mu_X = \mu_V$:

A test of H_0: $\mu_X = \mu_V$ (or $\mu_X - \mu_V = 0$)

- When n_X and n_V are large enough that $\bar{X}$ and $\bar{V}$ are both about normally distributed, use (4-14) to find the 95% confidence interval for $\mu_X - \mu_V$.

This confidence interval extends from:

$$\bar{X} - \bar{V} - 1.96\sqrt{\frac{S_X^2}{n_X} + \frac{S_V^2}{n_V}} \quad \text{to} \quad \bar{X} - \bar{V} + 1.96\sqrt{\frac{S_X^2}{n_X} + \frac{S_V^2}{n_V}}$$

- $$\begin{cases} \text{If 0 is in the 95\% confidence interval, accept } H_0 \\ \text{If 0 is not in the 95\% confidence interval, reject } H_0 \end{cases} \tag{4-15}$$

Given that a 95% confidence interval for a parameter has a 95% chance of including its true value—whether that value is zero or anything else—the risk of accidental rejection of H_0 (α) is the usual 5% under this test. The logic of the test is familiar: if zero is in the range of plausible values for the quantity $\mu_X - \mu_V$, then the sampling error that separates $\bar{X}$ from μ_X and $\bar{V}$ from μ_V could explain why $\bar{X} \neq \bar{V}$ even though $\mu_X = \mu_V$. (We know, however, that "could explain" is a weaker phrase than "must explain.")

This author thinks that n_X and n_V are sufficiently large if both are at least 15. But some people prefer a threshold close to 30.

EXAMPLE 4.5

Health and Wealth

Suppose it is hypothesized that those Americans who support a certain health care reform have the same average income as those who oppose the reform. A random poll of 200 Americans is taken, and their incomes and views about the reform are recorded. The poll finds 88 Americans who favor the reform, and their family incomes have a sample mean of \$36,800 and sample standard deviation of \$9500. The sample mean of family incomes for the 112 Americans polled who oppose the reform is \$51,300, and the sample standard deviation is \$16,200.

Some questions:

(i) Can the 88 Americans in the survey who favored reform be viewed as a random sample of Americans who favor reform?

(ii) Same question for the 112 Americans who oppose reform.

(iii) Would a usual difference of means test on these data reject the hypothesis that, in the full population of Americans, those who favor reform have the same mean income as those who oppose it?

Solution

(i) Yes indeed. A random sample is one in which each member of the relevant population has the same chance of being selected. In N is the number of

Americans and the selection of 200 people is genuinely random, then each American who supports health care has the same probability of being chosen for the poll, namely, $200/N$.

(ii) Yes and for the same reason as in (i). Each American who opposes the health care reform also has the same $200/N$ chance of being selected for the poll.

(iii) We see that the sample means for the two groups differ by \$14,500 (\$51,300 – \$36,800). But the high sample standard deviations (S_x and S_V) imply considerable diversity of incomes within each group, a circumstance that makes the sample mean an imperfect guide to the true mean. Then again, we had reasonable sample sizes (n_X and n_V) for both supporters and opponents of health care. All these considerations are accorded serious weight in the difference of means test tied to (4-15).

Using $\bar{X} = 36{,}800$, $S_x = 9500$, $n_X = 88$, $\bar{V} = 51{,}300$, $S_V = 16{,}200$, and $n_Z = 112$, we find from (4-14) that $\mu_X - \mu_V$, the difference in mean incomes between those for the reform (subscript X) and those against the reform (V) would follow the approximate distribution:

$$\mu_X - \mu_V \approx N(-14{,}500,\ 1835)$$

The 95% confidence interval for $\mu_X - \mu_V$ would therefore extend from [–14,500 – 1.96 * 1835] to [–14,500 + 1.96 * 1835], which is from [–18,097 to –10,903]. In other words, the 95% confidence interval comes *nowhere close* to including zero. Under test (4-15), therefore, we reject the hypothesis that, had every American been questioned, the mean income for supporters of the reform would have been the same as that for opponents.

EXERCISE 4.5 In Example 4.5, perform the corresponding difference-of means test with a 1% risk of accidental rejection.

EXAMPLE 4.6

From the Heart (True Story)

To assess whether the drug Vytorin was more effective than Zocor in slowing plaque buildup in heart arteries (which raises heart attack risk), a controlled experiment was performed. A random sample of 325 people with known plaque buildup took Vytorin over a period of several months, while another 325 similar individuals took Zocor. None of those conducting or participating in the experiment knew which individuals where taking Vytorin and which Zocor; the experiment was thus said to be *double-blinded*. Vytorin would be considered the more effective treatment if the mean increase in plaque buildup under Vytorin was less than that under Zocor.

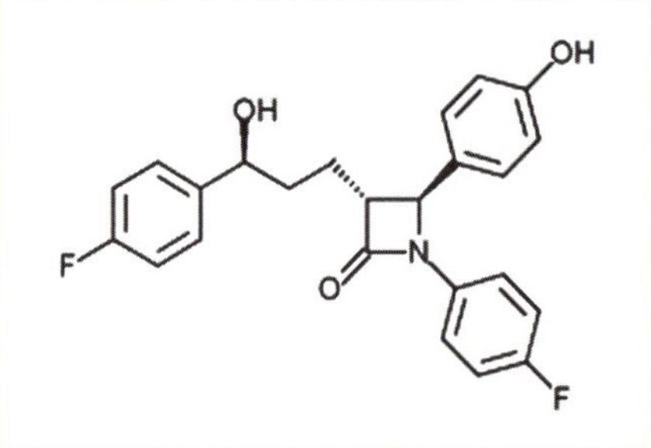

This Compound in Vytorin Supposedly Cut Heart Attack Risk

When the results were analyzed after the experiment ended, the mean change over the experimental period in plaque buildup was +0.0111 among the Vytorin patients and +0.0058 among those who took Zocor. (The more positive the result, the greater the plaque buildup, and thus the worse the outcome)

(i) Given these results in themselves, can we make any inferences about whether Vytorin was more effective than Zocor in slowing plaque buildup?
(ii) The sample standard deviation of plaque buildup was 0.067 among the Vytorin patients and 0.069 among the Zocor patients. Do the experimental results offer statistically significant evidence that Vytorin and Zocor differ in their effectiveness against plaque buildup?
(iii) Can we infer from the experimental results the probability that a particular patient would benefit from taking Zocor rather than Vytorin? Briefly discuss.

Solutions

(i) Yes, we can indeed make some limited inferences. When the sample means $\overline{X_V}$ and $\overline{X_Z}$ for, respectively, the Vytorin and Zocor groups are +.0111 and +.0058, it follows that $\overline{X_V} - \overline{X_Z} = +.0053$. That result means that plaque buildup was *higher* on average in the Vytorin patients than in the Zocor ones, and that the difference $\mu_V - \mu_Z$ (V = Vytorin; Z = Zocor) between the two population means has approximately a normal probability distribution with mean +.0053. Even without the sample standard deviations, we know that $P(\mu_V - \mu_Z > 0) > 50\%$. Thus, it is more likely than not that, in the broader population, Vytorin would be associated with more plaque buildup than Zocor. That outcome would not suggest that switching from Zocor to Vytorin would improve heart health.
(ii) The difference between Vytorin and Zocor is described as ***statistically significant*** if and only if the null hypothesis H_0: $\mu_Z = \mu_V$ is rejected. Here we have enough information to perform a difference-of-means test on this hypotheses. The population mean μ_V denotes the average outcome had all people with plaque buildup taken Vytorin under the conditions of the experiment,

and μ_Z is the corresponding mean had all such people taken Zocor. Here we can use (4-14) to say that:

$$\mu_V - \mu_Z \approx N\left(.0053, \sqrt{\frac{.0067^2}{325} + \frac{.0069^2}{325}}\right) = N(.0053, .0053)$$

meaning that the 95% confidence interval for $\mu_V - \mu_Z$ extends from .0053 – 1.96 * .0053 to .0053 + 1.96 * .0053, or from –0051 to .0157. We see that that 0 falls within the confidence interval, meaning that we cannot reject H_0 and therefore cannot conclude that there is statistically significant difference between Vytorin and Zocor. In words, we are saying that the observed difference in mean plaque buildup of .0053 is subject to such sampling error that we aren't convinced that there would be any net difference in the broader population.

(iii) No: we can't say very much about individual patients, because there are none in the study who took both Zocor and Vytorin. Suppose that, for argument's sake, we assume that the true value of $\mu_V - \mu_Z$ was .0053. That outcome could arise in theory because everyone fares .0053 worse in plaque buildup under Vytorin than Zocor; it could also arise because half of patients do better on Vytorin by .0053 and the other half do worse by .0159. In the first of these situations, 0% do better on Vytorin while, in the second, 50% do better. Average outcomes across groups of patients do not make clear what happened at the level of individual patients.

EXERCISE 4.6 Using the data in Example 4.6, estimate $P(\mu_V - \mu_Z > 0)$.

Now we turn to a venerable statistical test devised by Sir Ronald Fisher.

7.5 Fisher's Exact Test

Suppose that there are two groups, one with m items and the other with n. Suppose further that exactly j of these items are of interest, and that, under hypothesis H_0, these j items are *randomly distributed* over the two groups. By this we mean that random variable X, the number of items (out of the j) in the first group, has the same probability distribution as if j items were randomly selected without replacement from among the $m + n$ items in the two groups. Readers might recall from Chapter 1 that, under these circumstances, X would have a hypergeometric probability distribution. If random variable X is the number of the j items that fall in the first group, then formulas from Chapter 1 imply that:

$$P\left(X = k \middle| H_0 \text{ true}\right) = \frac{\binom{m}{k}\binom{n}{j-k}}{\binom{m+n}{j}} \tag{4-16}$$

Fisher's Exact Test involves dividing the $j + 1$ possible values of X that could arise (from 0 to j) into an acceptance range for H_0 and a rejection range. Typically, the probability of falling in the rejection region when H_0 is true is about 5%.

An example illustrates how Fisher's Exact Test works in practice.

EXAMPLE 4.7

Back to the Manholes

In the Manhattan manhole example in Chapter 1 (Example 1.15), we discussed how Professor Rudin and her team had ranked the 51,213 manholes in Manhattan based on the risk that they would suffer future fires or explosions. Given what actually happened in the year after the classification scheme was advanced, test the hypothesis that it was no better than random guesswork in identifying high-risk manholes. Focus on the 1,000 manholes that the Rudin team classified as the riskiest among the 51,213 in Manhattan.

Solution

Suppose that when Professor Rudin and her team had just released their risk assessments, some skeptic asserted that the scheme was useless and equivalent to random guesswork (hypothesis H_0). To test the "guesswork" hypothesis, the 51,213 manholes could be divided into two groups: (i) those classified by the Rudin team as the riskiest 1,000 of the manholes and (ii) the remaining 50,213 manholes. A test of H_0 could be based on how those manhole covers that suffer problems over the next year are divided between these two groups. If j is the total number of manholes with trouble in the next year, then H_0 implies they are randomly distributed over the 51,213 manholes in Manhattan, with no tendency to fall in the group that the Rudin team deemed especially risky. In other words, the number X that fall in the Rudin high-risk group would be hypergeometric with parameters j, $m = 1{,}000$, and $n = 50{,}213$, as specified in (4-16).

The setting is ideal for Fisher's Exact test. Once j is known, the test would take the general form:

$$\begin{cases} \text{Accept } H_0 \text{ if } X \le c \\ \text{Reject } H_0 \text{ if } X > c \end{cases} \tag{4-17}$$

where c would be chosen so that $P(X > c \mid H_0 \text{ true}) \approx .05$

This test meets the usual standard $\alpha = .05$, but it does so in a different way than the tests we considered previously. It sets the rejection region as the top 5%

of the X-distribution under the "guesswork" hypothesis whereas, in the past when we used 95% confidence intervals, we rejected if the outcome was in the bottom 2.5% of the distribution (below the 95% confidence interval) or the top 2.5% (above the interval). The reason we focus here on the top 5% is that these are outcomes in which the "Rudin 1000" group contained far more troubled manholes than would have been expected by chance alone. If that happens, it suggests the Rudin classifications had done better than guesswork in identifying dangerous manholes. On the other hand, if the Rudin high-risk group actually suffered fewer troubled manholes than would be expected under random guessing, that circumstance would strengthen the suspicion that the classification scheme was worthless. We are really testing H_0: *Rudin is no better than guesswork* against the alternative hypothesis H_a: *Rudin is better than guesswork*. We incline towards H_a rather than H_0 when X is unusually high, not when X is unusually low.

A year later, it was known that $j = 44$ (for 2009). To achieve a rejection region that starts at the 95th percentile of the hypergeometric distribution of X, we turn to an electronic hypergeometric table for $j = 44$, $m = 1000$, and $n = 50{,}213$. Given that the hypergeometric distribution is discrete, no single value of X is exactly equal to its 95th percentile. But the table tells us that:

$$P(X > 2 \mid H_0 \text{ true}) = P(X > 2 \mid X \text{ is } H(44, 1000, 50{,}213) = .0544$$

Thus, a Fisher's Exact Test very close to one based on the 95th percentile would be:

$$\begin{cases} \text{Accept } H_0 \text{ if } X \le 2 \\ \text{Reject } H_0 \text{ if } X > 2 \end{cases}$$

In terms of error probabilities, this test follows $\alpha = P(\text{reject } H_0 \mid H_0 \text{ true}) = .0544$.

As we know from Example 1.15, the actual value of X in the Manhattan manhole data was five. That outcome is comfortably in the rejection region. If the Rudin ranking was equivalent to guesswork, we would have expected only about 2% of troubled manhole covers to fall among the 1000 ranked riskiest (i.e.,1000/51,213). And 2% of 44 is about one. But the actual number of troubled covers in the top 1000 was five times that expected number, and it is rare to have such an excess by coincidence alone. Indeed, $X = 5$ was not only in the top 5% of the relevant hypergeometric distribution, but in the top 1/4 of 1%.

In short, the skeptic's hypothesis is rejected because the Rudin team did better than random guesses would suggest. Beating random guessing hardly means that the scheme was perfect: Of the 44 manholes that had trouble in 2009, 39 of them did not fall in the "Rudin 1000." But the Rudin scheme was better than what preceded it, and one should not belittle a forecasting tool because it is not a crystal ball.

EXERCISE 4.7 (i) Construct Fisher's Exact Test when X is the number of the 44 troubled manhole covers that fell in the riskiest 5,000 of the Rudin rankings. Choose the threshold of the rejection region so that the chance of accidental rejection is as close to 0.05 as possible.

(ii) What is the test outcome given that X was nine in the actual data?

4.6 The Chi-squared Test

"If I can make it there, I'll make it anywhere."

—*Frank Sinatra, "New York, New York"*

A distinguished member of the hypothesis-test elite is the Chi-square (χ^2) procedure. It is a bit more complicated than the tests we have seen up to now, but the reader who absorbs the main points about this test will have no trouble with the key concepts of hypothesis testing. If we can make it there, we'll make it anywhere.

Here we describe the Pearson variant of χ^2 test. It assumes that we are testing a specific null hypothesis H_0 against the generalized alternative H_a that H_0 is false. We assume that we have ***N independent data points*** that provide information about the accuracy of H_0. At the heart of the Chi-square procedure is the creation of ***a set of M categories***, defined in such manner that *every* data point must fall in *one and only one* of them. Thus, the categories cannot overlap, yet they must be sufficiently broad that no data point lacks a category into which it can fall.

Suppose, for example, that we are testing uniform birthdays (H_0). Then the categories might be the months of the year. Because the categories must cover all possibilities, they could not be only the ten months from January to October. And because they cannot overlap, the first category cannot be January/February while the second is February/March. When we say that the N data points are independent, we mean that knowing which category a particular point falls into says essentially nothing about the placement of any other point.

In broad terms, the χ^2-test compares the actual spread of data points across the categories with the spread that might have been expected under H_0. If the two spreads are sufficiently similar, H_0 is accepted while, if the spreads are sufficiently dissimilar, H_0 is rejected. To flesh out the test, we need to specify how the comparison between expected and actual spreads is performed, and how we interpret the outcome of the comparison. We do so below in stages, getting more detailed as we proceed.

At the outset, we note that the χ^2 test requires that the hypothesis be specific enough to allow the calculation of $\boldsymbol{p_i}$, the probability under H_0 that any given data point (out of the N) falls in category i:

$$p_i = P(\text{data point falls in category } i \mid H_0 \text{ true})$$

For example, if the hypothesis is uniform birthdays and category i is the ith month of the year, then $p_i = 1/12$ under H_0 for all data points. (Here we are neglecting slight differences in the lengths of different months.)

When we have categories and probabilities associated with each, what happens next? Before the data are inspected, let random variable a_i be the *total* number of the N data points that fall in category i. What can we say about a_i? If H_0 is true then, *prior to inspection of the data*, each of the N points independently has probability p_i of falling in category i and $1 - p_i$ of not doing so.* But that means we can specify the probability distribution of a_i: it would have a **binomial** distribution with parameters N and p_i. (Right? Each inspection of an individual data point is a trial which has probability p_i of resulting in category i. And there are N such inspections.) Then a familiar binomial distribution formula shows that the mean of a_i, denoted as e_i, would therefore follow:

$$e_i = Np_i \tag{4-18}$$

The χ^2 test calls for construction of the following chart:

Category	e_i, Expected Number of Data Points Under H_0	a_i, Actual Number of Data Points
1	Np_1	a_1
2	Np_2	a_2
...	...	...
...	...	...
M	Np_M	a_M

EXAMPLE 4.8 Toss Up

Suppose that it is hypothesized (H_0) that a coin is fair, and that it will be tossed 100 times to assess whether the hypothesis is plausible. Set up a χ^2 test of the hypothesis.

Solution

The test could have two categories: heads and tails. There would be $N = 100$ data points (one for each toss), each of which has probability 1/2 of falling in a given category. The test table would therefore be:

Category	e_i, Expected Number of Data Points Under H_0	a_i, Actual Number of Data Points
(1) Heads	$e_1 = Np_1 = 100 * \left(\frac{1}{2}\right) = 50$	a_1
(2) Tails	$e_2 = Np_2 = 100 * \left(\frac{1}{2}\right) = 50$	a_2

* We slightly modify this assumption later on, but don't worry about it now.

EXERCISE 4.8 Create a χ^2-table for testing the hypothesis that a six-sided die is fair, based on the outcomes of 60 tosses

Returning to the general chart, e_i is the expected number of data points in category i assuming the hypothesis is true, while a_i is the number that actually fell in category *i*. We have noted that, if H_0 is true, a_i would be the outcome of a binomial process with parameters N and p_i. That circumstance affects our interpretation of the complete χ^2 table. In the coin-toss setting above, for example, we do not anticipate that a_i will be exactly 50. Given the variability in binomial processes, we understand that the actual number of heads could vary modestly around 50 even if the coin is fair. But while the a_i's could differ from the e_i's even if H_0 were true, we hope that they will differ far more if H_0 is false. We look to the a_i's to send up a flare if the hypothesis is greatly wrong: If the coin is in fact heavily biased towards heads, we might wind up with (say) 88 heads out of 100. Such a huge degree of deviation from the expected 50 would make the hypothesis highly suspect.

The χ^2 test asks us to compare the second column of the table (what we expected on average) with the third column (what we got), and then make a decision whether to accept or reject H_0. If the deviations between the e_i's and a_i's are small, we might attribute them to meaningless fluctuations in binomial processes. Thus, we would be inclined to accept H_0. But if the deviations are large, then we worry that *the reason that what we saw differed so much from what we expected is that what we expected was wrong*. In that case, we might well reject H_0. In other words, we follow the Biblical injunction "ye shall go so far, and no further": While *some* oscillation of a_i's around e_i's is consistent with H_0, huge discrepancies between expected and actual outcomes cannot plausibly be ascribed to binomial variation.

But how do we decide which deviations between the expected and actual columns are small and unthreatening to H_0, and which are large and menacing? Stay tuned.

A Discrepancy Index

In a χ^2 test, we compute a numerical **index of discrepancy** between the e_i and a_i columns. (Discrepancy is a fancy word for disagreement.) In qualitative terms, this index takes on higher numerical values as the two columns diverge further. The general form of the test is:

$$\begin{cases} \text{Accept } H_0 \text{ if the discrepancy index does not exceed some threshold } c \\ \text{Reject } H_0 \text{ if the discrepancy index exceeds } c \end{cases}$$

In essence, any level of discrepancy up to c is attributed to binomial fluctuations around correctly-specified e_i's, and thus H_0 is accepted. Any discrepancy greater than c, on the other hand, is viewed as too large to be explained by chance fluctuations:

the reason the a_i's differed so much from the e_i's, it is surmised, is that H_0 is false and the e_i's are incorrect.

One can imagine lots of reasonable measures of how far the a_i's and the e_i's differ. For reasons that need not seem obvious to the reader, an especially attractive index denoted χ^2 follows:

$$\chi^2 = \frac{\sum_{i=1}^{M}\left(e_i - a_i\right)^2}{e_i} \tag{4-19}$$

The quantity χ^2 calculated under (4-19) is the **index of discrepancy** in the χ^2-test at hand. We will momentarily discuss what probability distribution of χ^2 takes on if H_0 is true. First, however, we should note the form of (4-19) gives the index several desirable features:

- ***$\chi^2 = 0$ when $e_i = a_i$ in all categories***

  It is nice that the index of discrepancy is zero when there is no discrepancy.

- ***χ^2 considers all M categories, and considers them equally.***

  That all categories should be considered rather than just some seems like common sense. And when $e_1 = e_7$, for example, the index gives the same weight to a given difference between e_1 and a_1 as to the same difference between e_7 and a_7. That makes sense: any divergent treatment would be hard to justify, especially when it is often arbitrary which category is numbered 1 and which 7.

- ***In squaring the differences between expected and actual outcomes, the index makes sure that large excesses of a_i over e_i in some categories do not cancel the effects of large deficits in others.***

  If a_i exceeds e_i by 500 in one category and falls below by 500 in another, both these outcomes could cast doubt on the hypothesis H_0. It would be unfortunate if the two deviations of opposite sign effectively cancelled each other in the overall index of discrepancy. When individual deviations get squared in the calculation, such cancellation cannot occur.

- ***Dividing the squared difference by e_i provides some perspective on how "big" the difference is.***

  A comparison between two cases illuminates this last point:

e_i	a_i	$(e_i - a_i)^2$	$(e_i - a_i)^2/e_i$
5	2	9	**1.8**
80	83	9	**0.11**

Intuitively, a difference of three between e_i and a_i seems more "negligible" when e_i is large than when it is small. In the first case above the difference of three is 60% of e_i (5 vs. 2), while it is only about 4% of e_i (80 vs. 83) in the second case. As the last column above indicates, the χ^2 formula takes account of that difference by dividing through by e_i.

The χ^2 test takes the general form:

$$\begin{cases} \text{Accept } H_0 \text{ if } \chi^2 \le c \\ \text{Reject } H_0 \text{ if } \chi^2 > c \end{cases} \tag{4-20}$$

But how do we choose c? To answer that question, we need ask probability distribution the discrepancy index χ^2 would follow if H_0 were true. Given the notation, it will hardly surprise readers that this distribution is a member of the χ^2 family. (A χ^2 random variable with k degrees of freedom is the sum of the squares of k independent random variables, each of which is $N(0,1)$.) In a χ^2-test with two categories, it is the case that:

When $M = 2$, χ^2 as defined in (4-19) is approximately χ^2-distributed with **one** degree of freedom when N is sufficiently large *and* H_0 is true. (4-21)

A derivation of (4-21) appears at the website.

Some characteristics of this distribution are:

The χ^2-*distribution* with one degree of freedom has:

- A mean of 1
- A standard deviation of $\sqrt{2}$
- A 95th percentile of 3.84
- A 99th percentile of 6.64

Other distribution characteristics of this χ^2-distribution (e.g., the median) can be obtained from an electronic calculator.

What do we mean by "when n is sufficiently large?" A rule-of-thumb is that we need both Np_1 and $N(1\text{-}p_1)$ to be at least five for the approximation in (4-21) to be reliable.

When we have more than two categories in a χ^2-test, the discrepancy score will also be approximately χ^2 distributed if H_0 is true. The number of degrees of freedom in the test will depend on several factors, the number of categories M being one of them.

To see how a χ^2 test is carried out, just perform the following exercise.

EXERCISE 4.9

Return to the two-category χ^2 test of the hypothesis H_0 that a coin is fair, based on 100 tosses.

(i) Suppose that the tosses yielded 57 heads. What would be the numerical value of the discrepancy score χ^2?

(ii) What value of c in (4-21) would yield $\alpha = .05$? (Hint: It's in the box just above.)

(iii) Would H_0 be rejected in the test in (ii) with the outcome 57 heads?

χ^2–Tests with More Than Two Categories

To get us started in considering χ^2-test with several categories, we introduce some data from MIT:

EXAMPLE 4.9

Born for MIT (Based on a true story)

Suppose that H_0 is the familiar uniform birthday hypothesis, and that 45 randomly chosen MIT students tell us their birthdays. To perform a χ^2 test of the hypothesis, we create a system of six nonoverlapping categories that will include all possible birthdays. These categories are: (1) the days in January and February, (2) March–April, (3) May–June, (4) July–August, (5) Sept–Oct, and (6) Nov–Dec.

(i) Formulate the χ^2-table for these categories. (Neglect the small differences in the numbers of days across the various two-month periods.)

(ii) The actual numbers of students in each category are:

Jan–Feb: 4	Mar–Apr: 12	May–June: 4
July–Aug: 8	Sept–Oct: 10	Nov–Dec: 7

With these data, calculate the discrepancy index.

(iii) Suppose (never mind why yet) that the discrepancy score here would follow a χ^2-distribution with five degrees of freedom if birthdays were in fact uniform. What threshold c would achieve $\alpha = .05$ in the χ-test?

(iv) Perform the χ-test with the MIT data. How should the results be interpreted?

Solution

(i) Neglecting slight differences in the lengths of these two-month periods, we conclude that, if H_0 is true, $p_i = 1/6$ for all six values of i. Thus, with 45 people, each $e_i = np_i = 45/6 = 7.5$. (Do not be upset that the mean e_i is not an integer although the a_i's must be. We are familiar with that circumstance; after all, the mean outcome when a fair six-sided die is tossed is 3.5, which can never come up on any toss.)

The χ^2-table here will be:

Category	Expected Number (e_i)	Actual Number (a_i)
Jan–Feb	7.5	
Mar–Apr	7.5	
May–June	7.5	
July–Aug	7.5	
Sept–Oct	7.5	
Nov–Dec	7.5	

(ii) With the actual birthday data, the table above becomes:

Category	Expected Number (e_i)	Actual Number (a_i)
Jan–Feb	7.5	4
Mar–Apr	7.5	12
May–June	7.5	4
July–Aug	7.5	7
Sept–Oct	7.5	10
Nov–Dec	7.5	7

The discrepancy index χ^2 for these numbers follows:

$$\chi^2 = \frac{\sum_{i=1}^{6}\left(e_i - a_i\right)^2}{e_i} = 6.87$$

(iii) An electronic table reveals that the 95th percentile of the χ^2-distribution with five degrees of freedom is 11.07. Therefore, the choice $c = 11.07$ in (4-20) would achieve $\alpha = .05$, because there is a 5% chance of "blundering" into the rejection region if H_0 is true.

(iv) Because $6.87 < 11.07$, the discrepancy score is within the acceptance region for H_0. What this means is that the observed variation in a_i's across categories is within the range of random fluctuations for a sample of size 45, assuming that birthdays are uniformly distributed. Thus, while we expect an average of 7.5 people per category, getting 12 in one of the six categories and getting 4 in two others is not terribly unusual, any more than getting six heads in ten tosses is terribly unusual for a fair coin. For that reason, someone who believed in H_0 can continue to do so in light of the test results. However, that the observed variation is consistent with uniform birthdays does not prove that birthdays are uniform. For example, there were three times as many births in the sample in Mar-Apr as in either Jan-Feb or May-June. The data at hand are consistent with uniform birthdays, but also with some other hypotheses that posit considerable non-uniformity.

EXERCISE 4.10 Suppose we learn later that two of the students in the class in Example 4.9 lied about their birthdays, but we have no idea which ones lied or how they did so. Might the *actual* birthdays among the 45 students (as opposed to the reported ones) have led to rejection of the uniform-birthday hypothesis? Discuss with specific reference to the data presented in Example 4.9.

Degrees of Freedom in a χ^2-test

As Example 4.9 indicates, the number of degrees of freedom in a χ^2-test is essential if we are to set the pass/fail threshold c. We have already encountered degrees of freedom in connection with t-tests, but determining the number for χ^2-tests is a bit more complicated. The general rule (as well as the justification for treating the discrepancy score as approximately following a χ^2-distribution) is based on theorems that are rarely if ever covered in an introduction to Statistics. For that reason, we do not discuss them here but simply report what the theorems say.

The number of degrees of freedom in a χ^2-test test (*hereafter* ψ) depends on three quantities:

M	the number of categories in the test,
M_P	the number of "puppet" (or redundant) categories in the test
Q	the number of distinct parameters (numerical constants) in the statement of H_0 that were estimated from the very data used in the χ^2-test.

We will define M_p and Q below. Given these quantities, ψ follows the rule:

Degrees of Freedom

$$\psi = M - M_p - Q \tag{4-22}$$

Among the quantities in (4-22), the number of categories M is obvious once the χ^2-test is specified. The number of "puppet" categories M_p relates to the circumstance that, in any χ^2-test, we know at the outset how many data points there are and sometimes even know how many come from particular sources (e.g., men and women). Thus, we need not be told all the a_i's to determine them all: some can be logically deduced from others.

In the six-category birthday test, for example, knowing both the total number of people asked their birthdays (N) and the number of people born in *five* of the two-month periods is enough to determine the number born in the sixth period: it is N minus the sum for the other five categories. In that sense, one of the six categories is a "puppet" of the others: the information it conveys is redundant given knowledge of what happened in the other five.

More generally, the number of puppet categories arises from a simple question. Given M, N and any supplementary information about the data sample known prior to examining the individual data, what is the **minimum** number of a_i's we need to determine *exactly* all the remaining a_i's? If we denote that number as m_p, then M_P, the number of "puppet" categories, follows:

$$M_p = M - m_p$$

In the birthday example, $M = 6$ and $m_p = 5$, so $M_P = 1$.

To see a case where M_P exceeds one, we consider a test of the null hypothesis: H_0: 30% of men and 30% of women smoke cigarettes.

Suppose our test data consist of yes/no answers to the question "Do you smoke cigarettes?" from random samples of 50 men and 60 women. We would presumably set up our χ^2 table as follows:

Category	e_i under H_0	observed a_i
M,S	$50 \times .3 = 15$	
M,NS	$50 \times .7 = 35$	
F,S	$60 \times .3 = 18$	
F,NS	$60 \times .7 = 42$	

(M = male, F = female, S = cigarette smoker, NS = not cigarette smoker)

The quantity m_p is the ***minimum*** number of a_i's we would need to complete the right column of the table. (We assume that these a_i's are strategically placed to be as informative as possible.) Because we know that 50 men and 60 women are in the database, we can finish the task with one of the "male" a_i's and one of the "female." (For example, if we knew that the test uncovered 18 male smokers and 40 female nonsmokers, we could fill in all four entries in the a_i column.) Thus, $m_p = 2$. It follows that $M_P = M - m_p = 4 - 2 = \mathbf{2}$.

Note that, in this test, the men are restricted to the first two categories, and the women to the other two. This arrangement differs from our original description of χ^2-tests, in which we assumed that each data point could fall in any category. Such a modification does not compromise the test, so long as we are careful in setting the number of "puppet" categories correctly to achieve the correct number of degrees of freedom.

The reader might be wondering why we even bother with puppet categories. If a category is a puppet of the others and provides no new information, why not simply delete it from the test? The simple answer is that we can't: The theorems that yield a χ^2-distribution for the discrepancy score assume that the redundant categories are retained. However, we take account of the extent of redundancy in setting degrees of freedom.

Because $\psi = M - M_p - Q$ and $M_p = M - m_p$, there is a shortcut to calculating ψ.

We can write:

$$\psi = M - M_p - Q = M - (M - m_p) - Q = \boldsymbol{m_p - Q} \tag{4-23}$$

where m_p = minimum number of a_i's needed to determine all the a_i's.

The last quantity we need to determine ψ is Q, the number of distinct constants in the statement of H_0 that were estimated from the very data used in the χ^2-test. We postpone discussing Q until after we complete an example, except to note that Q = *0 when* H_0 *was fully specified before the individual test data were examined*. That circumstance prevailed in the uniform birthdays test involving MIT data: we did not "peek" at the MIT data before advancing the hypothesis.

EXAMPLE 4.10 **Back to MIT**

We said in Example 4.09 that the discrepancy index had five degrees of freedom in the MIT birthdays test. Why five?

Solution

In the MIT test, we had six categories, one of which was a puppet, and no constants in the statement of the hypothesis that arose from the MIT data. Thus, $M = 6$, $M_P = 1$, and $Q = 0$. Under (4 22), we have $\psi = M - M_p - Q = 6 - 1 - 0 = 5$. (Alternately, $m_p = 5$, so $\psi = m_p - Q = 5 - 0 = 5$.) As noted earlier, we used ψ to set the pass-fail threshold for the test: it was the 95th percentile of the χ^2-distribution with $\psi = 5$ degrees of freedom.

EXERCISE 4.11 What were m_p, Q, and ψ in the test you devised in Exercise 4.8 about whether a six-sided die was fair?

The *Q* Factor

The last quantity in the degrees-of-freedom formula is Q. Often, the calculation of the p_i's in a χ^2 test depends on a few specific numbers in the statement of H_0. These numbers are called *parameters*. In the uniform birthdays case, for example, the key parameter is 1/365, the probability that a randomly-chosen person was born on *any* given day/month pairing (e.g. March 9). Ideally, these parameters would not be estimated from the very data used in the χ^2 test. In some instances, however, such separation between test data and the hypothesis cannot be achieved. (Examples of this difficulty are just paragraphs away.)

The quantity Q is the *number* of distinct parameters in the statement of the null hypothesis based ***directly*** on the data used in the test. The presumption is that these parameters were estimated from the test-data so as *to minimize the discrepancy index*. (As a practical matter, the maximum-likelihood estimates that we discussed in

Chapter 3 can be used.) Under this definition, Q must be zero if the hypothesis was fully advanced before the test data were examined. That circumstance obtained, of course, in the uniform-birthdays example.

Q is important because parameters in H_0 estimated from a data set tend by their very nature to increase the consistency between the hypothesis and the data. The more the analyst has "peeked" at and used data from the χ^2-test in spelling out his hypothesis, the less surprising it becomes that his hypothesis fits the data well. The χ^2 test recognizes this potential for an artificially-good test result, and tries to correct for it. In essence, the test takes the stance "look at the data if you must, but you're going to pay for it." The price is exacted as follows: More peeking leads to higher values of Q, which lowers the degrees of freedom ψ under (4-23). That reduction leads to lower c-thresholds and thus harder tests to pass.

We need an example to show why it is sometimes necessary to "peek" at χ^2 test data, and what "peeking factor" Q reflects what happened. Here's one.

EXAMPLE 4.11

Defective Work

A computer-chip CEO believes that the chance that a given chip is defective is the same regardless of which of the company's three factories produced it. More formally, she hypothesizes that:

> H_0: The probability a chip is defective is the same regardless of whether it was produced at factory A, B, or C.

But while the CEO assumes a common defect probability p, she does not know what p is.

An assistant goes out and samples 100 chips from the first factory and determines how many are defective. He similarly examines 200 chips from the second factory and 100 from the third, meaning that he examined 400 chips in total.

Discuss a χ^2 test that the CEO might perform of her hypothesis, and discuss why it cannot be performed without "peeking" at the data. How might she overcome the obstacles, and perform the test?

Solution

The CEO might eagerly start to fill out a χ^2 chart in preparation for a a χ^2 test:

Category	e_i	a_i
Defective Chips at Factory A		
Acceptable Chips at A		
Defective at B		
Acceptable at B		
Defective at C		
Acceptable at C		

Note that these six categories meet the requirement that every one of the 400 data points (i.e., chips sampled) will fall in one and only one category. And note that, in covering both condition (defective/acceptable) and location (A/B/C), the categories deal with both factors that are important in testing the hypothesis.

But there is a problem. Not knowing the common defect rate p at the three factories, how can the CEO assign numerical values to the e_i's? And, without e_i's, to what should the a_i's be compared?

To circumvent the difficulty, the CEO might then ask her assistant: Don't tell me where the defective chips were produced, but can you tell me how many defects there were out of the 400? Suppose that the answer she gets is 30. She notes that 30/400 works out to an overall defect rate of 7.5%. Indeed, having read our Chapter 3, she knows that 7.5% is the maximum likelihood estimate of the Bernoulli parameter p. She could then proceed to estimate the common defect rate and reformulate the hypothesis more precisely:

> H_0: The probability a chip is defective 7.5% regardless of whether it was produced at factory A, B, or C.

Now she is in a position to work out the e_i's.

Category	e_i	a_i
Defective Chips at Factory A	$100 \times .075 = 7.5$	
Acceptable Chips at A	$100 \times .925 = 92.5$	
Defective at B	$200 \times .075 = 15$	
Acceptable at B	$200 \times .925 = 185$	
Defective at C	$100 \times .075 = 7.5$	
Acceptable at C	$100 \times .925 = 92.5$	

Now suppose that the assistant reveals (for the first time) that 7 defective items were found at factory 1, 8 at factory 2, and 15 at factory 3. With these three a_i's, both we and the CEO can complete the entire table. (For example, with eight defects at factory 2, we can deduce that 192 of the 200 items sampled there were acceptable.) We can write:

Category	e_i	a_i
Defective Chips at Factory A	$100 \times .075 = 7.5$	7
Acceptable Chips at A	$100 \times .925 = 92.5$	93
Defective at B	$200 \times .075 = 15$	8
Acceptable at B	$200 \times .925 = 185$	192
Defective at C	$100 \times .075 = 7.5$	15
Acceptable at C	$100 \times .925 = 92.5$	85

For this table, the calculated discrepancy index χ^2 follows:

$$\chi^2 = \frac{\sum_{i=1}^{6}\left(e_i - a_i\right)^2}{e_i} = 11.68$$

The only remaining issue is how many degrees of freedom accompany this test. There are obviously $M = 6$ categories. There are *three* puppet categories because we needed to know only know three strategically-placed a_i's (one from each factory) to determine the remaining three. (We know this literally because we just found all six a_i's based on three of them.) In other words, $m_p = 3$.

What about Q? The "peeking factor" Q equals **one** here because, in setting the hypothesis, the CEO used the maximum likelihood estimate of one parameter p (the defect rate), namely, .075, based on the very data about 400 chips. (We do not count the 92.5% "acceptable" rate as a *distinct* parameter, because it is simply 100% – 7.5%.) Note that Q does not depend on the fact that the parameter estimate was .075: it relates solely to the *number* of distinct parameters estimated from the data used in the χ^2-test. That number is one.

The number of degrees of freedom therefore follows: $\psi = M - Mp - Q = m_p - Q = 6 - 3 - 1 = 2$. Because the 95th percentile of the χ^2–distribution with two degrees of freedom is 5.99, it is customary to see the pass/fail threshold c at 5.99. Thus, with $\chi^2 = 11.68$ here, we would *reject* the hypothesis with the data at hand. We would instead infer that the large differences in observed defect rates—which range from 4% at factory B (8/200) to 15% at factory C (15/100)—reflect real differences across the factories rather than mere sampling fluctuations.

If the CEO had not used a numerical value for the common defect rate p, the χ^2-test that was performed could not have gone forward. But what if her estimate of the common defect rate p was based on information other than the test data? Had she estimated p from data not used in the test (for example, the average defect rate for the previous year), then Q would have been zero, and there would have been three degrees of freedom rather than two. With three degrees of freedom, the pass/fail threshold c would have been 7.81 rather than 5.99. The price the CEO paid for "peeking" at the test data to estimate p was therefore about a 25% drop in the pass-fail threshold, which resulted in a more stringent test.

EXERCISE 4.12 Suppose that, instead of finding and using an estimate of p based on the test data in Example 4.11, the CEO had estimated p as .07 because that was the defect rate the previous year. What would change in the χ^2-test above? Would the accept/reject conclusion have changed?

How Many Categories?

We're almost finished with our introduction to the χ^2 test, but we have one more issue to pursue: How many categories should we use in a particular situation? In setting such categories, we face two conflicting pressures:

(1) ***If we limit ourselves to a few very large categories, there is a danger of so much aggregation that strange patterns could wind up getting concealed.***

Suppose, for example, that we wish to test the uniform-birthdays hypothesis with data from 100 randomly-chosen people. If we use the two categories January–June and July–December, then we would expect 50 birthdays to fall in each. Thus, the outcome $a_1 = 48$ and $a_2 = 52$ would easily pass a χ^2 test of the hypothesis. If, however, a_1 was 48 because 48 people were born in January and none in the next five months, a test that ignored that circumstance would look foolish. The test would have combined six monthly results that were all hostile to the uniformity hypothesis in a way that created an illusory harmony with it. In short, we don't want *too few* categories.

(2) ***At the same time, we don't want an excessive number of categories.***

The more categories we have, the more e_i's we must compute, the more a_i's we must count, and the more $(e_i - a_i)^2/e_i$'s we must calculate and add up. In an age of versatile calculators and ubiquitous computers, it might be overly lazy to press this point. But a more cogent consideration is that using the χ^2 table to set pass/fail thresholds in the test depends on some *asymptotic approximations* that are only reliable once the e_i's are sufficiently large. It has been known for many years that these approximations work well when *each e_i is at least 5*. That rule-of-thumb would imply that, given n data points in total, the maximum number of categories should be $n/5$.

In recent years, however, computer simulations have shown that the χ^2 table is highly accurate even when some e_i's fall below 5. For that reason, one need not exclude from a χ^2 test a natural and interesting category with an e_i of (say) 3. Still, we should understand that there is such a thing as *too many* categories, and having numerous e_i's well below 5 suggests that there are too many.

How does it all add up? We might as well start with the "e_i near 5" rule and try, on some plausible basis, to partition the possible outcomes into roughly $n/5$ categories. That a few of these categories have e_i's below 5 should not trouble us. As a practical matter, however, cutting back to (say) a dozen categories is often acceptable, because it is hard to make sense of a χ^2-table that looks interminable. Of course, in some situations (e.g., our cigarette-smokers example above), there are clear limits to the number of relevant categories.

Three Warnings Concerning the χ^2-Test

Our formal discussion of the χ^2-testing procedure is over. But, over the years, the author has noted certain mistakes that students often make when applying the χ^2-*test*. We discuss them now in the hope that readers are thereby "immunized" against these errors. It seems especially worthwhile to do so because some of the issues pertain to hypothesis testing in general.

Warning 1: The Categories the χ^2-Test Must Cover Every Point in the Data Set Used for the Test

What sometimes happens is that people instinctively discard redundant categories in setting up their χ^2-Tables. For example, suppose the hypothesis is that drug A is effective in 80% of the patients with a certain disease. The temptation is to compare the expected number of cures with the actual number, and to work with a one-category test. But that procedure is impermissible as a χ^2-test, because it lacks a category for those who used the drug and did not benefit (a group that, under the hypothesis, will constitute an expected 20% of the sample). True, the second category is redundant: If 86 out of 100 were cured, we can deduce that 14 of 100 were not cured. But the theorems underlying the χ^2-test assume that we will treat the redundant category as a puppet in calculating degrees of freedom rather than exclude it from the test.

Warning 2: Because it is possible to manipulate the χ^2-test result through the choice of categories, using a set of categories that seems strange will evoke suspicion of cheating. It is important, therefore, that the categories seem natural.

Statistical tests can be abused, as can anything else that brings knowledge and/or pleasure. It is not easy to engage in mischief with a straightforward procedure like the one-sample t-test (unless one simply fabricates the data), but the χ^2-test is more vulnerable. If we are to achieve more than a mechanical familiarity with hypothesis testing, we must not flinch from discussing the issue.

Suppose that a desperate analyst urgently wanted to **reject** the uniform birthday hypothesis with the MIT birthday data for 45 individuals (Example 4.9). He might, for example, work for a greeting card company, and hope to propose a change in production schedules that exploits the nonuniformity of birthdays to raise profits. But that scheme—which could result in a promotion—will only be adopted if birthdays are believed nonuniform. Is there any way he could fudge the categories to get a χ^2 result that appears to support his preferences?

With the MIT data in Example 4.9, he might get away with this unsavory scheme if he rearranged the categories as follows:

Category e(*i*) Under H_0	Observed a(*i*)	
Jan/Feb/May/June	15	8
Mar/Apr/Sept/Oct	15	22
Jul/Aug/Nov/Dec	15	15

What he did here was to combine the two periods that had few data points—Jan/Feb and May/June—into one large category that has a big deficit. He likewise hitched Mar/Apr and Sept/Oct together, to get one large category with substantial excess. Then he dumped the remaining months into one final category.

The χ^2 statistic for the data above is 6.53. Because $M = 3$, $M_P = 1$ (and $m_p = 2$) and $Q = 0$ (he's obviously not going to hint that he peeked at the data!), $\psi = 3 - 1 - 0 = 2$ here. For $\psi = 2$, the *c*-value corresponding to $\alpha = .05$ is 5.99. In other words, he could loudly proclaim that H_0 had been *rejected* under usual criteria.

Readers should not suspect that the author is advocating such chicanery. Among other things, it is stupid: even if he fooled the managers into rearranging the production schedule, the company would hardly benefit if the revision were based on meaningless fluctuations. But beyond the obvious point "don't do it," we need recognize that setting up a χ^2 test involves some discretion, which unprincipled individuals could use to get a desired outcome. When we are the recipients of χ^2 test results, therefore, we should inspect the categories before going along with the conclusion. Do they look as if they were chosen in advance, or do they look instead as gerrymandered as an old Massachusetts Congressional district? In the latter case, we have a right to be skeptical and to inquire politely about the choices made.

Warning 3: Don't Play the Percentages

Often, a hypothesis being tested via χ^2 specifies the *percentage* of observations that are expected to fall in each category. But the test must be based on comparing expected and actual *numbers* in various categories, not expected and actual percentages.

EXAMPLE 4.12 **New York Story**

Suppose that it is posited that 60% of readers of *The New Yorker* are college graduates. In a random survey or readers, 50% are college graduates.

(i) Illustrate an incorrect way to set up a χ^2-test of the 60% hypothesis based on the survey result.

(ii) Does the size of the random sample matter in conducting the χ^2-test properly? Might the accept/reject conclusion given the 50% result be different for some sample sizes than for others?
(iii) Discuss intuitively why working with percentages makes little sense.
(iv) For what sample sizes would the 50% result lead to acceptance of the 60% hypothesis in a χ^2-test with $\alpha = .05$, and for what sample sizes would it lead to rejection?

Solution

(i) Some people using χ^2 would compare 60% to the actual percentage obtained in a random sample. Thus if 50% of the readers in a random sample had graduated college, these people would set up χ^2-table as follows:

Category	Expected	Actual
College Graduate	60%	50%
Not College Graduate	40%	50%

They would then compute the discrepancy score as $(60-50)^2/60 + (40-50)^2/40 = 4.17$, and would reject based on one degree of freedom (and a pass/fail threshold of $c = 3.84$).
This exercise is *not* a legitimate χ^2-test. The test we have been discussing requires a comparison between expected and actual *numbers* in various categories, not percentages. The usual pass/fail thresholds are of no value when percentages are compared.

(ii) It is instructive, however, to explore why the χ^2-test will not play with percentages. Consider two possibilities: (i) a random sample of 20 readers was questioned in the test, and (ii) a random sample of 1000 was questioned. In both cases, suppose that 50% of the respondents were college graduates (as opposed to the hypothesized 60%).
If we work out the appropriate tables in the two cases, we reach:

	$n = 20$		$n = 1000$	
Category	e_i	a_i	e_i	a_i
College Grad	12	10	600	500
Not Grad	8	10	400	500
	$\chi^2 = 0.83$		$\chi^2 = 41.67$	

With one degree of freedom, the 60% hypothesis is easily accepted in the first case and emphatically rejected in the second.

(iii) The χ^2-test is making a sensible distinction here. When only 20 readers are questioned, random sampling error could easily explain the difference between the expected twelve graduates and the actual ten. Thus, a discrepancy of ten percentage points is not troubling. In a sample of 1000 readers, however, it would be extraordinary to find only 500 graduates if the 60% rate were accurate (and 600 were expected in the sample). In this instance, the discrepancy of ten percentage points is far greater than can plausibly be attributed to sampling error. The comparison 60% vs. 50% applies in both cases, but what to make of the comparison depends heavily on the *size* of the sample. Without making a big deal about it, the χ^2-test quietly takes account of sample size n. (Actually, that is why the test uses $(e_i - a_i)^2/e_i$ in the discrepancy index rather than $|e_i - a_i|/e_i$, where $|b|$ is the absolute value of b. The latter ratio is just percentage error.)

(iv) Let n be the sample size, and let e_c and a_c be the expected and actual numbers of college grads. Suppose that $e_c = .6n$ while $a_c = .5n$, corresponding to 60% expected and 50% actual. Then the χ^2-table for the *New Yorker* hypothesis would be:

Category	Expected Number	Actual Number
College Grad	$.6n$	$.5n$
Not College Grad	$.4n$	$.5n$

The discrepancy score χ^2 would then follow:

$$\chi^2 = \frac{\sum_{i=1}^{2}\left(e_i - a_i\right)^2}{e_i} = \frac{(.6n - .5n)^2}{.6n} + \frac{(.4n - .5n)^2}{.4n}$$

which simplifies to: $\chi^2 = .0417n$

In other words, the discrepancy score arising from the ten-percentage-point discrepancy is proportional to n, the size of the random sample. The higher the value of n, the more unhappy the test is with the difference between 60% and 50%. If we set $.0417n = 3.84$ (the pass–fail threshold at $\alpha = .05$ for one degree of freedom), we get $n = 92$. This outcome means that the hypothesis would be accepted despite a 60%/50% difference for n-values up to 92, and rejected for larger values.

The difficulty with a χ^2-test that works only with percentages is that it pays no attention whatever to the value of n, which is irrational. Don't let that happen to you!

We end this chapter with the example about *in vitro* fertilization that started it. If you understand the example, you have mastered all you were intended to learn about the χ^2-test.

EXAMPLE 4.13 In Vitro Fertilization

In the prologue to this chapter, we discussed a conversation between a doctor and an engineer that sets the stage for the following analysis. (That conversation actually did take place.) A Yale study of couples who underwent *in vitro* fertilization by Professor Edward Kaplan* and colleagues yielded the following data:

Trial Number (k)	Number Who Went Through Trial k	Number of Pregnancies on Trial k	Number of Exits After Failure on Trial k
1	571	75	158
2	338	36	129
3	173	12	68
4	93	4	48
5	41	2	18
6	21	0	11
7	10	0	4
8	6	0	2
9	4	0	4

SOURCE: Kaplan et al (1992) In all cases, women who got pregnant in trial k ceased further participation in the trials. "Exits after Failure on Trial k" are couples who failed a kth time and decided not to go on to a k + 1st attempt.

Two questions:

(i) Suppose that p is the probability of pregnancy in one trial of *in vitro* fertilization. Assuming that p is the same for *all* couples (and is unrelated to the number of unsuccessful trials a couple went through previously), what is a sensible estimate of p given the Yale data?

(ii) What do the data imply about the hypothesis H_0 that the probability that a trial of *in vitro* fertilization leads to pregnancy is the same for all couples and is independent of the trial number (x)?

Solution

As noted, the suspicion here is that the chance of success in *in vitro* fertilization goes down as the number of unsuccessful attempts increases. If so, then those who have experienced several disappointments face worse odds the next time. The reason the probability might go down is that the population might be heterogeneous, with some couples more likely than others to succeed with the technique. If so, the more successful couples might tend to achieve pregnancy early (and then "drop

* E. Kaplan, A. Hershlag, A. DeCherney,and G. Lavy (1992) "To Be or Not to Be? That is Conception: Managing *in vitro* Fertilization Programs," *Management Science* 38(9), Sept. 1992, pp.1217–1229

out" from the process), meaning that those who go on to later attempts are those less likely to succeed.

We should make sure we understand the table above. The first row tells us that 571 couples attempted in vitro fertilization for the first time at Yale New Haven Medical Center. Of these, 75 achieved pregnancy, while 158 of those who did not do so decided not to try a second time. Those numbers mean that 571 – 75 – 158 = 338 couples make a second attempt. Of these, 36 achieved pregnancy while 129 couples stopped without success after two attempts.

(i) If the probability p of success is independent of trial number (hypothesis H_0), then a sensible way to estimate p would be to divide the total number of successes by the total number of attempts. (That process yields the maximum-likelihood estimate of p.) The total number of successes is obtained by adding the numbers in the third column of the table, which yields 75 + 36 + 12 + 4 + 2 + 0 = 129. The total number of attempts is the sum of entries in the second column, which is 571 + 338 + 173 + 93 + 41 + 21 + 10 + 6 + 4 = 1257. We thus make the approximation: $p \approx 129/1257 = .103$. In other words, the chance of success is about 1 in 10 based on aggregating all the data.

(ii) To test the hypothesis H_0 that the probability of success is independent of trial number, we might posit that, for any attempt on any trial number, there is a 10.3% chance of success. But then we could perform a χ^2 test on the data, in which the categories are based on trial numbers. If there were n_k attempts at trial number k, then the expected number of successes based on the 10.3% rate would be $.103n_k$. The expected number of failures at the kth trial would be $(1 - .103)n_k = .897n_k$.

(iii) If $.103n_k$ is well below 5—meaning that n_k is well below 50—we have too few expected successes to allow an appropriate category in a χ^2 test. Noting that the n_k's for all $k > 5$ are far below the necessary level, we do well to combine several k-values above 5 to create the category "trial numbers at least six" and get a total of 41 attempts. At a 10.3% rate, we would expect 41 * .103 = 4.22 successes, which is close enough to five that we needn't worry about the shortfall.

We could then create a χ^2 table, noting as always that *every* data point must have a category in which it can fall. Thus, we cannot simply have the category "success on first trial;" we also need the complementary category "failed on first trial." The table would look as follows:

Category	Expected Number Under Hypothesis	Actual Number
Success on First Trial	571 * .103 = 58.8	75
Failure on First Trial	571 * (1 – .103) = 512.2	496
Success on Second Trial	338 * .103 = 34.8	36
Failure on Second Trial	338 * .897 = 303.2	302

Category	Expected Number Under Hypothesis	Actual Number
Success on Third Trial	$173 * .103 = 17.8$	12
Failure on Third Trial	155.2	161
Success on Fourth Trial	$93 * .103 = 9.6$	4
Failure on Fourth Trial	83.4	89
Success on Fifth Trial	$41 * .103 = 4.2$	2
Failure on Fifth Trial	36.8	39
Success on 6+ Trial	$41 * .103 = 4.2$	0
Failure on 6+ Trial	36.8	41

The discrepancy score $\chi^2 = \frac{\sum_{i=1}^{12}(e_i - a_i)^2}{e_i} = 16.73$ for these data.

How many degrees of freedom are there in this test? There are a total of $M = 12$ categories. In considering puppet categories, we note that 571 couples in total went through a first trial of in vitro fertilization, so we only need one of the a_i's for the first two categories (succeed on first trial/ fail on first trial) to determine the other. The same is true for all other pairs of success and failure at a given trial number. Thus, we only need *six* a_i's strategically placed to determine the other six. It follows that $m_p = 6$, and we can then use (4-23) to write $\psi = m_p - Q = 6 - Q$.

What about the factor Q? Here we estimated one distinct parameter in the statement of H_0 from the very data used in the χ^2 test: the success rate under H_0 of 10.3% on any trial (Again, the parameter .897 for the failure rate per trial is not a distinct parameter; it is simply $1 - .103$.) Thus, $Q = 1$.

In summary, the test has $\psi = m_p - Q = 6 - 1 = 5$ degrees of freedom. The pass/fail threshold c for $\alpha = .05$ is 11.07, so we would *reject* H_0 under usual standards.

Such rejection is not shocking if we consider how the actual success rate varied with trial number:

Trial Number	Actual Success Rate
1	13.1% (75/571)
2	10.7%
3	6.9%
4	4.3%
5	4.9%
6+	0%

Clearly, the likelihood of success is dropping appreciably as trial number increases. (The lone reversal of the decline at trial number 5 is based on a small sample: two successes in 41 attempts.) What the χ^2 test adds to this chart is forceful evidence that the observed decline cannot be dismissed as a coincidence.

EXERCISE 4.14

(i) Might the χ^2-test also reject H_0 if, contrary to suspicions, the chance of success increased dramatically on later attempts?

(ii) If you answered yes, propose a set of 129 successes out of 1257 attempts consistent with a success rate that increases with k and which would cause H_0 to fail the χ^2-test. For each trial k, assume that the sum of successes and dropouts is the same as in Example 4.13. Thus, there are still 338 couples who go on to trial 2.

(iii) If you answered no, repent for your error by completing part (ii).

Final Remarks

We have now discussed five different statistical tests:

The One-Sample t-Test

The Two-Sample t-Test

The Differences of Means Test For Moderate Samples

Fisher's Exact Test

The Pearson χ^2-Test

Although these tests differ in their details, they have the same general structure, which consists of:

- The choice of a *test statistic* that can be calculated from the test data
- Determination of the probability distribution of the test statistic *assuming that* the hypothesis being tested is true
- Choice of a rejection region for the hypothesis, which is typically set so that the probability of accidental rejection of the hypothesis (α)is 5%

The rejection region is typically situated at:

The highest 5% of test statistic's distribution under the hypothesis

OR The lowest 5% of test statistic's distribution under the hypothesis

OR The lowest 2.5% and the highest 2.5% of test statistic's distribution under the hypothesis

Which of these tests to use depends on the kind of hypothesis being tested. If we want to investigate whether two distribution have the same mean, it makes sense to perform a t-test when data points are few and a general difference of means test when data are more plentiful. Fisher's Exact Test is relevant when, under the null hypothesis, key data points are distributed between two groupings according to a hypergeometric distribution. A χ^2-test is useful if we can create categories in which the various data points must fall, and know the expected number of points in each category if the null hypothesis is true.

In subsequent chapters, we will encounter several other hypothesis tests. They too follow the general structure above.

Yet this introduction to hypothesis testing leaves several questions unanswered:

- Can we say something about the risk of accidentally accepting the hypothesis (β) beyond forever proclaiming that it can't be calculated?
- Can we go beyond the simple "accept/reject" dichotomy in discussing the compatibility between data and hypothesis? If a hypothesis is rejected, for example, did it just barely fail the test or did it fail colossally?
- We have repeatedly said that accepting a hypothesis means not that it is definitely true, but rather that it is among the hypotheses that are consistent with the available data. But how wide is the range of hypotheses that are consistent with a particular data set?

We can make considerable progress in dealing with all of these questions, but we postpone doing so until the next chapter. For now, we pursue a final series of exercises that can help us review the material presented thus far and consolidate our current gains.

The Takeaway Bar

Glossary

Hypothesis testing is a process that explores the consistency of a given hypothesis with a particular data set.

Classical hypothesis testing emphasizes the conditional probability of observing a result like the one that shows up in the test data, assuming that the hypothesis being tested is true.

The **rejection region** of a test of hypothesis H_0 is the set of outcomes in the test data that would lead to the rejection of H_0. (Ideally, the region is specified before the test data have been examined.)

The **acceptance region** of a test of hypothesis H_0 is the set of outcomes in the test data that would lead to the acceptance of H_0. (Same "ideally" as above.)

Type I Error in a test of hypothesis H_0 is rejecting H_0 under the test rules even though H_0 is true.

Type II Error in a test of hypothesis H_0 is accepting H_0 under the test rules even though H_0 is false.

The **test statistic** is a number meant to reflect the compatibility between hypothesis H_0 and the data used to test H_0

A **one sample t-test** explores whether a random sample from a normal distribution is consistent with a hypothesized value for the mean of that distribution.

A **two-sample t-test** uses random samples from two normal distributions to investigate whether the means of these distributions are the same.

A **large sample difference-of-means test** uses random samples from two distributions to investigate whether two distributions have the same mean. It invokes the Central Limit Theorem, and does not assume normality for the underlying distributions being compared.

Fisher's Exact Test explores whether data points having a certain characteristic are randomly distributed between two groups that together comprise the full data set.

A **chi-square (or χ^2) test** of hypothesis H_0 examines how the spread of actual data across various test categories compares with the spread that would be expected if the hypothesis were true.

The **index of discrepancy** in a chi-square test is a numerical measure of deviation between the spread of the actual data across the test categories and the spread that was expected under the hypothesis being tested.

Degrees of freedom in a chi-square test reveals which particular χ^2-distribution the index of discrepancy would follow if the hypothesis being tested is true. Degrees of freedom depend on the number of categories in the test, the number of redundant (or "puppet") categories, and the number of parameters in the statement of the hypothesis that were estimated from the very data used in the chi-square test.

Formulas

Type I Error Risk: $\alpha = P(\text{reject } H_0 | H_0 \text{ true})$
Type II Error Risk: $\beta = P(\text{accept } H_0 | H_0 \text{ false})$

One Sample t-test:

Hypothesis H_0: $\mu = d$ for some particular value d
Test Statistic:

$$t_{n-1} = \left(\bar{X} - \mu\right) / \left(S/\sqrt{n}\right)$$

where $\bar{X}$ is sample mean, S is sample standard deviation, and n is sample size
Distribution of Test Statistic if H_0 True: t-distribution $n - 1$ degrees of freedom
(Underlying distribution of X assumed normal)

Two Sample t-test:

Hypothesis H_0: $\mu_X = \mu_V$ for two random variables X and V
Assuming Equal Variances $\sigma_X^2 = \sigma_V^2$
Test Statistic:

$$t_{XV} = \frac{\bar{X} - \bar{V}}{S\sqrt{\left(\frac{1}{n_x} + \frac{1}{n_v}\right)}}$$

where $\bar{X}$ and $\bar{V}$ are sample means, S is the estimated common standard deviation, and n_X and n_V are sample sizes
Distribution of Test Statistic if H_0 true: t-distribution $n_X + n_V$ degrees of freedom

Unequal Variances

Test Statistic:

$$t_{XV} = \frac{\bar{X} - \bar{V}}{\sqrt{\left(\frac{S_X^2}{n_X} + \frac{S_V^2}{n_V}\right)}}$$

where S_X^2 and S_V^2 are estimates of, respectively, σ_X^2 and σ_V^2

Distribution of Test Statistic if H_0 true: t-distribution, ψ degrees of freedom with ψ given by (4-12)

(Underlying distributions of X and V assumed normal in these tests)

Large Sample Difference of Means Test:

Hypothesis H_0: $\mu_X = \mu_V$ for two random variables X and V
Test Statistic: $\bar{X} - \bar{V}$
Approximate Distribution of Test Statistic if H_0 true:

$$N\left(0, \sqrt{\left(\frac{S_X^2}{n_X} + \frac{S_V^2}{n_V}\right)}\right)$$

$\bar{X}$ and $\bar{V}$ are sample means, S_X^2 and S_V^2 are estimates of, respectively, σ_X^2 andσ_V^2, and n_X and n_V are sample sizes

(No assumption about X or V distributions if n_X and $n_V \geq 30$, but normal approximation generally works well so long as n_X and $n_V \geq 15$)

Fisher's Exact Test:

Hypothesis H_0: j items are randomly distributed between two groups
Test Statistic: X = number of items in first group
Distribution of Test Statistic if H_0 true: Hypergeometric with parameters (j,m,n)

$$P\left(X = k \middle| H_0 \text{ true}\right) = \frac{\binom{m}{k}\binom{n}{j-k}}{\binom{m+n}{j}} \text{ for } 0 \leq k \leq \min(j,m)$$

where m = size of first group and n = size of second group

Chi-Square Test:

(Usual) Hypothesis H_0: For N data points and M Test Categories:

- P(given data point falls in category $i | H_0$ true) = p_i
- Categories of different data points are independent

Under H_0, Number of data points in category i is binomial with mean $e_i = Np_i$
Test Statistic:

$$\chi^2 = \frac{\sum_{i=1}^{M}\left(e_i - a_i\right)^2}{e_i}$$

where a_i = actual number of data points in category i

Approximate Distribution of χ^2-Test Statistic if H_0 true:

$$\chi^2 \text{ with } m_p - Q \text{ degrees of freedom}$$

where m_p = minimum number of a_i's needed to determine all of them
Q = number of parameters in H_0 estimated from data used in the χ^2-test
(Generally favor categories with $e_i \geq 5$, though modest exceptions are not problematic)

Some Parting Thoughts

- Again, accepting a hypothesis does not mean that we are sure it is true. Rather, it means that the hypothesis is reasonably consistent with available data. It could be that other hypotheses that are very different are also consistent with the data.
- Rejecting a hypothesis means that available data seem inconsistent with it. That circumstance does not mean the hypothesis must be false; perhaps "bad luck" with a random data sample is the reason for the inconsistency. Also, the fact that a hypothesis is not literally true does not mean that it is not approximately true. Sometimes, the imperfections in a hypothesis are of *little practical importance* even if it (correctly) fails a statistical test.
- There is some flexibility in setting up χ^2 and other statistical tests, and that discretion can create opportunities for mischief (e.g., on setting the categories after looking at the data, to achieve a desired test result). For that reason, the statistical tests we perform might invite suspicion if they appear unusual without a plausible explanation.
- Please recall that a χ^2 test is based on comparing expected and actual *numbers* of data points in various categories, not expected and actual percentages of data points.
- The justification of the χ^2-test involves a deep theoretical analysis. We have focused primarily on how to perform a χ^2-test successfully, and on some intuitive discussion about why the procedure makes sense.

EZ Pass

1. Suppose that α, the chance of Type I error, is .02 in a particular hypothesis test. Then:
   (A) If the hypothesis is true, there is a 2% of accidentally rejecting it under the test rules.
   (B) If the hypothesis is false, there is a 2% of accidentally accepting it under the test rules.
   (C) Whether the hypothesis is true or false, the chance of reaching an incorrect conclusion in the test is 2%.
   (D) The chance of accidentally rejecting the hypothesis exceeds by .02 the chance of accidentally accepting it.

2. An example of a hypothesis test with $\beta = 1$ is:
   (A) One in which the data contain inaccuracies
   (B) One in which the hypothesis is always rejected
   (C) One in which the hypothesis is always accepted
   (D) One in which α must exceed .05
3. In Fisher's Exact Test, the test statistic assuming the hypothesis is true follows:
   (A) a binomial distribution
   (B) a geometric distribution
   (C) a hypergeometric distribution
   (D) a Poisson distribution
4. The underlying data used in a hypothesis test are assumed to come from a normal distribution in:
   (A) The Difference of Means Test for Large Samples
   (B) Fisher's Exact Test
   (C) The Pearson χ^2-Test
   (D) None of these
5. In a χ^2-test with k degrees of freedom, the pass–fail threshold c for $\alpha = .02$:
   (A) falls between the thresholds for $\alpha = .01$ and $\alpha = .05$ for all values of k
   (B) falls between the thresholds for $\alpha = .01$ and $\alpha = .05$ for some values of k
   (C) falls between the thresholds for $\alpha = .01$ and $\alpha = .05$ for no values of k
   (D) none of these
6. If a hypothesis is rejected in a χ^2-test in which the discrepancy score is 9.99 and $\alpha = .05$, then the number of degrees of freedom in the test:
   (A) cannot exceed four
   (B) must equal four
   (C) must exceed four
   (D) all of these

Further Exercises

EXERCISE 4.15 A noted legal scholar once said that, in courtroom verdicts, it is better that 100 guilty people should go free than that one innocent person should be convicted. Suppose that each trial starts with the null hypothesis H_0 that the person on trial is *innocent* and that, through its verdict, the jury either accepts or rejects H_0. Of Type I and Type II errors in hypothesis testing, which does the scholar consider the more serious in this context? Explain.

EXERCISE 4.16 Suppose that a test for a disease has a false positive rate of 7% and a false negative rate of 9%. The hypothesis H_0 is that the patient has the disease, and it will be accepted if the test result is positive and rejected if the test result is negative.

(i) What are the error probabilities α and β associated with this test procedure?

Suppose that, prior to the test, the doctor believes that the patient has an 80% chance of suffering the disease. Taking into account the error probabilities of the test:

(ii) What would his revised probability be that the patient has the disease if the test comes out positive? (Hint: Bayes' Theorem)

(iii) What would his revised probability be that the patient has the disease if the test comes out negative?

(iv) Do you see why some people believe that consideration of α and β might be inadequate in assessing the test results? Discuss briefly.

EXERCISE 4.17 (Adapted from *Real Statistics Using Excel*)

"A weight reduction program claims to be effective in treating obesity. To test this claim 12 people were put on the program and the number of pounds of weight gain/loss was recorded for each person after two years, as shown in the columns below. Can we conclude that the program is effective?

Subject	Weight Loss
1	23
2	15
3	–5
4	7
5	1
6	–10
7	12
8	–8
9	20
10	8
11	–2
12	–5

(Negative weight loss means weight gain.)

A common approach to dealing with this question is to test the hypothesis H_0: $\mu = 0$ which would imply that the program had no net effect. Only if that hypothesis is rejected because $\bar{X}$ exceeds zero by a sufficient amount would the program be declared effective.

(i) Assuming that the 12 people studied are a random sample from the broader obese population and that weight changes for those on the program follow a normal distribution, argue that an appropriate test for H_0 takes the form:

$$\begin{cases} \text{Accept } H_0 \text{ if } \bar{X} > c \\ \text{Reject } H_0 \text{ if } \bar{X} \le c \end{cases} \quad \text{for some } c > 0$$

(ii) For what value of c is $\alpha = .05$? (*Hint:* What is the 95th percentile of the t-distribution that pertains to this situation if H_0 is true? Given S and n, what is the 95th percentile of $\bar{X} - 0 = \bar{X}$?

(iii) What is the accept/reject outcome of the test with these data?

EXERCISE 4.18 (Adapted from *IBM SPSS Data Collection Library 7*)

During a given period, 128 people visit a museum which has a dinosaur exhibit and a human biology exhibit. 32 of these people majored in Biology in college, while the other 96 did not. 79 people out of the 128 preferred the dinosaur exhibit, while 49 preferred the display on human biology.

(i) Treating these 128 people as random samples from the two broader populations (Biology majors and others), formulate a Fisher's Exact Test of the hypothesis H_0 that Biology majors have the same preference split between the dinosaur and human exhibits as other people. Choose as test thresholds approximately the 2.5th and 97.5th percentiles of the distribution of X, the number of Biology majors who prefer the dinosaur exhibit if H_0 is true. (Hint: Use a hypergeometric calculator.)

(ii) Now perform the test with the information that, among the Biology majors, 14 preferred the dinosaurs and 18 preferred the humans.

EXERCISE 4.19 To assess whether a coin is fair, it will be tossed twelve times. Then a χ^2-test will be performed on the results.

(i) Set up the test. How many degrees of freedom does it have?

(ii) What pass/fail threshold c will achieve approximately a 5% chance of accidental rejection?

(iii) What numerical values could the χ^2 discrepancy score possibly take on in this test? (*Hint:* There are seven of them, one possibility being $\chi^2 = 0$.)

(iv) What is the probability that the score will take on each of these values, assuming that the coin is in fact fair? (*Hint:* Think binomially.)

(v) What is the mean value for the discrepancy score if the coin is fair?

(vi) If a hypothesis is correct, the mean for the discrepancy score in a χ^2 test should be approximately equal to the number of degrees of freedom in the test. How accurate is that approximation in this case?

EXERCISE 4.20 When 1510 randomly-chosen Americans killed in the Vietnam War were partitioned based on which decile of the national income distribution their families were from, the results were as follows:

Decile of Family Income	Percent of Fatalities
Lowest (tenth)	9.6 (%)
Second Lowest	13.1
Third Lowest	11.2
Fourth Lowest	11.4
Fifth Lowest	9.4
Fifth Highest	10.1
Fourth Highest	10.2
Third Highest	9.2
Second Highest	8.0
Highest	7.8

SOURCE: A. Barnett, T. Stanley, and M. Shore, "America's Vietnam Casualties: Victims of a Class War?" *Operations Research*, 40(5), pp. 856-866, Sept/Oct 1992.

Using these data and an appropriate χ^2-test, investigate the hypothesis that, over the entire US population, the per capita death rate in Vietnam was the same for all ten deciles of the distribution of family income. (Assume that all deciles had equal numbers of military-age men.)
(HINT: Remember what two quantities get compared in a χ^2-test. Do we compare the expected and actual *percentages* of data points in category i, or something else?)

EXERCISE 4.21 On April 14, 1912, HMS *Titanic* struck an iceberg in one of the most famous disasters in all history. The order of entry into the lifeboats was supposedly "women and children first," but it has been suggested that class was also relevant to the chance of surviving the calamity.

Some facts:

- Of the 511 women and children (hereafter W/C) on board, 353 were saved.
- There were 150 first-class W/C on board
- There were 117 second-class W/C on board
- There were 244 third-class W/C on board

(i) What is a sensible estimate of the overall probability of survival for W/C on the Titanic?

(ii) Construct a χ^2-test of the hypothesis H_0 that all W/C regardless of class has the same chance of survival. Identify the categories of your test, the expected numbers of data points in each, the number of degrees of freedom in the test, and the pass/fail threshold.

(iii) The actual numbers of W/C survivors were:

146 in first-class

104 in second-class

103 in third-class

Now, perform the χ^2-test described in (ii). What is the outcome?

EXERCISE 4.22 Mendel plays Sudoku and golf every day, and it is believed that his two performances on a given day are unrelated. More specifically, it is hypothesized that that the two events A and B below are independent:

A: Mendel is above his median performance in Sudoku

B: Mendel is above his median performance in golf

To test this hypothesis, Mendel's performances in each sport (time to complete in Sudoku, 18-hole score in golf) are recorded for 80 days. The performance on any day is then classified as above/below his 80-day median for golf, and also for Sudoku. (Assume that he is never exactly at the median.)

(i) Construct a (general) χ^2-test of this independence hypothesis, specifying the categories, expected number in each, and degrees of freedom for the test. (Neglect issues like day-to-day variation in the difficulty of Sudoku.)

(ii) Now suppose that we learn that Mendel was above the median in both golf and Sudoku on 22 of the 80 days. Perform the χ^2-test outlined above. (Yes: you have enough information.)

(iii) *Now* how many degrees of freedom do you think there are? (What is implied by what you did in (ii)?)

CHAPTER 5

Further Concepts in Hypothesis Testing

Alleged Age Discrimination (Based on a true story)

The older workers at the high-tech factory were grim and angry. Their employer had laid off five of the 33 workers in their division, and all of them were old. They told the lawyer that the pattern was not a coincidence and that they wanted to file an age-discrimination suit against the company. The lawyer asked them a few questions, including whether they had any non-statistical reason to believe that the company had focused its layoffs on older workers. When they reluctantly said no, he said that he was terribly sorry but they had no case.

(To be continued)

Here we focus less on expanding the set of statistical tests that we know than on exploring some new ways to think about hypothesis testing. We discuss the important concepts of statistical power and p-value, and consider when to take seriously the fact that a given hypothesis has passed a statistical test. We also ponder how to proceed when the hypothesis is subjected not to one statistical test but to a barrage of several. This is a brief chapter, but an important one.

5.1 Rejection Regions in Hypothesis Testing

In a χ^2-tests, the null hypothesis H_0 is rejected if the discrepancy score is unusually high. (i.e., at the right tail of its distribution under H_0). But in the one-sample t-test we discussed, the null hypothesis is rejected if the sample mean is either well above or well below the proposed mean μ (i.e., at either the left tail or right tail of its distribution under H_0). Why the difference, and what does the difference tell us about rejection regions in hypothesis tests?

We discussed the issue briefly in connection with Example 4.8 about the "manhole" hypothesis, but let's probe a bit more now. When we test a null hypothesis H_0, we need consider the question: If H_0 is false, what alternative hypothesis H_a do we believe would be true? H_a is usually not a specific alternative but rather a general one. In setting the rejection region for H_0, we focus on those outcomes which would be unusual under H_0 but *less so* under H_a. Those are the outcomes that do the most to reduce our confidence in H_0.

For example, for the "manhole" hypothesis H_0—which posits that the Rudin manhole rankings are no better than random guesses in identifying dangerous manholes—the alternative H_a of interest is that the rankings outperform random guesses as guides to dangerous manholes. If the test statistic X—the number of manholes among those ranked most dangerous which suffered difficulties—is considerably *higher* than would have been expected by sheer chance, then H_0 looks less plausible and H_a more so. The rejection region corresponding to this assessment is at the right tail of the distribution of X under H_0.

However, this example quietly implies that, even if the test statistic takes a highly unusual value under the null hypothesis, that circumstance would *not* argue for rejecting H_0 *if it is even more unusual to get that value under the alternative* H_a. For the Manhattan manholes, let X be the number of manholes out of the 44 that suffered fires/explosions that had been ranked in the riskiest 5000 (out of 51,219) by Rudin and her colleagues. What would be the proper accept/reject decision for H_0 had X turned out to be zero? Based on 44 random guesses about which manholes would have trouble, one would expect about 9.8% of the troubled manholes (5000/51,213) to be among the 5000 deemed riskiest, which works out to about four manholes out of 44. Getting $X = 0$ is surprising even under random guessing: $P(X = 0) = \dfrac{\binom{1000}{0}\binom{50{,}213}{44}}{\binom{51{,}213}{44}} = .016.$

But even though $X = 0$ is well out on the left tail of the "random guesses" distribution, it would be very weird to cite that outcome as evidence that the forecasts are better than random guesses (H_a). Because $X = 0$ damages H_a even more than H_0, it would not be in the rejection region for H_0.

Two important definitions:

- When H_0 is rejected if the test statistic is at one tail of its distribution under H_0 but not rejected at the other, the statistical test is said to be **one-sided**.
- When rejection occurs for outcomes out on either tail of the distribution, the test is said to be **two-sided**.

EXAMPLE 5.1 **To Repeat**

It is hypothesized that 64% of the consumers who try a certain product for the first time will purchase it again within one year. To assess the soundness of that "repeat rate" estimate, a random sample of 250 people who tried the product in a given month will be questioned one year later. The test statistic X is the number of those questioned who went on to a second purchase.

(i) Why would we expect that, under H_0, X would have a binomial distribution with parameters 250 and 0.64?

(ii) Set up a rejection region for H_0 with $\alpha \approx .05$ for each of the following generalized alternatives (H_a) to H_0:

(a) The percentage who try the product again is greater than 64%

(b) The percentage who try the product again is different from 64%.

Solution

(i) If p is the fraction of people who try the product who will buy it again within one year, then a randomly-chosen individual who tried the product will have probability p of a repeat purchase within a year and probability $1 - p$ of no such purchase. Different individuals selected at random can be assumed to decide independently. For these reasons and hypothesis H_0, random variable X—the number of the 250 individuals canvassed who report a further purchase within a year—would follow a binomial distribution with $n = 250$ trials and $p = .64$.

(iia) If H_a follows H_a: $p > .64$, then the outcomes that favor H_a over H_0 are X-values that are unusually high under H_0. Outcomes that are unusually low under H_0 favor H_0 over H_a. The rejection region should therefore focus on the upper tail of B(250, .64), the distribution of X under H_0. To achieve $\alpha = .05$, we would set the rejection region as all X-values above the 95th percentile of $B(250, .64)$. These are the 5% of outcomes, after all, that hurt H_0 the most in comparison with H_a.

A short exercise of trial-and-error with a binomial table shows that the test of the form:

$$\begin{cases} \text{Accept } H_0 \text{ if } X \le 172 \\ \text{Reject } H_0 \text{ if } X > 172 \end{cases}$$

has an α-value of .051, which is the closest we can get to $\alpha = .05$ given that X has a discrete distribution.

(iib) Here X-values that are unusually high or unusually low work against H_0 in comparison with H_a. Thus, it makes sense to reject when X is either at the upper or lower tail of $B(250, .64)$. To achieve $\alpha = .05$, the two-sided rejection region should cover essentially the bottom 2.5% and the top 2.5% of $B(250, .64)$. We learn from the binomial table that the test should be:

$$\begin{cases} \text{Accept } H_0 \text{ if } 146 \le X \le 174 \\ \text{Reject } H_0 \text{ if } X < 146 \text{ or } X > 174 \end{cases}$$

EXERCISE 5.1 Suppose that Mendel decides to test H_0 in Example 5.1 using a χ^2 procedure.

(i) Formulate the test with $\alpha = .05$, specifying the categories and the pass/fail threshold c.

(ii) Which of the two tests in part (ii) of Example 5.1 is most similar to your χ^2-test?

(iii) "The χ^2-test is better for comparing H_0 to the generalized alternative H_a: H_0 is false than to an alternative that specifies why H_0 is false."
Do you agree with this statement? Briefly discuss.

The ideas above will come to the fore again when we discuss p-values later in this chapter. But first we further explore the notion of alternatives to H_0 should it be false.

5.2 Some Power Points

In Chapter 4, we repeated to the point of monotony that we can typically calculate α, the risk of accidentally rejecting a hypothesis in a statistical test, but can very rarely calculate β, the risk of accidental acceptance. For that reason, test rules generally focus on keeping α at some low value, most often at 5%. But there is another approach to dealing with β, the "devil we don't know." Suppose we make the assumption that, if H_0 is false, some *very specific* alternative H_1 must be true. In that circumstance, calculating β is no more difficult than calculating α, as a simple example illustrates.

EXAMPLE 5.2 **Fair Enough?**

Suppose that it is desired to test H_0: *A coin is fair* against the alternative H_1: *the coin is twice as likely to come up heads as tails.* (It is assumed that either H_0 or H_1 must be true.) The coin will be tossed ten times and the test will be based on X, the number of heads over the ten tosses. The test rule is:

$$\begin{cases} \text{Accept } H_0 \text{ if } X \le 7 \\ \text{Reject } H_0 \text{ if } X \ge 8 \end{cases}$$

(i) What is the risk of accidentally rejecting of H_0 for this test?

(ii) What is the risk of accidentally accepting H_0?

Solution

Here we need recognize that, under H_0, random variable X is $B\left(10,\frac{1}{2}\right)$ while, under H_1, X is $B\left(10,\frac{2}{3}\right)$. We can then write:

(i) $\alpha = P\left(\text{rej } H_0 \middle| H_0 \text{ true}\right) = P\left(X \geq 8 \middle| X \text{ is } B\left(10, \frac{1}{2}\right)\right) = .055$ (from table)

(ii) $\beta = P\left(\text{acc } H_0 \middle| H_1 \text{ true}\right) = P\left(X \leq 7 \middle| X \text{ is } B\left(10, \frac{2}{3}\right)\right) = .700$

These answers are interesting. We see that is as easy to calculate β as α under this "either H_0 or H_1, nothing else" rule. We also see that, while α is close to the familiar "comfort level" of 5%, β is an unnerving 70%. This last outcome should not really surprise us: if X is $B\left(10, \frac{2}{3}\right)$, then the mean of X is 10*(2/3) = 6.67, so the range from 0 to 7 should "absorb" a large share of the probability about where X will fall.

EXERCISE 5.2. Suppose that the test of H_0 vs. H_1 in Example 5.2 will be conducted on the basis of 20 tosses, with the test rule:

$$\begin{cases} \text{Accept } H_0 \text{ if } X \leq 14 \\ \text{Reject } H_0 \text{ if } X \geq 15 \end{cases}$$

where X = number of heads over the 20 tosses

(i) Without calculations, do you think α under this test will be the same as, greater than, or less than α in Example 5.2? Briefly discuss your reasoning.
(ii) Now calculate α under this test.
(iii) Do you think β under this test will differ substantially, differ slightly, or not differ at all from β in Example 5.2? Again, offer a clue about your reasoning.
(iv) Now calculate β for this test.

To generalize from Example 5.2, we note that:

When H_1 is the *only* possible alternative to H_0, it must be true that:

$$\beta = P(\text{accept } H_0 | H_0 \text{ false}) = P(\text{accept } H_0 | H_1 \text{ true}) \quad (5\text{-}1)$$

The reader might be wondering: How do we know that, if H_0 is false, H_1 must be true? The answer has two parts. In some circumstances, we can argue on logical grounds that H_0 and H_1 are mutually exclusive and exhaustive possibilities: In a medical setting, for example, if H_0 is the hypothesis that a given patient has disease A and H_1 is the hypothesis that she doesn't, then either H_0 or H_1 must be true. In other cases, however, it is not at all clear that either H_0 or H_1 must be true: In Example 5.2, for instance, the likelihood a coin comes up fair could be 80% or 8% rather than one of the two probabilities associated with H_0 or H_1.

Hypothesis testing that chooses between H_0 and H_1 might best be thought of as raising a "what if?" question. If H_0 is false because H_1 is true, what would be the chance that we would be fooled under the test into accepting H_0? If we answer that question for a variety of possible alternatives H_1, we get some sense of when the test is especially vulnerable to Type II (β) error. If high vulnerability arises against alternatives H_1 that we sense might be true (even though we're not sure that they are), then we might worry that the test fails to protect us from the risk of accidental acceptance under realistic scenarios. That circumstance could well reduce our confidence in the test, and might encourage us to do something (e.g., collect more data) that would yield a more reliable test.

We titled this section " some power points," but have so far said nothing about power. Power concerns the optimistic counterpart of accidentally accepting null hypothesis H_0: it is the probability of correctly rejecting H_0 when the alternative H_1 is true:

Power (Against Alternative H_1)

$$= P(\text{reject } H_0 | H_0 \text{ false}) = P(\text{reject } H_0 | H_1 \text{ true}) \qquad (5\text{-}2)$$

when it is assumed that $P(H_0 \text{ true}) + P(H_1 \text{ true}) = 1$

The presumption is that H_1 is specific enough that the latter conditional probability can be calculated.

Because accidental acceptance and correct rejection of H_0 are complementary events, we can write:

$$P(\text{reject } H_0 | H_0 \text{ false}) + P(\text{accept } H_0 | H_0 \text{ false}) = 1 \qquad (5\text{-}3)$$

It follows from (5-1), (5-2), and (5-3) that:

In a test of H_0 against H_1:

$$\text{Power} = 1 - \beta \qquad (5\text{-}4)$$

where β is given in (5-1)

In Example 5.2 in which $\beta = .7$, the power of the test to reject the fair-coin hypothesis is only 30% when in reality heads are twice as likely as tails. This outcome implies that ten tosses rarely provide convincing evidence that a coin is unfair even when it is considerably so.

EXAMPLE 5.3 **A Gender-Biased Disease?**

It is uncertain whether a particular disease is equally likely to strike people of both genders, or instead affects one gender more than the other. The null hypothesis H_0 is "no gender difference," and, to test that hypothesis, Mendel will gather data about a random sample of 200 sufferers of the disease and will see how close the observed gender split is to 50/50.

If X is the number of males among the 200 disease sufferers, Mendel will test H_0 under the rule:

$$\begin{cases} \text{Accept } H_0 & \text{if } 90 \le X \le 110 \\ \text{Reject } H_0 & \text{Otherwise} \end{cases}$$

(i) What is α for this test?

(ii) What is the power of this test against the alternative H_a: *60% of the disease sufferers are male*?

(iii) Before performing this test, Mendel meets Minerva, who suggests that he evaluate H_0 with a usual χ^2-test with $\alpha = .05$. What would be the power of that test against H_a: *60% of Sufferers are male*?

(iv) Was Mendel's original test better than Minerva's proposed alternative? Briefly discuss.

Solution

(i) If H_0 is true, then each member of the sample independently has a 50% chance of being male. Thus, the number of males among the 200 disease sufferers would follow a binomial distribution with $n = 200$, and $p = 1/2$. Thus:

$$\alpha = P(\text{reject } H_0 | H_0 \text{ true}) = P(90 < X \text{ or } X > 110 | X \text{ is } B(200, 1/2)) = .1374$$

(ii) To find β, we note that, if H_1 is true, each disease sufferer has a 60% chance of being male, and thus X is $B(200,.6)$. But then we have:

$$\beta = P(\text{accept } H_0 | H_1 \text{ true}) = P(90 \le X \le 110 | X \text{ is } B(200,0.6)) = .0857$$

(Here we use the binomial calculator with $p = 0.6$, noting that:

$$P(90 \le X \le 110) = P(X \le 110) - P(X \le 89) = .0857 - .0000 = .0857.)$$

The power of Mendel's test = $1 - \beta = .9143$.

In summary, Mendel's test has about a 14% chance of erroneously suggesting a gender difference when there is none, but a 91% chance of correctly highlighting a gender difference when there is a 60/40 split between males and females.

(iii) We could formulate a χ^2 test of H_0 (i.e., a 50/50 split by gender) with two categories: male sufferer of the disease, and female sufferer. The χ^2-table would be:

Category	e_i	a_i
Male sufferer	100	X
Female sufferer	100	200 – X

where X= actual number of male sufferers

The test would have one degree of freedom ($M = 2$, $M_p = 1$, and $Q = 0$), meaning that the usual pass/fail threshold for $\alpha = .05$ would be 3.84. In terms of X, the discrepancy score would be:

$$\chi^2 = \sum_{i=1}^{2}\left(\left(e_i - a_i\right)/e_i\right)^2 = \frac{(X-100)^2}{100} + \frac{(200-X-100)^2}{100}$$

$$= 2\frac{(X-100)^2}{100}$$

Thus, the test rule would be:

$$\begin{cases} \text{Accept } H_0 \text{ if } 2\dfrac{(X-100)^2}{100} \le 3.84 \\ \text{Reject } H_0 \text{ if } 2\dfrac{(X-100)^2}{100} > 3.84 \end{cases}$$

What values of X would lead to acceptance of H_0 under the χ^2 test? With a spreadsheet or with algebra, we learn that $2(X - 100)^2/100 < 3.84$ when X falls in the range (87, 113), while $2(X - 100)^2/100 > 3.84$ if $X \le 86$ or $X \ge 114$. In other words, performing the χ^2 test is equivalent to using the test rule:

$$\begin{cases} \text{Accept } H_0 \text{ if } 87 \le X \le 113 \\ \text{Reject } H_0 \quad \text{Otherwise} \end{cases}$$

As noted, $\alpha = .05$ for this test. To find the power of this test against H_1 (60/40 split), we write:

$$\beta = P(\text{accept } H_0 | H_1 \text{ true}) = P(87 \le X \le 113 | X \text{ is } B(200{,}0.6))$$

$$= .1739 - .0000 = .1739.$$

The power of this test against $H_1 = 1 - \beta = .8261$.

(iv) Which of the two tests is "better" depends on one's priorities. Mendel's original test has $\alpha = .14$, while Minerva's is constructed to achieve $\alpha = .05$. However, against the "60% male" alternative, Mendel had a power of 91% while Minerva only had a power of 83%. In short, neither test is clearly better than the other.

EXERCISE 5.3

(i) In Example 5.3, what is the power of Mendel's original test against the alternative H_1: *70% of the Disease Sufferers are Male*?

(ii) Same for Minerva's test.

(iii) Would the power for Mendel's test and Minerva's be the same as above against H_1: *30% of the Disease Sufferers are Male?* Briefly discuss.

Power Curves

Suppose that in reality A% of the disease sufferers in Example 5.3 are male, and we want to calculate the power of Minerva's test as a function of A. Between Example 5.3 and Exercise 5.3, we have already obtained the results for $A = 60$ and 70, which must equal the power results for, respectively, $A = 40$ and $A = 30$. (Think about it.) What can we say for other values of A?

We can actually say quite a bit. Exercise 5.3 (if done correctly) demonstrates that the power is already close to 100% when $A = 30$ or 70; that being the case, we can infer that power is essentially 100% for the entire ranges 0–30 and 70–100. To find the power when A approaches 50%, we should consider what happens when A is exactly 50%. In that case, H_0 is correct, so that there is no issue of false acceptance. And the chance of correct acceptance when $A = 50\%$ equals $1 - \alpha$, which is 95%. If A differs very slightly from 50%, the chance of falling in the acceptance region is only slightly below 95%. Thus, β will itself be slightly below 95%, and the power $(1 - \beta)$ will only barely exceed 5%.

What this last outcome means is that the test has very little power to pick up differences from gender neutrality if those differences are very small. Then again, if they are so small, maybe it isn't terribly important to pick them up.

We could work out the power at each A-value using the method of Example 5.4, and then could plot power versus A for the range 0% to 100% (excepting the point $A = 50\%$ when H_0 is correct). Such a curve is called, not surprisingly, a *power curve*, Figure 5.1 approximates the power curve in this situation

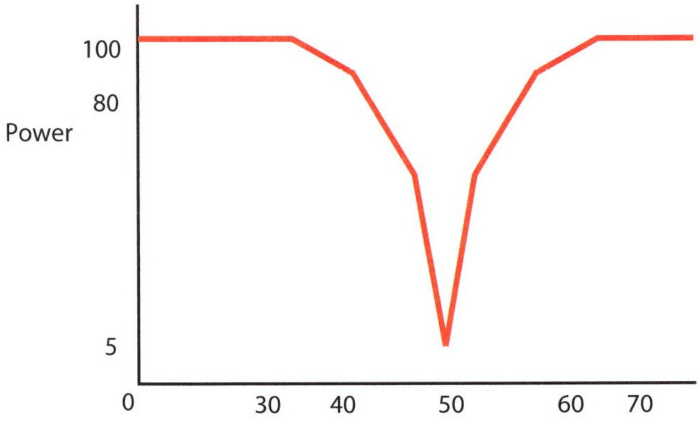

FIGURE 5.1 The Power of a χ^2 Test Concerning "Gender Neutrality" of a Disease, as a Function of Percentage of Males Among Disease Sufferers. (The outcome at exactly 50% is not a power calculation, because there the null hypothesis is correct.)

A power curve suggests the sensitivity of the test to violations of H_0 of various sizes. It is maximally illuminating when there is only one parameter of interest, which is hypothesized to take some particular value. When the hypothesis gets more complicated, the wide range of potential deviations from H_0 cannot easily be compressed to a single dimension. For example, if the null hypothesis is "uniform birthdays," then no one parameter can readily capture all potential forms of non-uniformity. In such settings, the concept of power curve is less helpful.

5.3 *p*-Values

Now we shift gears and consider an issue conceptually closer to α than to β. Standard hypothesis testing is often criticized because it partitions the broad spectrum of test outcomes into just two classes: "Accept H_0" and "Reject H_0." The more closely one watches the process at work, the more uneasy one is likely to get. For example, in a χ^2 test with 5 degrees of freedom and $\alpha = .05$, the accept/reject threshold (c) is 11.07. Thus, discrepancy scores of 0 and 11 are treated the same way (accept), as are the scores 11.2 and 90 (reject). Yet 11 and 11.2, despite their close proximity, are treated in opposite ways. More generally, statistical tests can produce a continuum of outcomes, ranging from extremely congenial to H_0 to extraordinarily hostile to it. To compress this diversity into two large ranges is a bit like replacing the 88 keys on the piano with two superkeys, each of which yields the average sound of 44 original ones.

Many analysts, dissatisfied by such reductionism, want a continuous measure of the compatibility between H_0 and the observed data. A measure often used is called the *p-value*. We introduce the concept informally in the example below, but will get more precise shortly thereafter.

EXAMPLE 5.4

Another Die Is Cast

Minerva purchases an expensive six-sided die from Zanzibar, having been assured that it is fair. She takes it home and tosses it four times, getting the sequence (3,3,5,3), in that order. Furious, she returns to Zanzibar's store to protest his sale of a non-random die. How compelling a case does she have?

Solution

We explore the question by considering a dialogue between Minerva (MV) and Zanzibar (Z).

MV: You're a crook and a scoundrel. Had you sold me a fair die, the probability would be only $(1/6)^4 = 1/1296$ of getting the outrageous outcome (3,3,5,3).

Z: But isn't that just as true for the sequence (2,4,6,1), which you would have interpreted as consistent with fairness? In fact, doesn't *every* individual outcome have probability 1/1296?

MV: Don't make it sound as if I got (2,4,6,1). Deal with the outcome I suffered.

Z: Well, would you have been just as upset with (3,5,3,3) (as with (3,3,5,3))?

MV: Yes

Z: Would you have been just as upset with (3,3,6,3)?

MV: Yes

Z: Would you have been just as upset with (4,4,6,4)?

MV: Yes

Z: Would you have been just as upset with (3,3,3,3)?

MV: No, you fool: I would have been more upset

Let's add this up, Zanzibar suggests. There are four possible sequences involving three 3's and a 5, all of which would have upset you as much as your actual outcome (3,3,5,3). There are another sixteen sequences involving three 3's and an X, where X is 1,2,4, or 6. So there are $16 + 4 = 20$ sequences with exactly three 3's that would have upset you as much as (3,3,5,3). And we should add 20 more sequences with exactly three 4's, another 20 with exactly three 1's, etc. Thus, there are $6 * 20 = 120$ sequences yielding exactly three of a kind, all of them as upsetting as what actually happened. These 120 sequences are mutually exclusive and each has a probability of 1/1296 for a fair die, meaning that:

$$P(\text{exactly three of a kind in four tosses of a fair die}) = 120/1296$$

The probability of getting four of a kind would be 6/1296, and that would have upset Minerva even more. Thus, the chance of an outcome at least as hostile to the hypothesis "the die is fair" as Minerva's was is $120/1296 + 6/1296 \approx 0.10$, *even assuming the die was fair.* So perhaps it was premature to conclude that the die was clearly biased. Zanzibar offered a refund at this point but Minerva, embarrassed that she had overstated the evidence of nonrandomness, left the store and founded Amazon.com.

EXERCISE 5.4 Suppose that Minerva had instead obtained the outcome (2,5,5,2) in Example 5.4. Assuming that the die actually was fair, what is the probability of an outcome at least as hostile to the hypothesis as this one? Assume that, in comparison with the double pair (2,5,5,2), "three of a kind" is considered more hostile while a single pair like (2,3,3,1) is less hostile.

Definition of *p*-value

What Zanzibar had done in Example 5.4 was to find the *p*-value of Minerva's result. To state a definition:

> The ***p*-value** in a statistical test is the probability—assuming that H_0 is true—of observing a test result at least as "hostile" to H_0 as the one that actually arose.

Because it is a probability, the *p*-value must fall in the range from 0 to 1. Zanzibar had determined that the *p*-value is about 10% for "three of a kind" out of four tosses of a die that is hypothesized to be fair. That outcome meant that, assuming fairness, the chance was about 1 in 10 of reaching an outcome as "hostile" to fairness as the one that Minerva got.

A Hierarchy of Hostility

We need be more specific about we mean about" hostility." We noted in Section 5.1 that we should identify some general hypothesis H_a that we believe true if H_0 is false. In many instances, H_a is just the opposite of H_0, meaning that H_a is simply "H_0 is false." But in some cases, we believe we know *why* H_0 is false if in fact it is, and we express that belief in the statement of H_a. For example, when H_0 is the hypothesis that an employer does not discriminate by gender in hiring, that alternative H_a could be that the employer discriminates against women. H_a is generally not as specific as H_1 that arises in the context of power calculations.

In defining "hostility" for a *p*-value calculation, we need specify at the outset the alternative hypothesis H_a. We take H_a into account when we rank the possible outcomes from the most hostile to H_0 to the least hostile, thereby creating a "hierarchy of hostility." In this hierarchy, the outcomes in the data that are *work against H_0 but work towards H_a* are the ones considered the most hostile to H_0. The outcomes that work most strongly against H_a are often considered the least hostile to H_0, even if they are not especially consistent with H_0 itself. (They hurt H_0's enemy H_a more than they do H_0.) In short, hostility towards H_0 is based largely whether the outcome tends to "tilt" us away from H_0 towards H_a, or instead pushes us further into the arms of H_0.

We will shortly encounter a job discrimination case, in which the null hypothesis H_0 is that there is no discrimination based on age, while the alternative H_a we have in mind is that older workers are discriminated against. If all the workers laid off are old, then the outcome is very consistent with H_a but not to H_0, so we assign it the highest hostility ranking for H_0. If all the workers laid off are young, by contrast, then the outcome is not especially congenial to H_0 (which posits no discrimination of *any* kind by age) but it is exceedingly inconsistent with H_a (which posits bias against the old). That outcome would be assigned low hostility to H_0.

Suppose that we plan to test uniform birthdays (H_0) by assembling 100 randomly-chosen people in a room, and only considering whether there is any sharing of birthdays. The alternative H_a is that the birthday distribution is not uniform over the year. Would we reject H_0 if there is no birthday sharing at all? As we discussed in Chapter

1 (Example 1.8), getting no sharing of birthdays among 100 people would be extraordinarily improbable under H_0. But if birthdays are not uniform (H_a), then they are concentrated in certain parts of the year, and getting no sharing is *even more* unlikely than under uniformity. Hence "no overlap" is actually more hostile to H_a than to H_0, and thus, paradoxically, the outcome "no overlap" would *increase* our confidence in H_0. (Think about it: it makes sense.)

We will gain more familiarity with the "hierarchy of hostility" as we work through some examples.

EXAMPLE 5.5

Hazardous Homes

Suppose that scientists at the Environmental Protection Agency hypothesize (H_0) that 24% of American homes have unhealthy levels of the gas radon. The alternative hypothesis H_a is that the percentage is other than 24%, either higher or lower. A random sample of 100 such homes will be selected and their radon levels measured. Let X be the number of sampled homes that exhibit excess radon.

Find the p-value of the result if:

(i) $X = 29$
(ii) $X = 98$
(iii) $X = 24$

Solution

Under the p-value concept, we wait until we find out what X is, and then find the probability of an outcome at least as hostile to H_0 as X, assuming that the 24% hypothesis is true.

(i) The outcome $X = 29$ is somewhat hostile to H_0 in that 29 is five more unhealthy homes than the 24 ($100 \times .24$) expected. But this excess of five homes could, of course, reflect nothing more than sampling error (i.e., the difference between an outcome within a random sample and the outcome for the full population). The p-value helps us assess the plausibility of treating sampling error as the reason for the excess.
To compute a p-value in this setting, we must decide which outcomes we consider "at least as hostile" to H_0 as the observed outcome 29. (Here H_a specifies that H_0 is false, either because the 24% estimate is too high or too low.) In identifying other outcomes as least as hostile to H_0 as 29, we start by describing 29 as equally hostile to itself (we will *always* include the observed outcome in the "at least as hostile" set). Then we identify $X = 19$ as *equally hostile* because, like $X = 29$, the outcome $X = 19$ is five away from the expected number 24. We interpret outcomes above 29 as *more hostile* to H_0 than is 29 itself; they are further away from 24 and do more to erode our confidence that H_0 is true. Similarly, outcomes below 19 are more hostile to H_0 than is $X = 19$, and thus more hostile than $X = 29$.

To summarize, outcomes at least as hostile to H_0 as the observed result $X = 29$ are:

(i) $X = 29$
(ii) $X = 19$
(iii) $X > 29$
(iv) $X < 19$

The p-value of H_0 given that $X = 29$ is the sum of the probabilities of these four (mutually exclusive) events *given that Ho is true*. Under H_0 and the concept of random sampling from a large population, each house chosen is treated as independently suffering a 24% chance of excess radon. Then X is *binomial* under H_0 with parameters $N = 100$ and $p = .24$ (i.e. X is $B(100,.24)$, and we can write:

$$p\text{-value} = P(X \leq 19 \mid X \text{ is } B(100,24)) + P(X \geq 29 \mid X \text{ is } B(100,24))$$

A binomial statistical package reveals that the sum of these two probabilities is **.29**. This number means that, if H_0 were true, the chance is nearly 3 in 10 that it would fare as "badly" in a test with 100 homes as it actually did (namely, getting an X-value five or more away from the "ideal" value 24). Given this relatively high probability, it would seem farfetched to seize on the test result as evidence that H_0 is preposterous.

(ii) The only outcomes as least as hostile to H_0 as the horrendous one $X = 98$ are:

$$X = 98, X = 99, \text{and } X = 100.$$

(Note that the outcome $X = 0$ is far closer to 24 than is $X = 98$, in consequence of which X=0 can be construed as less hostile to H_0 than is $X = 98$*.) The p-value in this situation would therefore follow:

$$p\text{-value} = P(X \geq 98 \mid X \text{ is B}(100,24))$$

That probability is so low that a usual binomial table will simply print out zeros. But using some approximations that apply in extreme situations, we reach the estimate:

$$p\text{-value} = P(X \geq 98 \mid X \text{ is } B(100,24)) \approx e^{-150}$$

Wow! What this startlingly low number is telling us is that, if the excess-radon rate was actually 24%, getting 98 or more homes with excess radon out of a random sample of 100 is so improbable as literally to be off the charts.

(iii) If $X = 24$ for 100 homes, we have hit the 24% rate right on the nose. If asked to identify the outcomes at least as hostile as this one, we have to list *every*

* To put the issue a bit more precisely, $X = 98$ is farther out as a percentile on the right tail of X's distribution than $X = 0$ is as a percentile of the left tail. Some people would say that we should set the p-value to $2P(X \geq 98)$ to "mimic" on the left tail the outcome equally hostile to that on the right tail, even though it corresponds to no actual left-tail outcome. As we will see, however, $P(X \geq 98)$ is so utterly minuscule that doubling it is of no consequence.

possible outcome, because this is the "friendliest" possible result. Under the definition of p-value, we would write:

$$p\text{-value} = P(X < 24 \text{ or } X = 24 \text{ or } X > 24 \mid x \text{ is } B(100,24))$$

Given that X must be somewhere, the p-value here would be 1.

To summarize these findings with a chart:

Outcome	p-value
24	1
29	0.29
98	0.0000

As these three cases suggest, the p-value starts at 1 at the outcome "friendliest" to the hypothesis, and declines as the observed outcome gets further away from the ideal one. But the decline is not precipitous: As the test result gets a bit more "hostile" to H_0 (e.g. as X moves from 32 to 33 in the excess-radon case), the p-value drops, but modestly. However, outcomes that make the hypothesis look ridiculous are associated with p-values exceedingly close to zero. The key advantage of the p-value is that it provides a ***continuous*** measure of how well the hypothesis fits with the available data; we no longer face an abrupt pass/fail shift in the test outcome at some accept/reject threshold. The p-value allows for nuances in assessing how "believable" H_0 is, distinguishing extremely favorable test outcomes from moderately-favorable ones from mildly-unfriendly ones from exceedingly hostile ones. With all these features, it is not surprising that the p-value has many admirers.

EXERCISE 5.5 In Example 5.5 above, find the p-values associated with each of the following outcomes:

(i) $X = 30$
(ii) $X = 35$
(iii) $X= 48$
(iv) $X = 18$

EXAMPLE 5.6

More Hazardous Homes?

In Example 5.6 above, the alternative to the 24% rate of excess radon (H_0) was some rate other than 24% (H_a). Thus, an observed rate far from 24% favors H_a over H_0 regardless of whether the rate is well above 24% or well below 24%. We therefore worked out the p-value looking in both directions from the ideal outcome of X = 24: the further away from 24, the more hostile to H_0.

But what if we were testing H_0: 24% against the suspicion (H_a) that the rate is *higher than* 24%? Against that alternative, what is the p-value associated with the outcomes:

(i) $X = 29$
(ii) $X = 98$
(iii) $X = 24$

Solution

We need to create a "hierarchy of hostility" that corresponds to the new alternative H_a. With that alternative, observed rates above 24% are considered hostile to H_0, but not those below 24%. An observed rate of (say) 15% doesn't make the 24% hypothesis look great, but it makes the hypothesis that 24% is too low (H_a) look even worse.

Here the outcome most hostile to H_0 based on 100 homes would be finding excess radon in all 100, the same as before. However, the least hostile outcome against H_a is now $X = 0$. (In other words, $X = 24$ is *not* the outcome that most favors H_0 against alternative H_a, because $X = 24$ does not make H_a seem absurd to the extent that $X = 0$ does.) The hostility ranking of a given outcome is based on two things: (i) whether it increases or reduces our confidence in H_0 relative to H_a and (ii) how strongly it increases or reduces our confidence. Applying this criterion in this situation, hostility grows steadily as X increases over the range (0,100).

Therefore:

(i) For $X = 29$, p-value = $P(X \geq 29 | X \text{ is } B(100,.24)) = .15$
(ii) For $X = 98$: p-value = $P(X \geq 98 | X \text{ is } B(100,.24)) \approx 0$
(iii) For $X = 24$: p-value = $P(X \geq 24 | X \text{ is } B(100,.24)) \approx .54$

Notice that the p-value for $X = 29$ has dropped from .29 against H_a: *Anything but 24%* to .15 against H_a: *24% too low*. A lower p-value means the outcome is less "congenial" to H_0. In both instances, we were off by five percentage points from hitting a 24% excess-radon rate, but the 29% rate when $X = 29$ leans toward the notion that 24% is "too low" (i.e., towards H_a). To be off by five points *and* in the direction that H_a anticipates makes us less sure about H_0 than just being off by five points. Here the p-value is telling us that, if H_0 were true, the chance is only about 1 in 7 of getting an outcome as unfavorable to H_0 (and thus favorable to H_a) as $X = 29$.

For $X = 24$, the p-value is still a formidable .54, but it has fallen from its value of one in Example 5.6. That happens because an outcome of (say) $X = 8$ would have done more to discredit the notion that 24% was too low (i.e., H_a) than does $X = 24$. In that sense, $X = 8$ is less hostile to H_0 than is $X = 24$.

EXERCISE 5.6 In Example 5.6, what is the p-value for the following outcomes?

(i) $X = 19$
(ii) $X = 25$
(iii) $X = 26$

p-values and the χ^2 Procedure

How can we relate the notion of p-values to the χ^2-mechanism that we discussed at such length in Chapter 4? In that setting, the test is based on the discrepancy score χ^2. A χ^2-test by its very nature is testing H_0 against the alternative H_a: H_0 *is false*. We have always reasoned that, the higher the discrepancy score, the less compatible (and thus more hostile) the outcome is to H_0. Thus, the "hierarchy of hostility" treats outcomes as increasingly hostile as the discrepancy score grows.

In a χ^2-test, the outcomes at least as hostile to H_0 as $\chi^2 = b$ are:

- $\chi^2 = b$
- $\chi^2 > b$

Thus, when $\chi^2 = b$ we would write:

$$p\text{-value} = P(\chi^2 \geq b \mid H_0 \text{ is true}) \qquad (5\text{-}5)$$

We can use an electronic χ^2-table to find the p-value under (5-5). First we need note the number of degrees of freedom in the test. Then, at the table, we:

- List the number of "degrees of freedom"
- List the discrepancy score b, leaving blank the line for "area right of" or whatever phrase the table uses for the probability of exceeding b
- Press the "compute" command, and watch the p-value appear on the blank line

EXAMPLE 5.7 **Hostility at MIT**

In the uniform-birthdays test in Example 4.9 with 45 data points and six categories, the calculated discrepancy score χ^2 was 6.87. What is the p-value of that outcome?

Solution

Under (5-5), we have:

$$p\text{-value} = P(\chi^2 \geq 6.87 \mid H_0 \text{ true})$$

If H_0 is true, then the discrepancy score in this test follows a χ^2–distribution with five degrees of freedom. Inputting the numbers five and 6.87 at the electronic table tells us that, for that distribution, the chance is .2305 of getting an outcome of 6.87 or higher. Thus the p-value of the outcome is .2305 (or .23 to the nearest hundredth). With about a 1-in-4 chance of reaching that discrepancy level even if birthdays are uniform, the data do not make the uniform-birthdays hypothesis look ludicrous.

What is nice about the p-value in a χ^2-test is that, if the discrepancy score goes up slightly, the p-value will only drop slightly. The p-value thus yields a continuous numerical rating of closeness between hypothesis and data, and not just a "pass/fail" assessment as in a standard hypothesis test.

EXERCISE 5.7 For a χ^2 test with the outcome $\chi^2 = 13.48$, what is the p-value of the outcome if the test has:

(i) five degrees of freedom
(ii) eight degrees of freedom
(iii) 16 degrees of freedom

5.4 p-Values And Hypothesis Testing

Paradise Lost?

With a hierarchy of hostility at hand, working out the p-value need not be difficult. But while the p-value is surely useful, it may not be the godsend it might first appear to be. As a practical matter, we must generally make *some* yes/no decision on whether we think H_0 viable. To do so on the basis of a p-value, we will probably wind up setting some formal or informal "go/no go" threshold (i.e., accept H_0 if the p-value is at least c; reject H_0 if the p-value is below c.) That puts us in an ironic situation: having used p-values to bypass some crude accept/reject dichotomy, we might furtively wind up resurrecting that dichotomy as a guide to action.

Because of this dilemma, p-values are sometimes incorporated into formal hypothesis tests rather than interpreted as replacements for them. To test the null hypothesis H_0, the analyst computes the p-value of the observed result; the test rule is typically:

$$\begin{cases} \text{Accept } H_0 \text{ if the computed } p\text{-value is above .05} \\ \text{Reject } H_0 \text{ if the computed } p\text{-value is below .05} \end{cases} \tag{5-6}$$

Does this test rule imply a 5% risk of accidental rejection (i.e., is $\alpha = .05$) ? Not necessarily. Suppose that H_0 is the hypothesis that a coin is fair, against the alternative

H_a that is it not fair. The coin is to be tossed three times. If X is the number of heads, then the p-value for each X value follows:

Value of X	p-value for fair-coin hypothesis
0	.25
1	1
2	1
3	.25

These p-values arise under the following calculations:

- p-value for $(X = 3) = P(\text{three heads} | \text{fair coin}) + P(\text{three tails} | \text{fair coin})$
  $= (1/2)^3 + (1/2)^3 = 1/4$
  (Note that three tails and three heads are equally hostile to H_0)

- p-value for $(X = 2) = 1$ because every outcome is "at least as hostile" to a fair coin as two heads and one tail
- p-value for $(X = 0) = p$-value for $(X=3)$; p-value for $(X = 1) = p$-value for $(X = 2)$

If we accept whenever the p-value is at least .05, then we *always accept* in this situation. Thus, $\alpha = P(\text{reject } H_0 | H_0 \text{ true}) = 0$, which is different from .05.

This example notwithstanding, the hypothesis test based on comparing the p-value to .05 *usually has a risk of accidental rejection close to .05* when one works with data samples that are at least moderately large. We will quietly assume in what follows that the p-value test in (5-6) has an α-value of essentially .05 unless otherwise indicated.)

A (real) job discrimination case against a high-tech firm that was "downsizing" illustrates hypothesis testing by p-value. In the U.S., it is illegal to discriminate by age when laying off workers, and courts particularly scrutinize how personnel decisions affect workers over age 40. A firm can be found guilty of age discrimination if (i) statistics offer strong evidence that its over-40 workers suffered disproportionately in its layoffs, and (ii) the firm cannot credibly rebut accusations that the pattern reflects "age animus."

EXAMPLE 5.8

Alleged Age Discrimination (Based on a true story)

In the situation described at the start of this chapter, 20 of the 33 employees of a high-tech company were over 40 years old and, in a layoff of five workers, all were over 40. Does this outcome constitute statistically significant evidence of "age animus" towards older workers?

Solution

As noted, the five terminated employees wanted to file a lawsuit against the company charging it with age discrimination. The company conceded that over-40 workers fared worse in the layoff than younger ones, but it strongly denied that age had anything to do with the layoff decisions. According to the firm, it was just a coincidence that the five workers least essential to the company's survival were all over 40.

In American courts, the question is sometimes asked: if the layoffs had been made *completely at random*, how likely is it that we would see a result as extreme as the one that actually arose? In essence, the courts ask for the p-value of the outcome against the null hypothesis H_0 that the layoffs were random. A model of k such layoffs would be an urn that contains a ball for each worker, from which k balls would be selected at random. In the present setting, the question was:

> If 5 of the 33 workers were selected for layoff **completely at random**, what is the probability that the age-related result would be as hostile to H_0: *No age discrimination* as the outcome that actually arose?

Here the outcome that arose was that all five laid off were old, and what was sought was the p-value for the outcome under a "layoff lottery." In this context, the alternate hypothesis was H_a: *Age discrimination against workers over 40.*

Given the phrasing of the question above, we are asked to imagine an urn with 33 balls, each containing the name of one of the branch's employees. If five balls are selected at random, what is the probability that over-40 workers would be at least as overrepresented among those chosen as they actually were in the layoff? In this instance, the answer to that question is the p-value of the observed result.

We should explain this last statement. If random variable X is the number of older workers laid off, then X must fall in the range (0,5); here $X = 5$. To find the p-value, we need first to identify the X-values at least as hostile to the random layoff theory as $X = 5$ (against H_a: *Discrimination against Older Workers*). The outcome that most strongly favors the alternative H_a over H_0 is $X = 5$, which means that $X = 5$ is rated "most hostile" in the hierarchy. On the other hand, the outcome $X = 0$—meaning that everyone laid off was young—does the most to advance H_0 over H_a, so that outcome is deemed the least hostile to H_0. (How can one say that layoffs were biased against the old when everyone laid off was young?) In the "hierarchy of hostility," the least hostile outcome for H_0 here is $X = 0$, the next least hostile outcome is $X = 1$, then comes $X = 2$, etc. Because $X = 5$ is viewed as the outcome most hostile to H_0, we have:

$$p\text{-value} = P(X = 5 \mid \text{random layoffs of five workers among the 33})$$

We can find this probability in two ways. With 20 workers over 40 and 13 below that age, the chance is 20/33 that the first person selected for termination in the layoff lottery would be over 40. Given that the first person selected was over

40, the chance that the second was also in that age range is 19/32. Continuing in this way in the spirit of sampling without replacement we find that:

$$P(\text{all five chosen over age 40}) = (20/33) * (19/32) * (18/31) * (17/30) * (16/29) = .07$$

We could also note that we are in the domain of the hypergeometric distribution. There are two groups: 13 young workers and 20 older ones, and we are selecting five people at random from among them. If X is the number who are old, then $P(X = 5)$ can be obtained from the hypergeometric formula (1-42) in Chapter 1:

$$P(X = 5) = \frac{\binom{20}{5}\binom{13}{0}}{\binom{33}{5}}$$

Turning to the electronic hypergeometric table, we once again get $P(X = 5) = .07$

Because the p-value of the outcome is .07, it exceeds the threshold of .05 in (5-6). Thus, under that standard rule, H_0 would be accepted. In essence, the observed outcome had enough of a chance of arising by pure coincidence that it would not be viewed as clear evidence of age discrimination. Under present standards, the firm would probably prevail in court unless the plaintiffs could offer additional evidence of age animus (e.g., a statement by the company president that he wants to "get rid of" the older workers). That is why the lawyers cited at the start of this chapter told the workers they had no case.

Discussion

The calculation above might seem contrived: no one in his right mind believes that the firm based its layoff decisions on an urn lottery. Yet the lottery might offer an adequate probabilistic model of the firm's decision process. Suppose that, as a first step in deciding whom to lay off, the company ranked the 33 employees based on how crucial their skills were to the company's future. If such skills are completely unrelated to age, then the person judged least essential—the first one to be laid off—has a 1 in 33 chance of being the oldest, a 1 in 33 chance of being the youngest, and a 1 in 33 chance of taking any other position in a ranking of employees by age. Thus, the chance that the "least essential" person would be old would be 20/33. Similarly, the second least essential person is equally likely to take any of the 32 remaining rankings, and would have a 19/32 chance of being old given that the "least essential" person were old. The key point is that, while the company is basing its decisions on employee skills, outcomes of the process can resemble those of an urn lottery if those skills *are randomly distributed with respect to age.*

If the firm had laid off more workers and the p-value of the outcome had fallen below .05, then the courts would be far less inclined to believe that the pattern is a mere coincidence. Still, however, the company could win the case. It might successfully argue that some skill related to its survival (e.g. dexterity with computers and other Information Age technology) is concentrated among the younger workers. Thus, while older workers suffered a higher-than-average layoff rate, the reason was not age animus but rather their deficits in a legitimate business-related skill. The plaintiffs might well object to such an explanation; by this stage, however, the argument is probably beyond Statistics and hence beyond the scope of this textbook.

EXERCISE 5.8 Suppose that the company in Example 5.8 had laid off six workers rather than five, all above age 40. Would that outcome yield a different accept/reject conclusion in the p-value test of the "no discrimination" hypothesis?

p-value Tests as Statistical Esperanto

Esperanto was created in 1887 to provide a universal second language that would allow speakers of English, Spanish, Russian, Chinese, and any other tongue to communicate more freely with one another. In some sense, hypothesis testing using p-values is a statistical version of Esperanto. We have already seen how a p-value test can arise in connection with the hypergeometric distribution (age discrimination), the binomial distribution (excess radon), and the χ^2 distribution (uniform birthdays). In these different cases, we had varying definitions of hostility to the hypothesis, and various ways of computing the chance of an outcome at least as hostile as the one that arose. But once we arrived at a p-value, the interpretation is the same regardless of how it was calculated. A p-value of .01, for example, would mean that, if the hypothesis of interest were true, the chance would be only 1% of getting an outcome at least as hostile to the hypothesis as the one that came up.

We will be discussing more statistical tests in subsequent chapters, and we will often provide the p-value associated with the outcome. p-values offer us a kind of "one-stop shopping," which allows us to interpret the outcome regardless of whether we have performed an F-test, a sign test, a χ^2 test, or any other procedure. And even when used in an accept/reject hypothesis test, the p-value gives some sense of how clearly the hypothesis passed (or failed) the test. In that way, it still offers a continuous measure of compatibility between the hypothesis and the test data.

5.5 Multiple Tests of the Same Hypothesis

Sometimes, we find it attractive to look at a hypothesis from different angles and test it more than one way. For example, we could test uniform birthdays by checking the spread of birthdays across the various months, but also by looking separately at whether the birthdays split about equally between odd and even days. Suppose, in an extreme case, that each month has 1/12 of birthdays, but that every birthday occurs on the 7th of the month. The uniformity test based on months would be totally insensitive to this deviation from uniformity, while the odd/even test would go ballistic, leading us properly to reject the hypothesis.

There is, however, a problem in testing a hypothesis in several ways. It is natural to say that, if we perform k tests, the hypothesis should pass every one to get a clean bill of health. But if there is a 5% risk of accidental rejection in each test, then, assuming independence, the chance of failing at least one of the tests by accident would be well above 5% (Right?). When $k = 6$, for example, the chance of at least one accidental failure would be .265 (i.e., $1 - 95^6$) if the six tests were independent.

Bonferroni's "Sudden Death" Test

How might we perform a series of k tests of a hypothesis H_0 and yet keep the overall α-value at .05? One very natural approach has been proposed by Bonferroni. If we want to use the "sudden death" rule (fail any one, and we reject at once), arrange each test so that *the risk of accidental rejection is .05/k*. Then:

$$\begin{cases} \text{Accept } H_0 \text{ if } H_0 \text{ passes all } k \text{ tests} \\ \text{Reject } H_0 \text{ if it fails one or more of the tests} \end{cases}$$

Under this regimen, the chance of accidental rejection cannot exceed 5%. Let A_i be the event that *H_0 fails the ith of the k tests even though H_0 is true*. Then, as indicated in Exercise 1.35, it must be the case that:

$$P(\text{reject } H_0 | H_0 \text{ true}) = P(A_1 \cup A_2 \cup \ \dots \ A_n) \leq \sum_{i=1}^{n} P(A_i)$$

But because $P(A_i) = .05/k$ for all i, we reach:

$$P(\text{reject } H_0 | H_0 \text{ true}) = P(A_1 \cup A_2 \cup \ \dots \ A_n) \leq k\left(\frac{.05}{k}\right) = .05$$

Thus, the risk of accidental rejection cannot exceed .05. Actually, it is likely to be very close to .05. If we are conducting different tests that focus on different aspects of the hypothesis, the tests are likely to be independent or close to independent. Then:

For k *independent* tests, each of which has $\alpha = .05/\text{k}$:

- the chance of passing all of them when H_0 is true is $(1 - .05/k)^k$
- the chance of failing at least one of them is $1 - (1 - .05/k)^k$.

The quantity $1 - (1 - .05/k)^k$ is almost exactly .05 for all positive integers, as the following chart suggests:

k	$1 - (1 - .05/k)^k$
1	.05
2	.049
3	.049
4	.049
5	.049
6	.049
10	.049
15	.049
30	.049

(All numbers in right-hand column rounded to nearest thousandth.)

EXAMPLE 5.9 Getting Even

Suppose that we have birthday data for 120 people, and wish to test two aspects of the uniform birthdays hypothesis H_0:

- Uniformity of birthdays over the months of the year
- Equal frequency of odd and even birthdays

(i) Formulate a pair of χ^2 tests that explore both aspects of H_0.
(ii) What should be the accept/reject thresholds for these tests be so as to achieve $\alpha \approx .05$ in a Bonferroni "sudden death" test?

Solution

(i) We could set up one χ^2 test of the hypothesis with twelve categories, one for each of the months. The test would have an e_i of ten for each month, and eleven degrees of freedom. We could do another "odd/even" χ^2 test with two categories: even birthdays (e.g., April 22) and odd birthdays (e.g., April 23). That test would have e_i's of 60, and one degree of freedom. (True: a month with 31 days has 16 odd and 15 even, but don't be petty.) Note that the two tests are essentially independent: if the hypothesis is true, then a randomly-observed excess of births in (say) April would not lead us to expect more births than usual on even dates.

(ii) But how would we combine the two test results? The Bonferroni approach would have us conduct each test with an α-value of $.05/2 = .025$, and require passing both. We therefore turn to the electronic χ^2-table, and quickly learn that:

- For 11 degrees of freedom, the pass/fail threshold for $\alpha = .025$ is 21.92.
- For one degree of freedom, the pass/fail threshold for $\alpha = .025$ is 5.02.

Our two-part test would therefore be:

- Compute the discrepancy index χ^2_M for the twelve-month test, and the discrepancy index χ^2_E for the odd/even test.
- Accept the uniform-birthdays hypothesis *if and only if*:

$$\chi^2_M \leq 21.9 \textit{ and } \chi^2_E \leq 5.02.$$

If birthdays are actually uniform, the chance is essentially 5% that we will conclude otherwise under this two-part test. If birthdays are not uniform, then we have two chances to get evidence on that point: an unusual spread of the data across the months, or a striking tendency of births to favor odd or even days. We might sense, therefore, we are reducing the risk of false acceptance (β). That is not automatically true, however: If birthday-frequencies differ across the months but not in the odd/even split, then our more "lenient" test for the monthly spread (with a pass–fail threshold of $c = 21.92$ for $\alpha = .025$ rather than $c = 19.68$ for $\alpha = .05$) would be less likely to pick up on the difference than a single test focused on the monthly spread.

Running multiple tests is feasible and often helpful, but it is not always necessary. For example, suppose that we suspect that birthdays are not uniform across the months and that, moreover, births are more common on odd than even days. With 120 people, we could set up a single χ^2 test of uniform birthdays with 24 categories: odd days in January, even days of January, odd days in February, etc. The discrepancy score should be elevated if either form of nonrandomness is a major factor in the birth pattern, meaning that one statistical test performs two tasks.

EXERCISE 5.9 Suppose we want to add a third test of uniform-birthdays, which is independent of the first two in Example 5.9.

(i) What would α be on each test if we want an overall chance rejection of 5%?

(ii) If the last test is a χ^2-test with two categories, what would be the pass/fail thresholds for the three tests?

5.6 The Range Of Acceptable Hypotheses

By now we probably understand that the mere fact that a hypothesis has passed a statistical test does not mean that it is surely correct. A power curve is one way to emphasize this point, by highlighting the risk of false acceptance as a function of the extent to which the hypothesis errs. Another approach is to explore the *range* of hypotheses that are all acceptably consistent with the data. If that range is large, we are vividly reminded that it is hazardous to seize upon any particular hypothesis and proclaim it demonstrably correct. Here we discuss this second approach.

EXAMPLE 5.10 **A Better Ad?**

A company is considering two banner ads at a certain web site for its product, and wants to compare their effectiveness. To make an assessment, it will have ad A appear at random 50% of the time when people go to the web site, and ad B the other 50%. Then it will compare the number of orders that arise over a two-week period from people exposed to each ad.

Suppose the managers are meeting *before* the results of the experiment are available, and different people advance four very different hypotheses about the relative effectiveness of the two ads:

H_1: Both ads are equal effective in stimulating orders

H_2: Ad A is twice as effective as Ad B

H_3: Ad A is three times as effective as Ad B

H_4: Ad A is four times as effective as Ad B

(i) Suppose that each of the four managers sets up in advance a χ^2 test of his or her hypothesis (while awaiting the results). What would we expect their tests to look like?

(ii) Now suppose that 30 people ordered the product over the two weeks, 18 of whom saw ad A. Which if any of these four hypotheses would be accepted under the tests in (i)?

(iii) To what extent can we discriminate among the four hypothesis based on the observed outcome in (ii)?

Solution

As they await the results, the managers could set up χ^2 tests of their various hypotheses. They could decide take the sales records for the items purchased over the two weeks and arrange them by (say) alphabetical order on the buyers' names. The person who believes that the two ads are equally effective (H_1) would assign the first buyer on the list a 50% chance of having seen ad A, and a 50% chance of seeing B. (Right?) He would make the same assumption for the second person independently and, indeed, for all others. If n people in total bought the product over the two weeks, he would assume that X, the number who saw A, is binomial $B(n,1/2)$.

From this point, it is easy to set up a χ^2 test of H_1 based on the following table (in terms of n, total sales over the two weeks):

Test for H_1:

Category	e_i	a_i
Bought having seen ad A	n/2	X
Bought having seen ad B	n/2	$n - X$

Once X is known, the test can proceed with $M = 2$ categories, $M_p = 1$ puppet category, and $Q = 0$, and thus one degree of freedom. (While he will use n when he learns it, that general information did not influence his hypothesis, any more than using the sample size n in a test of uniform birthdays is considered "peeking" at the data.)

The test of H_2 would be animated by the idea that, if ad A is twice as effective in stimulating sales as ad B, then the first buyer should have a 2/3 chance of having seen A, as would all others among the n buyers. The corresponding χ^2 table would be:

Test for H_2:

Category	e_i	a_i
Bought having seen ad A	$2n/3$	X
Bought having seen ad B	$n/3$	$n - X$

Similarly, the tables corresponding to the other two hypotheses would be:

Test for H_3

Category	e_i	a_i
Bought having seen ad A	$3n/4$	X
Bought having seen ad B	$n/4$	$n - X$

Test for H_4:

Category	e_i	a_i
Bought having seen ad A	$4n/5$	X
Bought having seen ad B	$n/5$	$n - X$

Like the test of H_1, all the others would have one degree of freedom ($M = 2$, $M_p = 1$, and $Q = 0$). For all the tests, the pass/fail threshold $c = 3.84$ at the usual 5% level.

(ii) With $n = 30$, $a_A = 18$ and $a_B = 12$, the table for testing H_1 becomes:

Category	e_i	a_i
Bought having seen ad A	15	18
Bought having seen ad B	15	12

The discrepancy score for H_1 is calculated as $\chi^2 = \frac{\sum_{i=1}^{2}(e_i - a_i)^2}{e_i} = \mathbf{1.20}$, which is well below the threshold of 3.84 . Thus, the hypothesis H_1 would be accepted under usual standards. In other words, even though sales were 1.5 times as high among viewers of ad *A* than among viewers of ad *B* (18 vs. 12), we could reasonably treat the difference as a mere fluctuation, and thus not attribute the difference to the superiority of ad *A*.

But it would be a mistake to stop there, and the other managers would surely protest if the advocate for H_1 announced "case closed." The fact that the ads ***may*** have been equally effective does not mean that they ***must*** have been equally effective. True: the observed difference in sales (18 vs. 12) could have be an artifact of statistical "noise," but statistical noise could just as easily have "camouflaged" the benefits of the ad *A* as exaggerated them. Thus, *A*'s true advantage over *B* might be greater than 18 to 12.

The table for H_2 becomes:

Category	e_i	a_i
Bought having seen ad *A*	20	18
Bought having seen ad *B*	10	12

The discrepancy score for this table is **0.6**, which easily calls for accepting H_2.

The corresponding discrepancy score χ^2 is 3.6 for H_3, which is close to the threshold of 3.84, but on the side of it calling for acceptance. For H_4, however, the discrepancy score is 7.5, which is in the rejection region. If *A* was really four times as effective as *B*, then it is surprising that 40% of the purchases were made by viewers of ad *B* (12/30) rather than something around 20%. Even with only 30 purchases in total, the difference between *B*'s share of purchases and its expected share is too large to be dismissed as a fluctuation.

(iii) The acceptability of all three of H_1, H_2, and H_3 means that the short-term statistics at hand support conjectures ranging from equal effectiveness to a factor-of-three advantage for ad *A*. When the data are compatible with such divergent hypotheses, it would be imprudent to test H_1 and, upon accepting it, to announce the ads are equally effective. The fallacy in that reasoning, as we see here, is that *consistency with a "no difference" hypothesis does not preclude consistency with an alternative hypothesis of strong differences*. Alas, such fallacious thinking is far from unknown in the use of hypothesis tests.

For a more precise sense of relative effectiveness of the two ads, the company might be wise to continue the experiment somewhat longer. If the ratio of ad *A* orders to ad *B* orders remains near 1.5 (i.e., near 18/12), then the range of acceptable hypotheses will eventually occupy a very narrow range around a ratio of 1.5.

One last point. Some readers might note the discrepancy index here was lowest for H_2, and wonder "Why not declare H_2 the "winner" in the χ^2

competition and stop talking about a range of winners?" At one level, we should reject this suggestion because we cannot change the rules of the game after the game is over: we had previously said that 3.84 was the pass/fail threshold, and should not in effect lower the threshold because of the observed results. In any case, the hypothesis that "best fits" the available data is not necessarily more accurate than other candidates that passed the test.

An example illustrates this last point. Suppose that a coin is to be tossed 100 times, and the Professor hypothesizes that the coin is fair. Mendel, more as a joke than anything else, contends that the coin has a 53% chance of coming up fair. If exactly 52 heads arise, then Mendel's expected number of 53 is only off by one, while the Professor was off by two. If they had set up χ^2 tests, Mendel's discrepancy score would be only about 1/4 as large as the Professor's (.04 vs. .16). But both of their hypotheses passed the χ^2 test with flying colors: If the Professor based his hypothesis on careful reasoning while Mendel did not, to favor Mendel because of the slightly better fit would be mechanical and artificial (to say nothing of indecent and rude).

The moral of the story? We should try to identify the range of hypotheses that are consistent with the data before settling on one of them. If the range is wide, then if we proclaim one hypothesis the "winner" we could be a good distance from the truth.

EXERCISE 5.10 Suppose that a fifth analyst joined the meeting in Example 5.10 and offered hypothesis H_5: **Ad B** is 1.25 times as effective as ad A in stimulating sales.

(i) Under H_5, what is the probability that someone who purchased the product saw ad A rather than B?

(ii) Does H_5 pass a two-category χ^2 test given that 18 of 30 purchasers saw ad A?

Down But Not Out: The Rejected Hypothesis

But much as a hypothesis that is accepted might not even be approximately true, a hypothesis that is rejected might not be meaningfully false. We have endlessly spoken of α, the chance of accidentally rejecting a correct hypothesis. Yet even if the rejection is appropriate, the imperfection of the hypothesis might be so small as to be of scant practical importance.

EXAMPLE 5.11 **Take Heart**

Suppose that a new medication is hypothesized to reduce the risk of a fatal heart attack by 40% within a certain population, from a ten-year risk of 10% to a risk of 6%. The drug has already been approved for widespread use, and 100,000 people who take it will be followed for a decade. The test of H_0: *Ten-year risk of 6%* will be analyzed via a χ^2 test as follows:

Category	e_i	a_i
Fatal Heart Attack	6000	
No Fatal Heart Attack	94,000	

(i) Suppose that the drug actually cuts the heart attack risk among these people to 6.2% over the ten-year period. What would be the accept/reject outcome of the test?

(ii) Do you believe that the test outcome shows that the drug was unsuccessful?

Solution

(i) The test will have one degree of freedom ($M = 2, M_p = 1, Q = 0$), and a usual pass–fail threshold of $c = 3.84$. If the fatal heart attack rate is 6.2% for the ten-year period, then $a_1 = 6200$ and $a_2 = 93{,}800$. The discrepancy score would follow:

$$\chi^2 = \frac{\sum_{i=1}^{2}\left(e_i - a_i\right)^2}{e_i} = \frac{200^2}{6000} + \frac{200^2}{94{,}000} = 7.09$$

Thus, we would reject H_0 under the test.

(ii) What happened here is that the drug reduced fatal heart attacks by 38% (from 10% to 6.2%). That is a bit below the hypothesized benefit of 40% and, given the huge sample size, even this difference of two percentage points is too large to attribute to chance alone. But does that mean the drug was unsuccessful, when it came very close to meeting expectations? To cite the outcome as evidence that the drug "failed" would seems excessive, and to imply that the success rate was far below 40% would seem unfair.

Especially when a hypothesis is tested against lots of data, it is being placed under a mathematical "microscope" which can detect even minor blemishes. Even if H_0 is rejected, it still might serve as an excellent *approximation* to what is going on. Even the laws of gravity are not exact: if an object is dropped from a rooftop, then its descent to the ground will be slower than expected because of air resistance. But no one has suggested that the laws of gravity are unhelpful because of this minor effect. Likewise, do not be too quick to dismiss a hypothesis solely because it failed a statistical test; look at the actual data to get a sense of how far off it was.

EXERCISE 5.11 For the χ^2 test in Example 5.11:

(i) What range of values of X would lead to acceptance of H_0, where X is the actual number of heart-attack sufferers? (Hint: Express the discrepancy score in terms of X.)

(ii) What is the probability that H_0 will be rejected if the medication actually reduces the risk of fatal heart attack by 38% instead of the hypothesized 40%? (By this we mean that the chance of a fatal heart attack is independently .062 for each person in the large sample.)

(iii) What is the power of the test of H_0 : 6% risk against the alternative H_a: 6.2% risk?

EXAMPLE 5.12

Mendel's Golden Goose?

Mendel had just moved to California in 2014, and he hoped to support himself in the Golden State through clever gambling in the California state lottery. But first, he had to assure himself that the numbers that arise in the Daily 4 drawings are actually random. In those drawings—which occur once a day, seven days a week—a *four digit* number in the range 0000 to 9999 is selected. To make an assessment about the randomness of the lottery, Mendel decided to test the winning numbers for all 365 days in 2013. When setting the tests, he had not looked at the numbers.

Mendel imagined four ways to evaluate the randomness of the lottery:

- He wanted to see whether the numbers were about uniformly distributed over the range 0000 to 9999. For example, he expects about the same number of selections between 0 and 999 as between 7000 and 7999.
- He expected odd and even numbers to arise with approximately equal frequency.
- He anticipated that the outcomes on consecutive days would be independent. For example, the fact that a high number came up on June 9 should neither decrease nor increase the chance of a high number on June 10.
- Mendel's astrologer told him that the Fibonacci numbers below 10,000 are especially likely to arise in the lottery. The Fibonacci numbers form a series that starts with 0 and 1, with each subsequent number the sum of the previous two. Thus, the sequence starts (0, 1, 1, 2, 3, 5, 8, 13, 21 …). Mendel wanted to assure himself that the astrologer is not on to something.

With these considerations in mind, Mendel wanted to construct a series of four tests of the hypothesis H_0 that the lottery is *completely random*, meaning that, each day, each of the numbers in the range 0–9999 independently has the same chance (of 1/10000) of coming up. He wanted the tests themselves to be essentially independent, and wanted the overall risk of accidental rejection of H_0 to be 5%. Had he consulted us, how might we have advised him to proceed?

Solution

If there are to be four independent tests with a total α-value of .05, then the Bonferroni approach suggests that each test should have an α-value of .0125. Why not construct the four tests one-at-a-time, and then assess whether the collection meets our requirements?

Uniform Spread of Outcomes

Here we could proceed in analogy with our hypothesis tests about uniform birthdays, and use the χ^2-methodology. We could divide the range from 0 to 9999 into subdivisions of equal size that seem natural. Perhaps a starting point would be the ten ranges: 0–999, 1000–1999, … ,9000–9999.

Given that we have 365 data points involving numbers drawn in 2010, we would expect 10% $*$ 365 = 36.5 outcomes in each of the ten ranges. We know that we could create many more than ten categories of equal size which would have e_i's at least five: one possibility would be 40 ranges each of length 250: 0–249, 250–499, 500–749, … . But suppose Mendel has made it clear that he does not want a χ^2 test with 40 categories: he wants to look at the Table and absorb easily what it says about the expected vs. actual spread of data points across categories. To humor him, we could stick with ten categories, and set up the table:

Range	Expected Number of 2013 Lottery Outcomes Under H_0	Actual Number of 2013 Lottery Outcomes
0–999	36.5	
1000–1999	36.5	
2000–2999	36.5	
3000–3999	36.5	
4000–4999	36.5	
5000–5999	36.5	
6000–6999	36.5	
7000–7999	36.5	
8000–8999	36.5	
9000–9999	36.5	

In the χ^2 test associated with this Table, the number of degrees of freedom would follow: $\psi = M - M_p - Q = 10 - 1 - 0 = 9$. In setting a pass–fail threshold, we want α to be .0125 rather than the usual .05; this is only the first of four tests we are devising for Mendel. We turn to an electronic χ^2 table and find that $c = 21.03$ fits the bill.

Odd/Even

This is an easy one: there are only two possible categories in a χ^2 test: odd number and even number. Given that half the possible numbers drawn are odd and the other half even, we would expect the 365 numbers selected in 2013 to follow:

Category	Expected Number of 2013 Outcomes Under H_0	Actual Number of 2013 Outcomes
Odd Number	182.5	
Even Number	182.5	

The test would have $\psi = M - M_p - Q = 2 - 1 - 0 =$ one degree of freedom. The electronic table informs us that pass/fail threshold for $\alpha = .0125$ is $c = 6.24$.

Independence on Consecutive Days

Recognizing Mendel's preference for simple tests over complicated ones, we might check for one major deviation from independence of consecutive outcomes. Suppose that we break the lottery outcomes into two ranges of equal size: numbers from 5000–9999 are classified as *high*, while those from 0–4999 are classified as *low*. We might say that a *high/low outcome* arises on a given date like June 10 if, over the two-day period June 9–June10, one of the outcomes is high and the other low. *Order does not matter*: we have a high/low outcome on June 10 if 3368 arose on the 9th and 9231 on the 10th *or* if 9231 arose on the 9th and 3368 on the 10th. If, however, we had 7402 on the 9th and 6333 on the 10th, we would not have a high/low outcome on the 10th.

To find the probability of a high/low outcome on any day (starting on January 2), we note that:

$$\begin{aligned} P(\text{high/low on day } k) &= P(\text{high on } k \text{ and low on } k-1) \\ &\quad + P(\text{low on } k \text{ and high on } k-1) \end{aligned}$$

Under H_0:

$$P(\text{high on } k \text{ and low on } k-1) = P(\text{high on } k)P(\text{low on } k-1) = (1/2)^2 = 1/4.$$

Likewise: $P(\text{low on } k \text{ and high on } k-1) = 1/4$
Therefore: $P(\text{high/low on day } k) = 1/2$

While it might not immediately be obvious, the high/low outcomes on two consecutive days are independent. Having a high/low outcome on (say) August 14—based on the numbers drawn on August 13 and 14—does not affect the chance of a high/low outcome on August 15—based on the drawings of August 14 and 15. The fact that the August 13 outcome disagreed with that on August 14 does not affect the chance that August 15 disagrees with August 14. That probability remains 50%, the same as it was before we heard of the August 13/14 disagreement.

In short, we have the essentially same χ^2 test table as for odd and even numbers.

Category	Expected Number of 2013 Outcomes Under H_0	Actual Number of 2013 Outcomes
High/Low Day	182	
Not a High/Low Day	182	

Note: The tally starts on January 2nd, meaning there are 364 days of interest rather than 365.

Once again, the pass/fail threshold to achieve $\alpha = .01$ is $c = 6.24$.

Fibonacci

There are a total of 20 Fibonnaci numbers below 10,000:

0, 1, 2, 3, 5, 8, 13, 21, 34, 55, 89, 144, 233, 377, 610, 987, 1597, 2584, 4181, 6765

Because 20 is 0.2% of 10,000, we would expect $365 * .002 = \mathbf{0.73}$ Fibonacci numbers to arise over a one-year period under random selection. Hardly enough to allow a χ^2-test. But we could try a p-value test, in which we reject H_0 if the computed p-value is below .0125. Given that the astrologer suspects that Fibonacci numbers will be overrepresented in the drawing (alternate hypothesis H_a), a "hierarchy of hostility" could specify that, the higher the number of Fibonacci numbers among the 365 selections in 2013, the more hostile the outcome is to the hypothesis of pure randomness. Thus, the relevant test would be:

- Let random variable X be the number of days during 2009 when Fibonacci numbers were drawn. Then, under the hypothesis of randomness, X should be *binomially* distributed with $n = 365$ and $p = .002$.
- If k Fibonacci numbers came up in 2013, then find $P(X \geq k \mid X \text{ is } B(365,.002)$. That probability is the p-value of the observed result.
- If p-value $\geq .0125$, then accept H_0; if p-value $< .0125$, reject H_0.

If we turn to the binomial table, we learn that the p-value $< .0125$ if k is four or greater.

Even if we accept H_0 in this Fibonacci test, it still has to pass the three other tests to get full acceptance. If we reject H_0 in the Fibonacci test, however, then the hypothesis is rejected at once under the "sudden death" Bonferroni procedure we are using. Thus, searching for four or more Fibonacci numbers is a good way to start.

Note that the four tests outlines are approximately independent under H_0 even if not perfectly so. Suppose, for example, that by chance there are 48 numbers in the range 2000–2999 rather than the expected 36.5. We would still expect under H_0 that these numbers are half odd and half even, meaning that an unusual result in the uniformity test does not raise the chance of a similar outcome in the odd/

even test. Because the Fibonnaci numbers below 10,000 are highly concentrated in the range 0–999, an excess of Fibonnaci numbers in the 2013 data might imply an excess in the 0–999 range in the uniformity test. But that is a very small effect that does not cast doubt on the overall testing program.

Now what? Are you curious about whether the 2013 California data pass all four tests? The author hopes so, for he points out that the numbers are available at the California Lottery website and he invites you to implement the tests on your own (Exercise 5.28).

EXERCISE 5.12

(i) Why not make up a (single) test of your own for the randomness of the Daily 4 lottery? (Don't worry if the test seems flaky: look at the test with Fibonacci numbers!)

(ii) How would you implement your test? What would be the test statistic, and what would be the rejection region?

The Takeaway Bar

Glossary

A **one-sided** statistical test rejects H_0 if the test statistic is far enough out on one (previously specified) tail of its distribution under H_0.

A **two-sided** statistical test rejects H_0 if the test statistic is far enough out on either tail of its distribution under H_0.

The **power** of a statistical test of hypothesis H_0 relative to the alternate hypothesis H_1 is the probability of *correctly* rejecting H_0 under the test rules, given that H_1 is true.

In a power calculation, the **alternate hypothesis H_1** is as specific as is the null hypothesis H_0, and it is assumed that H_1 is true if H_0 is false.

The ***p*-value** in a test of hypothesis H_0 is the probability of getting an outcome at least as hostile to H_0 as the one that actually arose, assuming that H_0 is in fact true.

In a p-value calculation, the **alternate hypothesis H_a** describes in general terms what we believe would be true if H_0 is false. Sometimes, H_a simply the hypothesis that H_0 is false; other times, H_a indicates why H_0 is false if in fact it is.

The **hierarchy of hostility** in a statistical test ranks the possible outcomes from most to least hostile to the hypothesis H_0 that is being considered. The ranking relates to whether the outcome increases or decreases the plausibility of H_0 relative to the alternate hypothesis H_a.

The **Bonferroni method** is a "sudden death" approach in which a hypothesis H_0 is subject to several independent tests, and is accepted only if it passes all of them. The individual tests are constructed so that the overall chance of accidentally rejecting H_0 (i.e., committing a Type I error) cannot exceed some specified level, often 5%.

The **range of acceptable hypotheses** indicates the diversity of hypothesis that would be accepted in statistical tests, based on a particular data set. If the range is wide, it means that the data are consistent with highly divergent possibilities.

The Formula Bar

Power of a test of H_0 against alternate hypothesis H_1:

$$\text{Power} = P(\text{reject } H_0 | H_1 \text{ true}) = 1 - P(\text{accept } H_0 | H_1 \text{ true}) = 1 - \beta$$

where $\beta = P(\text{accept } H_0 | H_1 \text{ true})$

p-value for hypothesis H_0 given observed outcome A

$$p\text{-value} = \sum_{Z_A} P\left(X = z_i | H_0 \text{ true}\right)$$

where the z_i's are the mutually exclusive outcomes in set Z_A, which contains those realizations of test statistic X at least as hostile to H_0 as the observed outcome A (under the prevailing "hierarchy of hostility")

Bonferroni Standard:

Suppose that hypothesis H_0 will be subject to k individual tests and will be accepted only if it passes all of them. Then, to keep the risk of Type I error from exceeding α, the risk α_k of accidental rejection in any given test is set according to:

$$\alpha_k = \alpha / k$$

Some Parting Thoughts:

- In general, the rejection region in a hypothesis test consists of outcomes that are unusual under H_0 but more typical under the alternative H_a that we think would be true if H_0 were false. Outcomes that are unusual under H_0 but even more so under H_a would not fall in the rejection region.
- The power of a statistical test of H_0 tells us the chance of correctly rejecting H_0 when it is wrong for a specific reason (i.e., alternate hypothesis H_1 is correct). If the power is low even against a very different hypothesis, then there is a real danger of accepting H_0 even if is very far wrong.
- Both p-values and power calculations involve alternatives to the null hypothesis H_0 that is being tested. Both they do so in different ways. In p-value calculations, the alternative H_a is quite general (e.g., H_0: $\mu = 0$ vs. H_a: $\mu \neq 0$) while, for power, the alternative hypothesis H_1 is specific enough to allow calculation of the probabilities of all possible test outcomes if H_1 is true(e.g., H_0: $\mu = 0$ vs. H_1: $\mu = 1$). The power of a test varies depending on which particular alternative to H_0 is under consideration.
- An advantage of conducting hypothesis tests with p-values is that the test statistic is transparent, and has the same general interpretation from test to test. The meaning of (say) χ^2 discrepancy score of 8.67 changes depending on how many degrees of freedom are present, but a p-value of .03 means the same thing regardless of the setting. Even if a p-value is simply compared to an accept/reject threshold like 0.05 in a formal test, it makes clear whether the hypothesis was easily accepted, barely accepted, barely rejected, or strongly rejected.
- A "sudden death" series of k tests of H_0, each with a risk α / k of accidental rejection, does not have an overall risk of accidental rejection of exactly α. (Type I errors in the individual tests, after all, are not mutually exclusive events.) As a practical matter, however, the risk very close to α if the tests are approximately independent.
- Identifying the range of acceptable hypotheses against a given data set is analogous to finding the confidence interval for a given parameter rather than stopping with a point estimate. It is an antidote to treating a given hypothesis as undisputed truth just because it passes a statistical test.

- Even if an observed difference is statistically significant, it might be of scant practical significance. With large amounts of data, even small "blemishes" in a hypothesis can cause its rejection. We should therefore look over the full test results in reaching judgments rather than become prisoners of p-values.

EZ Pass

1. In a two-sided statistical test, the hypothesis is rejected if the test statistic:
   (A) is far enough out on either tail of its distribution under the hypothesis
   (B) is near the middle of its distribution but on either side
   (C) is at the upper tail of its distribution
   (D) takes on two distinct values
2. If hypothesis H_0 is only slightly incorrect, then a test of H_0 with $\alpha = .05$ could be expected to:
   (A) have a β-value of .05
   (B) have a β-value of .50
   (C) have a β-value near .95
   (D) have a β-value near zero
3. To test the hypothesis H_0 that (exactly) 50% of Florida residents favor on-line gambling, 400 such residents will be chosen at random. If $\alpha = .05$ for a χ^2-test, its power against the alternative that 90% of Florida residents favor on-line gambling is:
   (A) Exactly .95
   (B) Less than .50
   (C) Greater than .95
   (D) Impossible to determine
4. If a hypothesis test in a medical journal is accompanied by the statement "p-value < .001," then:
   (A) the hypothesis has at least a 999 in 1000 chance of being true
   (B) the hypothesis has at most a 1 in 1000 chance of being true
   (C) if the hypothesis were true, the chance of getting an outcome at least as hostile to it as the one observed is more than 1 in 1000
   (D) if the hypothesis were true, the chance of getting an outcome at least as hostile to it as the one observed is less than 1 in 1000
5. When hypothesis H_0 is tested against alternative H_a, the calculated p-value is .06. Consider the two statements:
   (i) If H_0 were true, the chance is 94% that the test outcome would have been less hostile to it than the actual outcome
   (ii) If H_0 were true, the chance is 6% that the test outcome would have been more hostile to it than the actual outcome
   Which of these statements are true?
   (A) Both (i) and (ii)
   (B) (i) but not (ii)
   (C) (ii) but not (i)
   (D) Neither (i) nor (ii)

6. Suppose that a hypothesis H_0 is tested in two different ways, each with $\alpha = .05$, but the two tests are not independent. Then the chance of rejecting H_0 under at least one of the tests:
   (A) equals $2 * (.05) * (.95)$
   (B) equals $(.05)^2$
   (C) equals .10
   (D) cannot be determined from the given information

Further Exercises

EXERCISE 5.13 In UCLA's 2009 annual survey of students entering US colleges, 41.6% reported that cost was a "very important factor" in deciding which college to attend. Minerva hypothesizes that the statistic of 41.6% applies at the school where she serves as financial aid officer. To test this hypothesis, she decides to contacts 100 students at random via e-mail and asked them whether cost was a "very important factor" for them.

(i) Assuming that all 100 students will respond (honestly), formulate a χ^2 test of her hypothesis with a 5% risk of accidental rejection.
(ii) Suppose that, in reality, cost was a very important factor for 50% of students at the school. What is the probability that Minerva will erroneously accept the 41.6% statistic based on her test? (Hint: What outcomes would lead to rejection under the test in (i)?)
(iii) What is the power of her test to reject 41.6% when the true statistic for her school is 50%?

EXERCISE 5.14 In Exercise 1.6 of Chapter 1, we had the following description:

> "Everyone buying a pizza at JR's pizzeria gets a card that conceals either a J or an R. When the wax covering is rubbed off, the chance is 80% that a J will appear. Different cards can be treated as having independent letters. Once a person has both a J and R, she is entitled to a free pizza."

The Consumer Affairs Office of Dallas (where JR's is located) gets a complaint from a JR's customer, who alleges that the chance of getting an R is less than 20%, meaning that the pizzeria is engaging in fraud. The office decides to conduct a furtive investigation.

(i) If p is the probability of getting an R, what is the probability that it will take exactly k purchases before the customer is eligible for a free pizza? (Hint: There are two ways to win a pizza on the kth purchase.)

The investigator will buy up to ten pizzas on different occasions, and will check the letter on the card for each one. If he is eligible for a free pizza at the tenth purchase or earlier, he will drop the investigation and conclude the JR's is innocent. Otherwise, he will conclude that JR's is guilty and recommend that it be penalized.

(ii) What is the probability that he will erroneously conclude that JR's is committing fraud when in fact $p = 20\%$?

(iii) In terms of p, what is the probability that he will erroneously conclude that JR's is innocent for some particular $p < 20\%$? (Answer in terms of p.)

(iv) For a specific value of p in the range $0 \leq p < 20\%$, find the power of this statistical test to reject the hypothesis the JR's is innocent.

(v) Construct a *power* curve for $0 \leq p < 20\%$, which plots power vs. p under your formula from (iv).

EXERCISE 5.15 Fears are expressed that 28% of the swimsuits produced by a certain company are radioactive. To test this 28% hypothesis, Minerva takes a random sample of ten such swimsuits and measures their levels of radioactivity. The alternate hypothesis H_a is that the rate is greater than 28%.

Suppose that she finds that exactly three of the swimsuits are radioactive. What is the p-value of this outcome?

EXERCISE 5.16 In a jury of twelve members, one is Asian-American. The percentage of Asians among those eligible to be jurors is 15%. Under the hypothesis that there is no discrimination against Asian-Americans in jury selection, what is the p-value of this particular outcome? (The alternative H_a is that eligible Asian-Americans are less likely to chosen as jurors than other citizens.)

EXERCISE 5.17 It is hypothesized that, among movies that play in both the United States and the United Kingdom, there is no relationship between box office successes in the two countries. To test the hypothesis, the box office receipts of 54 recent movies in the two countries will be gathered. Each movie will be classified as in the top third/middle third/bottom third of receipts in each country.

(i) Devise a sensible nine-category χ^2-test of this hypothesis. What are the e_i's?

(ii) How many puppet categories are there in this test? (How many a_i's do you need to determine all nine?)

(iii) If the p-value of the test outcome is .14, what was the discrepancy score? (The alternative H_a is that box office receipts in the two countries are related in some way.)

(iv) Would the hypothesis have been rejected under a standard χ^2-test with 5% risk of accidental rejection?

EXERCISE 5.18 To test uniform birthdays (H_0), suppose that 60 people will be picked at random. The test statistic will be V, the number of those people who share a birthday with at least one other person selected.

(i) What is the hierarchy of hostility of possible outcomes, where the alternate hypothesis H_a is that birthdays are not uniformly distributed?
(ii) What is the p-value of the outcome $V = 0$?
(iii) What is the p-value of the outcome $V = 2$? (Hint: What is $P(V = 0)$ if H_0 is true?)

EXERCISE 5.19 The *New York Times* of March 9, 2010 reported that:

> "Over the 36 years Warren E. Burger and William H. Rehnquist served as chief justices, there were on average of three dissents from the bench each term. In the first four years of the court under Chief Justice John G. Roberts, the number rose by a quarter, to 3.75 (per term)."

Suppose that when Chief Justice Roberts took over, Mendel hypothesized that dissents from the bench under Roberts would occur under a Poisson process with a mean of three per year.

(i) Given the data above and the Poisson model, what is the p-value of the observed outcome for the first four years under Roberts? (Assume that the alternative H_a is that the mean number of annual dissents from the bench would rise under Roberts.)
(ii) Give an example of four yearly numbers of dissents (y_1, y_2, y_3, and y_4) that average out to 3.75 per term, but which would be highly unusual as outcomes of a Poisson process with rate three per year.
(iii) Was your p-value calculation in (i) sensitive to the deviation from a Poisson process that you raised in (ii)?
(iv) If you answered "no" to (iii) (as you should have), does that circumstance mean that a p-value can be misleading if based only on aggregate data? Discuss briefly.

EXERCISE 5.20 In the age-discrimination case in Example 5.8 (with 20 workers over age 40 and 13 below that age), suppose that 7 employees were laid off and 6 of them were over 40. We wish to compute the p-value of that outcome under the "urn lottery" model of randomly-generated layoffs. (The alternate hypothesis H_a is discrimination against older workers.)

(i) What is the overall chance that exactly 6 of the 7 laid off would be over 40? (Hint: Think hypergeometrically.)
(ii) What is the p-value of the observed outcome under the urn lottery?

EXERCISE 5.21 The founders of a new website for cats hypothesizes that 60% of the first-time visitors to the site will return within the next month. Others are more skeptical, and think the return rate will be below 60% (H_a). Thirty cats that visit the site will be followed, and the number who come back within a month will be recorded.

(i) If X is the number of cats who return over the next month, what is the "hierarchy of hostility" of outcomes to the null hypothesis?
(ii) If exactly 13 cats come back in the next month, what is the p-value of the outcome?
(iii) In a p-value test of H_0 with a .05 accept/reject threshold, what is the outcome?

EXERCISE 5.22 In the banner ad problem of Example 5.10, suppose that it is desired to test H_1: *Ads A and B are Equally Effective* against H_4: *Ad A Is Four Times as Effective as Ad B*. It is decided to conduct a test with a 5% risk of Type I error for H_1 that is based on X, the number of orders out of the next 30 received that were placed by customers who saw ad A.

(i) Assuming that an appropriate χ^2-test will be used, what is the power of the test of H_1 against the alternative H_4? (Hint: First find the values of X for which H_1 would be accepted under this test.)
(ii) Suppose instead that it is desired to calculate the p-value of the result "18 A's out of 30" for hypothesis H_1, against the generalized alternative H_a that ad A is more effective than ad B. What is the "hierarchy of hostility" of X-values in this test?
(iii) What is the p-value of the test result in (ii)? (Hint: Instead of a χ^2-test, why not use a binomial test based on the distribution of X under H_0?)

The next three exercises relate to the Excel File ANSKEY at the website, which contains *actual* answers to a (very well known) standardized multiple-choice test. There are five possible answers to each question: A, B, C, D, and E. Suppose that the company that prepares the test says that the answer keys to its exams are completely random in every way, and you want to test that hypothesis against the data in the file.

In the file ANSKEY, column A contains the first 49 answers, column B contains the next 49, and column C contains the last 49.

EXERCISE 5.23 (i) *Before looking at the ANSKEY data*, devise a test of the hypothesis that each answer has a 20% chance of being A, 20% of being B, etc. What kind of test are you proposing? What are its details, and what is the chance of Type I error?
(ii) Suppose that the first 20% of the answers are all A's, the next 20% are all B's, etc. Would this deviation from pure randomness show up in your test above?

(iii) Indeed, against the pattern in (ii), what is the probability that the test you performed would detect the deviation from pure randomness?
(iv) Now look at the actual data and perform the test. What do you conclude?

EXERCISE 5.24 Now we want a test of whether different answers in ANSKEY are independent. For $j \geq 2$, we say the answer to question j is "matched" if it is the same as the answer to question $j - 1$.

(i) What is the probability under pure randomness that the answer to question j is matched?
(ii) If the answer key is random, does the fact that question j is matched change the probability that question $j + 1$ is matched?
(iii) How many matches would you expect out of the 147 items on the answer key? (Hint: Under the definition, there can't be a match for the first answer.)
(iv) Please devise a statistical test on the randomness of the answer key based on the number of matches. (Strive for a risk of accidental rejection of 5%.)
(v) Now look at the data, and perform your test. What is your conclusion?

EXERCISE 5.25 Now we want to perform both tests on the ANSKEY data discussed in the last two questions, and to combine the results under the Bonferroni "sudden death" procedure.

(i) How would you revise your two test rules so that the total risk of accidental rejection of H_0 is 5%?
(ii) Given the data, what is the outcome of your new test with two distinct parts?

EXERCISE 5.26 Suppose that you plan to test a hypothesis in three different and independent ways, each of which involves a χ^2 test. The first test has eight degrees of freedom, the second test has four, and the last has two.

(i) Under the Bonferroni sudden death procedure, what should be the chance of accidental rejection under each test so that the total risk of accidental rejection is 5%?
(ii) What should be the pass/fail thresholds on each test to meet the conditions set out in (i)?
(iii) If the hypothesis is in fact true, what is the probability of failing all three of the individual χ^2 tests?
(iv) If the hypothesis is in fact true, what is the probability of failing at least two of the three of the individual χ^2 tests?

EXERCISE 5.27 It is hypothesized (H_0) that consumer reaction to a flashlight/toothbrush will be normally distributed on some satisfaction scale with mean 6 and standard deviation 1.2. Different consumers are assumed independent in their scores. The hypothesis will be tested in a "sudden death" manner as follows:

A first consumer's score will be obtained and:

- If the score is below 3.2 or above 8.8, H_0 will be rejected
- If the score is in the range (3.2, 8.8), a second consumer will be selected.

If a second consumer's score is obtained, then:

- If the score is below 3.2 or above 8.8, H_0 will be rejected
- If the score is in the range (3.2, 8.8), a third consumer will be selected

This process continues if necessary until a tenth consumer is selected:

- If the tenth consumer's score is below 3.2 or above 8.8, H_0 will be rejected.
- Otherwise, H_0 will be accepted.

(i) Under this test, what is the probability of rejecting H_0 when it is true?

(ii) Suppose that one desired to modify this sudden death test so that the overall chance of accidental rejection is 5%. What thresholds would replace 3.2 and 8.8, assuming that they must be the same distance from 6 in opposite directions?

(iii) Suppose that, in reality, consumer scores for this product are normally distributed with mean 4.5 and standard deviation 1.2. What is the probability that H_0 will erroneously be accepted under the original procedure with the thresholds 3.2 and 8.8?

EXERCISE 5.28 You knew this was coming. Please go to California lottery website and obtain with the 2013 California lottery data discussed in Exercise 5.12 ("Mendel's Golden Goose") and perform the four tests outlined there. (If you want to use a more recent year, that's fine.) What is the conclusion about randomness of the lottery under Mendel's procedure?

EXERCISE 5.29 Please add your fifth test from Exercise 5.12 to the mix. Now, perform a Bonferroni test on all five outcomes. Assume that your test was approximately independent of the other four.

CHAPTER 6

More Statistical Sampling

The 7% Solution (Based on a true story)

Over a one-year period, an insurance company's payouts over 20,000 claims were 7% larger than the company had projected. The effect on the company's bottom line was substantial, and some enraged stockholders filed a lawsuit, contending that the company's payout projections (and thus the overly-low premiums it had charged its customers) demonstrated gross incompetence. The company responded that it had suffered a temporary spasm of bad luck: the number of claims was not unusually large, but an unexpected proportion came from the upper tail of the long-term payout distribution. Oddly, however, it refused to provide any further information about either the distribution or the individual payouts.

"Could the company possibly be correct?" the lawyers representing the stockholders asked the statistician. "With a sample size of 20,000, how could there possibly be a sampling error of 7%?" "That's exactly the question we need to answer," the statistician replied, "and we have to do so without the information that could provide the answer."

To be continued

At the end of Chapter 3, we noted that, despite all we had accomplished in our introduction to statistical sampling, we left several key questions unanswered. Here we take up many of these questions. Much of this chapter's material does not usually appear in a first course in Statistics, so your instructor might choose not to cover all the topics. Because what we discuss here is important, however, you might find yourself returning to this chapter in the future to learn material that you do not cover now. The different sections of this chapter are essentially self-contained: the treatment of stratified random sampling, for example, does not depend on material in other sections, and they in turn do not depend on stratified random sampling.

We start our work by considering various threats to random sampling, as well as some possible remedies. Then we discuss a modification to the sampling formulas we've seen, and a different approach to drawing a data sample. We move on to consider bootstrapping, a modern statistical method that greatly enhances our ability to learn from random samples. Then we end with two challenging problems in sampling.

6.1 Random Sampling: Easier Said Than Done

We have an elegant theory for extrapolating from random data samples to full populations, but achieving a sample that is genuinely random is not always simple. Under random sampling, each member of the population independently has the same chance of being selected. But the very manner in which the sample is drawn can unintentionally favor some population members over others. And even if the initial process for selecting the sample is random, there may be a filtering process by which some of those selected "drop out" of the sample, yielding a final sample that is anything but random. In all these circumstances, we have suffered a **selection bias**. Below we discuss several selection biases from the real world.

Illustration 6.1a: Crime Wave

A police department wanted to know the average number of arrests over some period among those people it had arrested at least once. Nowadays, one can quickly get the exact answer from a spreadsheet of individual arrests but, back in the day, each arrest led to one card that was placed in a huge file. To estimate the average, the department picked n cards at random from the file and, for each individual selected, found the number of other cards in the file for that person. If a person initially chosen had three additional cards, for example, his total arrest number was recorded as four (3 + 1). The various arrest numbers were averaged to get a sample mean $\bar{X}$, which was used as the estimate of mean arrests per capita (μ) among those in the file.

There's a problem with that procedure. Suppose that Person A has been arrested seven times, while person B has been arrested once. Then A has seven cards in the file while B has one, meaning that A is far more likely to be selected for the "random" sample than B. When the sampling is biased towards "arrest prone" individuals, the sampling error $W = \bar{X} - \mu$ does not have mean zero; instead, there is a systematic tendency for $\bar{X}$ to exceed μ. Our formulas about random sampling error are worthless in such a situation.

Illustration 6.1b: Midway to Nowhere

We already encountered this example in Chapter 2, in connection with a poll about passenger satisfaction with Midway Airlines. To assess Midway's popularity Airlines among air travelers between New York and Chicago, the company took a poll aboard Midway Airlines planes. This sampling of New York-Chicago air travelers was not random: those who did not fly Midway had *zero* chance of being selected. This deviation from random sampling may explain why Midway did so much better in the poll that in actual traffic on the route.

Illustration 6.1c: Draft Choice

Back in 1969, when the United States still had compulsory military service (the "draft"), it was decided to select draftees via a birthday lottery. One ball for each of the 366 birthdays was placed in a large urn, and then the balls were removed from the urn one by one. The birthday selected first was assigned ranking number 1, meaning that people born on that day (which turned out to be September 14) were at the top of the list to be drafted. June 8 was the last number chosen; those born that day ranked 366 and were almost certain not to be drafted.

There was a problem, however, in the way the balls were placed into the urn. The balls for months towards the end of the year were concentrated at the top, while those for early months fell towards the bottom. Not surprisingly, the balls on the top tended to be selected sooner than the balls below them. Thus, not all birthdays had an equal chance of being ranked in the range 1–195 (the range from which draftees were selected in 1970): fully 26 of the 31 birthdays from December fell into that range, as compared to only 13 of the 31 from January. This apparent deviation from randomness was widely noted and sharply criticized.

Illustration 6.1d: Polls Apart

None of the many polls about political sentiment in (say) Australia can literally be described as random. If the poll is conducted by (say) telephone land-lines, then those who have only cell phones have zero chance of being selected. If the poll is conducted via the Internet, then older voters who never go on-line cannot be chosen. Attempts are made to correct for sampling biases, but they require assumptions. Suppose, for example, that, in an on-line poll, only 8% of respondents are above age 60, as opposed to 18% in the general population. Then one can "overweight" the over-60 results from the poll so that they count as 18% in an overall estimate. That procedure would work if one can assume that the views of older people who use the Internet are similar to those of older people who don't. But can one?

These four illustrations involved procedures that fail to achieve a random sample, but problems can arise even when the original sampling scheme is perfectly random. We turn now to a major hazard of that form.

Nonresponse Bias

Nonresponse bias is an occupational hazard of conducting surveys, because it frequently happens that some fraction of the people surveyed fail (for one reason or another) to provide responses. Often, such nonrespondents are ignored, while those who did answer are treated as a random sample from the full population. That approach is perilous.

Suppose, for example, that a university canvasses its graduates by mail about their recent professional achievements. A person who has just been promoted might complete the form promptly, while another who has just been imprisoned for embezzlement might be too depressed to respond. The tabulated results, therefore, could be artificially upbeat. More generally, a survey is hardly random when some members of the population are likelier to take part than others. The mathematics of random sampling has little relevance to such cases.

EXAMPLE 6.1

Home Front (Based on a True Story)

About two decades ago, a legal battle erupted between a mail-order house that sells clothing and the shipping company that delivers it. A central issue was what percentage of customers were receiving their packages at home rather than at work. (The higher the figure, the better for the shipper.) Yet that key quantity was not known for the period in dispute. To make an estimate of that "home delivery" proportion, the shipping company hired a surveying firm, which randomly chose 4000 of the clothing firm's customers from those who had received packages roughly three years earlier (a time frame pivotal to the lawsuit). The firm tried repeatedly to reach each of these 4000 people by telephone, but succeeded in only 35% of the cases. Of those customers successfully reached, 75% reported that they took delivery at residential addresses.

The shipper then argued that a random sample of $4000 \times .35 = 1400$ customers had yielded a 75% home-delivery rate. Applying usual sampling formulas like (3-38), it concluded that (.726, .774) was the 95% confidence interval for the home-delivery rate in the full population of customers. The shipper therefore asserted that a minimum of 72.6% of the packages delivered were destined for homes.

Is there any potential problem with this analysis?

Solution

There most certainly is a problem. The shipping company showed no curiosity about the 2600 customers out of 4000 who had not been reached in the random survey. But might the very fact that a customer wasn't accessible say something about where the package arrived? The surveyors, after all, had used three-year old phone numbers about places of delivery. If people are less likely to retain their work numbers for several years than their home numbers—as seems plausible, given that phone numbers are likelier to change when one changes jobs than when one changes homes—then the survey success rate among "commercial" customers may have been *lower* than for others. In addition, people at work may be especially unwilling to speak to strangers conducting surveys, or to respond to voicemail by such strangers.

These and other possibilities about the many unreached customers make it seem silly to treat 72.6% as a "lower bound" on the home-delivery rate. Of the 4000 people in the initial sample, only $4000 \times .35 \times .75 = 1050$ of them actually confirmed getting deliveries at home. It is theoretically possible, therefore, that as few as 26% (1050/4000) of the packages arrived at homes. That figure is surely too low, but the fact remains that the 75% estimate is subject to far greater uncertainty than the shipper suggested based on the 95% confidence interval. If the 1400 customers reached were unusually likely to have arranged for home deliveries, they can by no means can they be treated as a random sample for the purposes of the survey.

EXERCISE 6.1

Suppose that, in a compromise, the parties in the legal battle in Example 6.1 were willing to assume that the home-delivery rate was half as high among those not reached in the telephone survey as among those who were reached. What estimate of the overall home-delivery rate would be reasonable in light of that compromise?

Coping with Nonrandomness

Identifying cases in which randomness was not achieved does not imply that randomness cannot be achieved (or that selection biases cannot be corrected). We can demonstrate this point by revisiting the situations just discussed.

EXAMPLE 6.2

Double Counting?

Consistent with the situation in Illustration 6.1a, suppose that a random sample of 100 arrest cards out of many thousands are selected for a particular city and particular year, and that 60 of them they arose from individuals who were arrested once that year, and 40 who were arrested twice. (For those arrested twice, the second arrest card was also removed from the deck, meaning that no person could be chosen twice.) Given this outcome, what is a reasonable estimate of the mean number of arrests that year for those individuals with at least one arrest?

Solution

As noted, the threat to randomness here is that those with two arrests are essentially twice as likely to make it into the sample as those arrested only once. Suppose that f_i is the fraction of individuals in the file with exactly i arrests that year ($i = 1,2$). (We assume that $f_i = 0$ for $i \geq 3$, given the absence of any such individuals in the data sample.) Let N be the total number of individuals with arrest cards, and assume given the problem statement that $N \gg 100$. Then:

- The *total* number of arrest cards associated with "double arrestees" is $2Nf_2$
- The number of cards for single arrestees is Nf_1,
- The total number of cards is $2Nf_2 + Nf_1 = 2Nf_2 + N(1 - f_2)$.

Thus, the probability q_2 that the first card drawn is a from a double arrestee follows:

$$q_2 = \frac{2Nf_2}{2Nf_2 + N(1 - f_2)} = \frac{2f_2}{2f_2 + (1 - f_2)} \qquad (6\text{-}1)$$

Note that q_2 is independent of N. Because $N \gg 100$, (6-1) is also essentially the probability the second card drawn was from a double-arrestee, despite the deletion of the card or cards for the first person chosen. q_2 is also the expected fraction of the 100 cards drawn that come from double arrestees.

Here we know that 40% of the 100 cards drawn were from double arrestees. Treating .40 as an estimate of q_2, we can use (6-1) and simple algebra to approximate f_2 by f_2^* that is given by:

$$f_2^* = .25$$

What this estimate implies is that double-arrestees, who constitute about 25% of the population of those arrested, amassed about 40% of the arrests.

We can then estimate f_1 by $f_1^* = 1 - .25 = .75$. Thereupon, we can estimate μ, the average number of arrests among those individuals in the arrest file, by:

$$\mu \approx 1 * (.75) + 2 * (.25) = 1.25$$

EXERCISE 6.2 We will now argue that f_2^* in Example 6.2 is the maximum likelihood estimate of f_2^*.

(i) Let X be the number of double-arrestees among the 100 cards drawn. Explain why X is essentially $B(100, q_2)$ if $N \gg 100$.

(ii) Recalling a result from Chapter 3, explain why .4 is the maximum likelihood estimate of q_2 given the 40 double-arrestees came up.

(iii) Explain why the fact that .4 is the maximum likelihood of q_2 means that .25 is the maximum likelihood estimate of f_2.

The procedure in Example 6.1 can be generalized as an adjustment for the over-representation in the sample of certain population subgroups. It requires knowing in advance how the chance of selecting an individual varies from one subgroup to another. In Example 6.1, for example, we could deduce that a double-arrestee was twice as likely to be chosen as a single arrestee.

Among the other illustrations of nonrandom sampling that we discussed were the 1969 US draft lottery and the Midway Airlines advertisement. The US military draft is history, and so are such clumsy attempts to achieve randomness. In the United States and around from the world, lotteries are conducted every day for cash prizes, and there are elaborate methods to assure that the drawings are random. That we almost never hear of problems in these lotteries is quiet testimony to their success in achieving the randomness that bettors expect.

The deviation from randomness in Midway Airlines example is harder to rectify. If the relevant population is those people who fly between New York and Chicago, it is hard to see how Midway could find a random sample of all those travelers without the cooperation of its competitors (which is unlikely to be forthcoming). But does Midway really need a random sample? Could it not have asked its own passengers whether they had flown other airlines between New York and Chicago and, if they had and they prefer Midway, exactly why they prefer Midway? It could then have collected the reasons given and, without in any way suggesting that these testimonials arose from "random" travelers, presented those reasons in its advertisements.

What about the home delivery controversy in Example 6.1? Let's think.

EXAMPLE 6.3 **Home Improvement**

What could the shipping company have done in Example 6.1 to blunt the criticism that its calculations were biased because its sample of customers was not random?

Solution

Recall that only 1400 of the 4000 randomly-selected customers in Example 6.1 (35% of them) were reached in the telephone survey. Suppose that the company selected a random sample of (say) 100 of the 2600 nonrespondents, and then did whatever was necessary to find out about their deliveries. For example, the firm could have arranged for actual visits to the 100 mailing addresses. It could then make a credible estimate of f_{NH}, the fraction of nonrespondents who received home deliveries, and could then estimate f_H, the overall fraction of customers who received packages at home, by $\widetilde{f_H}$ which follows:

$$\widetilde{f_H} \approx .35 * .75 + .65\widetilde{f_{NH}}$$

where $\widetilde{f_{NH}}$ is the fraction of home-delivery customers in the new random sample. (As noted in Example 6.1, 75% of the 1400 customers (35%) who were originally reached had received their deliveries at home.)

If the new sampling was conducted properly, then $\widetilde{f_H}$ is an unbiased estimate of f_H.

EXERCISE 6.3 Suppose that $\widetilde{f_{NH}} = .45$ in Example 6.3, based on a random sample of 100 nonrespondents to the telephone survey. What is $\widetilde{f_H}$ in that case, and what is the standard deviation of the sampling error when using $\widetilde{f_H}$ to estimate f_H?

(Hint: In random sampling for a proportion in which the fraction "yes" is $\hat{p}$, what is the approximate variance of the sampling error associated with that estimate?)

6.2 The Finite Population Correction

Now we return to an issue that we touched on only briefly in Chapter 3. There we assumed that the sample size n in a random sampling exercise is very small relative to the population size N (or, effectively, that $N = \infty$). But what happens when an appreciable fraction of the population is included in the sample? Intuitively, we sense that the sampling result becomes *more accurate* in that case. Suppose, for example, that 40 families live in an apartment building and we want their mean income last year. If we obtain the incomes for a random sample of 38 of these families, then the sample mean should be very close to the population mean. After all, the sample contains 95% of the population (38/40): Even if the two families not sampled had different incomes from the rest, they are unlikely to change the overall mean by much.

We can address the issue quite specifically. Suppose that a population has mean μ and standard deviation σ, that the overall population size is N, and that a random sample is of size n is taken. If the population mean μ is to be estimated based on the sample mean $\bar{X}$, then:

When a random sample of size n is selected from a population of size N, the variance of the sampling error W when μ is estimated by $\bar{X}$ is given by:

$$\sigma^2(W) = \frac{\sigma^2}{n}\left[\frac{N-n}{N-1}\right] \tag{6-2}$$

The standard deviation of W therefore follows:

$$\sigma(W) = \left(\frac{\sigma}{\sqrt{n}}\right)\sqrt{\frac{N-n}{N-1}} \tag{6-3}$$

The quantity $\sqrt{N-n/N-1}$ in (9-3) is called the **finite population correction factor** because it adjusts the variance formula $\sigma^2(W) = (\sigma^2/n)$ that applies when the population size is effectively treated as infinite.

The derivation of (6-3) appears at the website. The key idea in the derivation is that, while the individual random selections can be treated as independent when $N = \infty$, they exhibit negative correlation when N is finite. If $X_1 > \mu$, for example, then $E(X_2) < \mu$ under sampling without replacement. (After all, the mix after X_1 is removed has an average less than μ.)

In estimating μ by $\bar{X}$, we generally do not know σ. In Chapter 3 where we effectively assumed that $N = \infty$, we proceeded to calculate a sample standard deviation S under (3-19). Then we assigned to the sampling error W the standard deviation $S_W = S/\sqrt{n}$. The adjustment to this formula when N is finite is the same as the one in (6-3):

If a random sample of size n is taken from a population of size N, the estimated standard deviation S(W) of the sampling error $W = \bar{X} - \mu$ follows:

$$S(W) = \left(\frac{S}{\sqrt{n}}\right)\sqrt{\frac{N-n}{N-1}} \tag{6-4}$$

where S is the sample standard deviation as defined by (3-24). The quantity $\sqrt{N-n/N-1}$ is once again called the ***finite population correction factor***, because it multiplies the earlier standard deviation of $(S/\sqrt{n})$ that applied when N was treated as infinite.

TABLE 6.1 The Finite Population Correction Factor for Some Combinations of Sample Size *n* and Population Size N

		Correction Factor: $\sqrt{\dfrac{N-n}{N-1}}$
n	*N* = 10, 000	*N* = 40,000
20	.999	.9998
40	.998	.9995
100	.995	.999
250	.987	.996
1000	.949	.987
2500	.866	.968
5000	.707	.935

Sometimes the correction is called the "finite sample correction factor," which hardly seems accurate given that we were always working with a finite sample. What is finite here is the *population* size N. Whenever $N \geq n > 1$, the correction factor is *less than one*, which means that we are reducing the standard deviation for W (and thus for μ) from that in the earlier case when $N = \infty$.

Because the finite population correction factor reduces the standard deviation assigned to W, it suggests greater precision in locating μ than we had before. Before we celebrate, however, we should assess how much we gain as n grows relative to N. Table 6.1 presents the correction factor for several combinations of n and N. We see that the correction factor is extremely close to one (meaning that it corrects practically nothing) even as the sample size goes into the hundreds. And even when the sample size is 5000—*half* the population size when N = 10,000—the uncertainty about the value of μ is 71% as large as it would be if N were in the trillions.

What happens when the random sample contains 1% of a large population? The calculation is straightforward:

When N is large and $n = .01N$, we can write:

$$\sqrt{\frac{N-n}{N-1}} = \sqrt{\frac{.99N}{N-1}} \approx \sqrt{.99} \approx .995 \qquad (6\text{-}5)$$

(We are making the approximation $N/(N-1) \approx 1$ for large N.)

We see from (6-5) that, when n/N is .01 or lower, the multiplicative correction for finite populations is essentially one, meaning that we can forget about it for most practical purposes. That outcome means that, unlike the sample size n, the exact value of population size N is generally *of little importance* in making inferences about μ. For a fixed σ^2, a sample of 100 yields essentially the same accuracy in a population of ten million as a population of 10,000. We offer an intuitive explanation in Example 6.4.

EXAMPLE 6.4

High Points

Suppose that the average height among 12 year-old boys is to be estimated from a random sample of 100 12 year-old boys in the city of Trenton, New Jersey (with 575 12-year boys). Random samples of the same size will be taken for the state of New Jersey (59,500 12 year-old boys), and the entire United States (2.09 million). Assume that the standard deviation σ of the heights among 12-year-old boys is the same in all three populations.

(i) In which of the three settings will the sampling error W have the lowest standard deviation?

(ii) Express the standard deviation of W for New Jersey as a multiple of σ_W for Trenton.

(iii) Express σ_W for the entire US as a multiple of σ_W for New Jersey.

(iv) Are you surprised by the answers to (ii) and (iii)? Briefly discuss.

Solution

(i) From (6-3), we know that the standard deviation $\sigma(W)$ follows:

$$\sigma(W) = \left(\frac{\sigma}{\sqrt{n}}\right)\sqrt{\frac{N-n}{N-1}}$$

where n = sample size, N = population size, σ = population standard deviation

Here $n = 100$ and σ is the same in all three populations.

Note that $(N - n)/(N - 1)$ can be rewritten as:

$$(N - n)/(N - 1) = \frac{(N-1)-(n-1)}{N-1} = 1 - \frac{n-1}{N-1}$$

With $n = 100$, $1 - (n - 1)/(N - 1) = 1 - 99/(N - 1)$. That quantity drops as N goes up, which means that $\sigma(W)$ is lower for the entire US than for Trenton or New Jersey.

(ii) The sampling error for Trenton boys ($N = 574$) has standard deviation given by:

$$\sigma_W(\text{Trenton}) = \left(\frac{\sigma}{\sqrt{100}}\right)\sqrt{1 - \frac{99}{574}} = .910\left(\frac{\sigma}{10}\right)$$

For New Jersey, we have:

$$\sigma_W(\text{New Jersey}) = \left(\frac{\sigma}{\sqrt{100}}\right)\sqrt{1 - \frac{99}{59499}} = .9992\left(\frac{\sigma}{10}\right)$$

Thus,

$$\frac{\sigma_W(\text{New Jersey})}{\sigma_W(\text{Trenton})} = \frac{.9992}{.91} = 1.098$$

(iii) The sampling error for the US follows:

$$\sigma_W(\text{US}) = \left(\frac{\sigma}{\sqrt{100}}\right)\sqrt{1 - \frac{99}{2089999}} = .9999\left(\frac{\sigma}{10}\right)$$

Therefore:

$$\frac{\sigma_W(\text{US})}{\sigma_W(\text{New Jersey})} = \frac{.9999}{.9992} = 1.001$$

(iv) Well, different people have different thresholds for surprise. But the random sample consisted of 100/575 or 17.4% of the 12 year-old boys in Trenton, but only 100/59500 or 0.2% of those boys for New Jersey. Yet, even though there were over 100 times as many 12 year-old boys in New Jersey as in Trenton, the sampling error for the entire state was only about 1.1 times that for Trenton (the state capitol). And even though there are 35 times as many 12 year-old boys in the entire United States as in New Jersey, the sampling error for the country was only one-tenth of one percentage point higher.
The upshot is that sampling error is highly insensitive to population size, except when N is fairly close to n. In the author's experience, this point is highly counterintuitive to people who have not studied Statistics (and even many who have). The real issue is that the outcomes for different members of the random sample are essentially independent so long as N is many times n. That being the case, the precise value of N is of little relevance to the degree of sampling error.

EXERCISE 6.4 For a fixed σ and sample size n, for what population size would $\sigma(W)$ be 80% as large as it would be if $N = \infty$?

With the finite-population correction, the probabilistic formulas about mean μ and proportion p are revised to the following:

For moderately large n,

$$\mu \approx N\left(\bar{X}, \left(\frac{S}{\sqrt{n}}\right) * \sqrt{\frac{N-n}{N-1}}\right) \tag{6-6}$$

where, as before, $S = \sqrt{\frac{\sum_{i=1}^{n}(X_i - \bar{X})^2}{n-1}}$

When $n \geq 20$ and $n\hat{p} \geq 5$ and $n(1-\hat{p}) \geq 5$

$$p \approx N\left(\hat{p}, \sqrt{\frac{\hat{p}(1-\hat{p})}{n-1}} * \sqrt{\frac{N-n}{N-1}}\right) \tag{6-7}$$

Readers might return to Chapter 3 if they don't recall what we mean when we treat μ and p as random variables. Put briefly, we know that they are numerical constants, but the probability distributions reflect our uncertainty about what their values are.

EXAMPLE 6.5

Home Runs

A firm worried about the quality of life of its 1000 employees decided to take a random sample of 40 and find out how long it takes those workers to commute from work to home. The sample mean $\bar{X}$ is 28 minutes and the sample standard deviation S is 9 minutes. Given these numbers:

(i) What is a 95% confidence interval for the mean commuting time (μ) for all 1000 workers?
(ii) Approximately what is the probability that the mean commuting time is over a half hour?
(iii) Approximately what percentage of workers have commuting times that exceed 40 minutes?

Solution

(i) Applying (6-6) with $n = 40$, $N = 1000$, $\bar{X} = 28$ and $S = 9$, we find that $\mu \approx N(28, 1.39)$, and thus that the 95% confidence interval for μ is approximately $(28 - 1.96 * 1.39, 28 + 1.96 * 1.39) = (25.3, 30.7)$ minutes.
(ii) Given that $\mu \approx N(28, 1.39)$, the normal table tells us that $P(\mu > 30) \approx .075$.
(iii) We don't have enough information to estimate this probability. While the location of μ has a normal distribution under the Central Limit Theorem, that circumstance says nothing about the shape of the original commuting-time distribution. And without the shape, we don't know what fraction of the distribution exceeds 40 minutes.
What we could do is go back to the 40 commuting times that we used to calculate $\bar{X}$, and see what proportion of them exceeded 40 minutes. Suppose, for example, that 10 of them did. Then, if p is the fraction of workers with commuting times above 40 minutes, we have $\hat{p} = 10/40 = .25$. We could use $\hat{p}$ as a reasonable first approximation of p (and could work out a confidence interval for p using (6-7)).

EXERCISE 6.5

If the sample size in Example 6.5 were 80 rather than 40 and everything else remained the same, what would be the distribution of μ?

6.3 Stratified Random Sampling

Instead of sampling completely at random from a population, it is sometimes desirable deliberately to break the population into groups, and then to take separate random samples within each group. The sizes of the various samples might not at all reflect the fraction of the population in each group: a group with 5% of the population might be the subject of 40% of the random sampling. Such a process—which is called **stratified random sampling**—might seem inherently unreasonable, but an example suggests why it might not be.

EXAMPLE 6.6

Drop the Charges? (Based on a true story)

A US hospital was accused of charging the government too much for its services to Medicare patients. To assess whether the accusation was justified, it was decided to take a sample of 100 billing records for individual visits to the hospital, and to audit each record for improper charges.

The hospital's patients fell into two distinct categories. About 90% were served in the emergency room, and were charged an average of \$500 apiece. The remaining 10% were admitted to the hospital and stayed at least one night; their average charge was \$5000. Though it was suspected that overcharging was present in both patient groups, there was no prior sense of how much there was, or whether one group suffered more (in percentage terms) than the other.

(i) Suppose that μ_E and σ_E are, respectively, the mean and standard deviation of the *percentage* overbilling for emergency room patients, while μ_H and σ_H are the corresponding quantities for patients admitted to the hospital. What is the total dollar amount of overbilling per year, assuming that N patients in total are served annually?

(ii) Suppose that a random sample of 100 patients are chosen, and that the hospitalized patients are used to estimate μ_H while the emergency room patients are used to estimate μ_E. Approximate the standard deviation of the sampling error in the estimate of total overbilling per year.

(iii) Can you imagine an alternative sampling scheme that is unbiased and might yield a lower standard deviation for total overbilling?

(iv) Approximately what would the standard deviation be under that scheme?

(v) If $\mu_E = \mu_H$, by approximately how much did the scheme in (iii) reduce the standard deviation compared to a simple random sample of size 100?

Solution

(i) We should note at the outset that percentage overcharge *per bill* is not the same as the percentage of cases in which Medicare was overcharged. If the correct bill was \$400, for example, and Medicare was charged \$500, the percentage overcharge would be 100/500 = 20% (which is the same as .20). If the correct bill was \$550 and \$500 was charged, then the "overcharge" would be $-50/500 = -10\%$.

Out of N patients, we have $.9N$ who were handled in the emergency room and $.1N$ who were hospitalized. Assuming (for simplicity) that \$500 and \$5000 were the exact charged amounts within these two groups of patients, then τ, the total dollar amount of overcharging, would follow:

$$\tau = .9N(500)\mu_E + .1N(5000)\mu_H = (450\mu_E + 500\mu_H)N$$

(ii) Of the 100 patients chosen at random, the number treated in the emergency room would be approximately 90 (based on its 90% patient share), while the number hospitalized would be roughly ten. Suppose that $\overline{X_E}$ is the sample mean of percentage overcharges for emergency-room patients, while $\overline{X_H}$ is the sample mean for hospitalized patients.

Then the estimate $\hat{\tau}$ of total dollar overcharge would follow:

$$\hat{\tau} = .9N(500)\overline{X_E} + .1N(5000)\overline{X_H} = (450\overline{X_E} + 500\overline{X_H})N$$

The sampling error that W_τ that attends the estimate $\hat{\tau}$ would be:

$$W_\tau = 450N(\overline{X_E} - \mu_E) + 500N(\overline{X_H} - \mu_H)$$

Treating the sampling errors associated with $\overline{X_E}$ and $\overline{X_H}$ as independent (as seems perfectly reasonable), we have:

$$\sigma^2(W_\tau) = (450N)^2\sigma^2(W_E) + (500N)^2\sigma^2(W_H) \tag{6-8}$$

where $W_E = \overline{X_E} - \mu_E$ and $W_H = \overline{X_H} - \mu_H$

Because $\overline{X_E}$ is based on approximately 90 patients while $\overline{X_H}$ is based on approximately ten, we can invoke (3-24) to write:

$$\sigma^2(W_E) \approx \frac{S_E^2}{90} \text{ and } \sigma^2(W_H) \approx \frac{S_H^2}{10}$$

where S_E and S_H are, respectively, the sample standard deviations of the emergency-room and hospital overcharges

Therefore (6-8) becomes:

$$\sigma^2(W_\tau) \approx (2250S_E^2 + 25000S_H^2)N^2$$

and we reach:

$$\sigma(W_\tau) \approx N\sqrt{(2250S_E^2 + 25000S_H^2)} \tag{6-9}$$

(Random Sample of 100 patients)

(iii) The overall statistics mean that 10 of every 100 patients were admitted to the hospital, and their combined charge was 10 * 5000 = \$50,000. The other 90 were emergency-room patients, and their total charge was 90 * 500 = \$45,000. Thus, the two patient groups—though very different in size—contributed about the same dollar volume.

Yet our estimate about overcharging in the hospitalized group is based on a sample of only about 10, which means it could be accompanied by sizable sampling error. This seems troubling, given that these patients might be a substantial source of total *dollar* overcharging. One potential way to improve accuracy with a total sample size of 100 is to replace the "one patient, one vote" aspect of pure random sampling with something closer to "one dollar, one vote." Given that each hospitalized patient involves about ten times the dollar charge as each emergency-room counterpart, perhaps a hospitalized patient should have roughly ten times the chance of entering the sample as a lower-cost emergency room patient. A simple strategy that approximately achieves this 10:1 ratio is:

- Choose a random sample of 50 emergency-room patients, and use them to estimate μ_E
- Choose a random sample of 50 hospitalized patients, and use them to estimate μ_H

Note that the chance an emergency-room patient is sampled is 50/.9*N*, while the chance for a hospitalized patient is 50/.1*N*. The latter quantity is nine times the former, which is close to ten times. And the 50/50 split has the virtue of simplicity.

What we have done here is introduced a *stratified sampling* scheme in which the two strata were emergency room patients and hospitalized patients. We specified in advance the sample sizes in each, and were well aware that patients in one of the strata would be "oversampled" and the other "undersampled," as compared to pure random sampling with the same total sample size. The hope is that the estimate of overcharging that arises from this stratified sampling would have a lower standard deviation than $\sigma(W_\tau)$ in (6-9).

If we shifted from a 90-10 scheme to a 50-50 scheme, we would not expect S_E^2 or S_H^2 to change on average. After all, the expected degree of *variability* among the observed measurements should not change with the number of measurements: recall from (3-19) that, S_E^2 is an unbiased estimator of σ_E^2 regardless of the sample size. With both patient samples of size 50, we have:

$$\sigma^2(W_E) \approx \frac{S_E^2}{50} \text{ and } \sigma^2(W_H) \approx \frac{S_H^2}{50}$$

Proceeding as in (ii), we reach:

$$\sigma(W_\tau) \approx N\sqrt{(4050S_E^2 + 5000S_H^2)} \qquad (6\text{-}10)$$

(50/50 split between emergency room and hospitalized patients)

Comparing (6-9) to (6-10), we see that the coefficient of S_E^2 has gone up by a factor of (4050/2250) = 1.8, while the far larger coefficient of S_H^2 has fallen by a factor of five (from 25000 to 5000). We would generally expect that these two opposite changes of different size will yield a reduction in the standard deviation of sampling error.

(v) If $\sigma_E = \sigma_H$, then we would expect that $S_E^2 \approx S_H^2$ for both the 90/10 and 50/50 splits. As noted, we would also expect that S_E^2 is about the same whether 90 emergency-room patients are considered or 50; similarly, S_H^2 would not change on average if 50 hospitalized patients are considered rather than 10. Using (6-9), we reach the approximation that, under random sampling with $S_E^2 \approx S_H^2$:

$$\sigma(W_\tau) \approx N\sqrt{(2250S_E^2 + 25000S_H^2)} \approx NS_E\sqrt{2250 + 25000} \approx 165NS_E$$

With (6-10), we estimate that, under the 50/50 stratified sampling:

$$\sigma(W_\tau) \approx N\sqrt{(4050S_E^2 + 5000S_H^2)} \approx NS_E\sqrt{4050 + 5000} \approx 95NS_E$$

Because S_E is about the same in both cases, we estimate that $\sigma(W_\tau)$ under the stratified scheme is only about 95/165 of that under random sampling. That means that stratification has reduced a key statistic about sampling error by about 42%. What has happened is that the gain in accuracy for hospitalized patients when the sample size goes from 10 to 50 far outweighs the drop in accuracy when the emergency-room sample size drops from 90 to 50. Described that way, the pattern makes intuitive sense.

EXERCISE 6.6 Suppose in Example 6.6 that $\sigma_H \approx 1.5\sigma_E$ and total dollar billings are the same for both groups of patients. Show that a stratified sampling scheme with 60 patients sampled among those hospitalized and 40 among those treated in the emergency room achieves lower $\sigma(W_\tau)$ than either pure random sampling of size 100 or a 50/50 stratified sampling scheme.

In Example 6.6, we used an *ad hoc* rule to set the sample sizes in the two strata. But might we devise a more systematic procedure for deciding how much sampling to do in each stratum? We consider the issue at the website, using an approach that resembles the way we found the minimum-variance unbiased estimator in Chapter 3.

When Should We Stratify?

Stratified sampling is most useful when random sampling might yield too little information about subsets of the data of special importance. That criterion was easily satisfied in the hospital overbilling example, where the small minority of patients admitted to the hospital generated half of the billing amounts. In other circumstances, however, stratified sampling might be more trouble than it is worth. The first question the analyst must face is: on what dimensions should the data be stratified? And, once stratification starts, when does it stop? If one stratifies the voting population based on age, should one also do so on gender? On ethnicity? On place of residence? And what fraction of the sampling resources should be devoted to each stratum? Very soon, the whole enterprise can become unwieldy.

And, for many purposes, pure random sampling works surprisingly well. Suppose, for example, that one wants to know what American adults think about a cap-and-trade scheme to reduce emissions of carbon dioxide. As we know, a random sample of 1500 adults should reflect the national viewpoint to within three percentage points. True, the results from this sampling about residents of Wyoming or older Korean Americans will be highly volatile, because we will have at most a handful of respondents in these categories. But if we want a national result, the sampling fluctuations in various categories of Americans largely cancel out, leaving us an aggregate result that is trustworthy.

In short, stratification is a potentially useful option. But, like most options, it should not always be exercised.

6.4 The Bootstrap Method

The statistical sampling theory we have been discussing is highly effective in some settings but of limited relevance in others. However, modern computers allow a quite different approach to sampling, tied to the extensive use of statistical simulation. The technique is called the *bootstrap*, because the available data set creates other data samples by "pulling on its own bootstraps." What the bootstrap method does is use extensive simulation to approximate a probability distribution for the statistic of interest. It applies an axiom that is simply-stated but exceedingly clever.

Suppose that we have a random sample of n observations $(x_1, x_2, \ldots ,x_n)$ from some large population Θ that has a number associated with each of its members. Some function $g(\Theta)$ of all the population numbers (e.g. their median) is of interest. As usual in random sampling, the aim is to obtain some information about $g(\Theta)$ given the n observations. The bootstrap approaches this task by considering two questions:

(i) Given $(x_1, x_2, \ldots, x_n)$, what is a reasonable approximation to the distribution of all numbers over the entire population Θ?

(ii) How does $g(z_1,z_2, \ldots, z_n)$ behave under repeated random samplings of size n (i.e., the z_i's) from the approximate distribution identified in (i)?

We elaborate on these questions below, using simple examples to illustrate the key points.

An Approximate Distribution

Assuming for simplicity that the x_i's all differ, a natural guess based *solely* on the x_i's about how the numbers are distributed over Θ (i.e., of random variable X, where X is the number for a randomly-chosen member of the population) follows:

$$P(X = x_i) = \frac{1}{n} \quad for\ i = 1,..,n$$

$$\sum_z P(X = z) = 0 \text{ for } z \notin (x_1, \dots, x_n) \tag{6-11}$$

where ∉ means "not contained in"

This distribution has the desirable properties that it assigns equal probability to the n numbers that each arose once in the sampling, and gives equal but far lower probabilities (namely, zeroes) to the numbers that did not arise. Admittedly, it is hard to believe that the assumption $\sum_z P(X = z) = 0$ for $z \notin (x_1, \dots, x_n)$ is exactly correct. However, the discrete distribution (6-11) seems easy to defend as a *first approximation* to the true distribution.

In the event that some of the numbers are repeated among the x_i's, then (6-11) is modified in a natural way: If a particular number b comes up k times within $(x_1, x_2, \dots, x_n)$, then we write:

$$P(X = b) = k/n \tag{6-11A}$$

Those numbers that came up once continue to be assigned probability $1/n$.

EXAMPLE 6.7

Time to Win?

Since 1912, the Boston Red Sox have won the baseball World Series in the following years:

1912 1915 1916 1918 2004 2007 2013

Suppose that random variable X is the time between successive Red Sox wins in the World Series.

(i) What probability distribution of form (6-11/11A) would arise from these data?
(ii) Do you think that distribution describes the time between 2013 and the next Red Sox victory?

Solution

(i) The interval between the first and second win is three years (1915–1912), between the second and third, one year (1916–1915), and so on: Note that two intervals out of the six are of length three. The approximate distribution reflecting this pattern is:

$$X = \begin{cases} 1 \text{ w.p. } \dfrac{1}{6} \\ 2 \text{ w.p. } \dfrac{1}{6} \\ 3 \text{ w.p. } \dfrac{1}{3} \\ 6 \text{ w.p. } \dfrac{1}{6} \\ 86 \text{ w.p. } \dfrac{1}{6} \end{cases}$$

(ii) Well, not necessarily, because the future is not simply the continuation of the past. But it does appear that there were two periods separated by 86 years in which the Red Sox won frequently. We might be in the midst of one of those periods now (it is November 2013 as of this writing), in which case the historical pattern might suggest that another victory is near. But the 86-year drought between 1918 and 2004 suggests that, if things get bad, they might get very bad.

EXERCISE 6.7 The New York Yankees, archrivals of the Red Sox, won the World Series in the years:

1923	1927	1928	1932	1936	1937	1938	1939	1941	1943	1947	1949
1950	1951	1952	1953	1956	1958	1961	1962	1977	1978	1996	1998
1999	2000	2009									

Find the discrete distribution of form (6-11/11A) for Z, the interval between successive Yankee wins in the World Series. (Start with the interval that begins in 1923 and end with the interval ending in 2009.)

Actually, the distribution (6-11) is the "maximum likelihood" distribution of X. We establish this point at the website. What that means is that, under those choices of parameters (i.e., of $1/n$'s, b/n's , and 0's), the observed results $(x_1, x_2, \ldots, x_n)$ would have the highest chance of arising. While maximum likelihood parameter estimates need not be correct, the appeal of the approximate distribution (6-11/11A) is increased when the observed results would be less likely under any other distribution.

What Bootstrapping Does

The bootstrap method combines the ideas of an approximate distribution and repeated random sampling from that distribution (which is called ***resampling***). Given a set of observed outcomes $(x_1, x_2, \ldots, x_n)$—all of which differ—we first create the discrete distribution that assigns probability $1/n$ to each observed value. Then, using a random number generator, we repeatedly draw *random samples of size n* from that distribution, and take note which specific sample is drawn each time. In each such random sampling (i.e., in each resampling), any of the n individual selections has a $1/n$ chance of assuming each value in the approximate distribution, *independent of any other selections*. For example, if the original sample is of size six and the observed outcomes are (1, 28, 2, 12, 6, 55), we take random samples of size six in which each of the six observations independently has probability 1/6 of taking on each of these values. In the first such sample, the numbers that come up could be (28, 6, 6, 55, 2, 1), meaning that there can be repetition in the resampling even when there was none in the original sample. In statistical parlance, we assume that the resampling is performed ***with replacement***.

An example reminds us that how the results of an individual resampling can diverge from the original numbers that are the basis of the resampling.

EXAMPLE 6.8

Six for Six

If a fair six-sided die is tossed six times, what is the probability that the outcomes are all different?

Solution

When a six-sided die is tossed six times, there are six possible outcomes on each toss, and thus a total of 6^6 overall outcomes. (Here different orderings of the same numbers—like (1,2,3,4,5,6) and (6,5,4,3,2,1)—are treated as different.) As for getting six different outcomes on the tosses, there are six possible outcomes for the first toss, for each of which there are five *differing* outcomes on the second toss. When the outcomes differ on the first two tosses, four possible outcomes on the third toss would differ from the preceding two. Continuing in this way, we see that there are a total of $6 * 5 * 4 * \ldots 1 = 6!$ final outcomes that involve no repetition of any individual outcome. Because all 6^6 possible outcomes are equally likely, we have:

P(six different outcomes on six fair-die tosses) = 6!/66 = 720/46656 = .0154 $\approx$ 1/65

This calculation shows that some repetition within the six tosses is extremely likely. Of course, the specific pattern of repetition would probably vary from one six-toss sequence to another.

Now consider six independent outcomes from the distribution that assigns equal probability to the numbers (1, 28, 2, 12, 6, 55). From the same reasoning as just used, there is only about 1 in 65 chance that each integer will arise exactly once. (Right?) That general observation is important in considering the resampling results under bootstrapping.

EXERCISE 6.8 Suppose that nine independent random selections are made from a discrete distribution, with nine possible outcomes that are all equally probable. What is the probability that all the selections yield different outcomes?

As suggested in Example 6.8 and Exercise 6.8, it is rare that the individual random sequences of size n drawn in bootstrap-resampling will contain all n of the original numbers that underlie the resampling. Moreover, the different sequences will show variability among themselves, which will create variability in any statistic calculated on a sample-by-sample basis. For example, suppose that we are interested in the mean of a certain distribution, for which we have the six observed values (1, 28, 2, 12, 6, 55). Then the first bootstrap resample might be (28, 6, 6, 55, 2, 1), which would have a sample mean of 16.5. The second resample might be (2, 6, 28, 12, 2, 28), which has sample mean 13. If we create 1000 bootstrap samples, we will have 1000 sequences of the form $(z_{1j}, z_{2j}, \ldots, z_{6j})$, where z_{ij} is the ith observation in the jth sequence ($1 \le i \le 6$; $1 \le j \le 1000$). Each of these 1000 samples would have a sample mean associated with it.

Then what? We would create a *histogram* for the 1000 values of the statistic of interest that were obtained through resampling. Working from that histogram, the bootstrap technique would provide estimates of familiar sampling results like the 95% confidence interval for the statistic. In simplest form, bootstrapping might literally use the histogram, and approximate the 95% CI as the range from roughly the 25th lowest of the 1000 observed values for the statistic to the 25th highest. In other instances, bootstrapping might adjust this particular interval, though often only slightly.

EXAMPLE 6.9 **All Is Fair?**

Suppose that we want to estimate the mean outcome on the toss of a 20-sided die, and perform eight tosses that yield the sequence (18,15,14,11,8,6,20,12). We do *not* assume that the die is fair, and thus do not consider it obvious that the mean is 10.5.

(i) What is the bootstrap's approximate distribution for X, the outcome on an individual toss?

(ii) Perform a bootstrap analysis with 5000 samples with replacement from that distribution. What is the bootstrap point estimate of the mean for the true distribution?

(iii) What is a 95% bootstrap interval for the mean?

(iv) What is the cumulative distribution function (CDF) for the sample means in the 5000 resamplings from the approximate distribution?

(v) Suppose that we had treated the eight original outcomes as arising from a normal distribution, and had proceeded in the traditional way. What would have been the 95% confidence interval for the mean?

Solution

(i) Using (6-11), we would reach the approximate distribution:

$$X = \begin{cases} 6 \text{ w.p. } \frac{1}{8} \\ 8 \text{ w.p. } \frac{1}{8} \\ 11 \text{ w.p. } \frac{1}{8} \\ 12 \text{ w.p. } \frac{1}{8} \\ 14 \text{ w.p. } \frac{1}{8} \\ 15 \text{ w.p. } \frac{1}{8} \\ 18 \text{ w.p. } \frac{1}{8} \\ 20 \text{ w.p. } \frac{1}{8} \end{cases}$$

(ii) Now we face a practical question: how do we perform a bootstrap analysis? Most if not all serious statistical packages make the task straightforward. We would be asked for:

- the original data sample (in this case (18, 15, 14, 11, 8, 6, 20, 12))
- the desired number of bootstrap "resamplings" with replacement from the approximate distribution corresponding to the data sample
- what statistic we would like calculated for each resampling (often, we have a drop-down menu for common statistics like the mean or median)
- what information we would like about the statistic's behavior over the course of the resamplings. (e.g. the average of the outcomes, their standard deviation, their histogram)

Once we provide this information, the results arrive within seconds. Not surprisingly, bootstrap methods only became popular when high-speed computing was routinely available.
When the author requested 5000 resamplings from (18,15,14,11,8,6,20,12) each of sample size eight, the package responded as follows:
The average of the 1000 sample means was 12.998. This is exceedingly close to the mean of 13 for the approximate distribution in (i). (As we'll discuss, that outcome is not surprising.)

(iii) The output reveals that the "standard" 95% bootstrap interval for the resampling results extended from 9.875 to 16.125. (This outcome means that approximately 2.5% of the 5000 sample means that arose in the resampling were below 9.875, and approximately 2.5% above 16.125. (Given that each of the sample means had to be a multiple of 1/8, it makes sense that the thresholds are themselves multiples of 1/8.)

In this situation, *the 95% bootstrap interval becomes a 95% confidence interval for μ, essentially without alteration.* We'll consider why shortly, much as we'll consider why in some instances the bootstrap interval is adjusted to yield a 95% confidence interval for the statistic of interest.

(iv) Conveniently, the package displays the CDF of the sample means based on 5000 random samples of size eight:

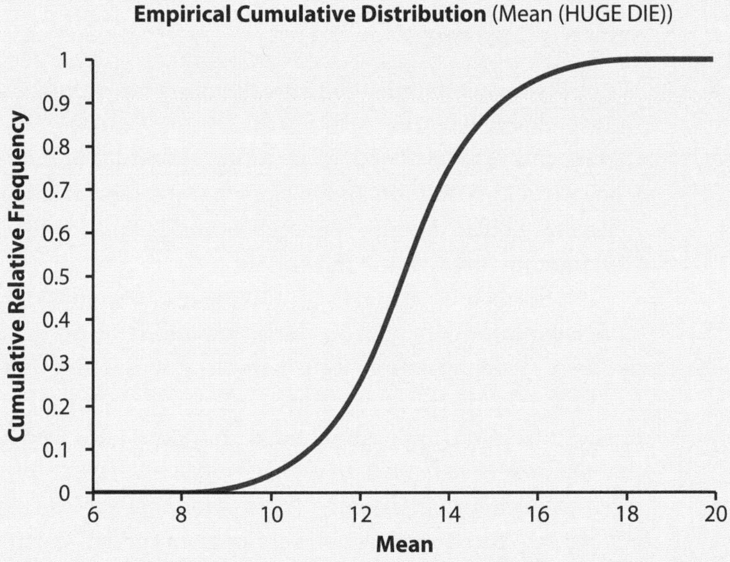

The curve at each value reflects the fraction of resampling results at or below that value. For example, the curve suggests that about 25% of the resamplings yielded a sample mean at or below 12. Note that the CDF grows substantially between about 11 and 15, meaning that heavy bulk of sample means arose there.

(v) Had we assumed that the data (18,15,14,11,8,6,20,12) were generated by a normal distribution, we would have found a 95% confidence interval for μ using the t-distribution that we met in Section 3.7. The sample mean $\bar{X} = 13$ here, and the sample standard deviation $S = 4.75$. We would therefore have assumed that the quantity t defined by:

$$t = (\bar{X} - \mu)/(S/\sqrt{n}) = (13 - \mu)/(4.75/\sqrt{8}) = .595(13 - \mu)$$

had a t-distribution with $n - 1 = 7$ degrees of freedom. We would then have learned from (3-27) and an electronic table that the 95% confidence interval for t extended from −2.37 to 2.37, meaning that the 95% confidence for μ extended from 9.02 to 16.98.

While this 95% confidence interval is wider than that arising from the bootstrap exercise (9.875 to 16.125), we should not be shocked by this difference. The two calculations involved different assumptions about the approximate distribution that generated the data (discrete for bootstrap, normal for the *t*-procedure), so we would not have expected the same confidence interval. All factors equal, the bootstrap approximate distribution based on the actual data might be more defensible than an assumed normal distribution that cannot be verified given only eight data points.

Why the Bootstrap?

While we have offered some introduction to how the bootstrap works, we have not indicated why it works. At one level, we could answer "because it does." Both a deep theory and extensive experimentation (in which initial data samples were drawn from a true distribution and various parameters for the distribution were then estimated via bootstrapping) converge on the conclusion that bootstrapping works very well in many (though not all) situations.

But the technique makes intuitive sense if we consider how it works. It starts with a plausible approximation for the distribution that produced the original data sample of size n. Then, to estimate the sampling variability that attends samples of size n from the true distribution, it uses the *observed* variability among repeated samples of size n from the approximate distribution. There seems no prior reason to believe that *this* ***observed variability*** *among different samples of size n from the approximate distribution* is either far larger or far smaller *than the* ***true variability*** *for samples of size n in the true distribution*, which cannot be observed. We might therefore approximate "true variability" by "observed variability" in assembling confidence intervals and other indicators about the statistic of interest.

We should be more specific. For the approximate distribution arising from the data points (18,15,14,11,8,6,20,12), the 5000 bootstrap simulations yielded a 95% range for the sample means from 9.875 to 16.125. Thus, there was about a 95% chance that the mean in a random sample of size eight in the approximate uniform distribution was within 3.125 of the correct mean of 13. The bootstrap approximation is that, in the unknown actual distribution that generated the data, there was also about a 95% chance that a sample mean based on a sample of size eight would be within 3.125 either way of the true mean. Thus we should move 3.125 in each direction from our original sample mean (namely, 13, based on the eight observed values) to get a 95% confidence interval for μ. Doing so yields the range we have already seen, from 13 – 3.125 to 13 + 3.125, or from 9.875 to 16.125.

The Plug-In Principle

Given a random sample from a population and some statistic of interest, it is natural to estimate that statistic for the population by its value for the sample. With a random sample of size nine, for example, we might estimate the population mean μ by the sample mean $\overline{X}$; similarly, we might estimate the interquartile range by the difference between the sample's third largest and third smallest observations. Such an approach has been named (or nicknamed) the *plug-in principle*. But it is often unclear whether the plug-in estimate is unbiased, or what level of sampling error attends the plug-in estimates. Unless we make some assumptions about the underlying distribution for the population (e.g., that it is normal), we might be unable based on existing theory to assign margins of error with any confidence. A great advantage of the bootstrap method is that it provides a basis for assigning margins of error, based on what happens in the repeated resamplings.

What the bootstrap method does is:

(i) note the ***point estimate*** of a parameter of interest when the plug-in principle is applied to the original data sample
and:
(ii) make an estimate ***based on resampling*** about the degree of bias and sampling error that attends the plug-in point estimate

EXERCISE 6.9

Given the five data points (33, 6, 21, 12, 3) from a distribution, use the plug-in principle to estimate the following statistics about the distribution:

- the mean
- the standard deviation (use n rather than $n - 1$)
- the median
- the lower quartile
- the upper quartile
- the interquartile range

Bootstrap-Adjusted Point Estimates

The bootstrap method does *not* assume that the plug-in estimate for a parameter is always an unbiased estimate of that parameter. Indeed, in some situations, bootstrapping provides potent evidence that the plug-in estimate is biased, as the following example illustrates.

EXAMPLE 6.10

Happy Median?

Given the seven data points: 21 6 48 33 3 66 12

(i) What approximation to the distribution that generated these points is used in bootstrapping?
(ii) What is the plug-in estimate of the median?
(iii) Suppose that the aim is to estimate the median of the distribution that generated the data points. How might one proceed under bootstrapping?
(iv) When 5000 resamplings were taken from the approximate distribution, the average of their individual medians was 23.2. Does that outcome suggest that using the sample median to estimate the true median might be biased?

(v) If you answered "yes" in (ii), what characteristics of the data suggest why the sample median might be biased for this particular set of seven numbers? (If you answered "no," you answered wrong.)

(vi) What adjustment might lead to an unbiased estimate of the median?

Solution

(i) The approximate distribution used in bootstrapping would be:

$$X = \begin{cases} 3 \text{ w.p. } \frac{1}{7} \\ 6 \text{ w.p. } \frac{1}{7} \\ 12 \text{ w.p. } \frac{1}{7} \\ 21 \text{ w.p. } \frac{1}{7} \\ 33 \text{ w.p. } \frac{1}{7} \\ 48 \text{ w.p. } \frac{1}{7} \\ 66 \text{ w.p. } \frac{1}{7} \end{cases}$$

(ii) The plug-in estimate of the median here is 21, the fourth largest number.

(iii) Under bootstrapping, one would take a large number of random samples of size seven with replacement from the approximate distribution in (i). Then one would find the median for each such resampling, which is the fourth largest (or fourth smallest) of the seven sample points.

(iv) The plug-in median—and the correct median in the approximate distribution—is 21. Because we learned that the average of 5000 estimates of the median exceeds that amount by 2.2 (23.2 – 21), there is indeed the suggestion of bias. (The average of 5000 medians should have been much closer to 21 if the estimation was genuinely unbiased.) Under the approximate distribution, therefore, the fourth largest outcome in a random sample of size seven tends to overestimate the median.

(v) Well, consider two extreme cases: all seven numbers selected in a resampling are at 3 (the lowest value), and all seven are at 66 (the highest). Each of the outcomes would have probability $(1/7)^7$. In the first case, the observed median of 3 falls below the actual median in the approximate distribution by 18 (i.e., by 21 – 3). Yet in the second case, the median of 66 exceeds the actual median by 45 (i.e., 66 – 21). Thus, there is an upward "tilt" associated with the extreme outcomes.

Or suppose that the seven points in a resampling are scattered about equally among the three lowest data points (3,6,12), or instead among the three highest (33,48,66). Both these patterns would have the same probability under the approximate distribution. In the first case, the observed median would be six, which is 15 below the approximate distribution's actual median (i.e., 21 – 6) while, in the second, the median would exceed 21 by 27 (i.e., 48 – 21). Again, there is evidence of an upward "tilt."

These discrepancies reflect that fact that the highest three numbers in the approximate distribution exceed the median of 21 by a larger amount than the three lowest numbers fall below it. The approximate distribute has a longer "right tail" than "left tail." For that reason, the average of the 5000 medians that arise in resampling tends upward compared to 21, which generates an upward bias in the plug-in estimate.

(vi) For the approximate distribution, the resampling suggests the sample median in a sample of size seven is on average about 2.2 above the actual median. Bootstrapping assumes that what happens in the approximate distribution is essentially what happens in the true distribution. The bootstrap outcome is therefore taken to imply that, in the unknown distribution that yielded the seven original data points, the median in an observed sample of size seven also exceeds the true median by 2.2 on average. (In effect, bootstrapping quietly assumes that the longer right tail that materialized in the approximate distribution also exists in the true distribution.) To get an unbiased estimate of the median in the true distribution, therefore, bootstrapping suggests that we should *reduce* the observed median among the seven original data points by 2.2. The resulting estimate is 21 – 2.2 = 18.8.

We have to remember that we have only one sample of size seven from the true distribution. Through studying the sampling behavior among many samples of size seven in the approximate distribution, we assume that we learn about the bias and volatility that inhere in a single sample of size seven from the true distribution. If the resamplings show an upward bias—as they did here—we assume that a bias of the same sign and approximate magnitude affected our original sample of size seven. It is therefore logical to use the bootstrap results to get an approximate adjustment to counteract that bias.

EXERCISE 6.10 Suppose that the three data points 1, 14, and 38 arise at random from a continuous distribution.

(i) Under the plug-in principle, what is the median of that distribution?

(ii) In a single resampling from the approximate bootstrap distribution, what is the probability that the median will be 38? (Hint: Let random variable X be the number of times the number 38 comes up in the resampling. What is the distribution of X?)

(iii) Approximately what will be the average of the individual medians that arise over 5000 resamplings? (Use probabilistic reasoning and not a bootstrap package to answer this question.)
(iv) Approximately what is the bias in using the sample median as an estimate of the true median?
(v) Why is any bootstrap analysis based on three data points of limited reliability?

Bootstrap-Adjusted Confidence Intervals

In some instances, the "raw" bootstrap confidence intervals—extending, for example, from the 26th to 974th outcomes among 1000 resamplings—might not offer a proper confidence interval for the parameter under study. A hypothetical example illustrates the issue.

EXAMPLE 6.11 Upside Down

Suppose that the correct value of a certain parameter in the approximate bootstrap distribution is 54, and 95% of the estimates of the parameter over 5000 resamplings range from 51 to 59. (The approximate distribution involves 16 distinct values, but we need not worry about the details.)

(i) Should we use the range from 51 to 59 as the 95% confidence interval for the parameter in the true distribution?
(ii) If not, what alternate confidence interval would be better?

Solution

In answering these questions, we might pursue the following logic:

- In the *approximate* distribution, the observed outcome fell between three below the correct parameter value (51 vs. 54) and five above that value (59 vs. 54) 95% of the time.
- The bootstrapping assumption is that:
  In the true distribution, there is also a 95% chance that the observed value is between three below and five above the true parameter value.
- While 54 is the correct parameter value in the approximate uniform distribution, it is instead an *estimate* of the parameter in the true distribution, based on a single random sample.
- We then ask: Under what circumstances is an estimate of 54 in the true distribution between three below and five above the true parameter value?
- Well, 54 falls three below the true parameter value when the true parameter is 54 + 3 = 57, and it is five above the true parameter when that parameter is 54 – 5 = 49.

- Thus, we might treat (49, 57) as the 95% confidence interval for the true parameter, and *not* (51,60).

In short:

$$\begin{cases}\text{The range from 51 to 60 is not an ideal confidence interval.}\\ \text{The range from 49 to 57 is more consistent with the logic of bootstrapping.}\end{cases}$$

EXERCISE 6.11 True or False: Suppose that the 95% range of bootstrap estimates of a parameter—ranging from the 2.5th percentile of the estimates to the 97.5th percentile—extends from λ below to the correct parameter γ in the approximate bootstrap distribution to λ above. Then the interval from $\gamma - \lambda$ to $\gamma + \lambda$ is a reasonable 95% confidence interval for the parameter value in the true distribution.

As Example 6.10 suggests, the histogram of outcomes arising from bootstrapping might be asymmetric, with positive deviations around the correct outcome in the approximate distribution being larger than negative ones. This asymmetry—like the bias discussed earlier—makes it prudent to adjust the raw bootstrap results to achieve more defensible confidence intervals.

How do we make these "prudent" adjustments? We will not discuss the issue here, but will instead defer to the wisdom of those who have explored this matter in depth. We can do so easily in practice by simply accepting the "bootstrap corrected" confidence interval that is typically offered by the statistical package, alongside the "raw" confidence interval based on percentiles of the bootstrap distribution. In many instances, the adjusted confidence interval is quite similar to the original one because the circumstances that warrant adjustment were not apparent in the bootstrap results. In the next example, we present both the raw and adjusted bootstrap results about confidence intervals.

EXAMPLE 6.12 **Wet Boots**

To illustrate the use of the bootstrap method with actual data, we turn to annual precipitation levels in New York's Central Park for the 143 years from 1869 to 2011. (The data set appears at the website that accompanies this book.) We will select 13 of these years at random, record the amount of precipitation each year, and then perform a bootstrap analysis on the sample of size 13 to make inferences about the population of size 143. An advantage of this approach is that we know the correct answer for the 143 years, and thus can assess whether the parameter estimates are accurate to within the margin of error assigned to them.

We focus on three statistics about the population:

- mean annual rainfall over the period
- mean absolute deviation of rainfall levels around the 143-year average (an alternative to standard deviation as a measure of variability)
- the 25th percentile of the distribution of annual rainfall

(i) How might we select a random sample of 13 years from 1869 to 2011?
(ii) Would the last 13 years (i.e., 1999–2011) provide a random sample of years from 1869 to 2011?
(iii) What point estimates of the three quantities of interest arise under the plug-in principle with the 13 data points?
(iv) What point estimates and 95% confidence intervals for these quantities arise from a bootstrap analysis with 1000 resamplings?
(v) How do the results in (iv) compare with the actual statistics for the 143-year period?

Solution

(i) One approach to achieving a random sample is to assign each year from 1869 to 2011 an integer from 1 to 143 (1869 = 1; 2011 = 143), and then go to RANDOM.ORG to get a set of 13 integers randomly selected from the range 1–143. Doing so yields the outcomes:

Random Integer	Year	New York Rainfall
40	1908	41.05 (inches)
125	1993	44.25
13	1881	36.26
118	1986	42.95
17	1885	35.37
29	1897	44.55
72	1940	45.07
11	1879	39.01
104	1972	67.03
71	1939	38.55
75	1943	36.74
143	2011	72.81
46	1914	40.17

(ii) In a word, no. In a random sample, all years from 1869 to 2011 would have an equal chance of making it into the sample. Choosing the last 13 years of the period obviously deviates from this criterion. Nor is the distinction a mere technicality: if there is any time-trend in rainfall levels, then choosing the end of the period could yield biased and unrepresentative results for the full period.

(iii) For these 13 data points:

- The sample mean is 44.91 inches.
- The sample mean absolute deviation Q is defined by:

$$Q = \frac{1}{n}\sum_{i=1}^{n}\left|r_i - \bar{X}\right|$$

where r_i = rainfall in the ith year sampled

Here

$$Q = \frac{1}{13}\sum_{i=1}^{13}\left|r_i - 44.91\right|$$

The numerical value of Q for the 13 data points is 7.72.

- Among 13 data points, the 25th percentile is the fourth lowest, which exceeds three of the points and falls below nine of them. Alternately, the 25th percentile is the median of the seven observations at or below the median, which is again the fourth lowest observation. Here the 25th percentile of the data points is 38.55.

(iv) The reported bootstrap results for 1000 resamplings the author performed are:

Mean:	
Average of the resampling estimates	44.78 inches
Average in original sample of size 13	44.91
95% Confidence Interval for the 143-year mean:	
Raw	(39.75, 51.43)
Adjusted	(40.11, 52.56)

We see that bootstrapping is essentially unbiased: the 1000 resamplings yielded means that were on average only 0.13 inches below the mean in the original sample (44.78 vs. 44.91). (This finding is no surprise: the individual resampling estimates of the mean are necessarily unbiased estimates of the actual mean in the approximate distribution.) Interestingly, the two confidence intervals edge upward a bit: they extend about five below the point estimate of 44.91 to about seven above it. That happens because the 13 original data points are not symmetric around 44.91: the values above 44.91 exceed it to a greater extent than those under 44.91 fall below it.

Note that the bootstrap is *not* telling us what we might learn from combining 1000 samples of size 13 (and thus in effect creating a sample of size 13,000). Rather, it is suggesting what we might infer based on a single sample of size 13. The variability around 44.78 that emerged in the 1000 resamplings of size 13 is an estimate of the sampling variability when the mean of the true distribution is estimated from a single random sample of 13 observations (as it is here).

For mean absolute deviation, we obtain:

Mean Absolute Deviation	
Average of the estimates:	7.56 inches
Average in original sample	7.72
95% Confidence Interval:	
Raw	(2.41, 14.14)
Adjusted	(2.66, 14.60

Once again, the estimation is nearly unbiased: 7.56 inches was the average of the 1000 calculated mean absolute deviations in the resampling, while 7.72 was the plug-in estimate for the original 13 data points. But we are taken aback by the widths of the two 95% confidence intervals, which both range from about 2.5 to 14.5. These outcomes suggest that the mean absolute deviation in the true distribution might be far lower or far greater than what we see in the original sample. What accounts for this strange result?

We understand this outcome better if we look at the 13 observations we have: 11 of them fall in a small range from the mid-30's to the mid-40's, while two of them hover around 70 (67.03 and 72.81). We can infer that:

- When we take a random resample of size 13 from the approximate distribution, the probability is $(11/13)^{13} = .114$ that none of the observations will be 67.03 or 72.81, in which case the mean absolute deviation among the 13 closely-bunched selections in the 30's and 40's will be far below 7.72. Because this happened in roughly 11% of the resamplings, then the 2.5th percentile of the resampling results will itself be far below 7.72.
- Turning to the binomial table with $n = 13$ and $p = 2/13$, we learn that the chance is .127 of getting at least four large observations (of 67.03 or 72.81) in a given resampling. When that happens, the 13 outcomes exhibit high polarization—with several around 70 and also several around 40—and a mean absolute deviation well above 7.72. It follows that the 97.5th percentile of the resampling results will far exceed 7.72.

In other words, the original estimate of mean absolute deviation is fragile, because we are uncertain about what fraction of outcomes fall near the two original data points far above the main mass of outcomes. The observed fraction is 2/13 = 15% in the original sample, but the true fraction might be 35% or 5%. The wide confidence interval obtained via bootstrapping quite properly warns us that our parameter estimate is unstable.

(iii) For the 25th percentile, we reach:

25th Percentile	
Average of the resampling estimates:	38.39 inches
Estimate in original sample	38.55
95% Confidence Interval:	
Raw	(36.26, 41.05)
Adjusted	(35.37, 40.17)

We note that the raw 95% confidence interval for the 25th percentile extends from one value in the original 13 observations (36.26) to another (41.05). That circumstance was to be expected: on every resampling, the fourth outcome from the lowest is necessarily among the original observations (though perhaps not 38.55, the fourth lowest of the 13 initial data points.) The bootstrap adjustment shifts the confidence interval down a bit.

It might seem surprising is that the confidence interval for the 25th percentile is far smaller than that for the mean absolute deviation (having a width of about five rather than twelve). The explanation for the difference is that the factor that widened the confidence interval for mean absolute deviation—namely, uncertainly about what proportion of the true distribution near the small "very high" cluster around 70 that arose among the 13 observed points—is almost irrelevant to the 25th percentile. Maybe the fraction of the true distribution that is near 70 is not 2/13 but rather 3% or 35%, but it hard to imagine given the data that the 25th percentile of the true distribution is near 70. For practical purposes, therefore, bootstrapping treats the 25th percentile of the true distribution as somewhere within the narrow range of measurements between 35 and 45 that constitute 85.6% (13/15) of the original data sample. Indeed, in the 1000 resamplings, *not once* was the 25th percentile that arose either 67.03 or 72.81.

(v) For the entire set of 143 years (the population):
- The mean rainfall was 45.29 inches.
- The mean absolute deviation of the annual rainfall amounts was 6.31.
- The 25th percentile of the rainfall amounts was 39.35 inches

All three of these statistics are well within their 95% confidence intervals based on bootstrapping the data for the random sample of 13 years.

EXERCISE 6.12 Suppose a bootstrap analysis is to be performed on the 13 data points in Example 6.12

(i) Given the results in Example 6.11 about the mean absolute deviation and the 25th percentile, what might you expect about the width of the bootstrap 95% confidence interval for the median. Briefly explain.

(ii) Using a statistical package that has a bootstrapping capability, perform 1000 resamplings from the distribution, and find an approximate 95% confidence interval for median rainfall in New York over 1869–2011.

(iii) Now find the actual median rainfall for New York over 1869–2011 from the full data set for the 143 years.

Bootstrapping is most helpful when we are otherwise at a loss to estimate the sampling error that attends a point estimate of an important statistic. Such a situation is the subject of the next two examples.

EXAMPLE 6.13

Free Parking? (Based on a true story)

A city that has parking meters on many of its streets wanted to estimate how much money it was losing because people were not paying the meters. Ideally, it would select a random sample of cars parked at meters at random times, and determine for each whether the meter read "violation" at the moment the car was approached. Then it would estimate p, the proportion of parked cars in violation of the meter, by $\hat{p}$, the fraction of such cars within the random sample.

However, such a sampling scheme was not practical. Even if the city knew exactly which meters had cars parked at them, a random selection among those cars might first pick a vehicle nine miles North of the city center, and then pick a second vehicle three miles South. Moving among the vehicles thus sampled would be hopelessly inefficient and expensive. Another scheme was necessary.

What the city decided to do instead was to pick metered *blocks* at random (e.g., 33rd Street between 3rd and 4th Avenues). Then, at a random time during a given interval (e.g., noon–1 PM), a surveyor would arrive at each such block and count both the number of cars parked there and the number of cars in violation of the meters. If a total of n blocks were sampled, and z_i was the number of cars parked on block i and and x_i the number of "violation" cars there, then the overall "scofflaw rate" p would be estimated by $\breve{p}$ that follows:

$$\breve{p} = \sum_{i=1}^{n} x_i \, / \sum_{i=1}^{n} z_i \qquad (6\text{-}12)$$

(i) In words, what approximation is being made in (6-12)?

(ii) Why does that approximation seem a reasonable one?

(iii) In working out the margin of error, would it be reasonable to apply the usual formulas as if $\breve{p}$ were estimated by the fraction of violations within a purely random sample of $\sum_{i=1}^{n} z_i$ different cars?

Solution

(i) In (6-12), $\sum_{i=1}^{n} x_i$ is the total number of cars found in violation on the n blocks sampled, while $\sum_{i=1}^{n} z_i$ is the total number of cars on those blocks. Thus, $\breve{p}$ is the fraction of all observed cars that were in violation, so (6-12) estimates the true fraction in violation by the observed fraction over the combined randomly-sampled blocks.

(ii) The approximation seem reasonable because, if the blocks were sampled randomly, their overall "scofflaw" rate under (6-12) would seem a plausible approximation to the true rate. Furthermore, the calculation pays more attention to blocks with many parked cars than to blocks with few, as a simple example makes clear. Suppose that two blocks are sampled, block A with two parked cars and one scofflaw (i.e., in violation), while block B has 16 parked cars and two scofflaws. Then the scofflaw rate was 50% on block A and 12.5% on block B, but $\breve{p} = (1 + 2)/(2 + 16) = 3/18 = 16.7\%$, meaning that the rate on the busier street got far more weight in the calculation than that for the other street. Such differential weighting makes sense, and seems a strength of the calculation.

(iii) The assumption that we are dealing with a random sample of $\sum_{i=1}^{n} z_i$ cars would be hazardous, for that assumption implies that the "scofflaw" statuses of different cars would be independent. But here the first z_1 cars considered are all on the same block, and the scofflaw rate might be strongly correlated with location. As an extreme case, suppose that the blocks in a city fall into two categories: those on which everyone pays the meter, and those on which no one pays. Then if the first car on the first block selected is a scofflaw, the chance that the second one is also a scofflaw is 100% and not p, the citywide average scofflaw rate.

Moreover, the failure of independence is not just a technical issue. Suppose that all the N blocks in a given city have exactly ten cars, and everyone pays the meter on half the blocks while no one pays on the other half. If we take two blocks at random, do we really have a random sample of size 20? Or do we actually have a sample of size two out of N? The sampling error can be considerably greater in the second case—which is what we actually face—than in the first.

EXERCISE 6.13 We said in Example 6.13 that $\breve{p}$ is a reasonable estimate of p, but did not say that it was an unbiased one. Is $\breve{p}$ an unbiased estimator of p? Either come up with an example in which $\breve{p}$ is biased or devise a simple proof that it must be unbiased.

(Hint: Imagine a city that has three blocks with parking meters, two of which will be sampled at random. Can you find a set of (x_i, z_i)'s for which $E(\breve{p}) \neq p$? The example can be simple.)

EXAMPLE 6.14

Park Ranges

In the situation of Example 6.13, suppose that the city samples 12 metered blocks at random, and reaches the following data of the form (x_i, z_i):

(4,12)	(3,8)	(3,9)	(3,16)	(2,7)	(4,15)
(4,10)	(3,15)	(2,6)	(1,12)	(3,8)	(2,14)

(i) What is the city's estimate $\breve{p}$ of the citywide "scofflaw" rate p?
(ii) Why is this a situation where bootstrapping seems desirable?
(iii) In a bootstrap analysis, what would be the approximate distribution of (X,Z) across metered blocks?
(iv) In words, how would a bootstrap analysis with 1000 resamplings proceed?
(v) Perform such an analysis, and assess whether the estimation procedure suffers an appreciable bias.
(vi) What is a 95% confidence interval for p based on percentiles in the bootstrap distribution of $\hat{p}$?
(vii) Referring to the histogram of the bootstrap statistic under resampling, discuss why adjusting the bootstrap estimate of the "raw" confidence interval is not important here.
(viii) Given the bootstrap results, approximate the probability that the true "scofflaw" rate is 20% or smaller.

Solution

(i) Applying (6-12) to these data, we obtain $\breve{p} = 34/132 = .258$.
(ii) As noted in Example 6.13, we cannot automatically treat different cars on the same block as independent in terms of their "scofflaw" status. Furthermore, the key statistic is the quotient X/Z of two random variables, $X = \sum_{i=1}^{n} X_i$ and $Z = \sum_{i=1}^{n} X_i$. (Here X_i, the number of cars in violation on the ith block selected, can be considered a random variable prior to the drawing.) Especially because $X_i \le Z_i$ for all i, the two variables X and Z might well be correlated. We have no prior theory how the ratio of these two correlated random variables would behave, meaning that we cannot reliably estimate the degree of sampling variability to which $\breve{p}$ is exposed.
Despite these complexities, bootstrapping generates a useful approximate sampling distribution for the test statistic (i.e. the "scofflaw" rate). Working with that distribution, we can accompany our point estimate of the "scofflaw" rate—arising from the plug-in principle applied to the original data—with a defensible estimate of its bias and margin of error based on bootstrapping.
(ii) Over the metered blocks in the city, the joint probability distribution of scofflaws (X_i) and parked cars (Z_i) on the ith block chosen would be discrete and limited to the nonnegative integers. The approximate distribution that corresponds to the data at hand would be:

$$P\left(X_i = j \cap Z_i = k\right) = \begin{cases} \dfrac{1}{12} & \text{for each } (j,k) \text{ pair that arose in the sample} \\ 0 & \text{for all other possible } (j,k) \text{ pairs} \end{cases}$$

For example, the pair (4,12) would be assigned probability 1/12, while (4,9) would be assigned zero.

(iii) In a bootstrap analysis with 1000 resamplings, each resampling would consist of 12 independent selections (with replacement) from the approximate distribution in (ii). Then a scofflaw rate would be calculated under (6-12) based on the 12 outcomes. Once the resamplings are complete, the distribution of scofflaw rates that were obtained would be construed as reflecting the extent to which the observed scofflaw rate in the original sample of size 12 is subject to sampling variability.

When the 1000 resamplings are performed, the histogram of scofflaw rates that arose was the following:

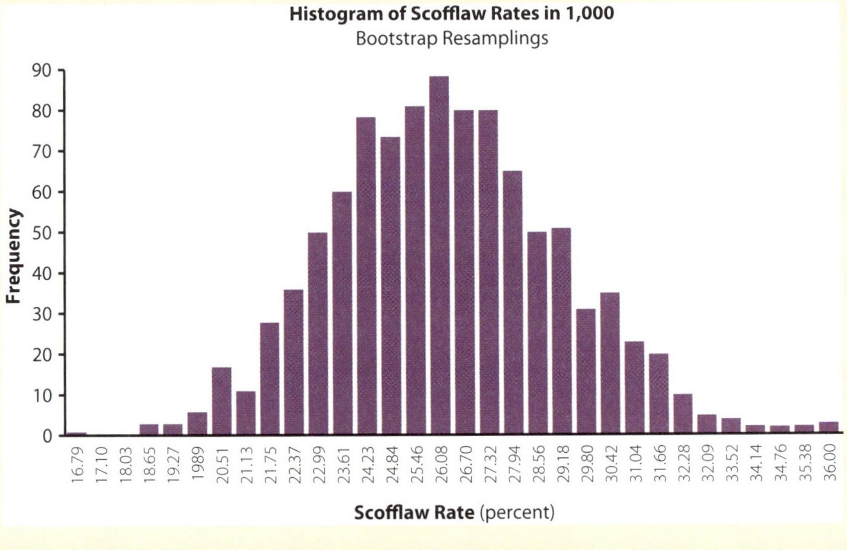

The mean of these 1000 scofflaw rates was 25.94%, and 95% of the estimates fell in the range from 20.44% to 31.68%. (That is the interval from the 26th smallest to the 26th largest outcomes: It excluded the lowest and highest 25 outcomes, which together constituted 50/1000 = 5% of the outcomes that arose.)

Given that the scofflaw rate in the original sample of 12 blocks was 25.8% under the plug-in principle, we see no real evidence of bias based on the bootstrapping procedure, which yielded an average outcome of 25.94%.

(v) The resamplings revealed that, under the approximate distribution, 95% of the 1000 individual values of $\breve{p}$ that arose in resampling fell between 5.36 percentage points below and 5.88 points above the value 25.8%, which is the plug-in estimate for the approximate distribution. Bootstrapping therefore hypothesizes that, for the entire city, there is about a 95% chance that the observed $\breve{p}$ of 25.8% is accurate to within these margins of error. That circumstance creates an approximate 95% confidence interval from 19.92% to 31.16% (i.e., from 25.8 – 5.88 to 25.8 + 5.36).

Note that we *subtract* 5.88 from 25.8 rather than add it to 25.8. We do this because there is a 2.5% chance that the plug-in estimate of 25.8 is too high by more than 5.88 percentage points. But under what circumstance is the plug-in estimate too high by more than 5.88 points? Answer: When the actual scofflaw rate is 25.8 – 5.88 = 19.92% or lower.

(vi) An adjustment to the original bootstrap confidence interval is important when (i) the average resampling statistic differs appreciably from the statistic based on the original sample and/or (ii) the histogram of resampling outcomes shows marked asymmetry. Neither of these circumstances applies here.

(vii) The true "scofflaw" rate would be 20% or lower if sampling error led the plug-in "scofflaw" rate—which was 25.8%—to be too high by at least 5.8%. In the 1000 resamplings from the approximate distribution, the outcome exceeded the correct ratio of 25.8% by 5.8 percentage points or more whenever it was at least 25.8 + 5.8 = 31.6%. That happened 30 times, or in 3.0% of the resamplings. For that reason, 3.0% is a reasonable estimate of the probability that the true city-wide scofflaw rate is 20% (25.8 – 5.8) or lower. Of course, we are not seriously suggesting that the probability is 3.0% rather than 3.1%, but we are confident that the probability is not (say) 40%.

EXERCISE 6.14 Based on the histogram of resampling results in Example 6.14 and some reasonable approximations, estimate a 90% confidence interval for the citywide scofflaw rate. Then estimate the probability that the city-wide scofflaw rate is at least 30%. (Hint: Note that each bar in the histogram covers an interval of size 0.62 percentage points. If you had to make a guess, what fraction of the outcomes in a given bar would you say are above (say) the halfway mark of the bar? The 3/4 mark?)

When Should We Bootstrap?

Actually, bootstrapping can be applied in any circumstance, even in cases in which we can find confidence intervals based on (say) the Central Limit Theorem. In those instances, we should get results from bootstrapping that are similar to those we would obtain on other grounds.

However, bootstrap results can be misleading if the original sample on which we bootstrap is itself unusual. After all, bootstrapping will never go beyond the original data points. In particular, a bootstrap exercise based on a small amount of data is vulnerable to sheer bad luck in choosing the sample, and the resulting uncertainty could well exceed anything suggested by the calculated bootstrap 95% confidence interval. Bootstrapping is reasonable and often helpful, but it is not magic.

6.5 Two More Advanced Sampling Problems

The two problems below are a bit advanced for an introductory course, but students might benefit if they read the solutions and not fret that they might not have reached the answers on their own. Such problems should NEVER appear on an exam.

We first turn to the dilemma that started this chapter.

EXAMPLE 6.15 **The 7% Solution**

To repeat what we said earlier, an insurance company estimated that its various payouts for a given type of claim would have a mean value of μ (though individual payouts would vary around the average). Over a one-year period, its actual payouts on 20,000 different claims averaged 7% above the projected mean. Some enraged stockholders filed a lawsuit against the company, contending that so much excess in such a large sample of claims demonstrated gross incompetence. The company's response was that it had suffered a temporary spasm of bad luck: an usually large number of claims came from the upper tail of the payout distribution.

How plausible is the company's explanation? Assume that no additional information is available. (This example is based on a true story, and there really was no additional information.)

Solution

In essence, the company was saying that there is a probability distribution for the payout for an individual claim. It implied that the 20,000 claims filed that year constitute a random sample of size 20,000 from that long-term distribution. The "sample mean" arising from those claims would deviate from the true long-term mean because of sampling error, and that sampling error—the company asserted—was enough to explain a 7% deviation from the long-term mean μ that the company had correctly specified.

How might we evaluate that assertion, absent further information that would indicate the level of sampling error in the observed data? Here we are not using $\bar{X}$ to estimate μ: the company asserts that it already knows μ. Rather, the relevant question is how far $\bar{X}$ can deviate from μ because of random fluctuations. Back in Chapter 1, we discussed the distribution of the sampling error $W = \bar{X} - \mu$, and concluded that, for even moderately large samples, $W \approx N\left(0, \frac{\sigma}{\sqrt{n}}\right)$ where σ is the standard deviation of the original distribution. It follows that a 95% confidence interval for W would be $\left(-\frac{1.96\sigma}{\sqrt{n}}, \frac{1.96\sigma}{\sqrt{n}}\right)$. Because $W = \bar{X} - \mu = 1.07\mu - \mu = .07\mu$ here, we could conclude that the company's claim is plausible so long as $.07\mu$ falls in the 95% confidence interval for W, based on the distribution of W when $n = 20{,}000$.

If we knew σ, we could directly find the 95% CI for W and evaluate the company's assertion. But we do not have the information that would allow us to determine σ. So what do we do?

The company's assertion that the 7% deviation is not statistically significant would work so long as $.07\mu$ falls in the 95% CI, however barely. Clearly, $.07\mu$ is on the positive side of zero, and it would squeak into the confidence interval if:

$$.07\mu = \frac{1.96\sigma}{\sqrt{20{,}000}} \tag{6-13}$$

Because $\sqrt{20{,}000} = 141.4$, (6-13) is equivalent to:

$$\mu = .198\sigma \text{ which implies that } \sigma \approx 5\mu$$

For the company's assertion to be plausible, in other words, we need a payout distribution with a standard deviation something like five times the mean.

Is it reasonable to suggest that $\sigma \approx 5\mu$ in the actual payout distribution? It's not hard to imagine a distribution with standard deviation-to-mean ratio of 5:1 if V is $N(1,5)$, for example, then $\sigma = 5\mu$. But this last distribution entails a large probability of a negative outcome, while the claims distribution by its nature falls completely in the positive range. (Claims that are denied are not included in the distribution.) Can we imagine a *positive* distribution with $\sigma \approx 5\mu$ that might plausibly resemble the actual claims distribution?

Well, let's try. Suppose that claims fall into two categories: a large percentage of small claims and a small percentage of very large claims. (Under these circumstances, the standard deviation of the distribution would be large but the mean relatively small.) To be more specific, suppose that the claims follow a simple two-point distribution:

Type of Claim	Percentage of Cases	Payout
Small	97%	1
Large	3%	Q

The 3% figure is not sacrosanct: It's just a number we use to get the exercise started. This 97/3 split suggests that about 1 out of 30 claims is large. Here, we are defining the small claim as 1 cash unit. Q therefore expresses the larger payout as a multiple of the smaller one.

In this discrete distribution, the mean μ is given by:

$$\mu = .97 * 1 + .03 * Q = .97 + .03Q$$

while the variance follows:

$$\sigma^2 = .97[.03(Q - 1)]^2 + .03[.97(Q - 1)]^2 = .029(Q - 1)^2$$

(In the formula for σ^2, we are using the facts that $(.97 + .03Q) - 1 = .03(Q - 1)$, while $Q - (.97 + .03Q) = .97(Q - 1)$.)

Taking the square root of this expression for σ^2, we see that $\sigma = .17(Q - 1)$, and hence that:

$$\sigma/\mu = .17(Q - 1)/(.97 + .03Q) \tag{6-14}$$

If we set σ/μ in (6-14) equal to 5 and solve for Q, we reach $Q \approx 252$. In other words, a payout distribution with standard deviation five times its mean is:

$$\text{Payout} = \begin{cases} 1 \text{ w.p. } .97 \\ 252 \text{ w.p. } .03 \end{cases} \tag{6-15}$$

(We can readily verify that $\mu \approx 8.5$ while $\sigma \approx 42.5$ here, and thus that $\frac{\sigma}{\mu} = 5$.

Under this distribution, the infrequent large claims would dominate the company's payouts. Out of every 100 claims, one would expect roughly 97 of size one (and total payout 97) and 3 of size 252 (total payout $3 * 252 = 756$). Of the overall payout of $97 + 756 = 853$, therefore, 756 (or about 89%) is absorbed by large claims. Thus, what really drives year-to-year variation in payouts in this situation is *not* the total number of claims (20,000) but the number of *large* claims (roughly3% of 20,000, or about 600). The actual number of large payouts would be binomial with the distribution B(20,000, .03) and, in that distribution, getting at least 642 large payouts rather than the mean number of 600 (i.e., getting 7% more than expected) is not a terribly rare event.

Discussion

We have just constructed one example of a claims distribution under which the company's 7% excess payout is within the range of plausible sampling error. But could the real claims distribution be remotely as lopsided as this one? Well, suppose that we are dealing with something like fire insurance. A small fraction of the claims arise when a house is destroyed, while the heavy majority might arise from relatively minor damage from fires that were quickly brought under control. In that case, a polarized distribution with the general character of (6-15) might be realistic (though it would certainly not be a two-point distribution).

Furthermore, this analysis assumes that the 20,000 claims are *independent* selections from the long-term claims distribution, and thus that there were 20,000 independent pieces of evidence about the company's payout model. But suppose you were asked to estimate your expenses over the next calendar year, for 200 items ranging from toothpaste to tuition. You might do so assuming an annual inflation rate of (say) 1%. But what if some financial anomaly raises the inflation rate to 7%? Then nearly all of your 200 forecasts might wind up too low, and a person looking at them separately might conclude you can't be trusted with money. But you didn't make 200 independent mistakes; you made one error that distorted all the forecasts the same way.

Analogously, isn't it possible that the company underestimated some single source of volatility in payouts (e.g. a hurricane), and for *that* reason paid out 7% more than anticipated? Of course it is possible, much as a polarized payout distribution is possible. But the company offered neither of these reasons why it paid out 7% more than it had foreseen. Perhaps the company was silent because neither explanation fit the situation at hand; perhaps instead one of the explanations was relevant but the company saw no need to offer it. Either way, the assertion that management was incompetent is hard to rebut given its highly inadequate defense.

EXERCISE 6.15 As in "the 7% solution" in Example 6.15, find *some* two-point positive probability distribution with a standard deviation that is four times its mean. More specifically, suppose that the two points are 1 and Q (where $Q > 1$). Assume that probability 4% is assigned to Q. What value should Q take on?

EXAMPLE 6.16 **Free Riders?**

A certain city sells monthly passes that allow unlimited rides on its mass transit system, and wanted to estimate the average number of rides per week among holders of these passes. There were two proposals about how to do so:

- Select 500 pass holders at random on a given date, and ask them how many rides they took in the previous week (Monday through Sunday). Use the sample mean to estimate the population mean.
- Select two independent groups of pass holders at random, each of size 250. Within the first group, ask about the number of weekday rides (Monday–Friday). Within the second, ask about the weekend (Saturday and Sunday). Find the sample means for each group, and add them together.

(i) Will both methods yield unbiased estimates of the mean number of rides per pass holder?

(ii) Can we assess which of these estimates has the greater variance of sampling error?

(iii) Are the any problems that could affect the reliability of both of these methods?

Solution

(i) Let random variable X be the number of weekly trips (during the week of interest) for a randomly-chosen pass holder, and let random variables X_d and

X_e be, respectively, the number of weekday and weekend trips for a randomly chosen pass holder. With these definitions:

$$X = X_d + X_e$$

Under the first method above, 500 values of X are obtained, and their sample mean $\bar{X}$ is found. Under the second method, 250 values of X_d are found and their sample mean $\overline{X_d}$ is calculated. Then $\overline{X_e}$ is calculated based on X_e-values for another sample of 250 pass holders.

If we define μ as the mean of X, and μ_d and μ_e as the means of X_d, and X_e, respectively, then:

- $\bar{X}$ is an unbiased estimator of μ, based on usual sampling theory.
- $\overline{X_d}$ is an unbiased estimator of μ_d, while $\overline{X_e}$ is an unbiased estimator of μ_e. It follows that $\overline{X_d} + \overline{X_e}$ is an unbiased estimator of $\mu_d + \mu_e$. But because:

$$\mu = \mu_d + \mu_e$$

we see that $\overline{X_d} + \overline{X_e}$ is also an unbiased estimator of μ.

Thus, both methods are unbiased.

(ii) Suppose that σ^2 is the variance of X, σ_d^2 is the variance of X_d, and σ_e^2 is the variance of X_e. We know from Chapter 3 that the variance of the sampling error in estimating μ by $\bar{X}$ follows:

$$\sigma^2\left(\bar{X} - \mu\right) = \sigma^2/500$$

(We would estimate σ^2 by S^2 in the actual sampling, but σ^2 is the true determinant in the volatility of $\bar{X}$.)

We likewise know that:

$$\sigma^2\left(\overline{X_d} - \mu_d\right) = \sigma_d^2/250$$
$$\sigma^2\left(\overline{X_e} - \mu_e\right) = \sigma_e^2/250$$

Because $\overline{X_d}$ and $\overline{X_e}$ are based on different samples, they can be treated as independent. Then we can use the rule "variance of a sum is the sum of variances" to write:

$$\sigma^2\left(\left(\overline{X_d} - \mu_d\right) + \left(\overline{X_e} - \mu_e\right)\right) = \sigma^2\left(\overline{X_d} - \mu_d\right) + \sigma^2\left(\overline{X_e} - \mu_e\right)$$
$$= \frac{\sigma_d^2}{250} + \frac{\sigma_e^2}{250}$$

But also

$$\sigma^2\left(\left(\overline{X_d} - \mu_d\right) + \left(\overline{X_e} - \mu_e\right)\right) = \sigma^2\left(\overline{X_d} + \overline{X_e} - \left(\mu_d + \mu_e\right)\right)$$
$$= \sigma^2\left(\overline{X_d} + \overline{X_e} - \mu\right)$$

Thus

$$\sigma^2\left(\overline{X}_d + \overline{X}_e - \mu\right) = \frac{\sigma_d^2}{250} + \frac{\sigma_e^2}{250} = \frac{\left(\sigma_d^2 + \sigma_e^2\right)}{250}$$

So the comparative question for the two sampling methods becomes:

$$\text{Is } \frac{\left(\sigma_d^2 + \sigma_e^2\right)}{250} \text{ greater than } \frac{\sigma^2}{500}?$$

The answer to this question would be clear if $\sigma_d^2 + \sigma_e^2 = \sigma^2$. In that case, we would be comparing $\frac{\sigma^2}{250}$ with $\frac{\sigma^2}{500}$, and it's obvious that the first quantity is the larger. But can we assume that $\sigma_d^2 + \sigma_e^2 = \sigma^2$? We could do so if X_d and X_e were independent, but it is not certain that they are. It could be, for example, that someone with an unusually high X_d-value uses mass transit for more than the home-to-work round-trip during the week. That circumstance might suggest that the person doesn't have a car, in which case his X_e-value might also be unusually high. In short, X_d and X_e might be correlated, which means we should use the Chapter 1 formula (1-95):

$$\sigma^2(X) = \sigma^2\left(X_d + X_e\right) = \sigma^2\left(X_d\right) + \sigma^2\left(X_e\right) + 2\rho_{de}\sigma\left(X_d\right)\sigma\left(X_e\right) \qquad (6\text{-}16)$$

where ρ_{de} is the correlation coefficient of X_d and X_e
Given the definitions of σ^2, σ_d^2 and σ_d^2, we can rewrite (6-16) as:

$$\sigma^2 = \sigma_d^2 + \sigma_e^2 + 2\rho_{de}\sigma_d\sigma_e$$

Thus:

$$\frac{\sigma^2}{500} = \frac{\left(\sigma_d^2 + \sigma_e^2\right)}{500} + \rho_{de}\frac{\sigma_d\sigma_e}{250}$$

and the comparative question between $\frac{\sigma^2}{500}$ and $\frac{(\sigma_d^2 + \sigma_e^2)}{250}$ becomes:

Is
$$\frac{\left(\sigma_d^2 + \sigma_e^2\right)}{500} + \rho_{de}\frac{\sigma_d\sigma_e}{250} > \frac{\left(\sigma_d^2 + \sigma_e^2\right)}{250}?$$

Which becomes:

Is
$$\rho_{de}\frac{\sigma_d\sigma_e}{250} > \frac{\left(\sigma_d^2 + \sigma_e^2\right)}{500}?$$

Or:

Is
$$2\rho_{de}\sigma_d\sigma_e > \sigma_d^2 + \sigma_e^2? \qquad (6\text{-}17)$$

We're almost there. Because $\rho_{de} \leq 1$, the quantity on the left of (6-17) has a maximum value of $2\sigma_d\sigma_e$. But because:

$$\left(\sigma_d - \sigma_e\right)^2 = \sigma_d^2 + \sigma_e^2 - 2\sigma_d\sigma_e \geq 0$$

it follows that:

$$\sigma_d^2 + \sigma_e^2 \geq 2\sigma_d\sigma_e$$

For that reason, inspection of (6-17) reveals that, even when weekend and weekday usage have a perfect positive correlation and thus $\rho_{de} = 1$, the answer to question (6-16 has to be "no." And what is true when $\rho_{de} = 1$ is equally true when ρ_{de} is less than one. In short, there is *no way* that asking 500 people about their weekly travels will entail sampling error with higher variance than asking 250 about weekdays and 250 others about weekends.

Indeed, if $\rho_{de} < 1$ (as seems all but certain), then the first scheme is strictly preferable to the second in terms of the variance of sampling error (and thus, also for the standard deviation of sampling error or the width of the 95% confidence interval).

(iii) There are at least two problems with both sampling schemes. If the sampling is done for one particular week, we don't know whether the results apply to other weeks. It could be, for example, that there are more weekend trips in the summer months than in the frigid winter. There is also the issue that respondents do not have perfect memories, and might sometimes misstate how many trips they took the previous Tuesday. There could even be misinterpretations of the question, and some respondents might treat a home/work round-trip as one ride rather than two. A survey that is well-designed and administered can reduce the risk of forgetfulness and misinterpretation, but is unlikely fully to eliminate it.

Such imperfections do not mean that the survey will be uninformative. But they do mean that random sampling error is not the only source of imprecision in the results.

EXERCISE 6.16 Suppose that the second proposal in Example 6.16 is amended to involve asking 500 pass holders about their weekday rides and another 500 independently about their weekend rides.

(i) Under what condition will this revised scheme achieve lower sampling-error variance than the first scheme, which asks 500 people about their total weekly rides?

(ii) Do you think it plausible that the condition in (i) would be satisfied? Briefly discuss.

Glossary

Selection bias arises when the data sample analyzed is not random even though it was intended to be. Thus, contrary to intention, some members of the population had a higher chance of entering the sample than others.

Nonresponse bias can arise when only some of those individuals asked to participate in a survey actually do so. Unless the responses of participants can be shown closely to resemble those of nonparticipants, the survey results could differ substantially from those that would have arisen had everyone taken part. This systematic bias has nothing to do with random sampling error, so the accuracy formulas associated with random sampling cannot correct for the problem.

The **finite population correction** to the standard error of a sample mean $\bar{X}$ takes explicit account of the population size N as well as the sample size n. So long as the sample size is a small fraction of the total population size (e.g, $n/N \leq .01$), the correction formula demonstrates that the exact value of N is essentially irrelevant to sampling accuracy.

Stratified random sampling is a method of choosing random samples within different strata of a population. In many but not all applications of stratified sampling, the sample sizes in certain strata are larger fractions of the total sample size than the populations in those strata are of the total population (i.e., some strata are "overrepresented" in the sampling, much as other strata are "underrepresented.") The aim of stratified random sampling is to achieve greater overall accuracy in sampling results than would arise in simple random sampling (in which all data points in all strata have the same chance of being selected).

Bootstrapping is a method for approximating the degree of bias and sampling error that attend a parameter estimate based on n data points. The bootstrap assessment is based on repeated samplings of size n without replacement (resamplings) from a discrete approximate distribution that "mimics" the data points at hand.

The **bootstrap approximate distribution** takes the n data points at hand and, if no number is repeated, assigns probability $1/n$ to each of the values that arose. If a particular number occurred k times for $k > 1$, then the distribution assigns probability k/n to it. The bootstrap distribution is discrete even if the unknown distribution that generated the n data points is continuous.

The **bootstrap approximation** is that the bias and sampling variability of the statistic of interest—as indicated through repeated resamplings of size n in the approximate bootstrap distribution—is a good proxy for the bias and variability of a single sample of size n from the true distribution. The information obtained through bootstrapping is therefore applied to the single sample of size n that started the bootstrapping exercise.

Formulas

Sampling with a Finite Population:

If $W = \bar{X} - \mu$, then:

$$\sigma(W) = \left(\frac{\sigma}{\sqrt{n}}\right)\sqrt{\frac{N-n}{N-1}}$$

where N = population size, n = size of random sample, σ = standard deviation within the full population

The quantity $\sqrt{\frac{N-n}{N-1}}$ in is called the ***finite population correction factor*** because it adjusts the standard deviation formula $\sigma(W) = \sigma/\sqrt{n}$ that applies when the population size is infinite.

When σ is estimated by $S = \sqrt{\frac{(X_i - \bar{X})^2}{n-1}}$, then $\sigma(W)$ is approximated by $S(W)$ that follows:

$$S(W) = \left(\frac{S}{\sqrt{n}}\right)\sqrt{\frac{N-n}{N-1}}$$

The Bootstrap Approximate Distribution

If the random sample from the true distribution of X consists of the distinct points $(x_1, \ldots, x_n)$, then the approximate distribution for X is:

$$P(X = a) = \begin{cases} \frac{1}{n} \text{if } a \in (x_1, \ldots, x_n) \\ 0 \text{ otherwise} \end{cases}$$

($\in$ means "is contained in)

If the same value a comes up k times in $(x_1, \ldots, x_n)$, then the approximate distribution is modified to:

$$P(X = a) = \frac{k}{n}$$

Bias in a Point Estimate:

If $\breve{\xi}$ is the estimate of a quantity of interest ξ under the plug-in principle and $\bar{\xi}$ is the average plug-in estimate of ξ under many repeated resamplings, then the bias associated with $\breve{\xi}$ is estimated under:

$$\text{Bias} \approx \breve{\xi} - \bar{\xi}$$

Some Parting Thoughts

- When a data sample that is meant to be random actually is not, that is bad news. When such selection bias occurs, extrapolating results from the sample to the population is problematic, and quantifying the potential distortion can be difficult if not impossible. A sensible way to treat selection bias is to avoid it.
- The finite population correction offers mathematical proof that, in most practical situations, the size of the population is far less relevant to sampling error than the size of the sample.
- When conducted sensibly, stratified random sampling can yield lesser sampling error than simple random sampling with the same total sample size. However, that potential advantage can become a drawback if one "undersamples" in strata where a larger sample would have been far more informative. Simple random sample works very well in many contexts, so stratification is sometimes more trouble than it's worth.
- Bootstrapping goes beyond earlier sampling theory, in that it allows the estimation of confidence intervals even in situations not covered by theory. It uses modern computing to conduct many resamplings in an approximate distribution that is believed to resemble the actual distribution that generated the available data. It's a great idea.
- However, bootstrapping does not always offer reliable results. If the original sample is small and the approximate distribution differs appreciably from the true distribution, lots of resampling won't reduce the problem. Then again, no methods are especially reliable when data points are sparse.

EZ Pass

1. An airline conducts an electronic poll of passengers that asks "How was your last flight with us?" That poll could produce biased results if:
   (A) All passengers respond to the poll.
   (B) The proportion of dissatisfied passengers who respond is the same as the proportion of satisfied passengers
   (C) Dissatisfied passengers respond at a different rate from satisfied passengers
   (D) None of these
2. To estimate the mean μ in a large population, a random sample consisting of 2% of the population is obtained and the sample mean $\bar{X}$ is calculated. The finite population correction factor for the standard error associated with $\bar{X}$ is closest to:
   (A) 0.99
   (B) 0.98
   (C) 1.01
   (D) 1.02

3. When a median of a distribution is estimated from a random data sample of size 11 (as the sixth largest observation), the estimator is:
   (A) always unbiased
   (B) sometimes unbiased
   (C) never unbiased
   (D) none of these
4. Five points are selected at random from a continuous distribution: 21, 36, 15, 28, and 26. If these points are used as the basis of 1000 resamplings in a bootstrapping exercise, the sample variances within the individual resamplings will average approximately:
   (A) 50
   (B) 7
   (C) .05
   (D) .22
5. The payouts μ and σ for individual insurance claims are known, and the 95% confidence interval for $\bar{X}$, the average for the next 4000 payouts, has an upper limit of 1.04μ. Given this information, the ratio of σ to μ is closest to:
   (A) one
   (B) two
   (C) four
   (D) cannot be determined from the given information
6. To assess the caliber of its customer service, Trader Jack's will interview at random customers who shop in its store on February 23. It will treat the respondents as a random sample of customers who shop at the store in February.

   This sampling scheme:
   (A) is unbiased, because all February customers have an equal chance of being interviewed
   (B) is biased, because frequent customers are more likely to appear on a given day than infrequent customers
   (C) is unbiased unless February 23 is a Thursday
   (D) is unbiased if restricted to customers who did not shop at Trader Jack's earlier in February

Further Exercises

EXERCISE 6.17 Suppose that the 100 randomly-chosen arrest cards in Example 6.2 had instead yielded 15 individuals with a single arrest that year, 40 with two arrests, and 45 with three arrests. What are reasonable estimates of the f_i's, the fraction of individuals in the arrest files with exactly i arrests that year (i = 1,2,3)? What is the corresponding estimate of the average number of arrests that year for individuals with at least one? (Yes: three linear equations and three unknowns.)

EXERCISE 6.18 In a telemarketing survey, 60% of those selected were successfully reached, and 70% of them were enthusiastic about the new product described. The other 40% were not reached.

(i) Based on these data, what are the maximum and minimum rates of enthusiasm for *all* of those who were sought for the survey. (Assume that the 70% figure for those who were reached has very little sampling error.)

(ii) A supplementary survey among those not initially reached found 40% enthusiastic about the product. Make a reasonable estimate of the product's "enthusiasm level" for all those sought for the survey.

EXERCISE 6.19 In an attempt to reach 200 randomly-chosen viewers who watch "The Good Wife," 120 are located, and 60% of them say it is the best television series of all time. In a follow up survey to get a random sample of those 80 viewers who were *not* reached, all are located, and 50% take the same view.

(i) Given these numbers, make an estimate of the overall fraction of "The Good Wife" viewers who consider it the best series of all time.

(ii) True or False: The margin of sampling error for your estimate (i.e. the width of the 95% confidence interval around its midpoint) is at least .07.

EXERCISE 6.20 A financial analyst wants to examine a random sample of 100 mortgages out of 900 that defaulted. He also wants to study a random sample of 100 mortgages out of 39,000 that did not default. In each case, he will estimate numerically the credit worthiness of the mortgage holders, based on information that they provided at the time they took out the mortgages.

Suppose that the sample means in the two groups are $\overline{X}$ (defaulted) and $\overline{Z}$ (did not default), and that the standard deviation S is the same in both groups.

(i) Which group has a wider 95% confidence interval for the true mean in its full population?

(ii) By what percentage does the width of the wider confidence interval exceed the narrower one?

(iii) If $\overline{Z}$ exceeds $\overline{X}$ by S/2, would one conclude under a difference of means test that the population means differ to a statistically significant extent?

EXERCISE 6.21 In a poll about whether students at a certain college want an "Inner Self" major, the school newspaper tries to get a random sample of students. After the poll is finished, it compares the characteristics of the students canvassed to those of the full student population, with the following results:

Characteristic	Sample (200 students)	Full Student Body
Male	42%	46%
Female	58%	54%
Freshmen	24%	28%
Sophomores	28%	25%
Juniors	23%	25%
Sophomores	25%	22%

(i) Perform two distinct tests on the hypothesis that the students questioned were a random sample of the student body. Assume that the gender test and the class-level test are independent. Arrange that the total risk of accidental rejection is 5%.

(ii) Give a numerical example of a major deviation from randomness that would not be picked up by these tests. (*Hint:* Suppose there were far too many female juniors and far too few female sophomores, while the pattern for males was the exact opposite.)

EXERCISE 6.22 Suppose that the newspaper in Exercise 6.21 is satisfied that it picked students at random, but is concerned that the male/female split in the sample is 42/58 rather than the correct 46/54. It therefore decides to estimate student support for "Inner Self" not by $\hat{p}$, the fraction who support it in the poll, but rather by $\breve{p}$ which follows:

$$\breve{p} = .46\widehat{p_m} + .54\widehat{p_f}$$

where $\widehat{p_m}$ = support for the new major among males in the sample $\widehat{p_f}$ = support for the new major among females in the sample

(i) Suppose that $\widehat{p_m}$ = .72 and $\widehat{p_f}$ = .56. What is $\breve{p}$, and what would $\hat{p}$ be had the results not been broken down by gender?

(ii) Let W be the sampling error associated with using $\breve{p}$ to estimate campus support for the new major. What is the mean of W? What is a sensible estimate of the standard deviation of W given the data in (i)?

(iii) Assuming that W has roughly a normal distribution (as it should), what is a 95% confidence interval for p, the fraction of people on campus who support the new major?

(iv) Should the newspaper routinely apply this gender-based adjustment to the sample results, rather than simply estimate p by $\hat{p}$? Discuss briefly.

(v) If the newspaper is adjusting the results by gender in this way, should it also adjust for the proportions of freshmen, sophomores, juniors, and seniors? Do you see how the whole business is getting complicated, which might be a good case for using $\hat{p}$ after all?

EXERCISE 6.23 Stratification is not always a good thing. To estimate a population mean, come up with a stratified sample scheme that divides a population into two parts and yields a *larger* standard error than a simple random sample with the same total number of data points. (Hint: Consider a case where the sampling is concentrated in the stratum in which it is less needed.)

EXERCISE 6.24 Suppose that n points are independently selected from an unknown probability distribution.

(i) Show that the sample mean $\bar{X}$ is an unbiased estimate of the true mean μ. (*Hint:* you need merely replicate the derivation we covered in Chapter 4.)

(ii) Show that, in the first resampling from the approximate bootstrap distribution, the average of the data points that arise will also have mean μ.

(iii) Show that, if the sample means in the repeated resamplings are averaged together, that average also has mean μ.

EXERCISE 6.25 Bootstrapping offers a natural approach for estimating a 95% confidence interval for the standard deviation of a probability distribution. Suppose that 12 observations from a distribution are available:

94 88 70 26 22 21 96 39 84 90 38 35

(i) What is the approximate bootstrap distribution given these numbers?

(ii) What estimate of the standard deviation arises from the plug-in principle?

(iii) Take 1000 resamplings from this distribution and find the mean of the individual standard deviations. Does the outcome suggest that the plug-in estimate is biased?

(iv) What are both the "raw" and bootstrap-adjusted 95% confidence intervals for the standard deviation?

(v) In fact, the 12 numbers above were generated from RANDOM.ORG, each of which is supposedly an independent pick from the uniform distribution on the integers from 1 – 100 (i.e., each pick has probability .01 of taking on any given integer in the 1–100 range). What is the true standard deviation of that distribution?

(vi) Do the bootstrapping results offer compelling evidence that the standard deviation in the distribution that generated the 12 numbers differs from that in the distribution that RANDOM.ORG says generated the numbers? Briefly discuss.

EXERCISE 6.26 Suppose that 100 math teachers who followed vegetarian diets were compared with 150 others with the same age/gender mix who ate meat, fish, and fowl. Over a 25-year period, 34 of the meat-eaters suffered serious heart problems, as opposed to 11 in the vegetarian group.

(i) Based on these data, make an estimate of the relative risk of heart disease for the meat-eaters compared to the vegetarians. (Use the definition of relative risk that was introduced in Chapter 2.)

(ii) Consider the broader populations of vegetarian math teachers and non-vegetarian math teachers with the same age/gender mix as those who were studied. Suppose $X = 0$ for an individual means no serious heart problems, while $X = 1$ means serious heart problems. What are the approximate bootstrap distributions for X in the two populations? (Hint: What are the 100 X-values for the vegetarians? What discrete distribution on 0-1 matches the observed pattern?)

(iii) Now conduct 1000 resamplings, each of which yields a relative risk estimate for the two groups. Find a bootstrap-adjusted 95% confidence interval for the relative risk factor.

(iv) Under the approximate distribution for vegetarians, the number out of 100 who suffer serious heart problems would be $B(100,.11)$. What is the corresponding binomial distribution for meat-eaters?

(v) Could the resampling have been conducted by taking 1000 ratios of the form Z/V, where V is $B(100,.11)$ and Z is the meat-eaters' distribution in (iv)? Briefly discuss.

CHAPTER 7

Analysis of Variance

Wine Sellers: A Trilogy

"We've got it," Minerva announced, looking over the small vineyard that she and Mendel had bought in the Napa Valley. "Our taste test shows that people think that our wine tastes better than the wines of our two chief competitors." "Does it really?," Mendel asked with a touch of sarcasm. "All the results suffer from sampling error, and most of the people who tasted our wine tried it after tasting the other two. And research shows that, given three choices, people are unusually likely to favor the last of the three." "You're right to say the test had some blemishes," responded Minerva, wondering how Mendel could state his objections so coherently. "But we have enough data to adjust for its imperfections. So have a glass and relax."

To be continued

Often we have data samples from different populations, and wonder whether they differ in their means and, if so, how. The statistical sampling theory that we studied in Chapters 3 and 6 provides a useful framework for thinking about that issue. In this chapter, we focus on a particular technique related to sampling theory called **analysis of variance (ANOVA)**. Sometimes it is preferable to assess differences in populations via the method of regression analysis, but we postpone introducing regression models until Chapters 8 and 9.

A Hypothesis of Equal Means

Analysis of variance might seem a misleading name for the technique we are studying, for it actually concerns itself with relationships among various means. (However, it does consider the variation among the means, and in that sense does involve variance.) In the simplest model, it is hypothesized that several means are equal, and the question is whether available data are consistent with the hypothesis. As an introduction to ANOVA, we concentrate first on two distinct ways of testing for equality-of-means.

We assume that we have m large populations, and that every member of each population has a number associated with him, her, or it. Suppose that μ_i is the average of the numbers for population i, and σ_i^2 the variance. In the standard ANOVA model—the subject of this discussion—there are there are three strong assumptions:

- ***A random sample of data of the same size n*** is available for each population.
- All m populations have ***the same variance*** (i.e. that $\sigma_i^2 = \sigma^2$ for all i).
- The numbers in each population follow a ***normal-distribution***.

Taken together, these assumptions mean that each randomly-selected data point in the ith sample is an independent pick from $N(\mu_i, \sigma)$. To put it another way, it is assumed that X_{ij}, the jth data point in sample i, is given by:

$$X_{ij} = \mu_i + \varepsilon_{ij} \tag{7-1}$$

where ε_{ij} is $N(0, \sigma)$, and different ε_{ij}'s are independent random variables. ((7-1) arises because adding a constant to a normal random variable yields another normal variable with the mean shifted by that constant.) Under (7-1), each of the m random samples of size n comes from a bell-shaped curve with the same width, but it is not certain that the means of these curves must be the same.

The question is whether the data are consistent with the equal-means hypothesis H_0 that follows:

$$H_0 : \mu_1 = \mu_1 \ldots = \mu_m = \mu$$

7.1 The Standard One-Way ANOVA Test

Here we present the logic of a familiar test of equal means, which is called the one-way ANOVA test. We have a total of nm data points in our m random samples each of size; let X_{ij} be the jth data point in sample i. As discussed in Chapter 3, we can estimate the mean μ_i in the ith population by the sample mean $\overline{X_i}$ and can estimate the variance σ_i^2 in an unbiased manner by S_i^2 which follows:

$$S_i^2 = \sum_{j=1}^{n} \frac{\left(X_{ij} - \overline{X_i}\right)^2}{n-1} \tag{7-2}$$

Two Ways to Estimate the Common Variance

The First Way

Because all the σ_i^2 take on the common value σ^2, each of the m S_i^2-values is an unbiased estimate of the same quantity. All these estimates are subject to sampling error. A sensible way to synthesize the m outcomes to estimate σ^2 is to take their simple average S^2:

$$S^2 = \frac{\sum_{i=1}^{m} S_i^2}{m}$$

What do we mean when we say that S^2 is a "sensible" estimate of σ^2? First of all, we can establish at once that S^2 is unbiased. After all:

$$E\left(S^2\right) = \frac{\sum_{i=1}^{m} E\left(S_i^2\right)}{m}$$

In chapter 3, we showed that $E(S_i^2) = \sigma^2$. Therefore:

$$E\left(S^2\right) = \frac{\sum_{i=1}^{m} E\left(S_i^2\right)}{m} = \left(\frac{1}{m}\right)\left(m\sigma^2\right) = \sigma^2$$

so S^2 is itself an unbiased estimator of σ^2,

Not only that, but it is the *minimum variance linear unbiased estimate*. By that we mean that, of all linear combinations of the S_i^2 that are unbiased estimators of σ^2, S^2 has the lowest variance. The sampling error that affects all the S_i^2 does the least "damage" to S^2 when the S_i^2's are averaged together. A proof of that statement appears on the disk that accompanies this book.

In summary:

When *m* populations have a common variance σ^2, and unbiased estimates of σ^2 (S_i^2's) are made from random samples of size *n* from each of these populations, a minimum-variance unbiased estimator S^2 of σ^2 is given by:

$$S^2 = \frac{\sum_{i=1}^{m} S_i^2}{m} \tag{7-3}$$

where $S_i^2 = \sum_{j=1}^{n} \frac{(X_{ij} - \overline{X}_i)^2}{n-1}$ and X_{ij} is the *i*th data point from the *j*th population

The quantity $(n-1)\sum_{i=1}^{m} S_i^2$ is typically described in ANOVA parlance as the ***within-group sum of squares*** in the data set, and is often denoted SS_W. We see from (7-3) that:

$$S^2 = \frac{\sum_{i=1}^{m} S_i^2}{m} = SS_W/(m(n-1)) = SS_W/(mn-m)$$

The variance estimator S^2 in (7-3) can also be written as:

$$S^2 = \frac{SS_W}{(mn-m)} \tag{7-4}$$

where $SS_W = (n-1)\sum_{i=1}^{m} S_i^2$ is the **within group sum of squares** in the data set

A Second Way to Estimate the Common Variance

If the equality of means hypothesis is true, however, there is another way to estimate σ^2 that is even simpler. All m samples would come from normal distributions with the same mean μ as well as the same variance. We could therefore pool the nm data points into one big "supersample." Their arithmetical average $\bar{X}$ would be called the **grand mean** of the observations, and is the estimator of μ discussed at length in Chapter 3. And an unbiased estimator W^2 of σ^2 follows:

$$W^2 = \frac{\sum_{i=1}^{n}\sum_{j=1}^{m}\left(X_{ij} - \bar{X}\right)^2}{mn-1} \tag{7-5}$$

(For reasons discussed in Chapter 3, we divide by $mn - 1$ rather than mn to achieve unbiasedness.)

If all m populations share the same mean and variance, an unbiased estimate of σ^2 can be obtained by pooling the samples from the various populations into one "supersample" and using the spread of the data points around the grand mean $\bar{X}$ of all the data points, according to:

$$W^2 = \frac{\sum_{i=1}^{n}\sum_{j=1}^{m}\left(X_{ij} - \bar{X}\right)^2}{mn-1} \tag{7-6}$$

where m is the number of samples and n the sample size in each one

Given that S^2 and W^2 are both unbiased estimates of σ^2 under H_0, we would expect that the ratio $U = W^2/S^2$ would be close to *one* if H_0 were true. The ratio could differ from one because both numerator and denominator are imperfect estimates of σ^2; too big a discrepancy, however, could raise concerns about the accuracy of H_0. That idea is central to the one-way ANOVA test, but we need develop it a bit more.

What If the Means Are Not Equal?

An important distinction arises here. Even if the various means differ, S^2 is still viable as an estimate of the common variance σ^2. (Each S_i^2, after all, is based solely on the spread within the ith sample.) But the same cannot be said of W^2. Suppose that H_0 is false, and that μ_i is a good distance from μ. Then the squared spread of the *i*-measurements around $\overline{X}$, the proxy for μ, could be far greater than that around $\overline{X}_i$, the proxy for μ_i. Figure 7.1 illustrates what the pattern might look like:

In situations like this, the spread of population-i measurements around the "alien" value $\overline{X}$ can be far greater than the spread among themselves. Indeed, it can be shown algebraically (Exercise 7.15) that:

$$\sum_{j=1}^{n}\left(X_{ij} - \overline{X}\right)^2 = \sum_{j=1}^{n}\left(X_{ij} - \overline{X}_i\right)^2 + n\left(\overline{X}_i - \overline{X}\right)^2 \tag{7-7}$$

When $\overline{X}_i$ is far away from $\overline{X}$ and thus $\left(\overline{X}_i - \overline{X}\right)^2$ is large, (7-7) implies that the contribution of the ith sample to formula (7-6) for W^2 is far greater than S_i^2, the local estimate of σ^2. Assuming similar problems with other samples, W^2 could *exceed* σ^2 by a substantial margin.

A Potential Test of the Equal Means Hypothesis

With all these considerations in mind, we see that an intuitively-appealing test of H_0 would take the form:

- Calculate $U = W^2/S^2$
- Accept H_0 if $U \le c$; reject H_0 if $U \ge c$.
- Choose c so that $\alpha = P(\text{reject } H_0 | H_0 \text{ true}) = 0.05$.

This test would exploit the fact that failure of equal means would inflate W^2 but not S^2 (i.e., would increase U). Unfortunately, determining the value of the test threshold c is not straightforward for this test. We can circumvent this difficulty, however, if we use S^2 and W^2 to find still a third unbiased estimate of σ^2 that is amenable to a statistical test with thresholds that can readily be determined. Be patient.

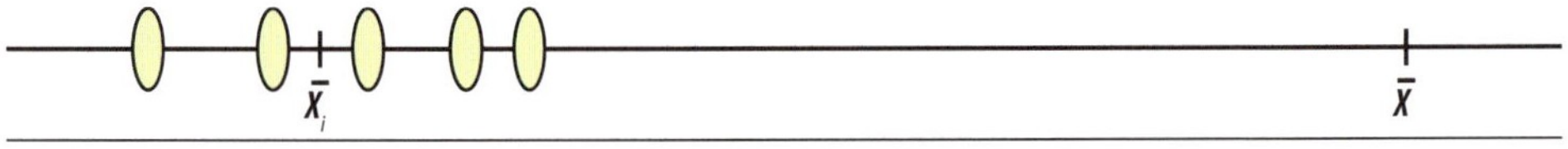

FIGURE 7.1 The sample points—represented by the monstrous yellow dots—deviate far more from the grand mean $\overline{X}$ than from their local sample mean $\overline{X}_i$.

Yet a Third Expression for the Common Variance

To reiterate (7-7), we have:

$$\sum_{j=1}^{n}\left(X_{ij}-\overline{X}\right)^2 = \sum_{j=1}^{n}\left(X_{ij}-\overline{X}_i\right)^2 + n\left(\overline{X}_i-\overline{X}\right)^2$$

But (7-1) implies that:

$$\sum_{j=1}^{n}\left(X_{ij}-\overline{X}_i\right)^2 = (n-1)S_i^2$$

Summing up over i in (7-7), we get:

$$\sum_{i=1}^{m}\sum_{j=1}^{n}\left(X_{ij}-\overline{X}\right)^2 = (n-1)\sum_{i=1}^{m}S_i^2 + n\sum_{i=1}^{m}\left(\overline{X}_i-\overline{X}\right)^2$$

Taking expected values of both sides of this equation, we get:

$$E\left(\sum_{i=1}^{m}\sum_{j=1}^{n}\left(X_{ij}-\overline{X}\right)^2\right) = (n-1)\sum_{i=1}^{m}E\left(S_i^2\right) + n\sum_{i=1}^{m}E\left(\left(\overline{X}_i-\overline{X}\right)^2\right) \quad (7\text{-}8)$$

But then (7-6) and (7-1) allow us to rewrite (7-8) *under* H_0 as:

$$(mn-1)\sigma^2 = m(n-1)\sigma^2 + n\sum_{i=1}^{m}E\left(\left(\overline{X}_i-\overline{X}\right)^2\right)$$

Then some quick algebra yields:

$$\left(\frac{n}{m-1}\right)\sum_{i=1}^{m}E\left(\left(\overline{X}_i-\overline{X}\right)^2\right) = \sigma^2 \quad (7\text{-}9)$$

The quantity $SS_B = n\sum_{i=1}^{m}\left(\overline{X}_i-\overline{X}\right)^2$ is often described as the ***between group sum of squares*** in the data set. Note that it includes the sample size n, and that its expected value under (7-9) is $(m-1)\sigma^2$.

We have just shown that:

If the μ_i's are all equal, then:

$$V^2 = \left(\frac{n}{m-1}\right)\sum_{i=1}^{m}\left(\overline{X}_i-\overline{X}\right)^2 \quad (7\text{-}10)$$

is an unbiased estimator of the common variance σ^2

The quantity $SS_B = n\sum_{i=1}^{m}\left(\overline{X}_i - \overline{X}\right)^2$ is called the "**between group sum of squares**"
Because $V^2 = SS_B/(m-1)$, $SS_B/(m-1)$ is also an unbiased estimator of σ^2.

Of course, the assumption that the means are equal is crucial is (7-10), because it depends on (7-6) which is predicated on equal means. If the means are not equal, then Figure 7.1 becomes relevant, and the between group sum of squares can grow for reasons that have nothing to do with the common variance σ^2.

The one-way ANOVA test of H_0 is based on a comparison between our first and third estimates of σ^2, expressed in (7-3) and (7-10). The test statistic F is given by:

$$F = \frac{V^2}{S^2} = \left(\frac{\dfrac{SS_B}{m-1}}{\dfrac{SS_W}{nm-m}}\right) = \left(\frac{SS_B}{SS_W}\right)\left(\frac{m(n-1)}{m-1}\right)$$

Under equal means, the random variable F should be about one, give or take sampling fluctuations. But a failure of equal means tends to push W^2 above the true σ^2, while S^2 is unaffected by the failure. That circumstance should push F above one. The upshot is that a one-sided test would be desirable for H_0: Equal Means, in which H_0 is accepted so long as $F \le c$, where c is some threshold, while H_0 is rejected if $F > c$.

To summarize:

The test statistic F in the one-way ANOVA test of equal means follows:

$$F = [SS_B/SS_W][m(n-1)/(m-1)] \tag{7-11}$$

where: m is the number of data samples and n the sample size in each one
SS_B is the between group sum of squares
SS_W is the within group sum of squares

The form of the one-way ANOVA test of H_0: $\mu_1 = \mu_2 \ldots \mu_k$ is:

$$\begin{cases} \text{Accept } H_0 \text{ if } F \le c \\ \text{Reject } H_0 \text{ if } F > c \end{cases}$$

where c is chosen so that $\alpha = P(F > c \mid H_0 \text{ true})$ is set at a pre-specified value, often .05

EXAMPLE 7.1 **Give Me an F…**

Suppose that the sample means and sample standard deviations from random samples of size 40 from five distributions are as follows:

Sample	Sample Mean	Sample Variance
1	5.43	2.66
2	3.99	3.08
3	4.86	2.57
4	5.65	3.38
5	4.80	2.60

(i) What is the grand mean $\bar{X}$ of all 200 observations?
(ii) What are S^2 and SS_W here?
(iii) What are V^2 and SS_B here?
(iv) What is the F-statistic?
(v) In general terms, what does this F-statistic mean?

Solution

(i) The sum of the 40 measurements from the first sample is 5.43*40 = 217.2. For all five samples the sum is 40*(5.43+3.99+4.86+5.65+4.8) = 989.2. The grand mean $\bar{X}$ is therefore 989.2/200 = 4.946. (We could also have found the grand mean here by simply averaging the five sample means, because all are based on the same number of data points.
(ii) S^2 is the simple average of the five sample variances under (7-3). Therefore, $S^2 = 14.29/5 = 2.858$. Under (7-4), $SS_W = (n-1)\sum_{i=1}^{m} S_i^2 = 39 * 14.29 = 557.31$.
(iii) Under (7-10), $V^2 = \left(\frac{n}{m-1}\right)\sum_{i=1}^{m}\left(\bar{X}_i - \bar{X}\right)^2$. Here $n = 40$, $m = 5$, $\bar{X} = 4.946$, and the X_i's are as shown above. Working through the algebra, we get $V^2 = 16.725$. Because SS_B is the numerator of V^2, we find that $SS_B = 66.90$.
(iv) F is the ratio of V^2 to S^2. Here $F = V^2/S^2 = 16.725/2.858 = 5.85$.
(v) An F-ratio of 5.85 means that the estimate of the common variance σ^2 based on the spread of sample means—an estimate that is only valid if all the individual μ_i's are the same—is nearly six times as high as the estimate that is unbiased even if the individual means differ. That excess over one seems to cast doubt on the equal-means hypothesis H_0. But we need to be more specific about what we mean by "near one" before we reach any definitive judgment, meaning that we have to explore what distribution the F-ratio would take on if H_0 is true. (We will do so shortly but, to avoid unbearable suspense, the author points out that the ratio of 5.85 here calls for emphatic rejection of the equal-means hypothesis.)

EXERCISE 7.1 If the means of the various distributions in Example 7.1 were in fact equal:

(i) Would we expect $\left(\overline{X}_1 - \overline{X}\right)^2$ to be larger than, smaller than, or about equal to S_1^2? Briefly but specifically discuss. (Hint: How would we expect $\left(\overline{X}_1 - \overline{X}\right)^2$ to compare to σ^2?)

(ii) What about $n\left(\overline{X}_1 - \overline{X}\right)^2$ as compared to S_1^2?

(iii) Does the ANOVA F-test come closer to comparing $n\left(\overline{X}_1 - \overline{X}\right)^2$ to S_1^2 than to comparing $\left(\overline{X}_1 - \overline{X}\right)^2$ to S_1^2?

(iv) Is the comparison the test makes the more reasonable of the two?

The Pass/Fail Threshold

We face a familiar question: how do we choose a pass/fail threshold in the ANOVA F-test to achieve that α-value we seek? If H_0 is true, both V^2 and S^2 have mean σ^2 but both exhibit sampling variability around their means. And the quantity $F = V^2/S^2$ is a ratio of the two random variables, and the behavior of such a ratio is typically complicated.

The F-distribution

Under the strong assumptions of the standard ANOVA model, however—normality and a common variance—the distribution of F is known when the various means of equal. While we will not work through the details here, F would follow (guess what?) an F probability distribution. Like the χ^2-family of distributions, F-distributions are characterized by degrees of freedom; the difference is that there are two degree-of-freedom statistics for the F-distribution rather than one for the χ^2. In the present setting, the F-ratio follows an F distribution with $(m - 1, n(m - 1))$ degrees of freedom.

F-distributions have unimodal but asymmetric density functions that vary with their degrees of freedom, but all are restricted to the nonnegative values. Some illustrative density curves appear in Figure 7.2. Note that all the curves in Figure 7.2 imply a considerable chance of an outcome below one and a very high probability of an outcome below 2.5.

The relation between the F and χ^2 distributions goes beyond the fact that both involve degrees of freedom. The F-distribution arises when one takes the ratio of two independent χ^2 random variables; its degrees of freedom are those of the χ^2-variable in the numerator and the one in the denominator, in that order. For example, an F-random variable with (2,100) degrees of freedom is the ratio of a χ^2-variable with two degrees of freedom to another with 100 degrees of freedom.

Because the F-distribution is important in ANOVA and elsewhere, electronic tables reveal any desired percentile of an F-distribution once its degrees of freedom are specified. For example, for $F(10,90)$, the table informs us that the 90th percentile is 1.67, the 95th percentile is 1.94, and the 99th percentile is 2.52. Should we want the probability of falling below 2.00 for $F(10,90)$, the table readily yields the outcome of .96.

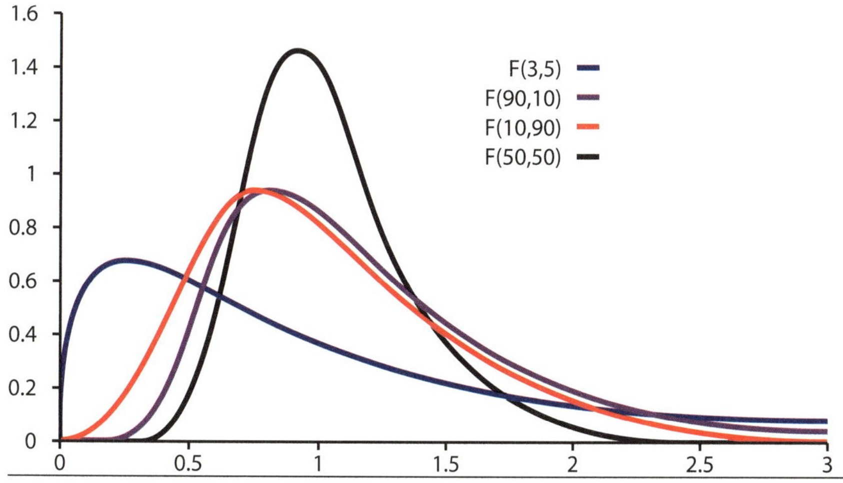

FIGURE 7.2 Density Functions for Some F distributions with varying degrees of freedom

Now we are in a position to determine c in an ANOVA F-test. We simply go to an F-table and find the $1 - \alpha$ fractile (i.e., the $100(1 - \alpha)$ percentile) of $F(m - 1, m(n - 1))$, and accept if an only if the calculated value of V^2/S^2 does not exceed that threshold. (For $\alpha = .05$, we find the 95th percentile or .95 fractile.) In summary:

- When H_0: $\mu_1 = \mu_1 \ldots \mu_m = \mu$ is true among m normal distributions with a common variance, the ratio $F = W^2/S^2$ has an F-distribution with $(m - 1, m(n - 1))$ degrees of freedom.
- A test of H_0 with Type-I error risk of α accepts H_0 if $F \leq c$, where c is the $1 - \alpha$ fractile of $F(m - 1, m(n - 1))$, and rejects if $F > c$. (7-12)

As noted, we assume both normality and a common variance as background conditions in the ANOVA F-test. Extensive simulations have shown robustness in the one-way ANOVA test, so neither the normality nor the equal-variance assumption is generally a "deal breaker" even when not actually correct. However, we are referring to moderate rather than extreme deviations from these assumptions.

EXAMPLE 7.2

A Musical Interlude

The following problem appeared on the web, thanks to some anonymous donor from Northern Kentucky University:

Many businesses have music piped into the work areas to improve the environment. At a company an experiment is performed to compare different types of music. Three types of music—country, rock, and classical—are tried, each on four randomly selected days. Each day the productivity, measured by the number of items produced, is recorded. The results appear below.

Country	Rock	Classical
857	791	824
801	753	847
795	781	881
842	776	865

(i) Can we conclude from this information that the mean number of items produced differs for at least two of the three types of music? Use α = .05.
(ii) Which type(s) of music seem to be best?
(iii) Which type(s) of music seem to be worst?

Solution

(i) In this exercise, we are being asked to assume that the productivity under a particular musical backdrop follows a normal distribution, and that day-to-day variability in productivity has the same variance for all three kinds of music. Obviously, we are in little position to verify (or contradict) that assumption with only four data points for each musical backdrop. Less stringently, we need assume that any deviations from normality and/or equal variance are not so great as to invalidate the (fairly robust) *F*-test.
There are of course statistical packages that perform the one-way ANOVA test, but we might work this one out by hand for practice. (Even here, the lazy author used an Excel spreadsheet.) We find directly that:

$$\overline{X_C} = 823.75, \quad \overline{X_B} = 775.25, \quad \text{and} \quad \overline{X_{CL}} = 854.25$$

where *C* refers to country music, *R* to rock, and *CL* to classical.
It follows that: $\overline{X}$ = (823.75 + 775.25 + 854.25)/3 = 817.75
Then, under (7-10), the between group sum-of-squares SS_B follows :

$$SS_B = 4 * ((823.75 - 817.75)^2 + (775.25 - 817.75)^2 + (854.25 - 817.75)^2) = 12{,}698$$

The within group sum of squares SS_W under (7-4) adds the squared variations around the sample mean for each type of music. For example, the squared variation SS_C for country music is given by:

$$SS_C = (857 - 823.75)^2 + (801 - 823.75)^2 + (795 - 823.75)^2 + (842 - 823.75)^2 = 2782.75$$

We likewise obtain SS_R = 776.75 and SS_{CL} = 1798.75. Hence SS_W = 5358.25.
We can then calculate the *F*-ratio according to:

$$F = SS_B/(m-1)/[SS_W/(m(n-1))] \text{ where } m = 3 \text{ and } n = 4 \text{ here.}$$

We obtain: F = (12,698/2)/(5358.25/9) = 10.66
What this *F*-statistic means is that, if H_0 is correct and the three means are truly equal, one of the two estimates of the common variance was nearly

eleven times as large as the other. Could such a big discrepancy be only a coincidence, caused by random fluctuations in the three small data samples? For α = .05, the pass/fail threshold for the ANOVA test of equal means is 4.26 here, because the F-ratio under H_0 would have an F-distribution with $(m - 1, m(n - 1) = (2,9)$ degrees of freedom. Because $F = 10.66$, the hypothesis H_0 that $\mu_C = \mu_R = \mu_{CL}$ would be rejected at the 5% level. Indeed, the statistical package tells us that the p-value of the result is .0042, meaning that the chance of getting a ratio of 10.66 or higher is estimated as about 1 in 238 under the equal-means hypothesis. In other words, it wasn't even close.

From these results, we would conclude that the three means cannot reasonably be treated as equal under the background assumptions of ANOVA. That could happen because all three means differ, or because two of the means are the same but the third mean differs from the other two.

(ii) Classical music achieved the highest mean productivity of the three types of music. But the question asks what type(s) of music had the best results, hinting that two types might be tied for first place. The question is inviting us to consider the possibility that a test comparing classical and country music would not find a significant difference between their mean responses. We accept the invitation in Exercise 7.2 below.

(iii) Given the data, rock music seems to be the worst: its mean response is the lowest, and every one of its four responses was lower than every one of the four responses for both classical and country music. Even given the small sample sizes, such a pattern would rarely arise by chance alone.

EXERCISE 7.2 Suppose that management views the results as inconclusive about whether classical music achieves greater productivity than country music, and therefore decides to try each type of music on six additional occasions. The additional results are the following:

COUNTRY	CLASSICAL
826	855
804	840
859	829
822	856
824	877
798	842

(i) Using all ten results about each type of music, perform an ANOVA F-test on the hypothesis that the mean responses for country and classical music are the same.

(ii) In Chapter 3, we discussed the *t*-test for difference of means under the same assumptions as the ANOVA *F*-test. Perform a *t*-test on the data here. Do you get the same accept/reject conclusions as in the ANOVA *F*-test?
(iii) Do you get the same *p*-value as in the *t*-test?

EXAMPLE 7.3

Keep the Change

We probably realize that the coins we use could have some bacteria on them, but might be less aware that the amount of bacteria depends on where we get the coins. But consider some data about two-pence coins in Britain, from a paper about the use of ANOVA in microbiology. Four randomly-chosen coins from a butcher's shop were analyzed for bacteria, as were four coins from a sandwich shop and a newsagent. The results were as follows:

Sure It's Safe to Touch This?

BUTCHER'S	SANDWICH	NEWSAGENT
140	2	40
108	21	5
76	0	5
400	42	0

Suppose there is interest in whether these data offer statistically significant evidence that the mean level of bacteria per coin differs in these three settings.

(i) Is this situation ideal for use of the ANOVA *F*-test?
(ii) In any case, perform the test. What are the results?
(iii) Does an informal look at the data suggest that the test yielded the correct conclusion despite its imperfections?

Solution

(i) The ANOVA test assumes that the data from each setting come from normal distributions with a common variance. The normality assumption seems suspect, given that 25% of the coins at the sandwich shop and the newsagent had no bacteria at all while negative levels of bacteria are impossible. The assumption of common variance seems … well, preposterous, given that the spread of outcomes at the butcher shop seems vastly greater than at the other two locales. If the viability of the *F*-test depends on the accuracy of these background assumptions, then it is not suited for the present case.

Yet, as noted, the author found this example in a paper that stressed the virtues of using ANOVA to investigate differences of means. The fact that the paper's authors—experienced users of ANOVA—thought the example useful to illustrate the technique emphasizes the point that even appreciable deviations from the background assumptions are not believed to compromise the test unduly. And these beliefs are based on extensive explorations about the "robustness" of the *F*-test.

(ii) The sample means and sample standard deviations are easy to calculate with only four data points from each setting:

$$\overline{X_B} = 181 \qquad \overline{X_W} = 16.25 \qquad \overline{X_N} = 12.5$$
$$S_B = 74.16 \qquad S_W = 19.60 \qquad S_N = 18.48$$

B = Butcher, *W* = Sandwich Shop, *N* = Newsagent

(The sample variances that we need are the squares of the sample standard deviations.)

From the statistical package, we learn that:

	SS	*df*	*MS*	*F*	*p*
Between:	74,065.16	2	37,032.58	4.889	0.037
Within:	68,173.47	9	7,574.83		
Total:	142,238.64	11			

This is an opportune time to decipher this output, which is standard for ANOVA programs:

- SS means *sum of squares*. "SS Between" is SS_B in our notation, while "SS Within" is SS_W. We can see that "SS Total" is the sum of SS_B and SS_W; it also corresponds to the sum of squared deviations of all mn data points around the grand mean $\bar{X}$.
- The quantity *df* means *degrees of freedom*, which are the quantities by which SS_B is divided to get V^2 and SS_W is divided to get S^2 in (7-10) and (7-4), respectively.
- *MS* means *mean sum of squares*, which corresponds to V^2 for "between" and S^2 for "within."
- *F* is the calculated *F*-ratio, which is "MS Between" divided by "MS Within."
- The quantity *p* is the *p-value* of the ratio under H_0.

Here the *p*-value implies that, if mean bacteria levels were the same in all three settings, the chance is only 3.7% of getting an *F*-ratio at least as high as 4.889. (As in a χ^2-test, the hostility of the outcome to H_0 grows as the test statistic gets larger.)

The *p*-value of .037 also indicates that the null hypothesis H_0 that all the means are equal is not sustained in a test with α = .05. Indeed, given the report from the statistical package, the usual way to perform the hypothesis test is to compare the p-value to α, and then reject H_0 if the p-value falls below α.

(iii) It is hard to look at the data and believe that the mean level of bacteria in coins from the butcher shop is the same as at the other two locations. All four butcher-shop measurements are greater than all eight of the other measurements. And not only are the four highest measurements from the butcher shop, but the smallest measurement from that shop is about twice as large as the largest from both the sandwich shop and the newsagent. Our "gut" sense is that the F-test yielded an accurate result here, and did so despite having to work with very small samples.

EXERCISE 7.3 Suppose that four sixpence coins were also gathered at random at the gym, and they showed bacteria levels of 122, 85, 60, and 113. Perform the ANOVA F-test on all four data samples (this one and the three in Example 7J.3). What is the result?

Unequal Sample Sizes

In many instances in which we wish to test the equal-means hypothesis, the sample sizes from the individual distributions vary. Fortunately, such variations do not compromise the ANOVA F-test, so long as we make appropriate adjustments. As before, we assume that all the data represent independent selections from normal distributions that share a common variance σ^2.

It is no surprise that, when the sample sizes vary, the minimum-variance unbiased estimate of σ^2 is no longer a simple average of the S_i^2's. Rather, arguments that we considered in Chapter 6 in conjunction with stratified sampling imply that we should give greater weight to the estimates arising from larger samples. Consistent with earlier results, the revised formula for S^2 is:

When the sample sizes are unequal, formula (7-3) for the minimum variance unbiased estimate S^2 of the common σ^2 becomes:

$$S^2 = \sum_{i=1}^{m} (n_i - 1) S_i^2 / (\tau - m) \tag{7-13}$$

where n_i = number of data points in sample i

$\tau = \sum_{i=1}^{m} n_i$ = total number of data points

m = number of samples

As before, $\sum_{i=1}^{m} (n_i - 1) S_i^2$ is called the *within group sum of squares* SS_w.

(Note that when all n_i's are the same, this expression is the same as (7-2).)

Similarly:

When the sample sizes are unequal, the estimator V^2 of the common σ^2 assuming equality of means is revised from (7-10) to:

$$V^2 = \sum_{i=1}^{m} n_i \left(\overline{X}_i - \overline{X}\right)^2 /(m-1) \qquad (7\text{-}14)$$

where n_i = number of data points in sample i
m = number of samples
$\sum_{i=1}^{m} n_i \left(\overline{X}_i - \overline{X}\right)^2$ is still called the *between group sum of squares*, SS_B.

We then proceed as before:

If the equal means hypothesis is true but sample sizes are unequal, then:

$$F = \frac{V^2}{S^2} \qquad (7\text{-}15)$$

where S^2 and V^2 are given by (7-13) and (7-14) respectively follows an F-distribution with $(m-1, \tau - m)$ degrees of freedom.

EXAMPLE 7.4

Time for a Bystander?

Some famous experiments in Psychology provided evidence for the "bystander" effect, which sought to explain why individuals did not help other people who were obviously in trouble. It was hypothesized that people might act swiftly if they are alone, but are less likely to do so if others are present who could help instead. (The idea is that, if others are not taking action, then maybe there is no need to do so.) And the larger the number of people doing nothing, the theory goes, the less likely a new person will intervene.

In one famous experiment, a student participant listened to someone speaking on a microphone from another room. This second person began describing personal problems in adjusting to New York City, and said that the stresses of living there sometimes led her to seizures. Then she became louder and less coherent, and said through increasing stammers that she was suffering a terrible seizure and desperately needed help. Then the microphone went off.

The seizure was fictional, and the aim of the experiment was to see whether the listener left the room to seek help and, if so, how long it took him to do so. (The genders of both listener and "victim" varied during the experiment; the author is assigning genders here simply to avoid clumsy phrases like "him or her.") Individual students who heard about the seizure fell into three groups:

- those who believed they were the only ones hearing the call for help
- those who believed that one other person also heard the call and
- those who believed that four other people heard the call.

The students participating in the experiment were alone in their rooms—supposedly for reasons related to privacy—meaning that no student knew how the other listeners were responding. (In actuality, there were no other listeners.) Thus, a student who thought that only he heard the plea would believe that no one would respond unless he did; a student who believed that there were four other listeners, on the other hand, might surmise that someone else would respond and therefore he didn't have to "get involved."

The primary measure of a listener's response to the simulated emergency was his *speed score*, which was defined as 100 times the reciprocal of his response time measured in seconds. For example, someone who sought help 80 seconds after the speaker began would get a speed score of 100*(1/80) = 1.25. (Apparently, it took about 70 seconds before it was reasonably clear that the speaker was suffering a seizure.) Those who did not respond within six minutes were assigned a speed score of 100*(1/360) = 0.28 and told they could leave. The overall results of the experiment were:

Number of Other Listeners	Number of Experimental Subjects*	Average Speed Score
0	13	0.87
1	26	0.72
4	13	0.51

*Different Students Who Believed that the Number of Other Listeners Who Heard the Call for Help was the Number to the Left

Source: J. M. Darely and B. Latane (1968), 'Bystander Intervention in Emergencies: Diffusion of Responsibility, *Journal of Personality and Social Psychology*, 8, pp. 377–383

Do these results offer statistically significant evidence of a relationship between the number of other listeners and the speed of response?

Solution

We see from the table above that the response was speediest when the student believed that only he heard the call, and slowest when he believed that he was one of five listeners. To investigate whether these differences were statistically significant, the authors of the original study conducted an ANOVA F-test. They reported the F-statistic they reached and the p-value corresponding to it, but they did not present in their paper the actual data for the 13 + 26 + 13 = 52 individuals whose responses they monitored. Below we conduct a "docu-drama," using individual response times that this author made up that were consistent with the average speed scores for the three groups, as well as other information about the response pattern that the authors provided. These "semi-fictitious" response times are useful for a learning exercise, but there is nothing fictitious about the role that ANOVA played in the real-life evaluation of the experimental results.

The "semi-fictitious" data appear below. The question is: Are these data consistent with the hypothesis that the mean speed score was the same regardless of how many other "listeners" heard the person having a seizure?

Speed Scores For Varying Numbers of Other "Listeners"

NO ONE	ONE OTHER	FOUR OTHER
0.96	0.82	0.42
0.94	0.88	0.28
0.62	0.55	0.8
1.11	0.28	0.41
0.79	0.9	0.28
0.83	0.86	0.28
0.59	0.8	0.82
0.97	1.02	0.84
0.84	0.85	0.83
0.87	0.28	0.28
0.99	0.87	0.56
1	0.85	0.28
0.8	0.28	0.55
	0.58	
	0.82	
	0.28	
	0.8	
	0.88	
	0.91	
	0.85	
	0.66	
	0.28	
	0.91	
	0.58	
	0.83	
	0.84	

These numbers match the average speed scores reported by Darely and Latane, as well as their statements on such topics as the fraction of participants in each group who did not summon help at all.

The statistical package that performed the ANOVA F-test yielded the following output:

	SS	*df*	*MS*	*F*	*p*
Between:	0.848	2	0.424	8.703	0.001
Within:	2.386	49	0.049		
Total:	3.234	51			

Note: "Between" and "Within" Calculations Use (7-13) and (7-14) respectively

We see that the *p*-value of the *F*-test was .001, meaning that the chance was only 1 in 1000 that the *F*-ratio would have been 8.703 or higher had the true means been equal (in which case the observed differences in speed scores would have been just fluctuations around the common mean). With equal means rejected, common sense suggests that the faster responses when participants thought they were alone (and slower ones when there were several other "listeners") support the view that people tend to "get involved" when convinced that if they don't act, no one will.

In Exercise 7.20 we will consider data about whether speed scores depended on the gender of the student participant, and/or the gender of the person who was apparently having a seizure.

EXERCISE 7.4 To make sure that you understand the output in Example 7.4, please do the following:

(i) Determine SS_B from the data for individual participants.
(ii) Determine SS_W from the data for individual participants.
(iii) Compute the *F*-ratio.
(iv) Verify that $P(F \geq 8.703) = .001$ (to the nearest thousandth)
(v) Find the 95th percentile of the appropriate *F*-distribution if the equal-means hypothesis were true.

7.2 Another Approach to Testing an Equal Means Hypothesis

A Limitation of the F-test

The *F*-test for equality of means in shares a limitation with the χ^2-test. If we reject a null hypothesis under a χ^2-test, the high discrepancy score does not make clear why the hypothesis failed. Similarly, if we reject the null hypothesis of equal means, the test result does not in itself indicate why the means differed. Was it the case that all the means were far apart, or might $m - 1$ of them have been the same while the last one was several miles away? The raw F-statistic does not answer that question.

Bonferroni and the Pairwise Tests

Given that limitation, the rejection of H_0 under the ANOVA F-test is often followed by "post hoc" procedures to assess which of the means compared actually differed.

Here, however, we focus instead on another way of testing H_0 that by its very nature indicates the "probable cause" that H_0 failed if it did. The procedure relates to our discussion in Chapter 5 about multiple tests of the same hypothesis and the correction identified with Bonferroni.

We are already familiar with difference-of-means tests when two means are being compared, having met them in Chapters 3. One approach to testing H_0: $\mu_1 = \mu_2 \ldots = \mu_m$ is to conduct a series of $\binom{m}{2}$ pairwise tests of equality: first $\mu_1 = \mu_2$, then $\mu_1 = \mu_3$, and all the way to $\mu_{m-1} = \mu_m$. Only if *all* of the pairwise hypotheses are accepted would we conclude that all the means are equal (i.e., we are conducting a "sudden death" test).

The procedure might seem highly tedious but, assuming a common variance σ^2, we will see momentarily that is easy to conduct the all the tests simultaneously. A more serious problem is that, when we conduct so many tests, the chance can grow disconcertingly large of getting at least one false rejection when H_0 is true. But we have already seen a way to prevent $\alpha = P(\text{reject } H_0 | H_0 \text{ true})$ from exceeding a specified value for the full series of tests. If there are k tests in total, we could arrange in the spirit of Bonferroni that each one have a Type-I error probability of α/k. We showed in Chapter 5 that—assuming H_0 is accepted only if all k tests call for acceptance—the overall risk of Type I error could not exceed α under any circumstances.

How could we implement that test regimen in the present setting? Recall that we are assuming that all the data samples come from normal distributions with a common variance σ^2, and that the only issue is whether they share a common mean μ. Suppose for simplicity that all the sample sizes are n. Then we know that $\overline{X}_i$ and $\overline{X}_j$, the sample means from, respectively, the ith and j^{th} of the distributions are independent normal random variables and that, under H_0, each has mean μ and variance σ^2/n. Then, under principles discussed in Chapter 3:

> If $\mu_i - \mu_j$, then $\overline{X}_i - \overline{X}_j$ would be normally distributed with mean *zero* and variance $2\sigma^2/n$.

If we knew σ, a test of the hypothesis H_{ij}: $\mu_i = \mu_j$ could take the form:

$$\begin{cases} \text{Accept } H_{ij} \text{ if } -c \le \overline{X}_i - \overline{X}_j \le c \\ \text{Reject } H_{ij} \text{ if } \overline{X}_i - \overline{X}_j > c \text{ or } \overline{X}_i - \overline{X}_j < -c \end{cases} \tag{7-16}$$

To keep the risk of accidental rejection down to some specified level α_{ij}, we would choose c so that $P\left(\overline{X}_i - \overline{X}_j > c \,\middle|\, \overline{X}_i - \overline{X}_j \text{ is } N\left(0, \sigma\sqrt{2/n}\right)\right) = \alpha_{ij}/2$. (Under that condition, $\overline{X}_i - \overline{X}_j$ would also have probability $\alpha_{ij}/2$ of falling below $-c$.) Finding the requisite value of c is a routine exercise with a normal table.

How would we choose α_{ij}? Well, the total number of pairwise tests would be $\binom{m}{2} = m(m-1)/2$, and there seems no reason to vary α_{ij} from one such test to another. Thus, to achieve a total Type-I error risk that does not exceed α, we could simply set $\alpha_{ij} = \alpha/\binom{m}{2} = 2\alpha/(m(m-1))$. For example, if $m = 6$ and $\alpha = .05$, we would set $\alpha_{ij} = .0033$.

But now we reach an important point. Because all the α_{ij}'s are the same and all tests have the same sample sizes (and the same assumed variance, $2\sigma^2/n$), *the pass-fail threshold c is the same in all the tests*. Thus, once we have c, there is an immediate way to see whether H_0 passed the series of tests:

- Find $\overline{X}_L$ the largest of the m sample means, then $\overline{X}_S$, the smallest, and see whether $\overline{X}_L - \overline{X}_S$) exceeds c. If not, then *all* the pairwise tests must lead to the outcome "accept H_{ij}," so H_0 is accepted. (Think about it.)
- If instead $\overline{X}_L - \overline{X}_S > c$, then we know that at least one pair of sample means is too far apart (i.e., $\overline{X}_L$ and $\overline{X}_S$) so H_0 is rejected.

But when we reject H_0 under the series of pairwise tests, we get some clear information about what went wrong. We can rapidly determine all the individual pairs of sample means that failed their tests: That happens whenever larger of the two exceeded the smaller by more than c. Thus, we get a sense of what alternative to the rejected hypothesis $\mu_1 = \mu_2 = \mu_m$ is suggested by the sample data. In other words, we get the "feedback" about why equal means failed that we do not get from the ANOVA F-test.

The discussion above focuses on two-sided pairwise tests, in which we reject H_{ij} when they differ by more than c regardless of whether $\overline{X}_i > \overline{X}_j$ or vice-versa. But suppose that the alternative H_{aij} to H_{ij} that we have in mind is that $\mu_i > \mu_j$. Then arguments in Chapter 5 suggest that we might do well to use a one-sided test of the form:

$$\begin{cases} \text{Accept } H_{ij} \text{ if } \overline{X}_i - \overline{X}_j \le d \\ \text{Reject } H_{ij} \text{ if } \overline{X}_i - \overline{X}_j > d \end{cases}$$

$$\text{where } P\left(\overline{X}_i - \overline{X}_j > d \,\middle|\, H_{ij} \text{ true}\right) = \alpha_{ij}$$

There is no problem in using one-sided pairwise tests in a Bonferroni-style test regimen, so long as we specify for each (i,j) pair which mean is expected to be larger if the means are not equal.

We've said a number of things in the last few paragraphs, so this is a good time put things together with an example.

EXAMPLE 7.5 **Four Makes Six**

Suppose that it is known that $\sigma = 1$, and we have the four sample means $\overline{X}_1 = 1.4$, $\overline{X}_2 = 1.9$, $\overline{X}_3 = 1.5$ and $\overline{X}_4 = 1.5$, all based on random samples of size 50. In this setting:

(i) How many different pairwise tests of the form $\mu_i = \mu_j$ can be conducted?
(ii) What should α_{ij} be so that the sum of the α_{ij}'s is .05?
(iii) What value of c would achieve the value of α_{ij} arising in (ii)? (We are using the two-sided test in (7-16).
(iv) What are the results in the individual pairwise tests?
(v) What conclusions might we reach here about which of the means (if any) are unequal?

Solution

(i) Here $m = 4$, so $\binom{m}{2} = m(m-1)/2 = 6$. So we can conduct six pairwise tests involving the four means.
(ii) To achieve $\alpha = .05$, each test would have $\alpha_{ij} = .05/6 = .0083$.
(iii) Under the hypothesis $\mu_i = \mu_j$, $\overline{X}_i - \overline{X}_j$ should be normally distributed with mean zero and standard deviation $\sigma\sqrt{2/n} = 1 * \sqrt{2/50} = .20$. We want to choose c in (7-16) so that:

$$P\left(\overline{X}_i - \overline{X}_j > c \,\middle|\, \overline{X}_i - \overline{X}_j \text{ is } N(0,.2)\right) = .0083/2 = .0042$$

In short, we want c to the 99.58%ile of the normal distribution $N(0,.20)$. Turning to the electronic table, we find $c = .527$.

(iv) The maximum difference among the sample means is between $\overline{X}_2$ (the largest) and $\overline{X}_1$ (the smallest), and it equals $1.9 - 1.4 = 0.5$. Because $.5 < .527$, the test of $\mu_1 = \mu_2$ yielded an outcome in the acceptance region. It follows that all the other pairwise tests reached the same results. Under the rules specified in advance, we would accept the hypothesis that all the means are equal. (You didn't expect that outcome, did you?)
(v) We would conclude that the level of sampling error that afflicts all the sample means is large enough to explain differences of the sizes that were observed. Someone who passionately believed in advance that the means were all equal need not abandon that view in light of the results. However, as we have said many times, accepting a null hypothesis H_0 does not mean that we are sure that it is true. Obviously, the data are also consistent with the possibility that (say) μ_2 is 25% larger than the other means.

EXERCISE 7.5

Suppose that in Example 7.5 there was a fifth sample mean $\overline{X_5} = 1.95$ based on a random sample of 50. Answer (i)–(v) for Example 7.5 in this revised situation. But, before doing so, make a guess about whether H_0 will be accepted in this new series of pairwise tests. And, at the end of the test, discuss very briefly why your hunch was right or wrong.

The Central Limit implies that, for $n \geq 15$ or so, $\overline{X_1} - \overline{X_2} \approx N\left(0, S\sqrt{2/n}\right)$ under H_0, where S is the data-based estimate of the common σ. In other words, we can use S in normal calculations the same way we used σ in Example 7.5.

EXAMPLE 7.6

Living Down to Expectations

To assess the extent to which self-confidence affects performance, an experiment is devised in which 150 young students are broken at random into three groups of size 50. All groups take a Math test. Those in the first group are (falsely) told before the test that they were selected as the excellent students in Math, so they shouldn't find the test especially difficult. Those in the second group were told nothing before the test, while those in the last group were told that it was known that they had trouble with Math, so they shouldn't be surprised if they found the exam difficult. (This example is based on actual experiments, though it doesn't replicate any particular one.)

The means and variances of test scores in the three groups were:

Group	Mean Score	Standard Deviation of Scores
Encouraged	81.33	14.58
Neutral	78.55	15.09
Discouraged	66.48	18.06

(i) Create a series of pairwise difference-of-means tests that have an overall α-value of .05.

(ii) Perform the tests. What is the result concerning the overall equal means hypothesis?

(iii) What do you infer from the pattern of acceptances and rejections of pairwise hypotheses?

Solution

(i) Given that there are $\binom{3}{2} = 3$ pairwise comparisons, we seek an α-value of .05/3 =.0167 for each test. We note that $n = 50$, and estimate the common variance σ^2 by:

$$S^2 = \frac{\sum_{i=1}^{3} S_i^2}{3} = \left(14.58^2 + 15.09^2 + 18.06^2\right)/3 = 255.48$$

Given the sample sizes of $n = 50$, we can approximate the $\overline{X_i} - \overline{X_j}$'s as zero-mean normally distributed under H_0: $\mu_E = \mu_N = \mu_D$ where μ_E = the mean score for encouraged students, etc. The standard deviation for each pairing is approximated as $S\sqrt{2/n} = \sqrt{255.48} * \sqrt{2/50} = 3.20$. Thus, each pairwise difference is $N(0, 3.2)$ under H_0. In any normal distribution, there is a 98.33% chance of falling within 2.13 standard deviations of the mean (one way or the other); in this case, that represents a range from 0 – 2.13 * 3.2 to 0 + 2.13 * 3.2, or from –6.82 to 6.82. It follows that the chance of falling outside that range is 1 – .9833 = .0167.

The pairwise tests all take the form:

Accept $\mu_i = \mu_j$ if	$-6.82 \le \overline{X_i} - \overline{X_j} \le 6.82$
Reject $\mu_i = \mu_j$	otherwise

In this experiment, the 50 students offered each level of encouragement might be considered a random sample of a population of young students offered that level. Thus, when the results are extrapolated to the broader population, there is an element of sampling error.

(ii) We see at once that two of the three comparisons involve sample means that differ by more than 6.82, so H_0 is rejected. Only the comparison for Encouraged and Neutral yields a result consistent with equality of means.

(iii) Given the results, it appears that "positive reinforcement" has little effect on scores compared to neutrality: although the average score among those encouraged was higher that those treated neutrally (by 81.33 – 78.55 = 2.78 points), the difference was small and well within the range of usual sampling error. In contrast, discouraging remarks were associated with a large drop in average score compared to both neutrality and encouragement (12–15 points). And sampling error is not a plausible explanation for the large disparity.

The strongest implication of the results is that students who are told that little is expected of them tend to "live down to expectations" and achieve less than they would have otherwise. Perhaps that happens because they are too ready to give up quickly when something seems hard, having received the prior message that it probably is too hard *for them*.

EXERCISE 7.6 Based on the sampling results in the "encouraged" and "neutral" groups, approximate the probability that encouragement did improve scores at least somewhat compared to neutrality (i.e., that the change in the population mean was greater than zero).

Unequal Sample Sizes

As we know from Chapter 7, unequal sample sizes do not hobble the t-test for equal means when normality and equal variance hold. We can likewise deal with unequal sample sizes in the present setting. For a pairwise test of H_{ij}: $\mu_i = \mu_j$, we can proceed as follows:

To test H_{ij}: $\mu_i = \mu_j$ one can exploit the fact that if t_{ij} is given by:

$$t_{ij} = \left(\overline{X}_i - \overline{X}_j\right) / S\sqrt{\left(\frac{1}{n_i} + \frac{1}{n_j}\right)} \qquad (7\text{-}17)$$

where S^2 is the estimate of the common variance σ^2 given in (7-13) and n_k the sample size from distribution k

Then, if H_{ij} is true, t_{ij} follows a t-distribution with $n_i + n_j - 2$ degrees of freedom

We can use (7-17) to set a one-sided or two-sided test of H_{ij} that has Type I error risk of α_{ij}. When n_i and n_j are both at least about 15, we can generally make the further approximation that $\overline{X}_i - \overline{X}_j \approx N\left(0, S\sqrt{\left(\frac{1}{n_i} + \frac{1}{n_j}\right)}\right)$.

Wine Sellers: A Trilogy

As noted at the start of this chapter, Mendel and Minerva bought a small winery in Northern California. They produced a Chardonnay, and wanted to know how consumers thought it tasted compared with two other California Chardonnays in the same price range. They therefore hired a market research firm to conduct taste tests involving the three wines in Northern California.

Given Minerva's background in Statistics, she had specific advice about how the taste tests should be conducted She saw merit in the usual test, in which randomly-selected people take sips of the three wines (with the order varying from person to person) and are asked to identify their favorite. But she stressed the importance of randomizing the order in which the wines are offered, because research shows that participants in such tests tend disproportionately to favor the last choice they sample. The firm assured her that it would use a random number generator to choose the ordering, but she decided that she would request the raw data and would test the results for randomness.

Mendel and Minerva decided to call their beverage M&M Chardonnay (and hoped that the candy company would not sue them for trademark infringement). Their prime competitors are Forest Hills Chardonnay (F) and White Surf Chardonnay (W). When the preference test was completed in Northern California, the firm reported that M&M had decisively outperformed the other wines, having been chosen the favorite by 50 of the 100 test participants. But Minerva decided to postpone any celebrations until she scrutinized the person-by-person results. These results are

TABLE 7.1 Taste Testing Results for Northern California

M, W, F	M, W, F	W, M, F	F, W, M	F, W, M
F, W, M	F, W, M	W, F, M	W, M, F	F, M, W
W, M, F	F, M, W	W, F, M	M, W, F	W, F, M
W, F, M	F, M, W	F, W, M	F, M, W	F, M, W
M, F, W	W, F, M	M, W, F	M, W, F	W, F, M
F, W, M	F, W, M	F, W, M	M, F, W	F, W, M
W, M, F	W, F, M	F, W, M	F, M, W	F, W, M
M, W, F	F, W, M	F, W, M	F, W, M	W, F, M
F, W, M	F, M, W	W, F, M	W, F, M	M, W, F
F, W, M	, F, M, W	M, F, W	W, F, M	W, F, M
W, F, M	W, M, F	M, W, F	F, M, W	F, W, M
F, W, M	W, F, M	F, W, M	W, M, F	M, W, F
F, M, W	M, F, W	W, M, F	F, W, M	F, W, M
W, F, M	F, W, M	M, F, W	W, M, F	F, W, M
W, M, F	F, W, M	M, F, W	F, M, W	W, F, M
M, W, F	F, W, M	F, W, M	F, W, M	M, W, F
F, M, W	F, W, M	W, F, M	M, F, W	F, W, M
F, W, M	W, F, M	W, M, F	W, F, M	F, W, M
W, F, M	F, W, M	M, W, F	F, W, M	W, M, F
F, W, M	M, F, W	W, M, F	W, F, M	W, F, M

presented below, with the brands listed in the order tested and the preferred brand in red. For example, the listing (W, M, F) means that White Surf was offered first, then M&M, and then Forest Hills, which was chosen as best by the respondent. Consistent with her prior plan, Minerva will investigate three questions:

(i) Did the marketing company actually randomize the order in which it presented the three beverages, as it said it would?

(ii) Was there clear evidence of a tendency of respondents to prefer the last wine they tasted, independent of any failure to randomize the order of selection?

(iii) All factors considered, did M&M outperform its competitors in the taste test?

We consider these questions below, in three separate examples.

EXAMPLE 7.7 **Disorder?**

Given the data in Table 7.1, test the hypothesis H_0 that the marketing company presented the three wines in random order to the test participants.

Solution

Although this chapter is about Analysis of Variance, we should hardly take the attitude that every question it presents need be analyzed via ANOVA. The overall aim of this book is to provide the reader with an arsenal of statistical weapons, which are deployed as appropriate to analyze data most effectively. In terms of random ordering of the wine choices, the χ^2-test seems to cry out as the procedure of choice.

There are six possible orderings in which a participant can taste the three wines:

(M, W, F) (M, F, W) (F, M, W) (W, M, F) (F, W, M) (W, F, M)

If the ordering was random, one would expect each of these orderings to occur on average in 1/6 of the 100 testings. But the χ^2 test table for the actual data looks as follows:

Pairing	Expected Frequency In 100 Tests	Actual Frequency
(M, W, F)	16.67	11
(M, F, W)	16.67	9
(F, M, W)	16.67	10
(W, M, F)	16.67	14
(F, W, M)	16.67	34
(W, F, M)	16.67	22

The discrepancy score for this test is given by: $\chi^2 = \sum_{i=1}^{6} \frac{(e_i - a_i)^2}{e_i} = 28.27$. The number of degrees of freedom in the test is five, because there are six categories, one "puppet," and no parameters in the statement of the hypothesis estimated from the test data (i.e., in the notation of Chapter 4, $M = 6$, $M_p = 1$, and $Q = 0$). For such a test, the pass/fail threshold c for $\alpha = .05$ is $c = 11.05$. Thus, the results are not easily consistent with random ordering. Disturbingly, we see that M&M was the last wine tasted in 56% of the tastings, meaning that any tendency to favor the last wine tasted creates an artificial advantage for M&M.

Given that the market research company apparently did not achieve random ordering, the next question is: Does it matter? Are the wine data consistent with earlier research, which suggests that people tend disproportionately to favor the last item they consider among those they are comparing?

EXAMPLE 7.8

Last is First?

Using the wine-testing data in Table 7.1, test the null hypothesis H_0 that order has no effect on preference, against the alternative H_a that a given wine is preferred more often when it is the third one tasted than when it is first or second.

Solution

If order of presentation does not matter, then one might initially think that the chance is 1/3 that the beverage tasted last will be the one preferred. But that need not be true if the three wines are not presented in random order. If the most popular wine tends to be the last one offered, then the third one tasted will be picked unusually often even if order has nothing to do with it. And, as noted in Example 7.7, the order of presentation here is not random.

Suppose that we define the following probabilities:

p_{Mi} = chance that a randomly-chosen wine drinker in Northern California would prefer M&M over Forest Hills and White Surf if M&M is the ith Chardonnay tasted among the three (i = 1,2,3)

p_{Fi} and p_{Wi} are the corresponding probabilities for, respectively, Forest Hills and White Surf

The estimates of these probabilities arising from the taste test (e.g., the $\widehat{p_{Mi}}$'s) are listed below.

TABLE 7.2 Preference probabilities by brand and order consumed in the wine tasting

Wine	Among All Tasters	When Tasted First	Tasted Second	Tasted Third
M & M	50%	33.3% (21)	30.4% (23)	64.3% (56)
Forest Hills	25%	17.8% (45)	16.1% (31)	50% (24)
White Surf	25%	23.5% (34)	19.6% (46)	40% (20)

Number in parenthesis is the sample size in each situation (e.g., in 21 of the tastings, M&M was presented first)

In the spirit of Bonferroni, we might consider six pairwise tests. One such test would focus on $\Delta_{13M} = \widehat{p_{M3}} - \widehat{p_{M1}}$, the difference between the expressed preference for M&M when it was tasted last from that when it was tasted first. The null hypothesis is H_{13M} (preference for M&M is the same whether tasted first or last), and the alternate hypothesis is H_a (M&M preferred more when tasted last than when tasted first). Because H_a suggests that $\Delta_{13M} > 0$, a one-sided test of H_{13M} seems desirable. Such a test would take the general form:

To test H_{13M}: $p_{m1} = p_{m3}$, calculate $\Delta_{13M} = \widehat{p_{M3}} - \widehat{p_{M1}}$ and:

$$\begin{cases} \text{Accept } H_{13M} \text{ if } \Delta_{13M} \le d \\ \text{Reject } H_{13M} \text{ if } \Delta_{13M} > d \end{cases}$$

Note that, the higher the value of Δ_{13M}, the greater its congeniality to the alternative hypothesis H_a relative to H_{13M}.

We would likewise formulate five other pairwise tests, for the hypotheses H_{23M}, H_{13F}, H_{23F}, H_{13W}, and H_{23W}. (Studying quantities like Δ_{12M} seems unimportant, because we have little interest in whether the second position in the ordering yields different preferences from the first.) Because we would be performing six tests in total, we might set $\alpha = .05/6 = .0083$ for each one to be sure that α for the full sequence does not exceed .05. (As usual, we accept H_0 (Order is Irrelevant) only if all six pairwise hypotheses are accepted.)

Hypothesis H_{13} and H_{23} taken together imply that $p_{M1} = p_{M2} = p_{M3} = p_M$. Suppose that $\widehat{p_M}$ is the overall fraction who preferred M&M among the 100 people who tried M&M. Then if these hypotheses are correct, calculations show that the sampling-error variance associated with each $\widehat{p_{Mi}}$ can be approximated by $\left(\frac{100}{99}\right)\frac{\widehat{p_M}\left(1-\widehat{p_M}\right)}{n_{Mi}}$ where n_{Mi} is the number of tasters for whom M&M is the ith wine they sipped. (Here we are applying a Chapter 4 result about the variance of $\widehat{p_M}$ as an estimator of p_M.) We use this variance estimate based on $\widehat{p_M}$ in the two pairwise comparisons involving M&M. In this special case, the general quantity $S\sqrt{\left(\frac{1}{n_i}+\frac{1}{n_j}\right)}$ in (7-17) is replaced in testing H_{13} by $\sqrt{\left(\frac{100}{99}\right)\widehat{p_M}\left(1-\widehat{p_M}\right)\left(\frac{1}{n_{M1}}+\frac{1}{n_{M3}}\right)}$.

Similarly, we use $\widehat{p_W}$ in the equivalent of $S\sqrt{\left(\frac{1}{n_i}+\frac{1}{n_j}\right)}$ in the two pairwise comparisons for White Surf, and use $\widehat{p_F}$ for Forest Hills.

For example, the fact $\widehat{p_M} = \frac{1}{2}$ means that H_{13} posits that $\Delta_{13M} = \widehat{p_{M3}} - \widehat{p_{M1}}$

$$\approx N\left(0, \sqrt{\left(\frac{100}{99}\right)\widehat{p_M}\left(1-\widehat{p_M}\right)\left(\frac{1}{n_{M1}}+\frac{1}{n_{M3}}\right)}\right) = N\left(0, \sqrt{\left(\frac{100}{99}\right).5 * .5 * \left(\frac{1}{21}+\frac{1}{56}\right)}\right) =$$

$N(0,.129)$. For the one-sided test of H_{13} with $\alpha_{13} = .0083$, we reject if Δ_{13M} exceeds the 99.17th percentile of $N(0,.131)$. The electronic table reports that this 99.17th percentile is .309. The test of H_{13} therefore becomes:

$$\begin{cases} \text{Accept } H_{13M} \text{ if } \Delta_{13M} \le .309 \\ \text{Reject } H_{13M} \text{ if } \Delta_{13M} > .309 \end{cases}$$

Because $\Delta_{13M} = \widehat{p_{M3}} - \widehat{p_{M1}} = .643 - .333 = .310$, we reject H_{13} (though only barely). However, comparable calculations reveal that H_{23} also fails, as do H_{13F} and H_{23F}. For that reason, the hypothesis that placement is irrelevant has already failed under the "sudden death" provision of the pairwise-test program.

Taken together, the significance tests support what seems apparent from the tabulated results: Each wine does roughly twice as well when tasted last as when tasted earlier. Thus, order does matter here, and the nonrandom ordering that put M&M last most of the time spuriously raised its margin of victory over its competitors.

Before we view these results as proving that people give positive preference to the last beverage tasted, we should remember the fundamental truth that correlation is not causality. Conceivably, the pattern arose not because people actively prefer the final beverage, but rather because they are biased against (say) choosing the first or second one. Perhaps this seems like a distinction without a difference, but choosing B over A because one loves B is not the same as doing so because one hates A.

EXERCISE 7.7 Perform the two pairwise tests for Forest Hills in Example 7.8 What are your conclusions?

EXAMPLE 7.9

A Win For M&M?

So far, we have made the following points about the Northern California test:

- The three wines were not presented in random order
- For all three wines, the preference was higher when it was tasted last than when it was tasted earlier. The overall pattern was statistically significant.
- M&M was the third wine tasted in 56% of the comparisons, as opposed to 24% for Forest Hills and 20% for White Surf

Under the circumstances, might the fact that M&M was preferred in 50% of the 100 tests (compared to 25% each for the other two wines) be misleading? If adjustments are made for "order bias" is there statistically significant evidence that M&M outperformed its competitors?

Solution

In answering this question, it makes sense to compare the different wines when placement is held constant. For example, how does M&M when tasted second compare to White Surf when tasted second? There can be six pairwise comparisons of interest: M&M in i^{th} place vs. White Surf in i^{th} place (i = 1,2,3), and M&M in i^{th} place vs. Forest Hills in i^{th} place. We are not especially interested in how Forest Hill compares to White Surf, so there is no compelling reason to do pairwise tests involving these two wines.

The six pairwise tests of interest are:

- p_{3M} vs. p_{3W}, based on $\Delta_{3MW} = \widehat{p_{3M}} - \widehat{p_{3W}}$
- p_{3M} vs. p_{3F}, based on $\Delta_{3MF} = \widehat{p_{3M}} - \widehat{p_{3F}}$
- p_{2M} vs. p_{2W}, based on $\Delta_{2MW} = \widehat{p_{2M}} - \widehat{p_{2W}}$
- p_{2M} vs. p_{2F}, based on $\Delta_{2MF} = \widehat{p_{2F}} - \widehat{p_{2F}}$
- p_{1M} vs. p_{1W}, based on $\Delta_{1MW} = \widehat{p_{1M}} - \widehat{p_{1W}}$
- p_{1M} vs. p_{1F}, based on $\Delta_{1MF} = \widehat{p_{1F}} - \widehat{p_{1F}}$

where $\widehat{p_{3M}}$ = fraction of tasters who tasted M&M last who preferred M&M, etc.

The null hypothesis H_0 is that M&M is equally preferred to Forest Hills and White Surf at the same position in the ordering. Proceeding the spirit of Bonferroni using six pairwise "sudden death" tests, we might choose α-risk in each test at .05/6=.0083 to prevent the overall chance of accidental rejection from exceeding .05.

But should the tests be one-sided or two-sided? A two-sided test seems preferable because, at the outset, we have no strong beliefs about how M&M compares in taste to its competitors. Also, we would presumably be just as interested in statistically significant evidence that (say) White Surf outperforms M&M as in significant evidence that favors M&M. Bad news is never welcome, but it could avert a disaster in the marketplace.

The test of H_{3MF}: $p_{3M} = p_{3F}$ would therefore take the form:

$$\begin{cases} \text{Accept } H_{3MF} \text{ if } -c \le \Delta_{3MF} \le c \\ \text{Reject } H_{13M} \text{ otherwise} \end{cases} \tag{7-18}$$

To find c, we first note that 55% of the 100 participants in the tasting favored the third wine that they sipped. Therefore, we use $\widehat{p_3}$ = .55 in the formulas for sampling variability in Δ_{3MF} under the overall hypothesis H_0, which posits that $p_{3M} = p_{3F} = p_{3W}$. In words, we are assuming that the overall 55% rate of last-place preference applies equally to all three wines, and assessing whether the observed differences in last-place preference rates can be construed as random fluctuations around 55%. In testing M&M vs. Forest Hills—in which the sample sizes in third place were, respectively, 56 and 24—the quantity Δ_{3MF} under H_{3MF} would follow:

$$\Delta_{3MF} \approx N\left(0, \sqrt{\left(\frac{100}{99}\right)\widehat{p_3}\left(1-\widehat{p_3}\right)\left(\frac{1}{n_{M3}}+\frac{1}{n_{F3}}\right)}\right) = N(0,.122)$$

where $\widehat{p_3}$ = .55, n_{M3} = 56, and n_{F3} = 24

Here we would choose c to be the 99.58th percentile of $N(0,.122)$ (and thus $-c$ is at the 00.42%ile of the distribution) to achieve α_{13} = .05/6 = .0083 (actually .0084, but let it go). We learn from the electronic table that c = .322.

From the table of results, we see that Δ_{3MF} = .643 – .5 = .143. With reject thresholds of –.322 and +.322 in (7-18), H_{3MF} is accepted. If we conduct the five other tests, we get the same outcome. In summary, the data do not offer clear evidence that

M&M performed differently from Forest Hills or White Surf when adjustments are made to correct for an "order of presentation" bias in favor of M&M. The test outcomes do not demonstrate that no difference exists, but they fall short of demonstrating that a difference does exist.

This outcome might seem unreasonable, given that M&M outperformed both other wines in all three placements by at least 14 percentage points. It is tempting to suggest that a better statistical test would use as a metric the *average* of the three differences between M&M and Forest Hills when order is held constant (i.e., $(\Delta_{3MF} + \Delta_{2MF} + \Delta_{1MF})/3$). However, we should not give in to the temptation to "pool" the results in retrospect. To change the rules of the test after the result is at hand is not sound statistical practice, and distorts the risk of accidentally rejecting the null hypothesis. The time to choose the test statistic is *before* any results are available, not afterwards.

A Comment about Bonferroni" Sudden Death" Testing

The Bonferroni approach to ANOVA is sometimes criticized on the grounds that the individual tests have such low α_k-values that they are extremely lenient. Thus, it is hard *not* to pass all the tests, so the risk of accidental acceptance is unnervingly high.

In this author's view, that concern is exaggerated. In a two-sided test with α = .05 involving a normal distribution, for example, the upper pass/fail threshold is 1.96 standard deviations above the mean. For the comparable test with α_5 = .01—corresponding to one of five tests in a Bonferroni "sudden death" procedure, the corresponding threshold is 2.58 standard deviations above the mean. The threshold is 2.80 standard deviations above the mean for α_{10} = .005, corresponding to ten Bonferroni tests. Thus, the acceptance region does not expand radically even as α drops by a factor of five or ten. If the null hypothesis is very far wrong, it is extremely likely to fail in the Bonferroni test regimen despite the procedure's alleged leniency.

EXERCISE 7.8 Suppose that each of the six pairwise tests in Example 7.9 had been conducted with an α-value of .05.

(i) Approximately what would have been the probability of at least one accidental rejection? (Treat the tests as roughly independent.)

(ii) Would all six tests have led to the conclusion "accept" given the data in Table 7.2?

7.3 Two-Way Analysis of Variance

So far, we have only considered two extreme hypotheses about the relationship among means of m different populations. We could simply assume that the means differ, and estimate each one based solely on a random sample from its population. Alternatively, we could make the opposite assumption that all the means are the same. One-way ANOVA offers a mechanism for testing this "equal means" hypothesis.

But there is the intermediate possibility that the means are not equal, but are related in some structured way. If the means are all the same, we effectively have one common parameter, namely, μ. If the means all differ, we need m parameters to describe the full set: $\mu_1, \mu_2, \ldots, \mu_m$. But perhaps we can specify the various means with fewer than m parameters but more than one. Suppose, for example, that we can classify the populations on two different dimensions, such as nationality and gender. In two-way ANOVA, we typically refer to those dimensions as ***factors***, and the different categories for each factor as ***levels***. (For example, if gender is the factor, then male and female would be the levels.) If we have random samples from each population, the mean for the population in category i on the first dimension and category j on the second might follow:

$$\mu_{ij} = \mu + \alpha_i + \beta_j \tag{7-19}$$

where: μ is the grand mean for all factor-level combinations

α_i is a general adjustment applied to all groups in level i on the first factor

β_j is a general adjustment applied to all groups in level j on the second factor

If there are h levels for the first factor and m for the second, the total number of factor-level combinations is hm. But if (7-19) prevails, we need only about $h + m$ parameters to determine all the means. (We say "about" for reasons we will discuss shortly.) Thus, if $h = m = 10$, we need about 20 parameters to specify the various μ_{ij}'s rather than 100.

Two-way analysis of variance focuses on whether a streamlined model like (7-19) offers a statistically sound representation of the set of means. For example, it sometimes investigates the hypothesis that X_{ijk}, the kth data point in the random sample from population (i,j) (level i on dimension one, j on two) follows:

$$X_{ijk} = \mu + \alpha_i + \beta_j + \varepsilon_{ijk} \tag{7-20}$$

where ε_{ijk} is $N(0,\sigma)$ for all i, j, k, with different ε_{ijk}'s independent

Like one-way ANOVA, Equation (7-20) implies that all the ε_{ijk} distributions are zero-mean normal with a common variance. Those aspects of the model are treated as given; the task of two-way ANOVA is to assess whether, under these assumptions, the linear formulation (7-20) provides a statistically-acceptable fit to the data. A more general model for X_{ijk} is:

$$X_{ijk} = \mu + \alpha_i + \beta_j + \gamma_{ij} + \varepsilon_{ijk} \tag{7-21}$$

where γ_{ij} is an ***interaction term*** that creates an additional adjustment beyond $\alpha_i + \beta_j$ to move from μ to μ_{ij}. Two-way ANOVA generally hypothesizes that $\gamma_{ij} = 0$ for all i and j (i.e., it posits a ***no interaction*** model). If we reject that hypothesis and must conclude that the γ_{ij}'s are nonzero, then we really have not moved beyond the general model that treats the μ_{ij}'s as different.

A Series of Models

Suppose that there are h levels of the first factor and m of the second, and random samples of size n for each of the hm combinations of factors. Let μ_{ij} be the mean for level i of factor 1 and level j of factor 2. If X_{ijk} is the kth data point from that combination (part of a random sample of size n), then a general model for X_{ijk} in the spirit of ANOVA could be:

$$X_{ijk} = \mu_{ij} + \varepsilon_{ijk}$$

where ε_{ijk} is $N(0, \sigma)$.

To estimate μ_{ij}, a simple approach would be to treat each (i,j) category separately, and to use the sampling theory in Chapter 3 to obtain a point estimate and confidence interval for μ_{ij}, as well as a distribution for the value of μ_{ij} given the data at hand. Though it's a bit cumbersome, we could just as well express μ_{ij} as:

$$\mu_{ij} = \mu + \alpha_i + \beta_j + \gamma_{ij} \tag{7-22}$$

where:

- μ is the grand mean for all categories
- α_i is a general adjustment for the mean, applied to all groups at level i on the first dimension
- β_j is a general adjustment for the mean, applied to all groups in category j on the second dimension
- γ_{ij} is a further adjustment for the particular combination (i,j)

Of course, if all the quantities on the right of (7-21) are nonzero, then we would have no fewer parameters if we simply dealt with the μ_{ij}'s separately.

But what if some of the quantities on the right of (7-22) are in fact zero? If $\alpha_i = \beta_j = \gamma_{ij} = 0$ for all i and j, then all the groups have a common mean μ, so pooling their data could yield a far more reliable estimate of μ than any of the individual samples on its own. If $\gamma_{ij} = 0$, then instead of having $h * m = hm$ different parameters to estimate (namely, μ_{ij} for $i = 1, \ldots h$ and $j = 1, \ldots m$), we would only need estimate about $h + m$, namely, μ, the α_i's and the β_j's. Two-way analysis of variance (two-way ANOVA) tests hypotheses that the μ_{ij}'s can be estimated from a streamlined model that assumes many zeros on the right of (7-22).

If it is hypothesized that $\mu_{ij} = \mu$ for all (i,j) pairs, then we are really in the domain of one-way ANOVA, and can perform the tests described earlier in this chapter using the hm data samples. Two-way ANOVA typically starts with the no interaction hypothesis H_I: $\gamma_{ij} = 0$, meaning that (7-21) reduces to (7-20):

$$X_{ijk} = \mu + \alpha_i + \beta_j + \varepsilon_{ijk} \qquad \text{(7-20 again)}$$

The no-interaction hypothesis implies that nothing idiosyncratic to the particular combination of i and j affects the mean there.

It is also possible to make the further stipulation that $\beta_j = 0$ for all j, in which case (7-20) reduces to:

$$X_{ijk} = \mu + \alpha_i + \varepsilon_{ijk} \qquad \text{(7-23)}$$

Under this model, μ_{ij} depends on the level of the first factor but not of the second.

Alternately, one could posit that $\alpha_i = 0$ for all i and thus that:

$$X_{ijk} = \mu + \beta_j + \varepsilon_{ijk} \qquad \text{(7-24)}$$

Here μ_{ij} depends on the level of the second factor but not of the first.

Two-way ANOVA allows tests of the models (7-20), (7-23), and (7-24), and the tests take the same general form as the one-way ANOVA test. Before discussing the procedure, however, we should say a bit more about the structure of the no-interaction model (7-20).

The No Interaction Model

In the no-interaction model with h levels of the first factor and m levels of the second, it is customary to define μ as the arithmetical average of the μ_{ij}'s:

$$\mu = \frac{\sum_{i=1}^{h}\sum_{j=1}^{m}\mu_{ij}}{hm} \qquad \text{(7-25)}$$

If there are n data points available for each factor-level combination, we can interpret μ as the answer to the question:

> "Suppose a data point X_{ijk} is chosen at random from among the hmn data points gathered at the various factor/level combinations. What is the expected value of X_{ijk}?"

In such a process, the data point has a $1/hm$ chance of being selected from each factor/level combination, and the conditional mean given that combination (i,j) is chosen is μ_{ij}. Then the conditional expectation rule tells us that the mean corresponding to the selection follows:

$$E(X_{ijk}) = (1/hm)\mu_{11} + (1/hm)\mu_{12} + \ldots (1/hm)\mu_{hm}.$$

But then (7-25) implies that $E(X_{ijk}) = \mu$.

In the no-interaction case, we have:

$$\mu = \frac{\sum_{i=1}^{h}\sum_{j=1}^{m}\mu_{ij}}{hm} = \frac{\sum_{i=1}^{h}\sum_{j=1}^{m}\left(\mu + \alpha_i + \beta_j\right)}{hm}$$
$$= \mu + \left(\frac{1}{hm}\right)\sum_{i=1}^{h} m\alpha_i + \left(\frac{1}{hm}\right)\sum_{j=1}^{m} h\beta_j \qquad (7\text{-}26)$$

Equation (7-26) is correct because α_i appears in all m of the μ_{ij} terms in which the first factor is at level i, while β_j appears in all h μ_{ij} terms in which the second factor is at level j. Note also that $\frac{hm\mu}{hm} = \mu$.

We can rewrite (7-26) as:

$$\left(\frac{1}{h}\right)\sum_{i=1}^{h}\alpha_i + \left(\frac{1}{m}\right)\sum_{j=1}^{m}\beta_j = 0 \qquad (7\text{-}27)$$

It is traditional to impose the condition $\sum_{i=1}^{h}\alpha_i = 0$, under which the adjustments of the mean μ_{ij} as the first factor varies across its levels average out to zero. But then (7-27) implies that $\sum_{j=1}^{m}\beta_j$ must also be 0. We will consider examples shortly.

To summarize:

For the no interaction model in which $\mu_{ijk} = \mu + \alpha_i + \beta_j$ and $\mu = \frac{\sum_{i=1}^{h}\sum_{j=1}^{m}\mu_{ij}}{hm}$, the traditional specification that $\sum_{i=1}^{h}\alpha_i = 0$ means $\sum_{j=1}^{m}\beta_j$ must also be 0. (7-28)

Because of (7-27), we only have $h - 1$ independent parameters among the α_i's (i.e., the last must be the negative of the sum of the others). Likewise, we only have $m - 1$ independent β_j's. Therefore, the no interaction model does not have $m + h + 1$ independent parameters (the 1 refers to μ); instead, it has only $1 + (m - 1) + (h - 1) = m + h - 1$. It was for that reason that we said earlier that this model had "about" $m + h$ parameters rather than exactly $m + h$.

Parameter Estimation in the No Interaction Model

How can we estimate the parameters for the no-interaction model from the X_{ijk}'s? It is surprisingly easy given (7-28); we proceed below under the assumption that the sample size takes the same value n for all groups.

Because $E(X_{ijk}) = \mu_{ijk}$, we can obtain an unbiased estimate of μ by taking the arithmetical average $\bar{X}$ of all the X_{ijk}'s. That is true because the grand sample mean $\bar{X}$ follows:

$$\bar{X} = \left(\frac{1}{hmn}\right)\sum_{i=1}^{h}\sum_{j=1}^{m}\sum_{k=1}^{n} X_{ijk}$$

Therefore:

$$E(\bar{X}) = \left(\frac{1}{hmn}\right)\sum_{i=1}^{h}\sum_{j=1}^{m}\sum_{k=1}^{n} E(X_{ijk})$$

With $E(X_{ijk}) = \mu_{ijk}$ and (7-25), we reach $E(\bar{X}) = \mu$.

As for the α_i's and β_j's, the conditions $\sum_{i=1}^{h}\alpha_i$ and $\sum_{j=1}^{m}\beta_j = 0$ allow a pleasant simplification. If we define $\overline{X_{1r}}$ to be the sample mean of all the mn data points from the first row of the (i,j) matrix (i.e., $i = 1$), we can write:

$$\overline{X_{1r}} = \left(\frac{1}{mn}\right)\sum_{j=1}^{m}\sum_{k=1}^{n} X_{1jk}$$

Thus:

$$E(\overline{X_{1r}}) = \left(\frac{1}{mn}\right)\sum_{j=1}^{m}\sum_{k=1}^{n} E(X_{1jk}) = \left(\frac{1}{mn}\right)\sum_{j=1}^{m}\sum_{k=1}^{n}(\mu + \alpha_1 + \beta_j)$$
$$= \mu + \alpha_1 + \left(\frac{1}{m}\right)\sum_{j=1}^{m}\beta_j$$

Because $\sum_{j=1}^{m}\beta_j = 0,$ we reach:

$$E(\overline{X_{1r}}) = \mu + \alpha_1 \tag{7-29}$$

Nice to know, but how does (7-29) help us estimate α_1? Because $E(\bar{X}) = \mu$, we have:

$$E(\overline{X_{1r}} - \bar{X}) = \alpha_1$$

Thus, we get an unbiased estimator of α_i by subtracting the grand sample mean from the average of the observations at level 1 of factor 1. Corresponding formulas offer unbiased estimators of the other α_i's and the β_j's.

In the no-interaction model $\mu_{ijk} = \mu + \alpha_i + \beta_j$ with random samples of size n from each (i,j) group, unbiased estimates of the parameters arise under the rules:

- An unbiased estimate of μ is the grand sample mean $\bar{X}$
- $\widehat{\alpha_\iota}$, an unbiased estimate of α_i is given by $\widehat{\alpha_\iota} = \overline{X_{ir}} - \bar{X}$ where $\overline{X_{1r}}$ is the average of all mn data points in the ith row of (i,j) matrix
- $\widehat{\beta_J}$, an unbiased estimate of β_j is given by $\widehat{\beta_J} = \overline{X_{jc}} - \bar{X}$ where $\overline{X_{jc}}$ is the average of all hn data points in the jth column of (i,j) matrix (7-30)

The no interaction model is also called a **main effects** model. The $\widehat{\alpha_\iota}$'s and $\widehat{\beta_J}$'s are interpreted as the "main effects" of being in individual rows or individual columns of the (i,j) matrix.

Before we turn to the specifics of an ANOVA test for no interactions, we should explore in intuitive terms what the no-interaction model is saying.

The Geometry of the No-Interaction Model

Suppose that we have two levels (high and low) for each of the two factors in a two-way ANOVA. Suppose too that we hold one of the factors constant, and graph what happens to the outcome variable as we move from high to low on the other factor. (By the ***outcome variable***, we generally mean the quantity $\overline{X_{ij}}$ that has mean μ_{ij} for combination (i,j).) For example, the outcome variable might be sales of a product, and the two factors might be the price charged and advertising on the product's behalf (each of which is either "high" or "low"). What might we expect the graph to look like?

The no interaction model takes the form:

$$\mu_{ij} = \mu + \alpha_i + \beta_j \quad (1 \le i,j \le 2)$$

where μ_{ij} = mean of the outcome variable for the (i,j) combination of factors

If we hold j constant at either 1 or 2, we can write:

$$\mu_{1j} = \mu + \alpha_1 + \beta_j$$

$$\mu_{2j} = \mu + \alpha_2 + \beta_j$$

which means that:

$$\mu_{1j} - \mu_{2j} = \alpha_1 - \alpha_2 \text{ (regardless of whether } j \text{ is 1 or 2)}$$

Therefore, the plot of outcome vs. Factor 1 with no interaction could look something like:

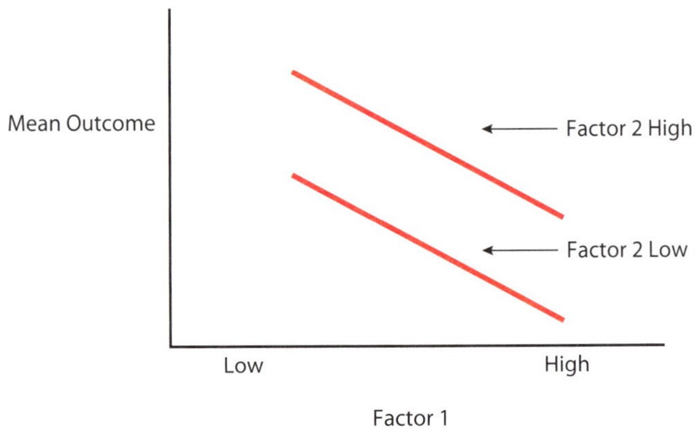

The point is that the two lines corresponding to the two levels of Factor 2 should be roughly parallel. We say "roughly" because we do not have the actual μ_{ij}'s; rather, we have the $\overline{X}_{ij}$'s that are all subject to sampling error. By the same reasoning, the two lines drawn when Factor 1 is held constant should also be roughly parallel.

These considerations mean that plotting an *interaction graph* (outcome vs. each factor, with the other factor held constant) gives a sense whether interactions are present. A strong deviation from parallelism would mean that the effect of changing one factor is quite different when the other factor is high than when it is low. But that difference by its very nature suggests that the two factors interact, and that the effect of a particular combination is not simply the sum of the two main effects.

When there are several levels of each factor, the geometric analysis of no-interaction gets more intricate. There is no great problem, however, if (say) Factor 1 can take on k values and Factor 2 can take on two. In that case, the k lines reflecting changes in the outcome variable when Factor 1 is held at a given level while Factor 2 goes from Low to High should all be about parallel.

EXAMPLE 7.10 **A No-Interaction Nightmare**

Suppose that average cholesterol levels are obtained for four groups of people in a certain community:

- Older Men (Age 40 and Up)
- Older Women
- Younger Men (Below Age 40)
- Younger Women

It is desired to approximate the relation between gender/age and cholesterol level by a main-effects model of the form:

$$\mu_{ij} = \mu + \alpha_i + \beta_j$$

where α_i = adjustment based on gender
β_j = adjustment based on age group

This is a rather weird community. Suppose that the sample means based on random samples of size 60 are:

	Men	Women
Older	180	120
Younger	120	180

(The various sample standard deviations are small.

(i) What would be $\bar{X}$, the $\widehat{\alpha_i}$'s, and the $\widehat{\beta_j}$'s given these data?
(ii) What would the interaction diagrams look like?
(iii) What might a person conclude if he was told the parameter estimates but not shown the underlying data ?
(iv) What is the moral of this story?

Solution

(i) $\bar{X}$ would be the average of the four sample means, which is 150. Because the row means and the column means are all 150, the four parameters $\widehat{\alpha_M}$, $\widehat{\alpha_F}$, $\widehat{\beta_O}$, $\widehat{\beta_Y}$ (M = male, F = female, O = older, Y = younger) would all be estimated as *zero* under (7-30).
(ii) The age vs. cholesterol lines for the two genders are:

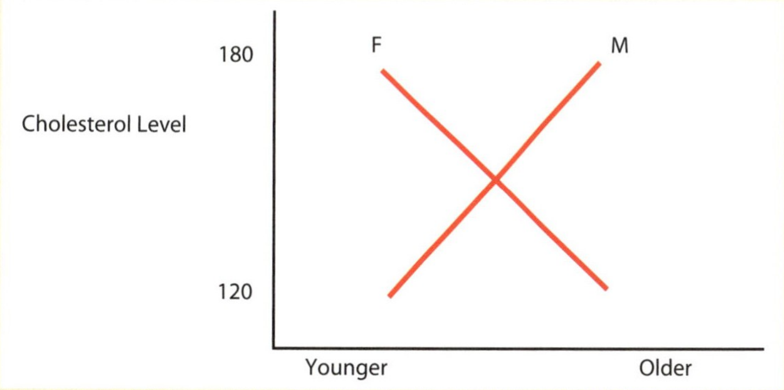

This pattern is hardly what one would expect under a no-interaction model, under which the two lines should be essentially parallel.

(iii) A person told that the grand mean was 150 and that the adjustments for age and gender were all zero might well believe that mean cholesterol level is the same across ages and genders. But that is not what the data show: cholesterol level depends on age for both genders, but the level increases with age for men yet decreases with age for women. Likewise, cholesterol in either age group very much varies with gender.

(iv) The moral is that, if the no-interaction model is wrong, its parameters can give a highly misleading impression about what is going on in various cells of the (i,j) matrix. For that reason, it might not yield a useful summary of the key patterns in the data.

EXERCISE 7.9 Consider the following two cases akin to Example 7.10:

Case I: The grand mean is 150, and the row mean for men is 170 while the column mean for younger people is 140.

Case II: The grand mean is 150 and the sample means are the same for younger men, younger women, and older women. The sample mean for older men, however, is higher by 60 than the means for the other three groups.

For each case:

(i) Offer sample means for the four categories that meet the stated conditions.
(ii) Draw an interaction diagram for those sample means
(iii) Estimate the main effects parameters for your data
(iv) Comment briefly on the adequacy of the main effects model

EXAMPLE 7.11 **Physical and Model Fitness (Based on a true story)**

In a courtroom proceeding, it was alleged that a physical fitness test for men hoping to be firefighters went too far in giving preference to military veterans. A main effects two-way ANOVA model was presented in evidence: its parameters suggested that, holding constant the level of preparation for the test, veterans passed at a rate 30 percentage points higher than nonveterans. Thus, the argument went, nonveterans were at a distinct disadvantage on the test, and one that they could not realistically hope to overcome.

The actual passing rates by veteran status and amount of preparation were (essentially) the following:

Preparation?	Veteran	Non-Veteran
No	95%	35%
Yes	97%	97%

For this exercise, assume that the sample sizes are the same for all four cells of the (i,j) matrix.

(i) Estimate the parameters of the main effects/no interaction model given these data.

(ii) How well does the model summarize the pattern in the data?

Solution

(i) The grand sample mean $\bar{X}$ is the average of the four passing rates, and is 81% = .81.

Given that the sample mean for the first row is 65% (which is 16% below the grand sample mean) and the sample mean for the second row is 16% higher than the grand sample mean, the row (preparation) parameters are $\widehat{\alpha_Y} = .16$ and $\widehat{\alpha_N} = -.16$. ($Y$ = yes; N = no)

The column parameters are $\widehat{\beta_V} = .15$ and $\widehat{\beta_{NV}} = -.15$ (V = veteran, etc.)

(ii) Because $\widehat{\beta_V} - \widehat{\beta_{NV}} = .30$, the main effects model suggests that veterans have a passing rate 30 percentage points higher than non-veterans at each level of preparation. Because $\widehat{\alpha_Y} - \widehat{\alpha_N} = .32$, the model also suggests that veterans who prepare a lot for the test pass at a rate 32 percentage points higher than those who prepare little, and that the same holds true for nonveterans.

Obviously, this is a poor summary of the underlying pattern. For veterans (who might already be in good physical shape), preparation scarcely seems necessary for success on the test. But that is apparently not the case for non-veterans. The data show that nonveterans who prepared did just as well as veterans who prepared, and that the gap between veterans and nonveterans was caused entirely by a low passing rate among nonveterans who did not prepare. The 30-point difference was a simple average of a gap of zero and a gap of 60.

If we use the main effects model to project passing rates in the four categories (e.g., the projection for veterans who don't prepare is $\bar{X} + \widehat{\alpha_N}$), we get:

Projected Passing Rates Under Main Effects Model, and Actual Passing Rates

Preparation?	Veteran	Non-Veteran
No	80% (95%)	50% (35%)
Yes	112% (97%)	82% (97%)

Beyond the absurdity of a 112% passing rate for veterans who prepare, this set of projections errs by 15 percentage points in every one of the four categories. Assuming a reasonable amount of data, such deviations cannot be reconciled with usual sampling fluctuations, implying that a no-interaction model is inadequate and that its parameter estimates cannot be taken seriously. In this instance, an *interaction* model of the form:

$$p_{ij} = .96 + \gamma_{ij}$$

$$\text{where } \gamma_{ij} = \begin{cases} 0 \text{ for veterans and prepared nonveterans} \\ -.60 \text{ for unprepared nonveterans} \end{cases}$$

would fit the data well, assuming that differences of one percentage point are well within the range of sampling error.

EXERCISE 7.10 Suppose that those defending the fitness test cited the table in Example 7.11 to argue that, if all nonveterans who took the test prepared for it, the difference in passing rates between veterans and nonveterans would disappear. Is there a potential flaw in this argument? Briefly discuss.

EXAMPLE 7.12

Of Ads and Mouths

A company that is offering a new product assumes that the reaction to the product among those who try it depends on *why* they tried it. The company believes that new users who buy the product because of advertising are more likely than others to react favorably (perhaps because the advertisement made clear what is special about the product). It also believes that those who try the product because of "word of mouth" from friends or relatives will be more favorably disposed for that reason, and that those who tried it because of both ads and word of mouth will be especially pleased with the product.

To quantify how satisfaction varies with the reason the new user tried the product, the company hired a market research firm and asked it to conduct a survey among four groups of new users, namely, those who were influenced by:

- Both advertising and word of mouth
- Advertising but not word of mouth
- Word of mouth but not advertising
- Neither advertising nor word of mouth

The market research firm was asked to get random samples of 100 new users in each of these four categories, and to ask each person in each sample to rate the product from 1to 10 (1 worst, 10 best) on an integer scale. The four groups will be assigned the subscripts 00 (neither ads nor mouths), 10 (ads not mouths), 01 (mouths not ads) and 11 (mouths and ads). When the results are at hand, the firm is asked to model satisfaction by a linear model of the form:

$$X_{ijk} = \mu + \alpha_i + \beta_j + \varepsilon_{ijk}$$

where X_{ijk} = satisfaction score of the kth person in group ij

μ = average score across the four groups (weighted equally)

α_i = an adjustment for the person's advertising exposure (i = 0 or 1)

β_j = an adjustment for the person's "words of mouth" exposure (j = 0 or 1)

ε_{ijk} = a zero-mean normal fluctuation to reflect diversity among people in the same category (with σ assumed the same in all four categories)

When the survey is complete, the firm comes back with the following data table:

Means and Standard Deviations of Satisfaction Scores By Exposure Category

Category	Mean Score	St Dev of Scores
00 (NA, NM)	6.24	1.32
01 (NA, M)	6.66	1.49
10 (A, NM)	6.52	1.20
11 (A,M)	8.82	1.19

All statistics are based on random samples of size 100.

A = advertising, M = word of mouth, N = No

Given these data:

(i) Make reasonable estimates of the various parameters in the linear model.

(ii) Using these estimates, approximate the satisfaction score in each of the four categories.

(iii) Do you think the differences between the approximations in (ii) and the various mean scores in the table can reasonably be ascribed to sampling error?

(iv) If not, what might you suggest for modeling the means for the four populations sampled?

Solution

(i) The average rating over 400 ratings is the grand sample mean $\bar{X}$, which is the average of the four category means. That average is 1/4(6.24 + 6.65 + 6.53 + 8.82) = 7.06.

We can express the results about means in matrix form as follows:

		Advertising		
		Yes	No	Row Mean
Word of Mouth	Yes	8.82	6.66	**7.74**
	No	6.52	6.24	**6.38**
Column Mean		**7.67**	**6.45**	
		Grand Mean **7.06**		

Given that the grand mean is 7.06, we see that the first column is on average 0.61 (7.67 – 7.06) above that value and the second column is 0.61 below. Thus, we choose $\widehat{\alpha_1}$ = 0.61 and $\widehat{\alpha_0}$ = –0.61 under (7-30) to reflect the average shifts from the grand mean based on the presence or absence of advertising. (The two parameters necessarily sum to zero, so there is really only one independent parameter about advertising.) For word of mouth, we note that the row mean is 0.68 above the grand mean when the new user had heard word of mouth and 0.68 below when he heard nothing. We therefore choose $\widehat{\beta_1}$ = 0.68 and $\widehat{\beta_0}$ = −.68.

The model therefore starts at 7.06, and adds or subtracts 0.61 based on yes/no for advertising and adds or subtracts 0.68 based on yes/no for word of mouth. We have thereby summarized the pattern with a three-parameter model.

As for the common variance σ^2, we would estimate it by S^2 which follows:

$$S^2 = 1/4(1.32^2 + 1.49^2 + 1.2^2 + 1.19^2) = 1.70$$

(ii) With these parameters, the model yields the following estimates for the category means:

$$\begin{aligned} \mu_{00} &\approx \bar{X} + \widehat{\alpha_0} + \widehat{\beta_0} = 7.06 - 0.61 - 0.68 = 5.77 \\ \mu_{10} &\approx 7.06 + 0.61 - 0.68 = 6.99 \\ \mu_{01} &\approx 7.06 - 0.61 + 0.68 = 7.13 \\ \mu_{11} &\approx 7.06 + 0.61 + 0.68 = 8.35 \end{aligned}$$

(iii) We might go back to sampling theory in Chapter 3 to compare these estimates to those implied by the raw data. In the category 00 (heard no mouths, saw no ads), the sample mean was 6.24 and the sample standard deviation was 1.32. Thus, the 95% confidence interval for μ_{00} would extend from:

$$6.24 - 1.96 * \left(1.32/\sqrt{100}\right) \text{ to } 6.24 - 1.96 * \left(1.32/\sqrt{100}\right) = 5.98 \text{ to } 6.50.$$

The model's estimate 5.77 does not fall within this interval: indeed, because $\mu_{00} \approx N(6.24, .132)$ given the sample data, 5.77 is approximately four standard deviations below the mean. The model, therefore, did not quite yield a plausible for estimate for μ_{00}. (Using the common estimate of σ, namely, $S = \sqrt{1.70} = 1.31$ as the standard deviation has no effect on this conclusion.) Furthermore, the analogous calculations for the other μ_{ij}'s yield very similar results. To answer question (iii), the differences between the model's estimates and the actual data can by no stretch of the imagination be ascribed to usual sampling error.

(iv) Here the inadequate linear model adds essentially nothing to what is apparent from the data summary. We see that advertising or word of mouth alone only modestly increased the mean response compared to neither of these, but that the combination of advertising and word of mouth was associated with a dramatic increase in satisfaction. In other words, there is an *interaction* between ads and mouths that makes their combined effect much greater than the sum of their individual effects. By its nature, the linear model does not allow for such interactions. That is a major shortcoming in this instance, and the simplification that reduced a four-parameter model (based on the individual sample means) to a model with three parameters misrepresented the data pattern in ways that are statistically significant and substantively important. Better, it would seem, to estimate the various μ_{ij}'s separately.

EXERCISE 7.11 Using Bonferroni pairwise tests, assess which pairs of sample means in Example 7.12 differ to a statistically significant extent. Then come up with a brief summary of the findings in simple English.

Two Way ANOVA F-Tests

Null hypotheses can be tested in two-way ANOVA with *F*-ratios that generalize the one we saw for one-way ANOVA. As before, we make an estimate S^2 of the common variance σ^2 based on the individual sample variances (here, the S_{ij}^2). Then we use as the numerator another quantity that also estimates σ^2 if the hypothesis is true but tends to overestimate it if the hypothesis is false. The test assesses whether the ratio exceeds one to a statistically-significant extent.

One important *F*-test concerns the consistency between the no interaction/main effects model and the test data. It is convenient to represent the situation with an $h \times m$ matrix (i.e., with h rows and m columns), in which the (i,j)th cell presents the sample mean when Factor 1 is at level i and Factor 2 at j. As before, we assume equal sample sizes (n) in all cells of the (i,j) matrix. The hypothesis H_I under scrutiny is:

H_I: In the general equation $\mu_{ij} = \mu + \alpha_i + \beta_j + \gamma_{ij}$, the quantity γ_{ij} is 0 for all i and j.

The test statistic is:

$$F_I = \frac{\dfrac{n\sum_{i=1}^{h}\sum_{j=1}^{m}\left(\overline{X}_{ij} - \left(\widehat{\alpha}_i + \widehat{\beta}_j + \overline{X}\right)^2\right)}{(h-1)(m-1)}}{S^2} \tag{7-31}$$

where $S^2 = \dfrac{\sum_{i=1}^{h}\sum_{j=1}^{m}\sum_{k=1}^{n}\left(X_{ijk} - \overline{X}_{ij}\right)^2}{(n-1)mh}$ and $\widehat{\alpha}_i$ is a data-based estimate of α_i under (7-30), etc.

What the summation term in the numerator does is compare actual sample means from various cells in the (i,j) matrix (the $\overline{X}_{ij}$'s) with the estimated cell means under the main-effects model (the $\widehat{\alpha}_i + \widehat{\beta}_j + \overline{X}$)'s). The comparative statistic is the squared difference between the two quantities. The denominator is the simple average of the sample variances from each of the cells. In other words, it is the familiar estimator of the common variance σ^2.

If H_0 is true, then it can be shown that F_I in (10-30) follows an F-distribution with $[(h-1)(m-1), (n-1)mh]$ degrees of freedom. Readers might recall χ^2 tests in which the number of degrees of freedom was reduced because parameters in the hypothesis were estimated from the test data. A similar reduction applies here: the fact that the $\widehat{\alpha}_i$'s, the $\widehat{\beta}_j$'s, and $\overline{X}$ were approximated under (7-30) is the reason that we have $(h-1)(m-1)$ degrees of freedom for the numerator even though we are summing squared quantities from hm different cells.

The quantity $n\sum_{i=1}^{h}\sum_{j=1}^{m}\left(\overline{X_{ij}}-\left(\widehat{\alpha}_i+\widehat{\beta}_j+\overline{X}\right)^2\right)$ is called the ***interaction sum of squares*** in two-way ANOVA, and is generally designated SS_I. The quantity $\sum_{i=1}^{h}\sum_{j=1}^{m}\sum_{k=1}^{n}\left(X_{ijk}-\overline{X_{ij}}\right)^2$ is generally called the ***error sum of squares***.

In short:

In two-way ANOVA, the no-interaction hypothesis H_I posits that $\gamma_{ij}=0$ for all i and j in the model $\mu_{ij}=\mu+\alpha_i+\beta_j+\gamma_{ij}$. If H_0 is true, then the ratio F_I that follows:

$$F_I = \frac{\dfrac{SS_I}{(h-1)(m-1)}}{\dfrac{SS_E}{(n-1)hm}}$$

where $n\sum_{i=1}^{h}\sum_{j=1}^{m}\left(\overline{X_{ij}}-\left(\widehat{\alpha}_i+\widehat{\beta}_j+\overline{X}\right)^2\right)$ and $SS_E=\sum_{i=1}^{h}\sum_{j=1}^{m}\sum_{k=1}^{n}\left(X_{ijk}-\overline{X_{ij}}\right)^2$

has an F-distribution with $[(h-1)(m-1), (n-1)hm]$ degrees of freedom (7-32)

The test of H_I then becomes:

$$\begin{cases}\text{Accept } H_I \text{ if } F_I \le c\\ \text{Reject } H_I \text{ if } F_I > c\end{cases}$$

where c is typically the 95th percentile of the F-distribution with $[(h-1)(m-1), (n-1)$ hm] degrees of freedom.

Even if the no-interaction hypothesis H_I fails, its estimates of the main effects (i.e., the $\widehat{\alpha}_i$'s and $\widehat{\beta}_j$'s are not rendered irrelevant. In the cholesterol example 7.10, the zero parameter estimates do reflect that facts that, within the data set, there was no net correlation between age or gender and cholesterol level. The main effects parameters answer the general question: On average, how far and in which direction do the sample means in the ith row or j^{th} column vary from the grand sample mean? We have always said that the mean of a distribution offers useful but limited information; the same is true of the main effects in two-way ANOVA.

Other Tests in Two Way ANOVA

In a two-way analysis of variance, it is possible that the γ_{ij}'s are not the only parameters equal to zero. It is possible that there are no "row effects," meaning that all the α_i's are zero (even if the β_j's are not). Instead, the "column effects" (the β_j's) might be zero. If all the α_i's, β_j's and γ_{ij}'s are zero, then all the means are equal.

Assuming that γ_{ij}'s are zero, a test for H_R: $\alpha_i = 0$ for all i arises if we consider what $\alpha_i = 0$ implies about the row means. Let $\overline{X_{ir}}$ be the average of all mn data points in the ith row of the matrix of sample means. If $\alpha_i = 0$, then we can write:

$$E\left(\overline{X_{ir}}\right) = \left(\frac{1}{m}\right)\sum_{j=1}^{m}\left(\mu + \beta_j\right) = m\mu/m + (1/m)\sum_{j=1}^{m}\beta_j = \mu + 0 = \mu$$

$$\text{because } \sum_{j=1}^{m}\beta_j = 0$$

Thus, all the row means equal the grand mean μ. In testing H_R we can therefore focus on the statistic F_R defined by

$$F_R = \left(hn\sum_{i=1}^{h}\left(\overline{X_{ir}} - \overline{X}\right)^2/(h-1)\right)/\left(SS_E/(n-1)hm\right)$$

where $\overline{X}$ is the grand sample mean of all nmh data points

Note that the $\overline{X_{ir}}$'s and $\overline{X}$ all have mean μ under H_R, but that the $\overline{X_{ir}}$'s generally suffer greater sampling error than $\overline{X}$ because they are based on smaller samples. The quantity $hn\sum_{i=1}^{h}\left(\overline{X_{ir}} - \overline{X}\right)^2$, which is called the ***between rows sum of squares*** and usually designated SS_R, is analogous to the between group sum of squares in one-way ANOVA. If H_R is true, then F_R would have an F-distribution with $(h-1, (n-1)hm)$ degrees of freedom.

To summarize:

Under the hypothesis H_R which states that $\alpha_i = 0$ for all i in the model $\mu_{ij} = \mu + \alpha_i + \beta_j$, the quantity F_R defined by:

$$F_R = (SS_R/(h-1)/(SS_E/((n-1)mh)) \qquad (7\text{-}33)$$

where $\overline{X_{ir}}$ = sample mean in jth row, $SS_R = hn\sum_{i=1}^{h}\left(\overline{X_{ir}} - \overline{X}\right)^2$ and

$$SS_E = \sum_{i=1}^{h}\sum_{j=1}^{m}\sum_{k=1}^{n}\left(X_{ijk} - \overline{X_{ij}}\right)^2$$

has an F-distribution with $(h-1, (n-1)hm)$ degrees of freedom

The same reasoning leads to the corresponding test of the hypothesis that the column means are all the same:

Under the hypothesis H_C which states that $\beta_j = 0$ for all j in the model $\mu_{ij} = \mu + \alpha_i + \beta_j$, the quantity F_C defined by:

$$F_C = (SS_C/(m-1)/(SS_E/((n-1)mh)) \qquad (7\text{-}34)$$

where $\overline{X_{ic}}$ = sample mean in j^{th} column, $SS_C = mn\sum_{i=1}^{h}\left(\overline{X_{jc}} - \bar{X}\right)^2$ and

$$SS_E = \sum_{i=1}^{h}\sum_{j=1}^{m}\sum_{k=1}^{n}\left(X_{ijk} - \overline{X_{ij}}\right)^2$$

where $\overline{X_{ic}}$ = sample mean in jth column,
has an F-distribution with $(m-1, (n-1)hm)$ degrees of freedom

The tests of H_R and H_C are based on how far out the calculated F-statistic falls on the upper tail of its distribution under the null hypothesis. In general, computer packages offers the p-value of the observed results, which is the chance under the hypothesis that the calculated F-statistic would be at least as high as the one that actually arose.

Computer packages for two-way ANOVA typically conduct all three of these tests and once and provide F_R, F_C, and F_I, along with their p-values. In the spirit of Bonferroni, we might want at least one p-value of .05/3 = .0167 or below before we reject the hypothesis that $\mu_{ij} = \mu$ for all i and j.

EXAMPLE 7.13 No Place Like Home?

It is widely believed that professional teams in American baseball tend to win more often when they play in their hometowns than when they play in the hometowns of their opponents. But is this "home field advantage" apparent in actual data about wins and losses? And, even if the advantage exists, does it do so to the same degree for all teams? Is it possible that some teams achieve a ten-point rise in their winning percentages in "home" games, while other teams only gain by two percentage points? Here we look into both of these questions with some data and a dose of ANOVA.

The table below presents the wins-and-losses record for six baseball teams in 2011, both for home games and "away" games. Each team chosen is the first one alphabetically (based on its home city) in its league and region. This choice of teams acts to minimize the number of games that the six teams play with one another, reducing the likelihood that the same game gets counted twice (as the home game for one team and the away game for another).

	Win-Loss Record, 2011	
Team	Home Games	Away Games
Baltimore Orioles	39-42	30-51
Chicago White Sox	36-45	43-38
Los Angeles Angels	45-36	41-40
Atlanta Braves	47-34	42-39
Chicago Cubs	39-42	32-49
Colorado Broncos (Denver)	48-43	35-46

Note: a-b means a wins and b losses. Observe that each team had exactly 81 home games and 81 away games.

(i) Given these data, is there statistically-significant evidence of a home-field advantage?

(ii) Is there statistically-significant evidence that the size of this advantage varies across teams?

Solution

Home-Field Advantage?

For our purposes, we will treat the games a particular team played in 2011 as *a random sample of its games over a longer period*. The first factor considered is team, and it has six different "levels," one corresponding to each team. The second factor is game location, and it has two levels, home and away.

Our first concern is whether the data are consistent with the hypothesis that there is no difference in winning percentages between home and away games is for all teams. We might start with the general form of the no-effects model:

$$p_{iH} = p_0 + \alpha_i + \beta_H$$

$$p_{iA} = p_0 + \alpha_i + \beta_A$$

where p_{iH} = probability that team i (i = 1, … 6) wins a home gain (H)

p_{iA} = probability that that team i wins an "away game" (A)

α_i = probability adjustment for team i

β_A = probability adjustment for "away" game;

β_A = adjustment for "home" game;

Here p_0 is the "grand mean" winning percentage for the six teams, which need not equal 1/2 because the teams selected might not win at exactly the average rate for all baseball teams.

The winning proportions corresponding to the table above are:

	Winning Proportion, 2011	
Team	Home Games	Away Games
Baltimore Orioles	.481	.370
Chicago White Sox	.444	.531
Los Angeles Angels	.556	.506
Atlanta Braves	.580	.519
Chicago Cubs	.481	.395
Colorado Broncos (Denver)	.593	.432

A glance at the table reveals that five of the six teams had a home field advantage in 2011, the exception being the Chicago White Six. But it isn't immediately clear whether the pattern is statistically significant.

In the parlance of two-way ANOVA, the null hypothesis H_0 of "no home field advantage" means that the there is no column effect in the table. In the no-effects model, we are testing $H_C : \beta_H = \beta_A$. (Because $\beta_H + \beta_A = 0$, we are actually testing $\beta_H = \beta_A = 0$.) To proceed, we note that:

$\bar{X}$ = overall proportion of wins = 477/972 = .491

$\overline{X_H}$ = overall proportion of wins in home games = 254/486 = .523

$\overline{X_A}$ = overall proportion of wins in away games = 223/486 = .459

$$SS_C = mn\sum_{j=1}^{m}\left(\overline{X_{jc}} - \bar{X}\right)^2 = 2 * 81 * \left((.523 - .491)^2 + (.459 - .491)^2\right) = .332$$

$$SS_E = \sum_{i=1}^{h}\sum_{j=1}^{m}\sum_{k=1}^{n}\left(X_{ijk} - \overline{X_{ij}}\right)^2 = 477 * (1 - .491)^2 + 495 * (0 - .491)^2 = 242.92$$

(Here each win is counted as a 1 and each loss as a 0.)

Therefore, F_C under (7-34) is given by:

$$F_C = SS_C/(m-1)/SS_E/(hm(n-1)) = .332/(242.92/960) = 1.31$$

Because F_C is $F(1,960)$ under H_0, we can turn to electronic table to find that the p-value of the outcome 1.31 is 0.25. Thus, we cannot reject the hypothesis of "no home field advantage" with these data. The net advantage for the six teams was about six percentage points in 2011, but that margin was not greater than the margin of usual sampling error.

Variation Across Teams in "Home Field Effect?"

Because we have not established convincingly the existence of a "home field" *advantage*, we will refer to a team's difference in success between its home and away games more neutrally as its "home field effect." Do the data offer statistically significant evidence that "home field effects" genuinely varied among the six teams studied? By "genuinely," we mean that the differences across teams that arose in 2011 reflect differences that would persist in the long run.

The "home field advantage" Δ_{Hi} for team i is given by:

$$\Delta_{Hi} = p_{iH} - p_{iA}$$

But are the 2011 data consistent with the hypothesis that Δ_{Hi} is independent of i?

If we think about it, we realize that we are in the domain of one-way ANOVA rather than two-way ANOVA. After all, we are testing H_0: $\Delta_{H1} = \Delta_{H2} \ldots = \Delta_{H6} = \Delta_H$, where Δ_H is the common mean. We can therefore use an ANOVA F-test based on the ratio:

$$F_\Delta = n\sum_{i=1}^{6}\left(\overline{\Delta_{H_i}} - \overline{\Delta_H}\right)^2 / (h-1)S^2$$

where $h = 6$, $n = 81$, and S^2 is the estimate of the common variance σ^2 based on the individual S_i^2's from the six teams.

To obtain S^2, we first find each of the S_i^2-terms. If team i won fraction $\widehat{p_{Ht}}$ of its home games, then its 0/1 variable about wins in home games has a sample variance of $\left[81\widehat{p_{Ht}}\left(1-\widehat{p_{Ht}}\right)^2 + 81\left(1-\widehat{p_{Ht}}\right)\left(0-\widehat{p_{Ht}}\right)^2\right]/80 = (81/80)\,\widehat{p_{Ht}}\left(1-\widehat{p_{Ht}}\right)$. Similarly, the sample variance for away games is $(81/80)\,\widehat{p_{At}}\left(1-\widehat{p_{At}}\right)$. As a difference of two (presumably) independent random variables, the home-field advantage Δ_{Hi} has a variance that is the *sum* of the variances for home and away games. For the Baltimore Orioles, for example, we have $\widehat{p_{Ht}}$ = .481 and $\widehat{p_{At}}$ = .370. Thus, the point estimate of its home field advantage is .111 and the sample variance of that estimate is .489.

Performing the calculations for all six teams, we reach the following chart for 2011:

Team	Home-Field Advantage ($\widehat{\Delta_{H_i}}$)	Sample Variance (S_i^2)
Baltimore Orioles	.111	.489
Chicago White Sox	–.086	.502
Los Angeles Angels	.049	.503
Atlanta Braves	.062	.499
Chicago Cubs	.086	.495
Colorado Broncos (Denver)	.037	.501

From these data, we find that the grand mean $\widehat{\Delta_H}$ is .043. When we conduct the ANOVA F-test with $n = 81$ and $h = 6$, we reach the output:

	SS	*df*	*MS*	*F*	*p*
Between:	1.907	5	0.381	0.766	0.575
Within:	239.143	480	0.498		
Total:	241.050	485			

The *p*-value of .575 means that the data are quite consistent with the hypothesis that the home-field advantage is the same for all teams in the long run. In essence, we are told that the six observed variations around the average advantage of 4.3 percentage points are consistent with random sampling error over one season, which comprises 81 home games and 81 away games. To put it another way, if all teams had home field advantages of 4.3 points in the long run, it would not be unusual for their observed advantages in individual years to vary around 4.3 points as much as we saw in 2011.

This outcome might seem surprising, given that the home-field advantages in 2011 ranged from 11.1 percentage points to a disadvantage of 8.6 points. But a calculation makes the result seem more plausible. Support that a given team has a 55% chance of winning each home game and a 50% chance of winning each away game. If $\widehat{p_H}$ is the proportion of games it wins in a set of 81 home games and $\widehat{p_A}$ the corresponding proportion for 81 away games, then sampling theory tells us that $\widehat{p_H} \approx N\left(.55, \sqrt{.55 * .45/81}\right) = N(.55, .055)$ while $\widehat{p_A} \approx N\left(.50, \sqrt{.5/81}\right) = N(.5, .056)$. (We are invoking the Chapter 3 result that $\hat{p} \approx N\left(p, \sqrt{p(1-p)/n}\right)$ where $n \geq 20$ and both np and $n(1-p) \geq 5$). Assuming independent outcomes in different games, we can invoke the result about the normality of a difference between independent normal random variables:

$$\widehat{p_H} - \widehat{p_A} \approx N\left(.55 - .50, \sqrt{.055^2 + .056^2}\right) = N(.05, .078).$$

Thus, a 95% probability interval for $\widehat{p_H} - \widehat{p_A}$ extends from $.05 - 1.96 * .078$ to $.05 + 1.96 * .078$, which is from −.10 to .20. A five-point edge in the long run, in other words, is not inconsistent with outcomes far away from that level in a particular season. It is notable that the width of that interval—30 percentage points—is greater than the maximum spread among our observed Δ_{Hi}'s. Given the limited sample sizes at hand, therefore, the ANOVA outcome is quite consistent with usual sampling variability.

In summary, we failed to find decisive evidence in the six-team 2011 data that there really is a home-field advantage, whether in general or for particular teams. Yet we say it again and again because it is true again and again: accepting the no difference hypothesis does not mean that we are sure there is no real difference. A review of decades worth of data might show that the 2011 pattern prevails year after year, in which case the pattern could not be viewed as a short-term fluctuation. What we can say now is that the 2011 data *in themselves* provide no convincing evidence of a home-field advantage or of cross-team diversity.

EXERCISE 7.12 Every month, the Gallup organization asks about 7000 American men and 7000 American women whether they approve of the performance of the US President. The results for President Barack Obama, starting with his first month in office (January 2009) and continuing at six-month intervals through January 2012 were the following:

	Approval for President Obama	
Month	Women	Men
January 2009	70 (%)	62 (%)
July 2009	60	54
January 2010	52	45
July 2010	48	43
January 2011	52	46
July 2011	46	41
January 2012	48	42

Source: The Gallup Organization The results are obviously rounded to the nearest percentage point.

A Gallup report that presented these results was titled "Gender Gap in Obama Approval Constant Since Term Began," and carried the sub-headline "Approval is Five to Six Percentage Points Higher Among Women than Men."

(i) For these data, test via ANOVA the hypothesis that President Obama's "gender gap" was the same in all the months considered (i.e., that the observed differences in the "gender gap" can reasonably be construed as sampling errors).

(ii) Recall the difference between statistical significance and practical significance that we discussed in Chapter 5. Does that difference seem relevant in this situation? Discuss briefly.

(iii) Suppose you were asked to test the hypothesis that there is no gender gap in support for President Obama (meaning that the observed gaps all result from random sampling error). What would you do (assuming that you wanted to proceed wisely), and what would you conclude?

Unequal Sample Sizes in Two Way ANOVA

We dealt without much difficulty with unequal sample sizes in one-way ANOVA, assuming that the assumptions of equal variance and normality continued to hold. In two-way ANOVA, the situation is less straightforward, as a simple example suggests.

Suppose that it is desired to estimate how the sales of a product vary with both its price and the amount of advertising on its behalf. In an experiment, there are two different price levels (high and low), and two different levels of advertising in different places (high and low). The hope is to model the relationship with a linear model of the form:

$$\mu_{ij} = \mu + \alpha_i + \beta_j$$

where μ = average sales per capita

α_{AH} =adjustment for high advertising and α_{AL} for high advertising

β_{PH} = adjustment for high price and β_{PL} for low price

Data are gathered for different advertising/price combinations but, prior to parameter estimation, the analyst notes a most stunning difference in sample sizes across cells of the 2x2 matrix of sample means. The observations are distributed as in Figure 7.3.

What do we make of such a pattern? We have equal amounts of data for the cases (A_H, P_L) and (A_L,P_H), but no data at all for (A_H,P_H) and (A_L,P_L). Suppose that observed sales were higher in the quadrant with low price/high advertising than in the quadrant with high price/low advertising. Given that outcome, was the increase in sales caused by lower price alone? Higher advertising alone? Some weighted average of high advertising and low price? If so, what was the weighting factor?

More mathematically, the sales level for high advertising and low price allows an estimate of $\mu + \alpha_{AH} + \beta_{PL}$, while the level when prices are high and advertising low allows an estimate of $\mu + \alpha_{AL} + \beta_{PM}$. Because $\beta_{PH} = -\beta_{PL}$ and $\alpha_{AH} = -\alpha_{AL}$, we can approximate that

$$S_{AH,PL} \approx \mu + \alpha_{AH} + \beta_{PL}$$

$$S_{AL,PH} \approx \mu - \alpha_{AH} - \beta_{PL} = \mu - (\alpha_{AH} + \beta_{PL})$$

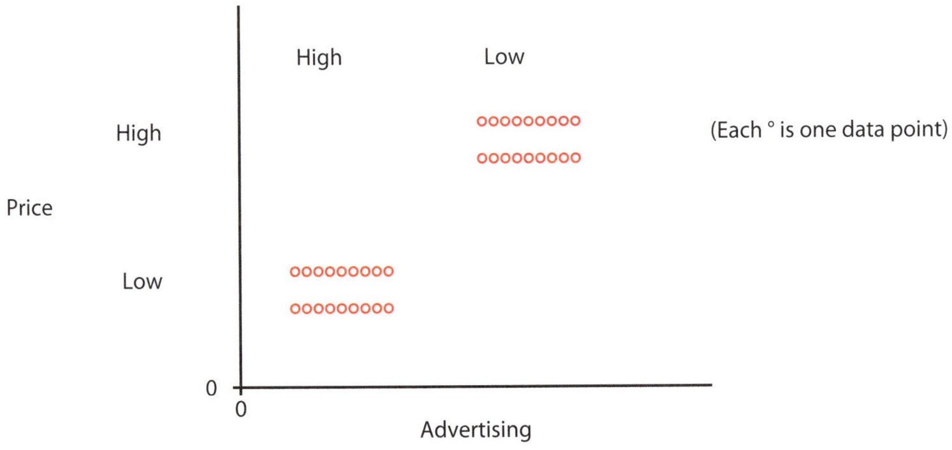

FIGURE 7.3 A Most Unfortunate Distribution of Data Points for Disentangling the Effects of Advertising and Price

With $S_{AH,PL}$ and $S_{AL,PH}$ at hand, we could estimate μ and the sum $\alpha_{AH} + \beta_{PL}$ but could do nothing beyond that. Having estimated μ by $(S_{AH,PL} + S_{AL,PH})/2$ all combinations of α_{AH} and β_{PL} that add up to $(S_{AH,PL} - S_{AL,PH})/2$ are equally consistent with the data.

For parameter estimation, there are clear reasons to prefer a *balanced* design with equal sample sizes for all four advertising/price combinations. Then we can vary advertising while holding price constant and vice-versa, and thereby be well-positioned to distinguish their separate effects on sales. If we have unequal samples sizes but nothing as extreme as Figure 7.3, then we retain some ability to partition the joint effect of advertising and price between them. But we typically pay some price in the reliability of the parameter estimates, an issue we will explore in detail in the forthcoming treatment of regression analysis.

There is no universal consensus on how to proceed in two-way ANOVA with unequal sample sizes. We should pay heed to the statistician who wrote that "This is a BIG problem, and one that statisticians have grappled with for years. One well-known statistician suggested that the only successful way to deal with unequal sample sizes is not to have any." Besides regression analysis, we can sometimes gain insight if we pretend the sample sizes are equal, and then see whether the results under that assumption—although obviously imperfect—make clear the results in the real life case. To illustrate that point, we go to the frigid North Atlantic.

EXAMPLE 7.14 That Night to Remember

The sinking of the *Titanic* is arguably the most famous accident in all history. Among the 2200 people aboard the ship when it hit an iceberg in April 1912, more than 1500 perished. Some readers of this book will have seen the award-winning movie *Titanic* released in 1997, though there were several previous movies and an enormous number of books about the doomed voyage.

In loading the lifeboats, the principle was supposed to be "women and children" first, so one would have expected the survival rate to vary by gender. Based on official rules, however, one would not expect vastly different survival rates for first-class, second-class, and steerage-class passengers. Here we investigate the survival pattern with the following data:

Survival Rates by Gender/Age and Class of Service on the RMS Titanic

Group	Survival Percentage	Number of Passengers
First Class Men	32.6	175
First Class W/C	97.3	150
Second Class Men	8.3	168
Second Class W/C	88.9	117
Third Class Men	16.2	462
Third Class W/C	42.2	244

W/C = Women and Children

(i) Create a main-effects model that offers a rule of thumb about survival as a function of class and status in the dichotomy "men/women and children."

(ii) Prepare some graphs that might suggest the presence or absence of interactions.

(iii) What would the phrase "no interaction" mean in the present context?

(iv) What would it mean to test the hypothesis that a "no interaction" model is consistent with the survival data?

(v) Test the main effects/no interaction model against the data in a reasonable way. Does it offer statistically significant evidence of an association between class and survival rate?

Solution

(i) A general model would assume that the survival probability p_{ij} for a given person in category (*i*,) would follow:

$$p_{ij} = p + \alpha_i + \beta_j + \gamma_{ij}$$

where $i = 1,2,3$ refers to the passenger's class of service
$j = 1$ for men and 0 for women/children
The main effects/no interaction model would assume that $\gamma_{ij} = 0$ for all i and j.
The parameters p, α_i, β_j, γ_{ij} would then be estimated from the survival data.

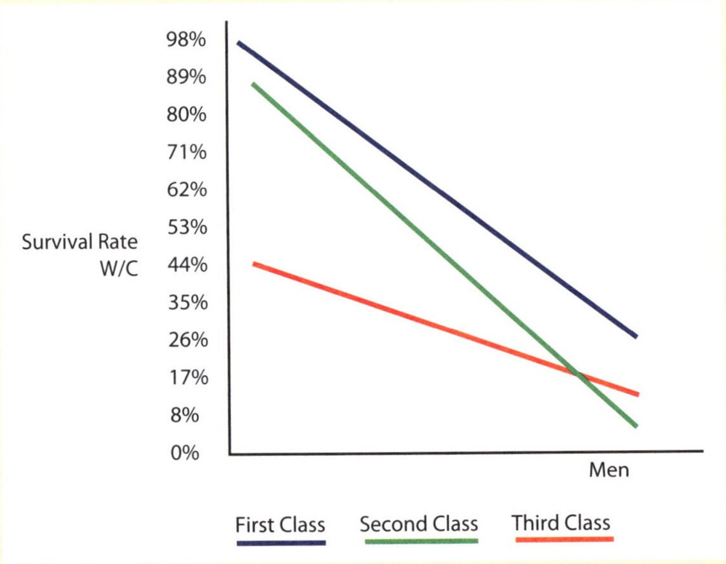

(ii) These interaction graphs show that the survival lines for the different classes do not decline in parallel when gender/age shifts from women/children to men. For that reason, we suspect the presence of interaction effects.

(iii) "No interaction" would mean that the survival rate in any gender/class combination can be modeled well by a formula that adds an average effect for gender to an average effect for class. There would be no substantial deviation from this additive linear model in any particular combination of gender and class.

(iv) The model estimates survival probabilities, and one can assess whether these probability estimates are consistent with actual survival rates. For example, if the model estimates a 50% chance of survival in a given gender/class combination and there were 100 people on board in that combination, the number of members who survived should be approximately $B(100, .5)$. If we had 53 survivors in that group, the data were clearly consistent with the 50% parameter estimate. If, however, there were but 23 survivors, then the estimate was almost surely too high.

(v) Obviously, we are nowhere near a situation of equal sample sizes in all cells. But suppose that we pretended that there were 117 members in each gender/class cell, while the survival rates remained at the levels in the table above. (We choose 117 because it is the smallest sample size for an actual gender/class combination.) Then we could perform a two-way ANOVA series of F-tests on the data.

The rationale for this drastic reduction in sample sizes is that the cuts subject five of the six individual survival rates to *more sampling error* (as reflected by $S/\sqrt{n}$) than they actually faced. For that reason, the ANOVA tests are artificially "lenient" towards accepting the null hypotheses being tested. If the null hypotheses are rejected under our tests, therefore, we can assume that they would have been rejected under tests that worked with the larger true sample sizes.

Treating all the sample sizes as 117, we would reach the following matrix:

	Class:			
	First	Second	Third	Row Means
Men	.326	.083	.162	.190
Women/Children	.973	.889	.422	.761
Column Means:	.650	.486	.292	
		Grand Mean: .476		

Given these data and "equal sample sizes," the parameter estimates of the main effects model are:

$$\hat{p} = .476; \quad \widehat{\alpha_M} = .190 - .476 = -.286; \quad \widehat{\alpha_F} = .761 - .476 = .286$$

$$\widehat{\beta_1} = .650 - .476 = .174; \quad \widehat{\beta_2} = .486 - .476 = .01; \quad \widehat{\beta_3} = .292 - .476 = -.184$$

In words, these main-effect parameter estimates imply that, within a given service class, women/children had survival probabilities that were approximately a 57 percentage points (28.6 + 28.6) higher than men. The estimates further suggest that, within a given gender/age category, first class passengers had survival probabilities that were 16.4 percentage points higher than passengers in second class (17.4 – 1.0) and 35.8 percentage points higher (17.4 – (–18.4)) than travelers in third class.

The projected survival probabilities based on these parameters and the no-interaction assumption that $\gamma_{ij} = 0$ for all i and j follow:

Projected (and Actual) Survival Probabilities By Gender/Class Under Main Effects Model

	Class:		
	First	Second	Third
Men	.365 (.326)	.201 (.083)	.007 (.162)
Women/Children	.935 (.973)	.771 (.889)	.577 (.422)

(Numbers in parentheses are actual survival probabilities)

The probability estimates seem fairly accurate for first class, but less so in second class and least so in third. Indeed, the model implies that only about one of the 117 men in third-class would survive, instead of the 117 $*$ .162 = 19 such men who did so. For this reason alone, we suspect that the no-interaction model will go down in defeat.

When we conduct two-way ANOVA on the matrix with six cells and $n = 117$, the tabulated results are:

Factor	Sum of Squares	Degrees of Freedom	F-stat	p-value
Gender/Age	(SS_R) 38.15	1	283.53	.0000
Class of Service	(SS_C) 22.55	2	83.79	.0000
Interaction	(SS_I) 9.23	2	34.30	.0000
Errors	(SS_E) 93.65	696		

These outcomes make obvious that both gender/age and class of service are related to survival rates, but they also make obvious that the main effects/no interaction model is inadequate in quantifying the pattern. In the original data, the most conspicuous deviation from the overall trends arises for women/children in third class: given the five other survival rates, one might have guessed a survival rate perhaps 80% for these passengers. Yet only 42.2% of women/children in third class actually survived.

Again, this analysis assumed away the huge differences in sample sizes, assigning to all six gender/class categories the sample size 117 of the smallest group. But the main-effects model failed even when we pretended there was more sampling error than there actually was, so it could never have passed reliable tests that used the real sample sizes.

Final Remark

As one might imagine, there are real-life ANOVA exercises that involve are three or more factors. Also, if there are h levels of factor 1, m levels of factor 2, and q levels of factor 3, it might not be feasible to sample all hmq combinations of factors and levels. Thus, one might want to sample from a subset of factor-level combinations to estimate the parameters of the ANOVA model, and then posit (hope?) that the model is accurate even for combinations from which no data were gathered. These are all interesting issues, and we discuss them in the web supplement to this book. Here, however, we end with some further exercises concerning one and two-way ANOVA.

The Takeaway Bar

Glossary

Analysis of Variance (ANOVA) is a method for investigating whether a set of m means can be summarized through a formula with fewer than m parameters.

The **grand mean** in ANOVA is the arithmetical average of the all the data points in the various samples at hand. If all the sample sizes are equal, the grand mean is the arithmetical average of the individual sample means.

One-way ANOVA considers whether a set of m distributions have the same mean.

The **F-distribution** in probability is a nonnegative distribution with two parameters called its degrees of freedom. (It is literally a ratio of two independent random variables that are both χ^2-distributed.) In one-way ANOVA, it is the ratio of two quantities that should both equal the common variance σ^2 if the null hypothesis is correct and the means being tested are all equal.

A **Bonferroni pairwise test** for equality of means is a series of $\binom{m}{2}$ difference-of-means tests, which are conducted for every possible pair of the m means that are hypothesized to be equal.

Two-way ANOVA models the means of "response variables" (μ_{ij}'s) as functions of two factors, each of which can take on several different levels.

A **main effect** in a two-way ANOVA model is the mean differences between the grand mean μ and the mean when an individual factor is held constant at a particular level.

A **main effect/no interactions** ANOVA model posits that that the mean when a particular factor/level combination is the sum of the grand mean and the individual main effects for that combination.

Formulas

One-Way Analysis of Variance:

$$H_0\colon \mu_1 = \mu_2 \ldots = \mu_m = \mu$$

To test H_0, calculate F-ratio defined by:

$$F = SS_B/(m-1)/(SS_W/(\tau - m))$$

where τ = total number of data points from the m distributions

n_i = sample size from distribution i ($i = 1, \ldots m$)

$\overline{X}_i$ = sample mean for the distribution-i sample

$\bar{X}$ = grand mean of all τ data points

$SS_B = \sum_{i=1}^{m} n_i(\bar{X}_i - \bar{X})^2$ is the *"between group sum of squares"*

$SS_W = \sum_{i=1}^{m}(n_i - 1)S_i^2$ is the *"within group sum of squares"*

$$S_i^2 = \sum_{j=1}^{n} \frac{(\overline{X_{ij}} - \overline{X_i})^2}{n-1}$$

- If H_0 is true, then the F-ratio follows an F-distribution with $(m - 1, \tau - m)$ degrees of freedom. (It is assumed that all m distributions are normal and have a common variance σ^2.)
- We reject H_0 if F exceeds the $1 - \alpha$ fractile of that F-distribution, where α is the risk of accidental rejection of H_0.

When all the sample sizes are equal (i.e., $n_i = n$ for all i), then we can write:

$$F = (SS_B/(m - 1))/(SS_W/m(n - 1))$$

where $SS_B = \sum_{i=1}^{m} n(\bar{X}_i - \bar{X})^2$ and $SS_W = \sum_{i=1}^{m}(n - 1)S_i^2$

Under H_0, F has an F-distribution with $(m - 1, m(n - 1))$ degrees of freedom

Pairwise Bonferroni Tests:

To test

$$H_0: \mu_1 = \mu_2 \ldots = \mu_m = \mu$$

- Create a series of $\binom{m}{2} = m(m - 1)/2$ pairwise tests of the hypothesis H_{ij} that $\mu_i = \mu_j$ for $1 \le i \le m$ and $i < j \le m$
- For each test, arrange that $P(\text{reject } H_{ij} | H_{ij} \text{ true}) = \alpha/(m(m - 1)/2)$. α is typically set equal to .05.
- If H_{ij} is true and the sample sizes are equal, then the quantity:

$$t_{ij} = (\bar{X}_i - \bar{X}_j)/(S\sqrt{2/n})$$

is approximately t-distributed with $2n - 2$ degrees of freedom

where n = common sample size

$\bar{X}_i$ = sample mean from the distribution-i sample

$S^2 = \frac{\sum_{i=1}^{m} S_i^2}{m}$; S_i^2 is the sample variance in the dist-i sample

Under H_{ij}: When n is moderately large, $\overline{X_i} - \overline{X_j} \approx N(0, S\sqrt{2/n})$

For differing but moderately large sample sizes,

$$\overline{X_i} - \overline{X_j} \approx N\left(0, S\sqrt{\left(\frac{1}{n_i} + \frac{1}{n_j}\right)}\right)$$

- A two-sided pairwise test of H_{ij} would reject if the outcome either is below the $(1/2)[\alpha/(m(m-1)/2)]$ fractile of the relevant distribution or is above the $1-(1/2)[\alpha/(m(m-1)/2)]$ fractile.
- A one-sided pairwise test of H_{ij} when the alternative H_a is $\mu_i < \mu_j$ would reject for an outcome in the bottom $\alpha/(m(m-1)/2)]$ fractile of the relevant distribution. If the alternative H_a is $\mu_i > \mu_j$, then the rejection region is the top $\alpha/(m(m-1)/2)]$ fractile.
- H_0 is accepted if and only if each H_{ij} is accepted

Two-Way Analysis of Variance

Here we assume that there are two factors that can affect the means, the first of which can take on h levels and the second of which assume m levels .

There is an hxm matrix of means, in which

$$\mu_{ij} = \text{mean when factor 1 is at level } i \text{ and factor 2 is at level } j$$

The general model is $\mu_{ij} = \mu + \alpha_i + \beta_j + \gamma_{ij}$

The quantities α_i and β_i are the main effects in the model, while γ_{ij} is an interaction term that applies only for the combination (i,j)

It is assumed that a set of sample means $\overline{X_{ij}}$ are at hand, all of which are based on a sample of size n.

Among the hypotheses that can be tested against the data are:

H_R: $\alpha_i = 0$ for all i and $\gamma_{ij} = 0$ for all i and j

H_C: $\beta_j = 0$ for all j and $\gamma_{ij} = 0$ for all i and j

H_I: $\gamma_{ij} = 0$ for all i and j

(R = Rows, C = Columns; I = Interactions)

The parameter estimates are:

$$\hat{\mu} = \overline{X} = \text{grand mean of all observations}$$

$$\widehat{\alpha_\iota} = \overline{X_{ir}} - \overline{X}$$

where $\overline{X_{ir}}$ = average of all observations in the ith row of the sample-means matrix

$$\widehat{\beta_J} = \overline{X_{jc}} - \bar{X}$$

where $\overline{X_{jc}}$ = average of all observations in the jth column of the sample-means matrix

$$\widehat{\mu_{iJ}} = \overline{X_{ij}} - \left(\widehat{\alpha_i} + \widehat{\beta_j} + \bar{X}\right)$$

To test H_R:

Compute the ratio:

$$F_R = (SS_R/(m-1))/(SS_E/(n-1)hm)$$

where $SS_R = hn\sum_{i=1}^{h}\left(\overline{X_{ir}} - \bar{X}\right)^2$ and $SS_E = \sum_{i=1}^{h}\sum_{j=1}^{m}\sum_{k=1}^{n}\left(X_{ijk} - \overline{X_{ij}}\right)^2$

SS_R is the between row sum of squares, while $SS_E/(n-1)$hm is an unbiased estimate of the common σ^2, independent of the values of α_i, β_j, or γ_{ij}.

If H_R is true, then F_R is $F(h-1, (n-1)hm)$, where h = number of rows in μ_{ij} matrix and m = number of columns.

- Reject H_0 if F_R exceeds the $1-\alpha$ fractile of $F(h-1, (n-1)hm)$

To test H_C:

- Compute the ratio

$$F_C = (SS_C/(m-1))/(SS_E/(n-1)hm)$$

where $SS_C = mn\sum_{i=1}^{h}\left(\overline{X_{jc}} - \bar{X}\right)^2$ and $SS_E = \sum_{i=1}^{h}\sum_{j=1}^{m}\sum_{k=1}^{n}\left(X_{ijk} - \overline{X_{ij}}\right)^2$

If H_C is true, then F_{RC} is $F(m-1, (n-1)hm$

- Reject H_0 if F_R exceeds the $1-\alpha$ fractile of $F(m-1, (n-1)hm)$

To Test H_I:

- Compute the F ratio: $F_I = \dfrac{SS_i/(h-1)(m-1)}{SS_E/((n-1)hm)}$

where $SS_I = n\sum_{i=1}^{h}\sum_{j=1}^{m}\overline{X_{ij}} - \left(\widehat{\alpha_\iota} + \widehat{\beta_J} + \bar{X}\right)^2$ and $SS_E = \sum_{i=1}^{h}\sum_{j=1}^{m}\sum_{k=1}^{n}\left(X_{ijk} - \overline{X_{ij}}\right)^2$

If H_I is true, then F_I has an F-distribution with $[(h-1)(m-1), (n-1)hm]$ degrees of freedom

- Reject H_I if F_I exceeds the $1-\alpha$ fractile of $F((h-1)(m-1), (n-1)hm)$

Some Parting Thoughts

- One-way ANOVA can never establish that several means are all equal. What it can do is demonstrate that the observed differences among sample means could plausibly be explained by statistical sampling error. But "could be" is not the same as "must be."
- All ANOVA F-tests use a ratio of two quantities and reflect a simple principle. Under the null hypothesis, both the numerator and denominator are estimates of the common variance σ^2; when the null hypothesis fails, however, the numerator should grow but not the denominator. Thus, ratios well above one raise questions about the viability of the null hypothesis.
- F-tests in one-way ANOVA can establish the falsity of an equal-means hypothesis, but they do not indicate why the hypothesis failed. There are "post mortems" that can be performed after failed *F*-tests, but there is a case for using Bonferroni pairwise tests instead of *F*-tests: the pairwise tests can be one-sided as well as two-sided, and by their very nature they reveal which pairs of mean differ at high levels of statistical significance (i.e. at low p-values).
- If strong interaction effects exist in ANOVA with two factors, then a main-effects model might be all but useless in approximating the means for particular factor/level combinations. Still, the main effects indicate the average change in mean compared to the overall mean when a particular factor takes a particular level. That aggregate information is often helpful.
- Virtually all standard ANOVA procedures assume that all the distributions compared have the same variance σ^2. It is rarely obvious that the variances must be the same, but moderate differences among the variances have been shown only minimally to affect the test results. With sample sizes large enough to justify use of the Central Limit Theorem, Bonferroni pairwise tests can be conducted without an assumption of common variance.
- Because they focus on means, ANOVA tests can foster the belief that the only major differences between distributions are those concerning their means. But other differences (e.g., in variances) might be just as important.

EZ Pass

1. If one is comparing four means in one-way ANOVA and has samples of size 40 from each population, the F-statistic under equal means has:
   (A) 4 and 160 degrees of freedom
   (B) 3 and 160 degrees of freedom
   (C) 4 and 156 degrees of freedom
   (D) 3 and 156 degrees of freedom
2. If Bonferroni pairwise tests involving seven means, the total number of comparisons is:

(A) 56
(B) 42
(C) 21
(D) 14

3. If we reject the hypothesis of equal means in a one-way ANOVA *F*-test, then we can infer that:
   (A) all the means are unequal
   (B) the largest mean is at least twice the smallest mean
   (C) the largest mean exceeds the smallest mean by at least twice the common standard deviation σ
   (D) none of these
4. In one-way ANOVA, the simple (arithmetical) average of the individual sample variances:
   (A) is always an unbiased estimate of the common variance σ^2
   (B) is unbiased unless the sample sizes are unequal
   (C) is biased but is the minimum-variance estimate of σ^2
   (D) is unbiased on Thursdays but not on other days
5. In an ANOVA with two factors and two levels, the sample means of the (nonnegative) response variables (i.e., the $\overline{X_{ij}}$) are twice as high when one factor is high and the other low as when both factors are low. When both factors are high, the sample mean is three times as high as when both are low. This pattern suggests:
   (A) that interaction effects are substantial
   (B) that a main effects model will fit the data well
   (C) that the grand mean is half the mean when both factors are high
   (D) none of these
6. In two-way ANOVA with two rows for Factor 1 and three columns for Factor 2, $\bar{X} = 82$, $\overline{X_{1r}} = 88$, $\overline{X_{1c}} = 80$, and $\overline{X_{3c}} = 86$. In the main-effects model, the estimate of μ_{22} given these data would be:
   (A) 74
   (B) 77
   (C) 82
   (D) cannot be determined from the given information

Further Exercises

EXERCISE 7.13 Some exercises with the electronic F-distribution table:

(i) Find the 95th percentile of F(2, 398).
(ii) Find the 99th percentile of F(8, 140).
(iii) The F-statistic is 1.88 in a one-way ANOVA F-test that compares five sample means with a sample size of 45 for each distribution. What is the p-value of the result?

EXERCISE 7.14 To test equality of means among several distributions, a random sample of size 100 is obtained from each. It is assumed that the variance is the same in all the distributions; the estimate S^2 of the common variance given the data is 4.

(i) To test the hypothesis using the Bonferroni method with a 5% overall risk of accidental rejection, a series of pairwise tests is performed. If each test has a Type-I error probability of 1/560 (.0018), how many means are being compared?
(ii) Specify each of the pairwise tests.
(iii) Suppose that the second largest of the calculated sample means is 8.44 and the second smallest is 7.48. Given this information, what is the *minimum* number of pairwise tests in (ii) in which the hypothesis $\mu_i = \mu_j$ is rejected? Explain your reasoning clearly.

EXERCISE 7.15 Verify Equation (7-7):

$$\sum_{j=1}^{n}\left(X_{ij} - \overline{X}\right)^2 = \sum_{j=1}^{n}\left(X_{ij} - \overline{X}_i\right)^2 + n\left(\overline{X}_i - \overline{X}\right)^2$$

Hint: $X_{ij} - \overline{X} = \left(X_{ij} - \overline{X}_i\right) + \left(\overline{X}_i - \overline{X}\right)$

EXERCISE 7.16 Sample means are obtained using random samples of size 101 from six distributions, and it is desired to test the null hypothesis H_0 that the six distributions have the same means. The sample variance S^2 is 1, and the grand mean is 6.

(i) List a set of six sample means for which H_0 would be rejected under an ANOVA F-test with α=.05 but would not be rejected under a series of pairwise Bonferroni tests with α = .05. (This needn't be complicated; just consider what each test focuses on.)
(ii) Find a set of six sample means for which H_0 would be rejected under a series of pairwise Bonferroni tests with α = .05. but not under an ANOVA F-test with α=.05. (Same hint as in (i)).
(iii) What is the moral of this story?

EXERCISE 7.17 (From the Web) "An experiment was performed to determine whether the annealing temperature of ductile iron affects its tensile strength. Five specimens were annealed at each of four temperatures. The tensile strength in (ksi) was measured for each temperature. The results are presented in the following table. Can you conclude that there are differences among the mean strengths?"

Temperature (C°)	Sample Values				
750	19.72	20.88	19.63	18.68	17.89
800	16.01	20.04	18.10	20.28	20.53
850	16.66	17.38	14.49	18.21	15.58
900	16.93	14.49	16.15	15.53	13.25

(i) Is this an exercise in one-way ANOVA or two-way ANOVA? Explain.

(ii) This exercise was meant to illustrate the use of ANOVA. What assumptions were being made about the distributions of tensile strengths at given temperatures?

(iii) Now, answer the question as stated above. Does your ability to process the data sensibly require an understanding of what annealing is?

EXERCISE 7.18 (From the web). "The Gourmet Cookie Company produces its best selling cookie, Chocolate Macadamia Delites, on all three of its cookie production lines and ovens. The quality control manager is concerned that the cookies on one or more of the lines weigh more than the cookies are supposed to weigh. The average weight of the cookies on all three lines should be 2.0 ounces. Using the (random) sample data from the table below, test the hypothesis at the .05 level of significance that there is no difference in cookie weights between the three cookie production lines."

Line 1	Line 2	Line 3
2.0	1.9	2.2
2.05	2.1	2.1
2.05	2.0	2.15
2.0	1.95	2.3
2.1	2.05	2.35
2.05	1.9	2.05
2.05	2.05	2.25
2.3	2.15	2.3
2.0	1.9	2.4
2.15		

(i) Answer the question above using an ANOVA method (as the problem intended).

(ii) Does your ANOVA approach ever use the fact that the average weight per cookie on each line "should be 2.0 ounces?"

(iii) Perform a test on the hypothesis that the mean cookie weight on all lines is in fact 2.0 ounces.

EXERCISE 7.19 The Pew Research Center conducts an annual random survey of attitudes in various countries around the world. In 2011, the United States had the following *favorability ratings* in the following countries in Europe:

COUNTRY	RATING	SAMPLE SIZE
BRITAIN	61	1000
FRANCE	75	1004
GERMANY	62	1001
SPAIN	64	1000
LITHUANIA	73	750
POLAND	70	750
UKRAINE	60	1000

The favorability rating is the percentage of respondents who said that they had favorable views of the United States. Obviously, the ratings were rounded to the nearest percentage point.

(i) Consider the null hypothesis H_0 that the favorability rating for the United States was the same in all seven countries in 2011 (and the variations in ratings across the countries reflect nothing more than random sampling error). Under that hypothesis, what is the estimate S^2 of the common variance? (Hint: The estimate S_i^2 is $\left(\frac{n}{n-1}\right)\widehat{p}_i\left(1-\widehat{p}_i\right)$, where $\hat{p}_i$ is the fraction "favorable" in country i.

(ii) Conduct the appropriate ANOVA F-test on these data. What do you conclude?

EXERCISE 7.20 Suppose that a follow-up experiment about bystanders is undertaken to amplify on the results of the study in Example 7.4 (This example is roughly based on actual data.) The gender of both the participant and the "victim" are varied, while the number of other "listeners" is either zero or one. The means and standard deviation of speed scores for the response are calculated for each combination, with the following results:

	Sample Speed Score:	
Combination (participant, victim, other listeners)	Mean	Standard Deviation
MM0	0.70	0.25
MM1	0.88	0.19
MF0	0.75	0.24
MF1	0.86	0.20

	Sample Speed Score:	
Combination (participant, victim, other listeners)	Mean	Standard Deviation
FM0	0.74	0.20
FM1	0.89	0.22
FF0	0.69	0.16
FF1	0.85	0.23

M = Male; F = Female 40 observations for each combination

(i) Suppose that the two factors of interest are (i) whether the participants are both of the same gender and (ii) whether there are zero or one other listeners. Find the sample means and standard deviations for each of the four groupings based on the data above. Please note that:

$$\sum_{i=1}^{m}(x_i - c)^2 = \sum_{i=1}^{m}(x_i - \bar{X})^2 + (\bar{X} - c)^2$$

(ii) It is hypothesized that the number of other listeners affects response time but not whether participant and victim are of the same gender. Conduct an ANOVA test of this hypothesis.

(iii) Are the data consistent with a no interaction/main effects model for these two factors?

(iv) Consider the alternate hypothesis that the number of other listeners and the gender of the participant matter but not the gender of the victim. Create a two-factor model based on number of listeners and gender of participants, and repeat (i)–(iii).

Introduction to Linear Regression

Car Less?

"I've figured out the subject of my masters' thesis," Minerva told Mendel. "I will write about urban population density and the use of mass transit. My theory is the fraction of residents who use mass transit goes up faster than linearly as population density increases. The greater the density, the fewer parking spaces per capita and the greater the road congestion, both of which discourage driving to work. In addition, greater population density increases the variety of public transit routes and the frequency of service they offer. So public transport gets more attractive as alternate transportation gets more repulsive."

"Sounds mildly interesting," Mendel mused. "But how do you propose to test your theory?" "I'm gathering data about population density and public transport usage in lots of American cities," Minerva answered, "and will devise a formula that summarizes the linkage. I suspect that usage varies with some power of population density greater than one, and I'll demonstrate that empirically. I'm hoping that the exact power will be known to history as Minerva's constant. Something akin to Moore's Law about computing power."

"What makes you think you can extract this "constant" from the messy data you'll have?" asked Mendel. "I know regression analysis," Minerva responded, "I can do anything."

To be continued

Often, one wants to understand how changes in one quantity induce changes in another. A manufacturer, for example, might want to know the relationship between the sales of a product and the expenditures on its advertising. To ascertain that higher advertising is associated with higher sales is to belabor the obvious; the real issue is whether the profits on the increased sales exceed the cost of the advertising that produced them.

Regression analysis is potentially of great value in this context, for the technique strives to develop a precise formula to relate the movements of the **dependent variable** to those of one or more **explanatory variables**. Of course, trying to do something is not the same as doing it successfully; the safe use of regression requires a good deal of thought and a good dose of skepticism. But the underlying theory is elegant, the procedures easy to use and, when conditions are auspicious, the results remarkably powerful. The method is worth the considerable attention that we will devote to it.

In this chapter, we limit ourselves to cases in which only one variable is systematically related to the dependent variable. In statistical parlance, we are considering only *bivariate* analyses. (Perhaps the simplest case is the relationship between the circumference of a circle and its radius.) It is obviously possible that several factors influence the quantity of interest, and we will generalize our analysis in Chapter 9 to address those cases. But our study of these more complex cases will be considerably easier if we understand how to perform bivariate regression analyses. The "default" assumption in many such bivariate analyses that the relationship between the dependent variable and the explanatory variable is linear; as we will see, however, some relationships that are initially nonlinear can be converted to linear form with an appropriate change of variables.

8.1 Sales vs. Advertising

Perhaps we should start our study of regression in the simple setting of sales vs. advertising. Suppose it is believed the per capita sales, S, are tied to per capita advertising, x, by a simple linear equation:

$$S = \alpha_0 + \beta_0 x \tag{8-1}$$

Here α_0 represents the base level of sales in the absence of any advertisements, while β_0 is the change in sales caused by each unit increase in the amount of advertising done.

Theoretical arguments might establish the plausibility of (8-1), but only rarely can they yield the precise values of α_0 and β_0 in a particular setting. To obtain these values, one would presumably have to gather some relevant data and then analyze them sensibly. Hypothetically at least, that task should be straightforward: one could divide the sales region into n disjoint parts and then consciously vary the level of advertising across them. After observing the sales in each district for some appropriate period, one would have n data points of the form (x_i,s_i), where s_i and x_i are the per capita sales and advertising, respectively, in the ith district. If, as assumed, the points actually fall on a straight line, finding its slope β_0 and its intercept α_0 would demand nothing more complicated than a ruler.

Unfortunately, the immaculate conditions of this simple experiment are almost impossible to achieve. While a firm might know how much it spent on advertising, it almost never knows precisely how much it sold. Sales estimates sometimes depend on data from a random subset of stores; hence they are subject to sampling error.

Moreover, the stores sampled provide imperfect statistics about their sales, involving estimates on subjects like losses and thefts as opposed to exact knowledge. It follows from such conditions that the *observed* S_i's can deviate from the true sales figures and thus that, instead of appearing along the straight line that describes the underlying relationship, the data points will usually exhibit some random scatter around it.

A more realistic model might recognize such a measurement problem by revising (8-1) to the more general form:

$$S_i = \alpha + \beta x_i + \varepsilon_i \quad (8\text{-}2)$$

where ε_i = the "random error" in the observed S_i-value

In this formulation, the true sales level is $\alpha + \beta x_i$ but the observed sales levels deviates from this quantity by random-variable ε_i. The observed sales could be considered the realization of random variable S_i.

Random variable ε_i is often assumed to have four properties:

(i) a continuous distribution
(ii) a mean of zero
(iii) symmetry around zero
(The notion here is that, if the errors are genuinely random, overestimating sales by b units is neither more nor less likely than underestimating by b.)
(iv) an assignment of greater probability to small errors than to larger ones.

Note that a zero-mean normal distribution would possess all four of these properties. For this and other reasons, it is common to assign ε_i the distribution $N(0, \sigma_i)$ (Figure 8.1).

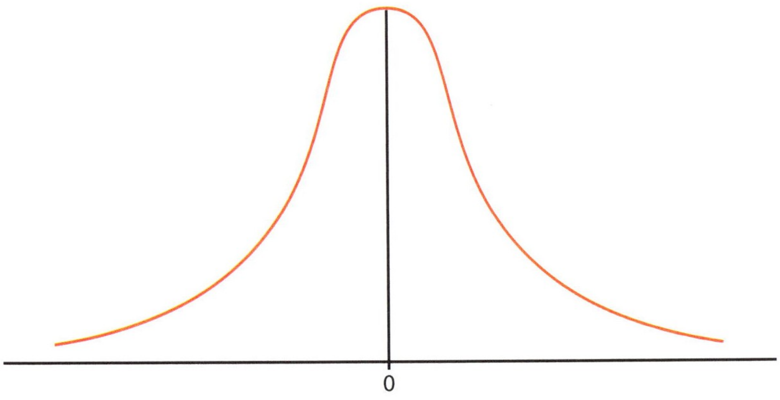

FIGURE 8.1 A zero-mean normal curve might be a good model for random error in an observed quantity

Frequently, though not always, two further assumptions about the ε_i's are quite realistic:

(v) The different ε_i 's can be treated as independent random variables.
(vi) There is no prior reason to believe that some of the random errors are larger or smaller than others. Thus, all of the σ_i's should be equated to some common value σ_ε. (This is known as the **homoscedasticity** axiom.)

When (v)–(vi) are incorporated into the sales-advertising model while (i)–(iv) are summarized by zero-mean normality, the overall relationship becomes:

$$S_i = \alpha_0 + \beta_0 x_i + \varepsilon_i \tag{8-3}$$

where ε_i is one sample from the distribution $N(0, \sigma_\varepsilon)$, independent of x_i and anything else.

In words, (8-3) is saying that a zero-mean "noise process" obscures the true linear relationship between S and x. For this reason, the estimation of α_0 and β_0 is no longer elementary; instead of arising from a simple graph, the two parameters must somehow be "retrieved" from data points not located where they actually should be. Providing a sensible mechanism for accomplishing such retrieval is the primary goal of linear regression analysis.

8.2 Regression Through the Origin

To introduce basic ideas about linear regression analysis, we concentrate first on the simple formulation:

$$Y_i = \beta_0 x_i + \varepsilon_i \tag{8-4}$$

where the ε_i's are independent samples from $N(0,\sigma_\varepsilon)$ for some σ_ε.

In this model, $Y = \beta_0 x$ is said to be the **true line** relating Y and x. Equation (8-4) is a special case of (8-3) in which α_0 is set equal to zero, a condition that requires the true line to pass through the origin. Sometimes that restriction is wholly appropriate: if, for example, Y is the risk of radiation-induced cancer and x the dose to which a person was exposed, then (0,0) must appear along their joint curve by definition. We will soon turn to the case where the intercept α_0 need not be zero, but will find our work there easier because we considered regression through the origin first. That circumstance largely explains why we devote so much attention to this special case.

In (8-4), the various ε's represent random errors (or *disturbances*) that distort the observed y-values, though—it is assumed—*not* the corresponding x's. Geometrically, the disturbances subject the various points to vertical buffeting around the line from whence they originally came. (Figure 8.2) *Throughout this section, we assume that model (8-4) correctly specifies the relationship between the x_i's and Y_i's.*

In this formulation, Y is said to be the **dependent** variable and x the **explanatory** variable. These terms reflect a strong assumption about cause and effect: the only

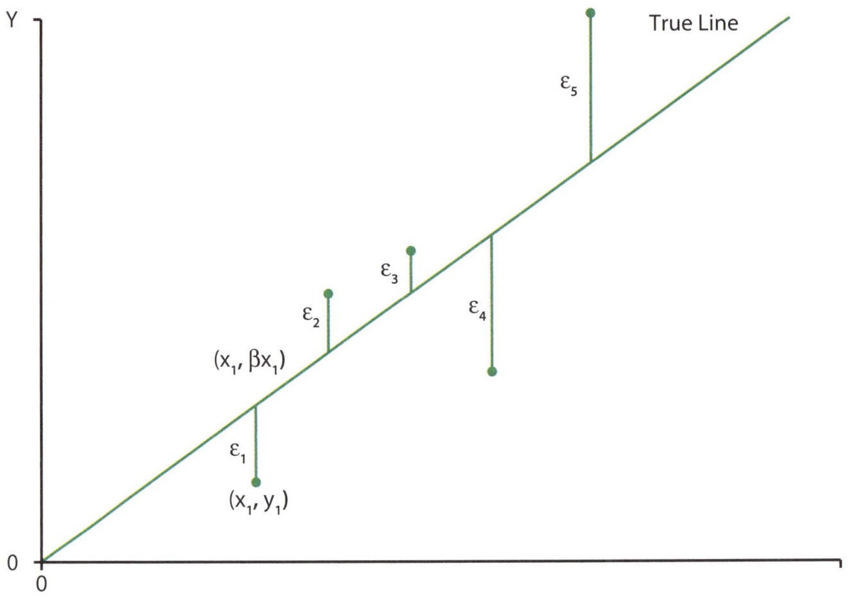

FIGURE 8.2 Random errors cause the points to scatter around the line to which they belong.

"power relationship" is that x affects Y, not that Y affects x or that some unidentified third force influences both Y and x simultaneously. Equation (8-4) is called a **linear regression** model. The word "regression" apparently arose because one of the earliest models of this form was used to quantify the regression-to-the-mean relationship between the heights of fathers and those of their sons.

Suppose we have a set of observed points (x_i, y_i) generated under (8-4). Though we might wish to deduce the slope β_0 of the true line from these data, the ε's have rendered that achievement unattainable. The problem is suggested by the three-point example in Figure 8.3a, in which $\varepsilon_i = 0$, ε_2 is highly negative, and ε_3 somewhat less so. If someone unaware of the values of the ε_i's s were asked to guess the true y-to-x

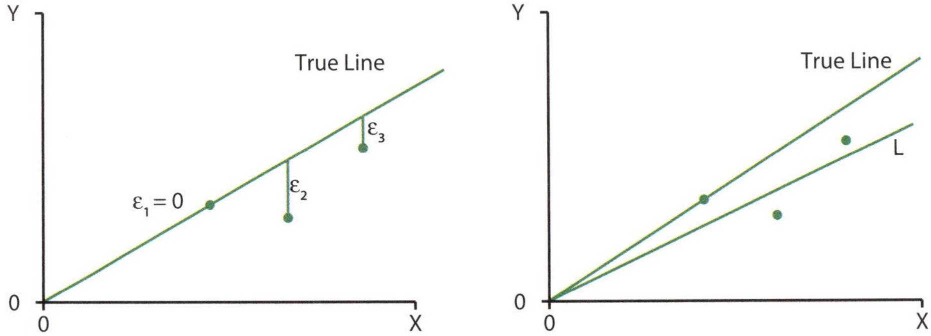

FIGURES 8.3A AND 3B Given this set of random errors in the observed y-values, one would probably underestimate the slope of the true line.

relationship, she would quite rationally prefer line L of Figure 8.3b to the one that is actually correct. Line L, after all, implies a more "natural" pattern of random measurement errors than the one that truly arose in this case. Yet L has a lower slope than the true line, and thus would lead to an estimate of β_0 that is too low.

It is the nature of random errors that, sometimes, they will combine to produce a misleading impression. We must simply accept the fact that, like estimating μ by $\bar{X}$, the estimation of β_0 based on vertically-perturbed data is an imperfect venture.

A Measure of Discrepancy

To say that we can't determine β_0 precisely is not to imply we can't approximate it. As a start to doing so, we might consider Figure 8.4, in which three lines are proposed as the "original source" of the data points plotted. Of the three choices, Line 2 clearly seems the soundest. Line 1 seems unpromising because all its vertical "misses" are large and, moreover, of the same sign. (Given n data points, the chance that *all* the independent zero-mean normal ε_i' s will be negative is $.5^n$.) Dissatisfying for the same general reasons as Line 1, Line 3 soars above the points, moving further away as x goes up. Line 2, by contrast, comes fairly close to the points, passing above some and below others.

More generally, suppose we consider the whole family of lines $y = \beta x$. We could start at the horizontal ($\beta = 0$) and gradually rotate our way through the first quadrant to the vertical. For a while, this rotation moves the line closer to the data; after some threshold corresponding roughly to Line 2, however, continued movement starts to take the line away. Quantifying this pattern could provide the embryo of a scheme for estimating β_0.

What would be helpful is some number to reflect the **discrepancy** between the n data points and any given line. The less satisfactorily does a line "fit" the points, the higher would this discrepancy number be. The true slope β_0 could be approximated by the **slope** of that line for which the index of discrepancy takes on its minimum value.

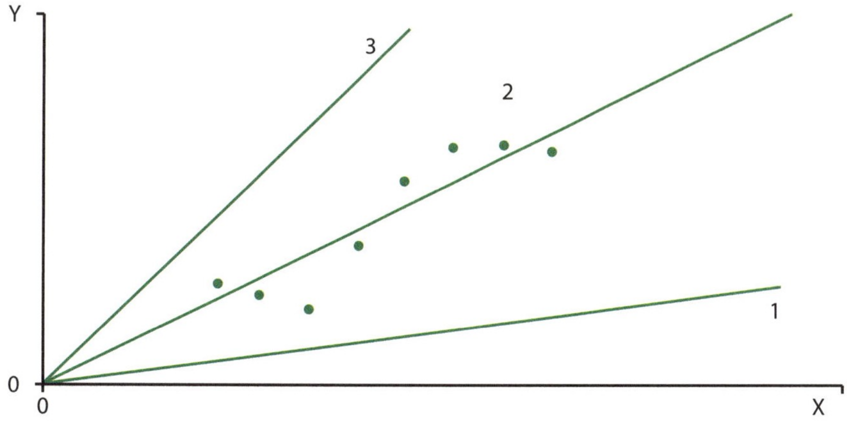

FIGURE 8.4 Three possible fits to the data points shown.

A Quadratic Index

The idea sounds good in the abstract, but how should discrepancy be measured? Presumably, a suitable index would depend on the sizes of the vertical disparities between points and line. (The focus on vertical distances is proper because, as noted, only the observed y_i's are assumed subject to random error.) Also relevant, perhaps, are the proportion of the points on each side of the line. One can imagine a large number of intelligent indices but, for reasons that need not seem at all obvious, a particularly useful one for the family $y = \beta x$ is:

$$L(\beta) = \sum_{i=1}^{n}\left(y_i - \beta x_i\right)^2 \tag{8-5}$$

In words, $L(\beta)$ sums the squares of the n vertical distances between each y_i and the y-coordinates of the point on the line $y = \beta x$ at $x = x_i$.

The Least Squares Slope Estimate

The quantity $r_i = y_i - \beta x_i$, is defined as the ith ***residual***. The residual r_i reflects both the vertical distance and direction from the ith point to the line $y = \beta x$: when the point is above the line, $r_i > 0$ while, when the line is above the point, $r_i < 0$. $L(\beta)$ in (8-5) is therefore the sum of squared residuals associated with $y = \beta x$. β^*, the value of β at which the sum-of-squares function in (8-5) is minimized, is called (not surprisingly) the **least-squares** estimate of β_0. In this section, we will focus on this β^*, when the line that is fit to the data must pass through the origin (as in 8-5).

Given that there are an infinite number of lines of the form $y = \beta x$, how can we find the one that minimizes the sum of squared residuals (hereafter SSR)? That task might seem forbidding to those who do not know Calculus, but quite routine to those who do. $L(\beta)$ is, after all, a quadratic function in β, and minimizing a quadratic function is not difficult. Finding $dL(\beta)/d\beta$ and setting it equal to zero yields:

The least squares estimate $\beta^*$of slope β in the model $Y_i = \beta x_i + \varepsilon_i$ follows:

$$\beta^* = \sum_{i=1}^{n} x_i y_i \Big/ \sum_{i=1}^{n} x_i^2 \tag{8-6}$$

Here the observed y_i's are the realizations of the Y_i's which are random variables.

EXAMPLE 8.1

Two Points Determine a Line

Students might remember the statement "two points determine a line" from their studies of algebra. At that time, what was meant was that, given two points (x_1,y_1) and (x_2,y_2), one can readily find the line $y = a + bx$ that passes through both of them.

But, in linear regression through the origin, the line $y = \beta^* x$ that best fits two data points need not pass through either.

Suppose that $Y_i = \beta_0 x + \varepsilon_i$, where ε_i is $N(0,\sigma_\varepsilon)$, and we have the two data points (1,3) and (20,40). Given these points:

(i) What is the least squares estimate of β_0?
(ii) What are the residuals at the two data points?

Solution

(i) Applying (8-6), we get:

$$\beta^* = \sum_{i=1}^{n} x_i y_i \Big/ \sum_{i=1}^{n} x_i^2 = (1*3 + 20*40)/\left(1^2 + 20^2\right) = 2.002$$

(ii) At $x_1 = 1$, the y-coordinate of the point is $y_1 = 3$, while the y-coordinate of the corresponding point on $y = 2.002x$ is $y_1^* = 2.002 * 1 = 2.002$. The residual r_1 is given by $y_1 - y_1^* = 0.998$. At $x_2 = 20$, $y_2 = 40$ and $y_2^* = 2.002 * 20 = 40.04$. Thus $r_2 = y_2 - y_2^* = -.04$.

EXERCISE 8.1 Suppose that the third point (6,24) is observed, in addition to the two in Example 8.1. If least-squares regression through the origin is to be performed:

(i) Before doing any calculations, make a guess of where β^* will be.
(ii) Now, find β^*. Are you surprised?
(iii) Find the residuals at the three data points.

A Post-Mortem on Example 8.1

A bit of thought about the β^*-estimate in Example 8.1 might make it seem peculiar. We know that the true line passes through (0,0) so, if we ignored the point (20,40) and based the estimate of β **solely** on (1,3), we would choose a slope of 3. If we instead looked exclusively at (20,40), we would set a slope of 2. Yet when both points are considered, the resulting β^* is nowhere near the average of 2 and 3; instead, the choice $\beta^* = 2.002$ means that virtually all the emphasis is given to the point *further* from the origin.

Scrutiny of (8-6) makes clear that this disparity is not an accident. The slope of the line from (0,0) to (x_i,y_i)—which represents the best guess of β based on (x_i,y_i) alone—is y_i/x_i. This quantity enters (8-6) if we rewrite the latter as:

$$\beta^* = \sum_{i=1}^{n} x_i y_i \Big/ \sum_{i=1}^{n} x_i^2 = \sum_{i=1}^{n} x_i^2 \left(y_i / x_i\right) \Big/ \sum_{i=1}^{n} x_i^2 \qquad (8\text{-}7)$$

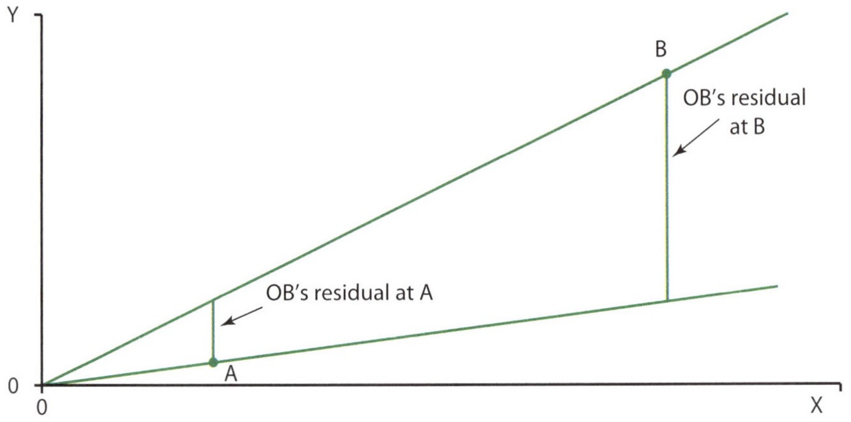

FIGURE 8.5 From the viewpoint of sum-of-squared residuals, those fitting points *A* and *B* with a line through the origin should be wary of displeasing *B*.

Because (y_i/x_i) gets weighted by x_i^2 in (8-7), it follows that the further a point falls from the vertical axis—whether to the right or the left—the more influential is its individual "guess" about β_0 on the value of β^*. The formula is responding to the fact that, if one gives a far-off point a slope estimate that it dislikes, it can retaliate by sending back an alarmingly large squared residual. (Figure 8.5)

But such unequal weighting, far from reflecting a mindless prejudice against close-in points, has a strong justification. It must be remembered that the ε_i's at *all* x-values have been assigned the same standard deviation. This assumption is less symmetric than it might at first appear. If, for instance $\sigma_\varepsilon = 1$ in the present situation, then what is a 95% confidence interval for the true y-value at $x = x_1$? It extends roughly two standard deviations in each direction from the observed y-value of 3 at (x_1, y_1) Therefore, it goes from about $3 - 2 * 1$ to $3 + 2 * 1$, namely, from 1 to 5. The analogous interval at $x = 20$ ranges from $40 - 2$ to $40 + 2$, or from 38 to 42. It follows that the *percentage* measurement error associated with the y-value at $x_1 = 1$—with the width of the confidence interval about 67% of the observed value of three—is much greater than that at $x_2 = 20$ (where it is roughly 5% of 40). It follows that the slope estimate based on (1,3) alone is subject to something like 67% uncertainty (1 to 5), while that based on (20,40) alone sustains 5% uncertainty ($38/20 = 1.9$ to $42/20 = 2.1$). Thus, (20,40) is considerably more reliable for the purpose of slope estimation than is (1,3). That the least-squares procedure takes account of this difference is not a drawback but a strength.

To be sure, people are sometimes unhappy about giving greater weight to distant points. They fear that such points are especially prone to inaccurate measurement, or that the relationship between y and x at the outer values might differ from within the "mainstream" of the data. But that concern is really about the viability of model (8-4)—which assumes that the same slope and error distribution prevails at all x-values—rather than about a least-squares technique that is contingent on the validity of (8-4). Within the framework for which it was intended, least-squares is a useful and defensible procedure.

Why Least Squares?

Minimizing squared vertical misses seems a reasonable goal in estimating the slope of the line that generated the data, but, as noted, one could come up with other reasonable goals. One could minimize (say) the sum of absolute residuals, or the sum of the fourth powers of the residuals. Why favor β^* over other estimates of β_0?

Actually, there are several reasons.

β^* is an **unbiased** estimate of β_0, meaning that $E(\beta^*) = \beta_0$. (8-8)

The proof of (8-8) appears at the website.

Actually β^* is not just unbiased, but is the "best linear unbiased estimate" of β_0, meaning that it suffers the lowest mean-squared estimation error in a broad class of candidates for estimating β_0.

Moreover, β_0 has another property that we discussed in Chapter 3.

β^* is the **maximum likelihood estimate** of β_0 (8-9)

The proof of (8-9) also appears at the website.

EXAMPLE 8.2

A Second Opinion about the Slope

Suppose that it is proposed to estimate β_0 in (8-4) by $\hat{\beta}$ which follows: $\hat{\beta} = \bar{Y} / \bar{X}$ where $\bar{Y}$ and $\bar{X}$ are, respectively, the averages of the y and x coordinates for the (x_i, y_i)'s.

(i) Is $\hat{\beta}$ an unbiased estimate of β_0?
(ii) Is $\hat{\beta}$ the maximum likelihood estimate of β_0?

Solution

(i) Because $\bar{Y} = \sum_{i=1}^{n} Y_i / n$ and $\bar{X} = \sum_{i=1}^{n} x_i / n,$ we can write:

$$\hat{\beta} = \bar{Y}/\bar{X} = \sum_{i=1}^{n} Y_i \Big/ \sum_{i=1}^{n} x_i$$

(Here the x_i's are known constants while the Y_i's are random variables because they include the ε_i's.)

Under (8-4), $E(Y_i) = \beta_0 x_i$, which means that:

$$E(\hat{\beta}) = E\left(\sum_{i=1}^{n} Y_i / \sum_{i=1}^{n} x_i\right) = \beta_0 \sum_{i=1}^{n} x_i / \sum_{i=1}^{n} x_i = \beta_0$$

Thus, $\hat{\beta}$ is an unbiased estimate of β_0. (We never said that β^* was the only unbiased estimate of β_0.)

(ii) In a word, no. We have already shown that β^*—which is different from $\hat{\beta}$ except in a rare coincidence—is the maximum likelihood estimate of β_0. (No: there can't be two maximum likelihood estimates, because we minimized a quadratic function to get the MLE, and that circumstance yields a unique solution: β^*)

EXERCISE 8.2 Find the variance of $\hat{\beta}$ in terms of σ_ε^2.

Slope Estimation Error With Least Squares

In Chapter 3, we spoke of the sampling error $W = \bar{X} - \mu$ that accompanied using $\bar{X}$ to estimate μ. Can we now speak about $W_\beta = \beta^* - \beta_0$, the ***slope estimation error*** that arises when β^* is used to estimate β_0? Yes we can, and we can say some fairly specific things. That is another advantage of working with the least-squares slope estimate.

First of all, we know that $W_\beta = \beta^* - \beta_0$ has mean zero, because $E(\beta^*) = \beta_0$. As for the variance of W_β, we can write:

$$W_\beta = \beta^* - \beta_0 = \left(\sum_{i=1}^{n} x_i Y_i / \sum_{i=1}^{n} x_i^2\right) - \beta_0 \tag{8-10}$$

All the quantities on the right-hand side of (8-10) are constants except the Y_i's. We note that W_β is of the form $W_\beta = \sum_{i=1}^{n} c_i Y_i - \beta_0$ (where $c_i = x_i \sum_{i=1}^{n} x_i^2$), and thus we can write:

$$\sigma^2(W_\beta) = \sum_{i=1}^{n} c_i^2 \sigma^2(Y_i) \tag{8-11}$$

Because $\sigma^2(Y_i) = \sigma_\varepsilon^2$, we have $\sigma^2(W) = \sigma_\varepsilon^2 \sum_{i=1}^{n} c_i^2$. But note that

$$\sum_{i=1}^{n} c_i^2 = \left(1/\sum_{i=1}^{n} x_i^2\right)^2 \sum_{i=1}^{n} x_i^2 = 1/\sum_{i=1}^{n} x_i^2$$

It follows that (8-11) can be written as:

$$\sigma^2(W_\beta) = \sigma_\varepsilon^2 / \sum_{i=1}^{n} x_i^2$$

If $W_\beta = \beta^* - \beta_0$, then:

$$\sigma^2(W_\beta) = \sigma_\varepsilon^2 / \sum_{i=1}^{n} x_i^2 \Rightarrow \sigma(W_\beta) = \sigma_\varepsilon / \sqrt{\sum_{i=1}^{n} x_i^2} \tag{8-12}$$

Moreover, we can readily go beyond specifying the mean and standard deviation of W_β. Because W_β is a linear combination of the Y_i's and the Y_i's are independent normal random variables, we know from Chapter 1 that W_β itself has a *normal* distribution. To put everything together, we have:

In regression through the origin, the **slope estimation error** $W_\beta = \beta^* - \beta_0$ that arises when β_0 is estimated by β^* follows a zero-mean ***normal*** distribution with standard deviation

$$\sigma(W_\beta) = \sigma_\varepsilon / \sqrt{\sum_{i=1}^{n} x_i^2} \tag{8-13}$$

It is interesting that $\sigma(W_\beta)$ does not depend directly on the number of data points. For example, $\sum_{i=1}^{n} x_i^2$ in (8-13) can be just as large because of a few points far from the origin as from many data points close to the origin. But that circumstance should not upset us, for we saw in the post-mortem to Example 8.1 that points further from the origin are subject to less random variability in percentage terms than those closer to zero. For that reason, we might learn as much about β_0 from a few "stable" points with high x_i's as from far-more-numerous "volatile" points near the origin.

EXAMPLE 8.3

Same Two Points

Suppose that we have the two points (1,3) and (20,40) from Example 8.1, and know that $\sigma_\varepsilon = 1$ in the homoscedastic model $Y_i = \beta_0 x_i + \varepsilon_i$ where ε_i is $N(0, \sigma_\varepsilon)$.

(i) What is the probability distribution for W_β?
(ii) Estimate the probability that β_0 is at least two.

Solution

(i) Because W_β is $N\left(0, \sigma_\varepsilon / \sqrt{\sum_{i=1}^{2} x_i^2}\right)$, the fact that $\sigma_\varepsilon = 1$ and $\sum_{i=1}^{2} x_i^2 = 401$ means that W_β is $N\left(0, 1/\sqrt{401}\right) = N(0,.050)$.
(ii) We found in Example 8.1 that $\beta^* = 803/401 = 2.002$. Thus, $\beta_0 \geq 2$ so long as $W_\beta = \beta^* - \beta_0 = 2.002 - \beta_0$ is .002 or less. Because W_β is $N(0,.05)$, we find from a normal table that $P(W_\beta \leq .002) = .516$. (We are tacitly assuming here that the observed β^* does not alter our distribution for W_β.)

EXERCISE 8.3 Suppose that we have the three points (1,3), (10,27) and (20,40). We continue to assume that $\sigma_\varepsilon = 1$ in the regression model $y_i = \beta_0 x_i + \varepsilon_i$. Answer the two questions in Example 8.3.

Estimating the Variance of the Slope Estimation Error

Recognizing the parallels between $S(W_\beta)$ here and the standard deviation of $W = \bar{X} - \mu$ for random sampling error, we might well be pleased with (8-12). As before, we can anticipate creating confidence intervals for the slope estimation error, and can move on to a probability distribution for β_0 akin to the one we developed for μ. We can specify, in other words, how much uncertainty the measurement errors (i.e., the ε_i's) have caused in the least-squares estimate of the slope. Next thing we know, there will be confidence intervals, and probabilistic statements about where β_0 is.

At this point, however, some readers will be reminded of our premature celebrations in Chapter 3. Not knowing the standard deviation σ in the population whose mean we sought, we could not apply formulas that were contingent on the value of σ. While we eventually overcame the difficulties with an approximation for σ (namely, S), we might ask: do we face a comparable difficulty here and, if so, how can we deal with it?

A procedure to estimate σ_ε^2 presents itself. Because each ε_i that we encounter has mean zero, its variance is the same as its mean-squared measurement error (Right?). Why not estimate the common σ_ε^2 in our homoscedastic model by observing the various squared deviations around the regression line and then taking their average V_ε^2? What we are discussing is averaging the squared residuals:

$$V_\varepsilon^2 = \sum_{i=1}^{n} r_i^2 / n \tag{8-14}$$

where $r_i = y_i - \beta^* x_i$ is the ith residual

The approximation here is that *the observed mean squared-spread around the regression line could serve as a good proxy for the true squared spread around the true line.*

But is V_ε^2 an unbiased estimator of σ_ε^2? If so, perhaps we can use V_ε^2 as a proxy for σ_ε^2 in an approximate version of (8-13). As show at the website, however, V_ε^2 suffers a downward bias:

$$E\left(V_\varepsilon^2\right) = \sigma_\varepsilon^2\left[(n-1)/n\right] \tag{8-15}$$

The intuitive explanation for the shortfall is similar to the one for the shortfall we faced when estimating σ^2 in statistical sampling (Chapter 3). If we have two points with ε_i's that are both positive, then that circumstance will push β^* up above β_0, meaning that the regression line will be artificially close to the data points. (Figure 8.6) Thus, the squared residuals that are observed will be less than the true squared

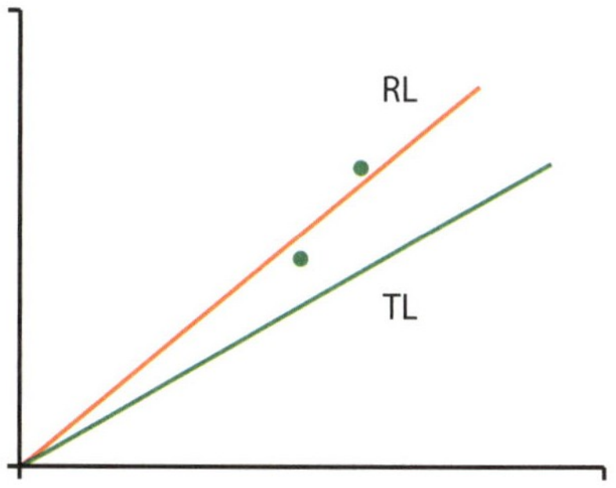

FIGURE 8.6 When random errors push the two observed points above the true line TL, the regression line RL is likewise higher so the spread around RL is less than around TL.

deviations around the true line. A similar phenomenon arises when both ε_i's are negative. Even though the calculated V_ε^2 will not always be below σ_ε^2, it will on average be so. Thus, V_ε^2 is subject to downward bias.

Under (8-14), the downward bias is tied to the factor $(n-1)/n$, which approaches one as n gets larger (i.e., the bias diminishes). That pattern makes intuitive sense: With lots of data points both above and below the true line, the regression line is less likely to "pander" to individual points.

As in estimating σ_x^2 in random sampling, the bias in V_ε^2 can be eliminated by a simple linear adjustment. Under (8-15), an unbiased estimate of σ_ε^2 is simply V_ε^2 multiplied by the correction factor $n/(n-1)$:

If $Y = \beta_0 x + \varepsilon$ where is $N(0,\sigma_\varepsilon)$, then S_ε^2, an unbiased estimator of σ_ε^2, is given by:

$$S_\varepsilon^2 = \frac{\sum_{i=1}^{n} r_i^2}{n-1} \tag{8-16}$$

where $r_i^2 = (y_i - \beta^* x_i)^2$ and β^* is the least-squares estimate of β_0 under regression through the origin with the n data points (x_i, y_i) for $i = (1, \dots, n)$.

EXAMPLE 8.4

Speed Kills?

Suppose that, among items produced by a machine, the fraction that are defective varies linearly with the speed at which the machine is run. (Note that the defect rate must be zero when the speed is zero: if nothing is produced, how can anything be defective?) To get more specific about the relationship, regression through the origin might be appropriate here.

If Y is the fraction defective and x the speed, the linear formula relating Y_i to x_i is:

$$Y_i = \beta_0 x_i + \varepsilon_i$$

The idea is that the fraction of defective items at speed x_i is subject to some random variation from one run to another, and that ε_i captures that variation Suppose that runs at nine different speeds generate the following data:

Speed	Fraction Defective
1	.029
1.25	.04
1.5	.05
1.75	.057
2	.062
2.25	.072
2.5	.077
2.75	.085
3	.093

(i) What is the least-squares estimate of β_0 given these data?
(ii) Assuming homoscedasticity and zero-mean normality for the ε_i's, what is an unbiased estimate of σ_ε^2?

Solution

(i) Direct application of (8-6) yields $\beta^* = \sum_{i=1}^{n} x_i y_i / \sum_{i=1}^{n} x_i^2 = .0313$
(ii) A chart of residuals is:

Speed	Actual Defect Rate	Defect Rate at Regression Line*	Residual
1	.029	.0313	–.0023
1.25	.04	.0392	.0008
1.5	.05	.047	.003
1.75	.057	.0548	.0022
2	.062	.0626	–.0006
2.25	.072	.0705	.0015
2.5	.077	.0783	–.0013
2.75	.085	.0861	–.0011
3	.093	.0939	–.0009

* Based on the estimate $y_i = .0313x_i$

An unbiased estimator of σ_ε^2 under (8-16) is $S_\varepsilon^2 = \sum_{i=1}^{9} r_i^2/8 = .00000322$. The corresponding estimate S_ε of the standard deviation of ε is $\sqrt{.00000322} = .00179$.

EXERCISE 8.4 Though it's not exactly correct (as we will soon see), suppose that one could approximate σ_ε by S_ε in (8-16) and then proceed with the relevant normal distribution identified in (8-13). What would be 95% and 99% confidence intervals for β_0 in Example 8.4 under the assumptions of the regression analysis?

A Probability Distribution for Slope Estimation Error

Under (8-7), our slope estimate β^* is a linear combination of independent random variables (the Y_i's). When n is sufficiently large, therefore, we can invoke the Central Limit Theorem to suggest that β^* and thus $W_\beta = \beta^* - \beta_0$ has a normal probability distribution. We have indicated that $E(W_\beta) = 0$, that $\sigma(W_\beta) = \sigma_\varepsilon / \sum_{i=1}^{n} x_i^2$, and that $\sigma(W_\beta)$ can be approximated by:

$$S_W = S_\varepsilon / \sum_{i=1}^{n} x_i^2 = \sqrt{r_i^2 / \left((n-1) * \sum_{i=1}^{n} x_i^2\right)}$$

For sufficiently large n, S_ε and σ_ε are close enough together that we can rewrite (8-13) as:

If $y = \beta_0 x + \varepsilon$ where is $N(0, \sigma_\varepsilon)$, then for sufficiently large n:

$$W_\beta = \beta^* - \beta_0 \approx N\left(0, \sqrt{r_i^2 / \left((n-1) * \sum_{i=1}^{n} x_i^2\right)}\right) \quad \text{(8-17)}$$

As usual, the question arises of what we mean by "sufficiently large?" For usual purposes, the approximation is very good when the number of data points is at least (say) two dozen, and quite good when the number of data points is at least about 15. We will provide evidence on this subject shortly.

A key assumption in making inferences about β_0 is *that the probability distribution of slope estimation error prior to the determination of β^* is the same as the distribution once β^* is known*. Suppose that, based on n and σ_ε, there was a 90% chance that W_β would fall between 5 and –5 before we knew β^*. Then it is assumed that there is still a 90% once chance we learn (say) that $\beta^* = 8$. And if the probability is 90% that β_0 is within five of $\beta^* = 8$, then there is a 90% chance that β_0 falls the range from three to 13.

Indeed, given both β^* and a probability distribution for W_β, we can speak of a probability distribution for where β_0 is located in the same sense that we previously spoke of a probability distribution for μ. (If your memory is hazy, please review our discussion on this last topic in Chapter 3.) More specifically, we can make the approximation that:

If $y = \beta_0 x + \varepsilon$ where ε is $N(0, \sigma_\varepsilon)$, then for sufficiently large n:

$$\beta_0 \approx N\left(\beta^*, \sqrt{r_i^2 / \left((n-1) * \sum_{i=1}^{n} x_i^2\right)}\right) \quad (8\text{-}18)$$

With (8-18) in hand, we can make some familiar statements about confidence intervals:

If $Y = \beta_0 x + \varepsilon$ where ε is $N(0, \sigma_\varepsilon)$, then for sufficiently large n:

- A 95% confidence interval for β_0 extends from about $\beta^* - 1.96S_\beta$ to $\beta^* + 1.96S_\beta$
- A 99% confidence interval for β_0 extends from about $\beta^* - 2.58S_\beta$ to $\beta^* + 2.58S_\beta$
- A 90% confidence interval for β_0 extends from about $\beta^* - 1.64S_\beta$ to $\beta^* + 1.64S_\beta$

$$\text{where } S_\beta = \sqrt{r_i^2 / \left((n-1) * \sum_{i=1}^{n} x_i^2\right)} \quad (8\text{-}19)$$

We can also use (8-19) to estimate the probability that β_0 falls in a particular range.

EXAMPLE 8.5

Sixty Minute

In a usual linear regression analysis through the origin with 60 data points and $\Sigma x_i^2 = 100$, we find that $\beta^* = 10$ and the sum of squared residuals is 1180. Given this information:

(i) What is an approximate 95% confidence interval for β_0?
(ii) What is the approximate probability that $\beta_0 > 11$?

Solution

(i) With this information, we can determine that $S_\beta = \sqrt{r_i^2 / \left((n-1) * \sum_{i=1}^{n} x_i^2\right)} = \sqrt{1180/(59 * 100)} = 0.447$.
Therefore, the approximate 95% CI for β_0 is $(\beta^* - 1.96\, S_\beta, \beta^* + 1.96\, S_\beta) = (9.12, 10.88)$

(ii) If $\beta_0 \approx N(\beta^*, s_\beta) = N(10, 0.447)$, then the normal table reveals that $P(\beta_0 > 11) \approx .0126$.

EXERCISE 8.5 In the situation of Example 8.5, what is the probability that estimated slope is within 5% of the true slope (i.e., that β^* is between $.95\beta_0$ and $1.05\beta_0$)? (Hint: For what range of β_0 would this event happen?)

Slope Estimation Error When n is Small

When n is small, we have no right to invoke the Central Limit Theorem to assume a normal probability distribution for W_β. But because of the special circumstance that the ε_i's are all zero-mean normal, we have access to the distribution for W_β even when data points are few. In Chapter 3, we encountered a random variable that was zero-mean normal, and we divided it by an unbiased estimate S of its standard deviation σ. Readers might recall that the ratio had a t-distribution and, if the estimate was based on n data points, the number of degrees of freedom was $n - 1$. But that is precisely the circumstance in which we find ourselves now.

It can be shown that:

Under the assumptions we have made about regression through the origin, the ratio t_β defined by:

$$t_\beta = W_\beta / S_\beta = \left(\beta^* - \beta_0\right) / \sqrt{r_i^2 / \left((n-1) * \sum_{i=1}^{n} x_i^2\right)} \tag{8-20}$$

has a t-distribution with $n - 1$ degrees of freedom.

A t-distribution with few degrees of freedom, you might recall, is zero-mean symmetric and has "fatter tails" than the distribution $N(0,1)$. As the number of degrees of freedom grows, the distribution gets closer and closer to $N(0,1)$.

It follows from (8-20) that $W_\beta = t_\beta S_\beta$. Because $\beta_0 = \beta^* - W_\beta$, we can proceed as in Chapter 3 to state that:

In usual linear regression through the origin with n data points:

- The A% Confidence interval for β_0 extends from

$$\beta^* - [\gamma_{A,n-1}]S_\beta \text{ to } \beta^* + [\gamma_{A,n-1}]S_\beta \tag{8-21}$$

where $\gamma_{A,n-1}$ is the **[(100 + A)/2] %ile** of the t-distribution with $n - 1$ degrees of freedom.

- We can convert a probabilistic question about β_0 to an equivalent probabilistic question about a t-distributed random variable with $n - 1$ degrees of freedom.

EXAMPLE 8.6

More Speed

Given the data in Example 8.4 (Speed Kills?):

(i) What is a 95% confidence interval for β_0?
(ii) What is the probability that $\beta_0 > .031$?

Solution

(i) Because $n = 9$, the 95% CI for β_0 extends from $\beta^* - \gamma_{95,8}\, S_\beta$ to $\beta^* + \gamma_{95,8}\, S_\beta$. For the data in Example 8.4, $\beta^* = .0313$. Also, $\sum_{i=1}^{10} x_i^2 = 39.75$, and $\sum_{i=1}^{10} r_i^2/8 = .00000322$, which means that $S_\beta = .00028$ under (8-19). From an electronic t-table, we find that $\gamma_{95,8} = 2.306$. Therefore, the 95% confidence interval for β_0 has a lower limit of $.0313 - 2.306 * .00028 = .0307$. The upper limit is .0319. Thus, the 95% CI for β_0 is (.0307,.0319). That is a very narrow range, indicating that we believe there was very little slope estimation error in this setting.

Note that $\gamma_{95,8}$ calls for finding the 97.5%ile for the t-distribution, not the 95th %ile. The author did not want to use $(A + 100)/2$ in a subscript, which is why he used the notation $\gamma_{A,n-1}$. But do not make the mistake of using the Ath percentile!

(ii) $P(\beta_0 > .031) = P([\beta^* - \beta_0]/S_\beta = (.0313 - \beta_0)/.00028 < .0003/.00028 = 1.07)$. Because $t_\beta = [\beta^* - \beta_0]/S_\beta$ is t-distributed with 8 degrees of freedom, the electronic table informs us that $P(t_\beta < 1.07) = .842$.

EXERCISE 8.6

Find the 99% confidence interval for β_0 in Example 8.4. How does that interval compare with the one that you calculated in Exercise 8.4 under a normal approximation?

More about the Normal Approximation for Slope Estimation Error

Now that we know the actual distribution of W_β at different values of n, we can assess when n is "sufficiently large" that the normal approximation holds. We see from Table 8.1 that the approximation does not fare especially well at $n = 6$: the chance of falling within three familiar confidence intervals is several percentage points lower under the correct t-distribution than under the normality approximation. For example, the 95% confidence interval under normality is actually a 89% confidence interval when $n = 6$. (Equivalently, the chance of falling outside the 1.96-standard deviation range is about twice as high under a t-distribution with five degrees of freedom as under a normal curve (11% vs. 5%). Once $n = 15$, the deviation between normal and t-distribution has dropped by about 2/3 compared to $n = 6$; once $n = 24$, the deviation has dropped by about 5/6. At $n = 50$, the deviation is a fraction of one percentage point for all three confidence intervals.

TABLE 8.1 A Performance Evaluation for the Normal Approximation about β_0 in Regression Through the Origin, with Various Numbers of Data Points

	Probability that β_0 Falls There:				
		Actual* For			
Interval	Normal Appx	$n = 6$	$n = 15$	$n = 24$	$n = 50$
$(\beta^* - 1.64S_\beta, \beta^* + 1.64S_\beta)$	.899	.838	.876	.886	.892
$(\beta^* - 1.96S_\beta, \beta^* + 1.96S_\beta)$	.950	.893	.930	.938	.944
$(\beta^* - 2.58S_\beta, \beta^* + 2.58S_\beta)$	.990	.951	.978	.984	.987

* Actual is based on t-distribution with $n - 1$ degrees of freedom

It is a matter of judgment when the normal approximation is "close enough." The table makes clear that there is no magic n-value at which the approximation shifts abruptly from horrendous to sublime. Actually, there is no need to use the normal approximation at any value of n, given that the t-distribution is readily accessible and is exactly correct under the assumptions of the regression. We speak of normality largely because of its intuitive familiarity, and because people tend to assume normality in connection with certain statistics that we will introduce shortly.

But how does linear regression change once we longer require the line to pass through the origin? That is the subject of the next section.

8.3 The General Linear Regression Model

In this section, we present several theoretical results for the general linear regression model. While offer a few numerical exercises to illustrate these results, we postpone until later sections any attempt to apply regression analysis in real-life situations. Thus, the applications-oriented reader should not fear based on this section that we have lost interest in applications.

The more general linear relationship between Y and x takes the form:

$$Y = \alpha_0 + \beta_0 x + \varepsilon \qquad (8\text{-}22)$$

where ε is $N(0, \sigma_\varepsilon)$ and the ε's for different data points are independent

Equation (8-22) implies that the line relating Y to x need not pass through the origin, a modification that often increases the realism of the analysis. In the case of sales vs. advertising, for example, it is plausible that advertising stimulates more sales, but hard to imagine that sales would disappear in the absence of advertising. Allowing a baseline level for Y when $x = 0$, namely α_0, avoids an assumption of zero sales that would be hard to defend.

We should say something more about the ε-term. Although ε is in theory zero-mean measurement error, in practice it has a wider interpretation. It is meant to take account of various factors—many of them unquantifiable—that get no explicit consideration in the regression model. For example, a formula to estimate the resale price of a used car—for a given year, make, and model—might consider the vehicle's mileage as a key explanatory variable. But it can rarely consider the relative bargain skills of the buyer and seller. If, however, one is willing to treat the difference in skills as creating zero-mean fluctuations of resale prices around a baseline average value—which is tied to a vehicle's objective characteristics—then a regression model like (8-22) can defensibly be used to estimate that baseline value.

The method of least-squares can be extended directly to this new setting. For any member of the family of lines $y = \alpha + \beta x$, its overall **vertical** discrepancy $L(\alpha,\beta)$ from the actual (x_i,y_i)'s could be reflected by the expression:

$$L(\alpha,\beta) = \sum_{i=1}^{n}\left(y_i - \left(\alpha + \beta x_i\right)\right)^2$$

As before, this formula takes difference between the observed y_i and the y-value corresponding to $x = x_i$ along the line $y = \alpha + \beta x$, namely, $\alpha + \beta x_i$. Then it squares the differences and sums up over the n data points.

How do we determine the least-squares slope and intercept estimates? Those readers who have studied multivariate Calculus know that the technique for finding maxima and minima in several variables is the direct generalization of the method used in one-variable Calculus: instead of setting a single derivative equal to zero, we set several individual "partial derivatives" equal to zero. In this setting, we work with two partial derivatives, and reach the following result:

When $y = \alpha_0 + \beta_0 x + \varepsilon$, the least-squares estimates of the slope and intercept follow:

$$\beta^* = \frac{\sum_{i=1}^{n} y_i\left(x_i - \bar{X}\right)}{\sum_{i=1}^{n}\left(x_i - \bar{X}\right)^2} \quad (8\text{-}23)$$

$$\alpha^* = \bar{Y} - \beta^*\bar{X} \quad (8\text{-}24)$$

where $\bar{Y} = \sum_{i=1}^{n} y_i / n$ and $\bar{X} = \sum_{i=1}^{n} x_i / n$

Equation (8-23) yields a slope estimate that is superficially similar to the slope estimate in (8-6) in regression through the origin: the only change is that $\bar{X}$ rather than zero is subtracted from the x_i's. Of course, the numerical slope estimates under the two approaches can differ drastically. Under (8-23), those points further from the middle of the data in their x-coordinates have greater influence on the slope estimate than points close to $\bar{X}$. (Indeed, if $x_i = \bar{X}$ for some i, then y_i has no effect whatever under (8-23) on the least-squares slope estimate.) The explanation for this pattern is similar to why points far from $x = 0$ in regression through the origin have greater weight in slope estimation than closer-in points.

EXERCISE 8.7 True or false: If $\bar{X} = 0$, the least-squares line in regression through the origin must be the same as the regression line that is not constrained to pass through the origin. Briefly but specifically discuss.

EXAMPLE 8.7

Three Points Determine a Line

Find the least-squares regression line for the points (1,4), (2,9), and (3,11).

Solution

Here $\bar{X} = 2$, and β^* is given by:

$$\beta^* = \frac{\sum_{i=1}^{3} y_i(x_i - 2)}{\sum_{i=1}^{3}(x_i - 2)^2}$$

Using the x_i's and y_i's here, $\sum_{i=1}^{3} y_i(x_i - 2) = 4 * (-1) + 9 * 0 + 1 * 11 = 7$, while $\sum_{i=1}^{3}(x_i - 2)^2 = 1^2 + 0^2 + 1^2 = 2$. Thus, $\beta^* = 3.5$.

Moreover, $\bar{Y} = (4 + 9 + 11)/3 = 8$, meaning that $\alpha^* = \bar{Y} - \beta^*\bar{X} = 8 - 3.5 * (2) = 1$. The least-squares regression line is therefore $y = 1 + 3.5x$.

EXERCISE 8.8 Suppose that the least-squares regression line for the points (5, 8), (12, y_2), and (19, 36) is $y = 14 + 2x$. Is this information enough to determine the value of y_2? If so, determine it; if not, specify what further information you would need to find y_2.

Other Expressions for the Least-Squares Slope Estimate

Two other formulations of (8-23) are sometimes of value in interpreting the expression for β^*.

$$\beta^* = \widehat{\rho_{xy}}\left(\frac{\widehat{\sigma_y}}{\widehat{\sigma_x}}\right) \tag{8-25}$$

where:

$\widehat{\rho_{xy}}$ is the correlation coefficient of the x_i's and y_i's

$\widehat{\sigma_x} = \sqrt{(x_i - \bar{X})^2/n}$ the standard deviation of the x_i's

$\widehat{\sigma_y} = \sqrt{(y_i - \bar{Y})^2/n}$ the standard deviation of the observed y_i's

The derivation of (8-25) appears at the website. Note that (8-25) implies that $\beta^* = 0$ when x and y are uncorrelated in the data set. We will soon find (8-25) of practical value, when we consider the capital asset pricing model in Finance.

EXERCISE 8.9 Show that the formula for β^* (8-24) can also be written as:

$$\beta^* = \frac{\sum_{i=1}^{n}(y_i - \bar{Y})(x_i - \bar{X})}{\sum_{i=1}^{n}(x_i - \bar{X})^2} \quad \text{(8-26)}$$

The least-squares procedure that leads to (8-23) and (8-24) is often called **ordinary least squares (OLS)** regression analysis. The reason for this wording is to distinguish this standard procedure from weighted least squares regression modeling, in which avoiding a large vertical "miss" at some data points is viewed as more important than doing so at others.

Some Truths about the Residuals.

In OLS regression, the ith residual r_i follows:

$$r_i = y_i - (\alpha^* + \beta^* x_i)$$

The sign of the residual indicates whether (x_i, y_i) is above or below the least squares regression line.

Certain properties of the residuals are of interest.

If r_i is the ith residual in OLS regression then:

$$\sum_{i=1}^{n} r_i = 0 \quad \text{(8-27)}$$

$$\sum_{i=1}^{n} r_i x_i = 0 \quad \text{(8-28)}$$

Showing that $\sum_{i=1}^{n} r_i = 0$ is a brief exercise in algebra. We write:

$$\sum_{i=1}^{n} r_i = \sum_{i=1}^{n}\left(y_i - \left(\alpha^* + \beta^* x_i\right)\right) = n\bar{Y} - n\alpha^* - n\beta^*\bar{X} = n\left(\bar{Y} - \alpha^* - \beta^*\bar{X}\right)$$

But because $\alpha^* = \bar{Y} - \beta^*\bar{X}$, we have $\bar{Y} - \alpha^* - \beta^*\bar{X} = 0$, and thus:

$$\sum_{i=1}^{n} r_i = 0$$

The derivation of (8-28) is somewhat lengthier, and appears at the website.

From these two rules about the residuals, we can establish another about $\widehat{\rho_{rx}}$, the coefficient of correlation between the residuals and values of the explanatory variables. Given (8-28) the mean of the product of the residuals and x-values (which equals $\sum_{i=1}^{n} x_i r_i / n$ is zero, while the mean of the residuals (which is $\sum_{i=1}^{n} r_i / n$ is also zero under (8-27). Therefore, (2-4) implies that the correlation coefficient $\widehat{\rho_{rx}}$ of x and z is necessarily zero.

In any OLS linear regression analysis, the coefficient of correlation $\widehat{\rho_{rx}}$ of the residuals and the explanatory variable **must be zero.** (8-29)

Statements (8-27)–(8-29) tell us something about the least-squares regression line. That the residuals average to zero means that the line cannot exceed all the data points; nor can it fall below all of them. That the correlation coefficient $\widehat{\rho_{rx}}$ is zero means that the residuals under OLS cannot systematically grow or shrink at either extreme of the x_i distribution. Even if, in reality, the standard deviation of ε (i.e., σ_ε) increases with x (i.e., if we face *heteroscedasticity* rather than the homoscedasticity that is assumed), that will not be apparent in the residuals around the least-squares line.

Incidentally, none of the properties in (8-27)–(8-29) need prevail in regression through the origin.

EXAMPLE 8.9

Zero for Three

For the OLS regression with the three points (1,4), (2,9), (3,11), show that the sum of residuals is zero.

Solution

We found in Example 8.7 that the OLS regression line for these points is $y = 1 + 3.5x$. The residual at $x = 1$ was therefore $4 - (1 + 3.5 * 1) = -0.5$. The residuals at $x = 2$ and 3 were, respectively, 1 and -0.5. Thus, $\sum_{i=1}^{n} r_i = 0$.

EXERCISE 8.10 Your turn. Show that $\sum_{i=1}^{n} x_i r_i$ and $\widehat{\rho_{rx}}$ are both zero in the situation of Example 8.9.

EXERCISE 8.11 Suppose that there are data points at $x = 1$, $x = 2$, and $x = 3$, and that $r_3 = 1$ in OLS regression. Given this information, what are r_1 and r_2? (Hint: You can write two linear equations in two unknowns.)

Some Properties of the Least-Squares Parameter Estimates

While there are no technical problems in transferring least-squares analysis from regression through the origin to more general linear regression, doing so will not be helpful unless we can make inferences about β_0 (and now α_0) that are similar to those we saw earlier. Fortunately, we lose none of the attractive properties of β^* when we allow the y-intercept of the regression line to differ from zero. As a start, it is not difficult to establish that *β^* is an unbiased estimate of β_0*, and that α^* is an unbiased estimate of α_0. We do so at the website.

It can further be shown that:

If $Y_i = \alpha_0 + \beta_0 x_i + \varepsilon_i$ where ε_i is $N(0,\sigma_\varepsilon)$ and different ε_i's are independent, then the least squares estimates α^* and β^* of, respectively, α_0 and β_0, are:

- the ***minimum-variance linear unbiased estimates*** of these quantities, and
- the ***maximum likelihood estimates*** of these quantities.

Under the circumstances, least-squares estimation appears just as useful in the general linear regression model as in the model with the intercept constrained to be zero. All the remains is considering whether the slope estimation error $W_\beta = \beta^* - \beta_0$ and its new counterpart, $W_\alpha = \alpha^* - \alpha_0$ have probability distributions that are tractable. Fortunately, they do.

Because the Y_i's are independent normal, it follows that β^* under (8-30)—as a linear combination of the Y_i's—must itself be normal. We already know that $E(\beta^*) = \beta_0$, and can directly show from (8-22) and (8-23) that $\sigma^2(\beta^*) = \sigma_\varepsilon^2 / \sum_{i=1}^{n} (x_i - \bar{X})^2$.

As for $W_\beta = \beta^* - \beta_0$, we know that $E(W_\beta) = \beta_0 - \beta_0 = 0$, and $\sigma^2(W_\beta) = \sigma^2(\beta^*)$. Putting it together, we have:

In the regression model $Y = \alpha_0 + \beta_0 x + \varepsilon$, the slope estimation error $W_\beta = \beta^* - \beta_0$ for the least-squares slope estimate β^* is normally distributed with mean zero and standard deviation

$$\sigma(W_\beta) = \sqrt{\sigma_\varepsilon^2 / \sum_{i=1}^{n} (x_i - \bar{X})^2} \qquad (8\text{-}30)$$

From this assessment, we can proceed as in Section 8.1 to speak of a probability distribution for the true slope β_0, given the observed value of β^* and the distribution of the estimation error W_β associated with β^*.

In the regression model $Y = \alpha_0 + \beta_0 x + \varepsilon$ with the n data points (x_i, y_i), we can treat the value of the unknown slope β_0 as a normally distributed random variable with mean β^* and standard deviation

$$\sqrt{\sigma_\varepsilon^2 / \sum_{i=1}^{n} (x_i - \bar{X})^2} \qquad (8\text{-}31)$$

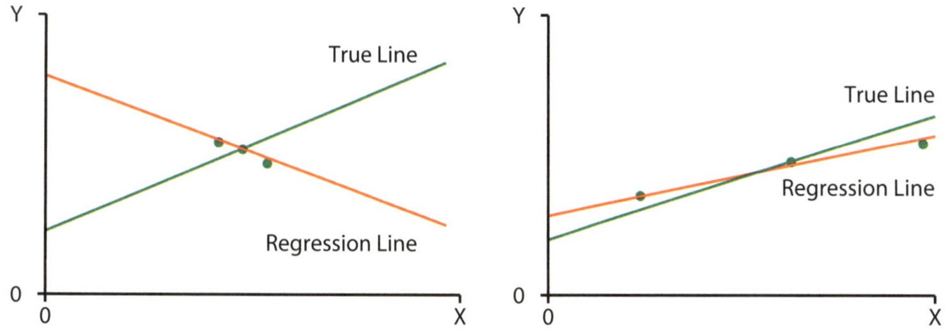

FIGURE 8.7 For a fixed number of data points, more spread in the x_i's is conducive to more accurate slope estimates.

We might reiterate something we first said in Chapter 3: some statisticians would not be comfortable with a statement like (8-31), which assigns to β_0 a probability distribution based on the observed value of β^* and the distribution of slope estimation error that inheres in β^*. This author's response to these concerns is the same as the one in Chapter 3.

We can rewrite the denominator $\sum_{i=1}^{n}(x_i - \bar{X})^2$ in the expression for $\sigma^2(W_\beta)$ as:

$$\sum_{i=1}^{n}(x_i - \bar{X})^2 = n\left(\frac{\sum_{i=1}^{n}(x_i - \bar{X})^2}{n}\right) = n\left(\widehat{\sigma_x}\right)^2$$

where $\left(\widehat{\sigma_x}\right)^2$ is defined as the mean-squared spread among the x_i's. Doing so reminds us that slope estimation error depends on both the number of data points (n) and the spread that they exhibit in the explanatory variable x. That dependence makes sense: as Figure 8.7 illustrates, having three data points close together—when the slope estimate is highly vulnerable to the caprices of random error ε—tends to yield less accurate slope estimates than three data points further apart. (In Figure 8.7, ε_i is the same in both graphs for $i = 1,2,3$.) Similarly, a few data points with high $\left(\widehat{\sigma_x}\right)^2$ can yield just as much accuracy as many data points with small $\left(\widehat{\sigma_x}\right)^2$.

What about the intercept estimate α^*? In parallel with the analysis about β^*, we can determine that:

$$E(\alpha^*) = \alpha_0$$

$$\sigma^2(\alpha^*) = \sigma_\varepsilon^2\left(\frac{1}{n} + \frac{(\bar{X})^2}{(n-1)\left(\widehat{\sigma_x}\right)^2}\right) \tag{8-32}$$

where $\left(\widehat{\sigma_x}\right)^2 = \sum_{i=1}^{n}(x_i - \bar{X})^2 / n$

And because α^* is a linear combination of the Y_i's (though a different one from β^*), α^* is also normally distributed. Putting everything together, we reach the following analogue of (8-31):

In the regression model $Y = \alpha_0 + \beta_0 x + \varepsilon$ with the n data points (x_i, y_i), we can treat the unknown intercept α_0 as a normally distributed random variable with mean α^* and standard deviation

$$\sigma(\alpha^*) = \sigma_\varepsilon \left(\frac{1}{n} + \frac{(\bar{X})^2}{(n-1)(\widehat{\sigma_x})^2} \right) \tag{8-33}$$

EXAMPLE 8.10

Hot News

Mendel recalls that there is a linear relationship between Fahrenheit and Centigrade temperatures, but he does not remember what it is. Suppose that he has two thermometers, one of which gives the exact Fahrenheit temperature while the other gives an estimate of the centigrade temperature that Mendel knows is subject to normal zero mean measurement error with some standard deviation σ_ε. (The errors for different readings are independent.) Unbeknownst to Mendel, $\sigma_\varepsilon = 0.5°$ Centigrade.

Mendel notes the temperatures on both thermometers at eight successive hours from 8 AM to 3 PM, when the Fahrenheit temperatures are as follows:

8 AM	56° F	12 Noon	68° F
9 AM	58° F	1 PM	72° F
10 AM	64° F	2 PM	69° F
11 AM	66° F	3 PM	66° F

He will perform an ordinary least squares linear regression that relates the observed centigrade temperature at these eight times to the exact Fahrenheit temperature

(i) What is the probability that his estimate of the slope of the centigrade/Fahrenheit relationship will be correct to within 5% of the true slope?

(ii) Will the probability be the same that his intercept estimate will be correct to within 5% of the true intercept?

(iii) Would you expect his slope estimation error and intercept estimation error to be correlated? If so, would you expect positive or negative correlation?

Solution

Obviously, this is a contrived problem. But it gives us an opportunity to think about some issues concerning the accuracy of linear regression results.

If x_i is the Farenheit temperature at the start of the ith hour he considers (i = 1,2,...,8) and y_i the observed Centigrade temperature, Mendel will estimate the slope β_0 in $Y = \alpha_0 + \beta_0 x + \varepsilon$ by β^*. We know from (8-30) that the slope estimation error that Mendel encounters will be $N\left(0, \sigma_\varepsilon^2 / \sum_{i=1}^{8}(x_i - \bar{X})^2\right)$. In the present situation, $\bar{X}$ = 67.88°F and $\sum_{i=1}^{8}(x_i - \bar{X})^2$ = 278.88, while $\sigma_\varepsilon^2 = (1/2)^2 = 1/4$. Therefore, W_β is $N(0, \sqrt{1/1111.5})$. = $N(0, .030)$.

The true Centigrade/Farenheit relationship is:

$$y = (5/9)(x - 32) = (5/9)x - 17.78.$$

(i) To get the slope estimate accurate to within 5% of the true slope, the absolute slope estimation error W_β must not exceed (5/9)(.05) = .028. Given that W_β is $N(0, .03)$, we find directly that $P(-.028 \le W_\beta \le .028) = .649$. In words, the chance is about two in three that the least squares slope estimate β^* will fall within .028 (or 5%) of the correct slope of 5/9.

(ii) Why would we think that? The distribution of W_α does depend on the x_i's, but not in the same way that W_β does. And the actual intercept is –17.78 and a 5% deviation from that number is different from a 5% deviation from the true slope of 5/9.

(iii) We know that $\alpha^* = \bar{Y} - \beta^*\bar{X}$. If measurement error yields a β^* is too high relative to β_0, this formula suggests that α^* will be too low relative to α_0 if $\bar{X} > 0$ and too high if $\bar{X} < 0$. Here $\bar{X} > 0$, so we might reasonably conjecture a negative correlation. True: if β^* is too high, then we might suspect that $\bar{Y}$ is also too high. But we will soon explore the issue further, and will confirm that our conjecture is correct.

EXERCISE 8.12 Work out the probability in the situation of Example 8.10 that the intercept estimate is within 5% of the true intercept. (Use (8-33).)

The Real World

As in regression through the origin, we confront the problem that the variance σ_ε^2 of the zero-mean error term is rarely known in advance. We handle this difficulty the same way we proceeded in Section 8.1: we first come up with an unbiased estimate of σ_ε^2 based on the residuals, and then use that estimate in setting a probability distribution for W_β and W_α.

With a great deal of algebraic patience, we can establish that:

S_ε^2, an unbiased estimate for σ_ε^2 when $Y = \alpha_0 + \beta_0 x + \varepsilon$ and ε is $N(0, \sigma_\varepsilon)$, follows:

$$S_\varepsilon^2 = \sum_{i=1}^{n} r_i^2 / (n - 2) \tag{8-34}$$

Algebra aside, it makes sense that we divide by $n - 2$ here rather than $n - 1$ (as we did in regression through the origin). If we have exactly two points (x_1,y_1) and (x_2,y_2), we can get a perfect fit to the data (and thus zero residuals) just by "connecting the dots," *regardless of the value of* σ_ε. It is thus comforting that the formula balks at offering a variance estimate when $n = 2$.

Slope Estimation Error With Few Data Points

As with regression through the origin, we can use S_ε^2 from (8-34) as an estimate for σ_ε^2 in the probability distributions for W_β and β_0 (and also W_α and α_0). When we have enough data points, we can continue to treat these variables as approximately normal in their distributions. For fewer data points, we can again turn to the *t*-distribution:

- For sufficiently large n, $W_\beta \approx N\left(0, \sqrt{S_\varepsilon^2 / \sum_{i=1}^{n} (x_i - \bar{X})^2}\right)$ under OLS regression on the linear model $Y = \alpha_0 + \beta_0 x + \varepsilon$, using S_ε^2 in (8-34).
- For any value of $n \geq 3$, $W_\beta / \sqrt{S_\varepsilon^2 / \sum_{i=1}^{n} (x_i - \bar{X})^2}$ follows a *t*-distribution with $n - 2$ degrees of freedom. (8-35)

Similarly:

- For sufficiently large n, $\beta_0 \approx N\left(\beta^*, \sqrt{S_\varepsilon^2 / \sum_{i=1}^{n} (x_i - \bar{X})^2}\right)$ under OLS regression on the linear model $Y = \alpha_0 + \beta_0 x + \varepsilon$, using S_ε^2 in (8-34).
- For any value of $n \geq 3$, $(\beta^* - \beta_0) / \sqrt{S_\varepsilon^2 / \sum_{i=1}^{n} (x_i - \bar{X})^2}$ follows a *t*-distribution with $n - 2$ degrees of freedom. (8-36)

With these statements, we can set confidence intervals for β_0 and estimate the probability that β_0 falls in any particular interval of interest. Analogous statements can be made about α_0.

Now we turn to some actual examples to see how these ideas can be applied. We assume that the reader has access to a statistical package that performs linear regressions on request, and that offers least-squares parameter estimates and confidence intervals for these parameters. The package will typically offer a good deal of additional information, but we will postpone discussion of such information until later in this chapter or the next chapter.

8.3 Some Linear Regression Analyses

EXAMPLE 8.11 **Delinquency Patterns in Philadelphia**

We consider first a regression analyses performed by this author some years ago. It has been argued that, in deciding whether to impose a prison sentence, a judge should consider not just what an offender has already done, but also what he might do if allowed to go free. This view has received some backing from the U.S. Supreme Court, which held in 1984 that some detention decisions about juvenile delinquents could be tied to projections of their future "dangerousness." Such a development inevitably gives rise to the question: to what extent is an offender's future criminal record foreshadowed by his existing one? We discuss a simple analysis on the subject.

The data we use were gathered at the University of Pennsylvania in what *Newsweek* described as "perhaps the most influential piece of criminal justice research in the last decade." (3/23/81.) The investigators assembled the juvenile police records of all males born in 1945 who lived in Philadelphia between ages 10 and 18. The published report about these youths paid particular attention to the *chronic offenders*, meaning those boys arrested at least five times. Because such boys amassed a grossly disproportionate share of all the juvenile arrests, they are a natural focus for any discussion of punishment policy.

Table 8.2 below, the basis of the present exercise, concerns all 164 youths who entered the chronic offender category (i.e. were arrested a fifth time) between the ages 15 and 16.5. The boys were broken into 16 groups of roughly equal size based on their individual rates of arrests *per year* early in their delinquent careers. Then their annual arrest rates were computed for their remaining juvenile careers (i.e, through age 18), with the fifth arrest treated as the dividing point between "early" and "late." All the statistics in the table refer to arrest rates while the offender was "on the streets;" periods when he was in detention and thus unable to commit crimes were deleted from the time line.

Using the table, answer the following two questions:

(i) How strong is the evidence that the early arrest rate x, and the later rate, Y, are positively related?

(ii) If a relationship exists between Y and x, how can one quantify it?

TABLE 8.2 Annual Arrest Rates of Juvenile "Chronic Offenders" Born in Philadelphia in 1945

Group	Early-Career Arrest Rate (x_i)	Later Arrest Rate (y_i)
1	.36	1.18
2	.43	1.06
3	.55	.97
4	.58	1.09
5	.80	1.03
6	.90	.94
7	1.02	1.55
8	1.08	1.15
9	1.20	1.39

Group	Early-Career Arrest Rate (x_i)	Later Arrest Rate (y_i)
10	1.33	1.38
11	1.60	1.32
12	1.86	1.18
13	1.98	1.83
14	3.32	2.21
15	4.22	2.37
16	6.51	2.88

(Arrest rates reflect number of bookings per year of' "street time.") These numbers are closely based on records of the Philadelphia Police Department. The fifth arrest is the dividing line between the early and later juvenile "criminal career."

Solution

One reasonable approach might be to model the Y-to-x relationship by the standard linear regression model: $Y = \alpha_0 + \beta_0 x + \varepsilon$. The ε-term would presumably reflect random variations in luck in avoiding arrest for crimes committed, as well as factors about individual offenders that were not included in the model.

Ordinary least-squares (OLS) regression analysis on the data in Table 8.2 yields the linear approximation:

$$Y = .908 + .325x$$

The regression printout reveals that the 95% confidence interval for β_0—which is tied to the t-distribution with $n - 2 = 14$ degrees of freedom under (8-35)—extends from .260 to .389. The 95% confidence interval for α_0 extends from .755 to 1.061. As for the specific questions:

(i) The correlation coefficient of x and y (which we can obtain without the need for regression analysis) is .944, which suggests a strong correlation. And the fact that the 95% confidence interval for β_0 does not include zero (and extends only narrowly around the β^*-value of .325) suggests that it would be implausible to treat the upward trend of the regression line with x as only a coincidence.

(ii) One can quantify the relationship via the OLS regression equation: the slope estimate of .325 means that, as the early arrest rate increases by one arrest per year, the later rate increases on average by about 1/3 per year. Looked at another way, the slope estimate of .325, means that two chronic offenders whose early arrest rates differed by some amount Δx would only differ on average by about $.325\Delta x$ in their later arrest rates. Thus, while the early behavior is not irrelevant to future behavior, earlier differences are on average diminished by something like a factor of three.

Indeed, the regression line (like the data themselves) suggest that something of a *regression to the mean* phenomenon is at work. Evaluating $y = .908 + .325x$ at the various values of x in Table 8.2 makes clear that those with the lowest early arrest rates tend to go up in their later rates (towards to overall average), while those with the highest rates tend to go down. That circumstance has some resemblance to the very reason that regression analysis is called regression analysis. As we noted, one of the earliest applications of the technique was in coming up with a rule relating the heights of fathers and sons.

EXERCISE 8.13

(i) Using the data in Table 8-2, find S_{α}^2, the unbiased estimate of σ_{α}^2. (Hint: Use (8-32) and (8-33).

(ii) Then, using the result in Example 8.11 and a normality approximation, find a 95% confidence interval for α_0.

EXAMPLE 8.12

Fare Game

It makes intuitive sense that, all factors being equal, more passengers will fly when fares are lower than when fares are higher. But at what speed does demand grow as fares decline?

Using a reasonable data set, conduct a reasonable linear regression analysis about "fare sensitivity" in the United States. Offer approximate answers to the questions:

(i) Is there statistically significant evidence that changes in average airfare change the passenger "participation rate" on an air route?

(ii) If so, how large is the estimated effect? Does it seem likely than a fare decrease "pays for itself" because it so greatly increases the number of passengers that total revenue goes up?

(iii) Given the uncertainty in the slope estimate, what is the range of possible effects for each dollar increase (or decrease) of fare?

Solution

The phrase "all factors being equal" in the first sentence of above is far from a formality. The decision to fly rather than drive obviously depends on the distance from origin to destination; for short trips, flying may save little time and be less convenient than traveling by road. Moreover, more passengers presumably travel between New York and Chicago—with a combined metropolitan population well over 20 million—than between Jackson, Mississippi and Lubbock, Texas, which together have fewer than one million residents. Unless one can sensibly deal with such factors, there seems no chance of getting a reliable estimate of the fare sensitivity of potential air travelers.

But the enterprise is not hopeless. The *Consumer Air Fares* report prepared by the US Department of Transportation offers quarterly data about the number of air passengers between various pairs of American cities, as well as information about the average fare paid by these passengers. In the first quarter of 2011, for example, an average of 461 passengers per day made the 612-mile air journey between Nashville, Tennessee and Tampa, Florida. One can imagine a strategy for using linear regression analysis to relate average airfare to passenger traffic by making comparisons across different air routes. The analysis could have the following characteristics:

- To avoid comparisons involving different travel distances, one could consider only trips of approximately the same length. There are, for example, 36 US air routes in the *Consumer Air Fares* report that are between 600 and 650 miles apart and have nonstop flights. Furthermore, one could use passenger traffic from the same year and quarter to avoid difficulties that could arise when data from different seasons are compared, or when data from different years (with different background economic conditions) are contrasted.
- To adjust for population differences in the cities on different routes, one could calculate a "participation rate," which is defined as the number of passengers per day divided by the sum of the metropolitan-area populations of the two cities on the route. That rate is, roughly speaking, the probability that a randomly-selected person from one of these cities flies the route on a randomly-chosen day. (We say "roughly speaking" because some people might fly the route twice in one day, in one direction in the morning and the other direction at night, and some people flying between A and B might live in neither city.)

Here we perform such an OLS regression analysis using data for the months April-June 2011 on US routes between 600 and 650 miles in length. Our regression equation takes the general form:

$$Y = \alpha_0 + \beta_0 x + \varepsilon \tag{8-37}$$

where Y = participation rate on a route, x = average fare paid on that route

ε is a zero-mean normal term meant to reflect special characteristics of a pair of cities that raise or lower air travel between them away from the overall trend line (8-37). OLS regression entails the homoscedasticity assumption that the ε's can be approximated as independent picks from the distribution $N(0, \sigma_\varepsilon)$. That assumption is not self-evidently plausible, but it might suffice for an approximate analysis.

We should acknowledge at the outset that the regression model is imperfect. It focuses on routes between 600 and 650 miles long, though fare sensitivity might be different on routes of 300 miles or 3000 miles. Moreover, the dependent variable "participation rate" might be misleading on routes to vacation destinations with relatively few permanent residents. To reduce that problem, we chose the second quarter of 2011 to avoid summer or winter vacations, and we deleted one route to Las Vegas because it is a classic "getaway" destination. And although nonstop flights exist on all the routes in our sample, on some routes there are dozens of nonstop flights each day while on others there is only one in each direction. On the routes over 600 miles long that we are studying, the time saving over driving is large even if the flight makes a stop. But the net advantage in speed and convenience does vary from route to route.

In principle, the zero-mean ε-terms in the regression model can accommodate such complicating factors. In practice, they might fall short.

The raw data that underlie the regression analysis appear at the website. Two examples of the data points are:

City 1	City 2	Average Fare Paid	Passengers per day	Combined Populations[a] (millions)
Boston	Detroit	273	956	10.85
Washington DC	Milwaukee	158	1261	9.85

[a] Sum of 2010 populations for the metropolitan areas of each city in the pair. (US Census)

Passenger numbers do *not* include travelers who changed planes at either city on longer journeys.

Source: *Domestic Airline Fares: Consumer Report* (US Dept. of Transportation, Second Quarter, 2011)

The participation rate Y on a given route is the ratio of the next-to-last column above to the last column. The data used in the regression analysis are:

TABLE 8.3 Average Fares and Daily Passengers per Million Residents in Spring 2011, on 35 US Air Routes Between 600 and 650 Miles Long

Route	Average Fare Paid (x)	Participation Rate (Y)
Boston-Detroit	273	88.1
Columbus-Minneapolis	265	50.8
Chicago-Richmond	265	34.3
New York-Grand Rapids	263	12.6
New York-Knoxville	253	10.6
Huntsville-Washington	253	66.4
Detroit-Kansas City	250	51.5
Atlanta-Grand Rapids	240	31.5
Dallas-Colorado Springs	236	28.9
Dallas-Nashville	234	85.9
Hartford-Charlotte	228	106.7
Boston-Columbus	222	45.5

Route	Average Fare Paid (*x*)	Participation Rate (*Y*)
St. Louis-New Orleans	217	57.8
New York-Greenville	214	11.6
New York-Charleston	214	30.2
Washington-Chicago	207	299.5
Nashville-Tampa	201	111.7
Charlotte-Miami	197	114.5
Chicago-Raleigh	193	98.7
Nashville-Kansas City	191	177.7
San Diego-Salt Lake City	189	66.5
Atlanta-Flint	189	66.3
Atlanta-Chicago	177	208.4
Austin-Kansas City	176	64.8
Phoenix-Sacramento	175	140.9
Seattle-Sacramento	170	180.3
Albuquerque-San Diego	164	83.9
Portland-Salt Lake City	163	203
Washington-Milwaukee	158	128
Phoenix-Reno	152	101.5
Boston-Raleigh	147	177.3
Denver-Boise	142	113.3
Denver-Phoenix	126	393.5
Myrtle Beach-Detroit	115	70.7
Denver-Tucson	115	157.7

X is in dollars; *Y* is measured in daily passengers per million combined population

The use of a regression package reveals that the least-squares regression model on data from the 35 cities above yields the equation:

$$Y = 313.63 - 1.04x$$

The calculated 95% confidence interval for β_0 extends from −1.58 to −.50.

What do these regression results mean? We might start by discussing what they do *not* mean, in connection with the intercept of 313.63 Taken literally, the equation implies that the participation rate will be 313.63 if $x = 0$ because the airline offered the flight free of charge. But we only have data about fares between \$115 and \$273, and lack any evidence that a formula about that range continues to apply well beyond the range. Thus, we do not know what would happen if fares dropped to (say) \$50, let alone to zero. In the same way, we have no empirical evidence about what would happen if fares reached \$500.

But the slope of −1.04 does suggest that higher fares are associated with lower participation rates, and the confidence interval for β_0 does not include the value zero. If the regression is treated as reliable, therefore, it does offer statistically significant evidence that higher fares do reduce demand for air service.

What about the specific coefficient of –1.04? The average fare on the 35 routes in Table 8.3 was \$199.26, and the estimated participation rate at that fare is 106.92. Using that rate, suppose that the fare on an "average" route were reduced by \$50 to \$149.26. Then the estimated participation rate would grow by –1.04 * –50 = 52, to about 159. In other words, a 25% reduction in fare yields approximately a 50% increase in passenger demand. Total revenue therefore changes by a factor of (1.5) * (.75) = 1.13, meaning that the fare reduction increases total revenue by 13% and therefore does pay for itself. Calculations for other fare changes within the range studied lead to similar conclusions.

This is not to say that the airlines should rush to reduce fares. Given that US flights are close to full now, the airline would presumably have to add extra flights to accommodate the sharp increase in passengers. It could well be that the costs of doing so might exceed the increase in revenue. Still, the regression analysis raises a point worth thinking about, which goes beyond the unsurprising news that passenger demand goes up as fares go down.

However, an inspection of the data used in the regression analysis raises questions about its value. The participation rates on the various routes range from a low of 11 per million residents (on New York-Knoxville, with fare \$253) to a high of 394 (on Denver-Phoenix, with fare \$126). Those two rates vary by a factor of 35. Does anyone seriously believe that halving the fare from \$253 to \$126 could cause a 35-fold increase in traffic? Perhaps the participation rate variable is (to say the least) not optimally defined, and perhaps factors not included in the model exert a large influence on demand patterns. To assume that such omitted effects behave like zero-mean noise (and thus are adequately captured by the ε-term) might be overly optimistic.

Reviewing the individual data points, we find that the Denver-Phoenix route—which had the third lowest one-way fare on the 35 routes studied—had by far the largest participation rate of all the routes studied. (As noted, passengers who change planes at Denver or Phoenix are not included in the statistic for that route.) We know that some data points in regression analysis can have far more influence on the results than others; is it possible that Denver-Phoenix is the main reason we got a negative and statistically significant value of β_0?

To approach the issue more systematically, we might first see whether any of the data points meet the Chapter 2 criterion for outliers in the dependent variable. The participation rates among the 35 cities have the following five-parameter summary:

L (maximum)	393.71
UQ (75th %ile)	149.17
M (median)	88.11
LQ (25th %ile)	66.45
S (minimum)	10.61

With these numbers, the interquartile range IQR= $UQ - LQ$ is 149.17 – 66.45 = 82.72. Measurements below LQ – 1.5 IQR and above UQ + 1.5 IQR are considered outliers under a convention we discussed in Chapter 2. By this standard, we have no outliers at the low end of the participation-rate distribution, but we have two at the upper end: Chicago-Washington (with participation rate of 300) and Denver-Phoenix (394).

A defensible next step is to run the regression again after we delete the two outliers. For the remaining 33 points, the revised regression equation is:

$$Y = 253.57 - .80x$$

The 95% confidence interval for β_0 extends from –1.18 to –0.43.

In short, the deletion of the outliers did reduce the (absolute) estimate of "fare sensitivity" by 23%, from –1.04 to –.80. But the confidence interval also shrank, and its upper end of –0.43 is nearly as far from zero as its counterpart in the original regression (–0.43 vs. –0.50). Thus, the statistical significance of the outcome is essentially unchanged. With the new slope estimate of –.80, a \$50 drop in fares from \$199.62 to \$149.62 would be expected to raise the participation rate on an average route by about 43%, from 93.68 (i.e., from 253.57 – .801 * (199.62)) to 133.68. But when a 25% fare drop raises demand by 43%, revenue changes by a factor of (.75 * 1.43) = 1.07 (i.e., goes up by 7%). Adding flights to raise revenue by 7% might well be uneconomic, although US airlines are far from strangers to uneconomic behavior.

We are not saying here that it is inherently desirable to delete outliers from the data set used in regression analysis. However, assessing the sensitivity of key results to the inclusion or exclusion of outliers is reasonable as part of a statistical investigation.

EXERCISE 8.14

Suppose that it is argued that New York City routes—the most populous in the table—should be deleted from the analysis, on the grounds that New Yorkers have abnormally little interest in traveling to places like Grand Rapids, and the citizens of Grand Rapids return the compliment. What would be the least-squares estimate of β_0 if the analysis were conducted without the New York routes? (You can do this without a statistical package, though it wouldn't hurt to get used to working with one.)

EXAMPLE 8.13

Heat Wave?

We hear a great deal about global warming; why not consider some data from New York City's Central Park? Since 1876, temperatures have been recorded every hour, and the yearly average of the 365 * 24 = 8760 readings (8784 in Leap Years) has been calculated.

(i) Using the data in Table 8.4 below, please perform a linear regression analysis of the form:

$$Y_t = \alpha_0 + \beta_0 t + \varepsilon$$

where Y_t = mean temperature in year t and t = time
Assume homoscedasticity (i.e., that ε_i is $N(0,\sigma)$ for all data points).

(ii) Discuss the key results.

Solution

For simplicity, we will group the data into five-year periods (except for the first period, which covers the six years 1876–1881). We will treat the first period as t = 1, and then increase t by one for each succeeding five-year period (e.g., 1882–86 is t = 2, and 1887–91 is t = 3). Proceeding in this way, we reach the data in Table 8.4.

TABLE 8.4 Mean Temperature in New York's Central Park for Five-Year Periods from 1876–2011

Period	Mean Temp	Period	Mean Temp
1	52.7	14	54.4
2	51.2	15	54.8
3	52.1	16	53
4	52.6	17	52.7
5	53.9	18	54.2
6	53.1	19	54.2
7	53.8	20	54.6
8	53.1	21	54.7
9	53	22	55.3
10	52.7	23	55.7
11	54.3	24	54.8
12	53.8	25	55.6
13	54.1	26	55.4
		27	55.6

Note: First period contained six years: 1876–1881. Temperature is in degrees Fahrenheit.

When we perform OLS regression on the data, we reach the following equation:

$$Y_t = 52.16 + .125t$$

95% confidence interval for α_0: 51.63 to 52.69

95% confidence interval for β_0: 0.091 to 0.158

Given that a unit increase in t is the same as moving forward by five years, the slope estimate β^* implies that mean temperature in New York has been increasing by an average of 1/8 degree Farenheit every five years (or, equivalently, by 1°F. every forty years). The confidence interval suggests that it would be farfetched to assert that the true value of β_0 is zero, and that the apparent upward trend in temperatures is but a random fluctuation. Nor does it appear that the trend line is responding only to very recent growth, as a graph makes clear:

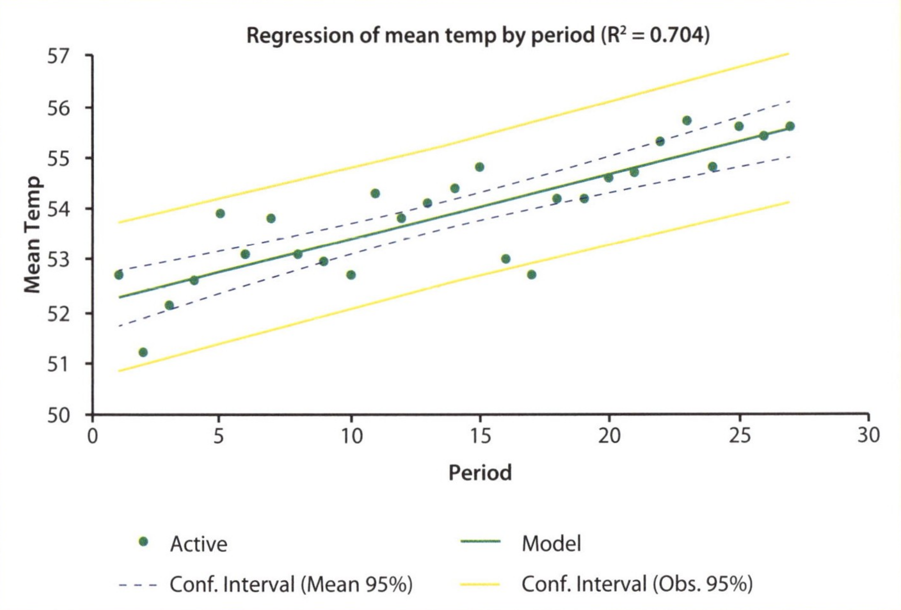

FIGURE 8.8 OLS Regression Line and New York Mean Temperatures

(For now, ignore R^2 and the confidence intervals in Figure 8.8.)

One cannot avoid the sense that the residuals in Figure 8.8 follow something of a cyclic pattern. For present purposes, we treat that outcome as a meteorologic curiosity. (We will discuss shortly what it means that $R^2 = 0.704$.)

EXERCISE 8.15 Using Table 8-4, perform OLS regression on only the data for the last ten periods (from 1962–2011). Now perform a separate regression on the first 17 periods. Do the slope estimates for the two time ranges suggest a consistent rate of growth in mean temperature per unit time? Discuss briefly.

8.4 Two Important Regression Statistics

The next items on our regression agenda are two famous summary statistics used in conjunction with the technique: the *t*-ratio and the R^2-value. These numbers relate to the statistical significance of the primary results and the credibility of the underlying model. Because both the statistics are often believed to convey more information than they actually do, we should scrutinize them with especially great care.

The *t*-statistic

Even if x has no effect on Y and thus $\beta_0 = 0$ in the linear model, it is most unlikely that β^* will itself be zero. Only an extraordinary coincidence would cause the measurement errors around the horizontal to juxtapose in such manner that the least-squares line has no tilt. In such circumstances, the mere fact that $\beta^* \neq 0$ is hardly definitive evidence that $\beta_0 \neq 0$.

Attention therefore turns to the question: when can we be confident that x *does* exert some influence on Y? A familiar and conservative response is to conjecture that $\beta_0 = 0$ and then test statistically the viability of that hypothesis.

In standard linear regression, a reasonable test of H_0: $\beta_0 = 0$ is :

Compute the 95% confidence interval (95% CI) for β_0 and:

- Accept H_0 if the 95% CI contains 0 (8-38)
- Reject H_0 if the 95% CI does not contain 0.

Because a 95% CI has a 19 in 20 chance of including the true parameter value, it has a one in 20 chance of failing to do so. Hence there is a 5% chance of rejecting H_0 under the test listed when $\beta_0 = 0$ or, in standard parlance, the Type I error probability for the listed procedure is the usual .05. We've been there, and done that.

The historic role of the *t*-statistic has been to hasten the implementation of this hypothesis test. Before proceeding to the definition, we note that, when n is large enough that the distribution for β_0 is approximately normal, the 95% CI for β_0 is $(\beta^* - 1.96S_\beta, \beta^* + 1.96S_\beta)$. If the interval is entirely in the positive range, then its lower end point $\beta^* - 1.96S_\beta$ exceeds 0. Likewise, if $\beta^* + 1.96S_\beta < 0$ if the interval is completely negative.

The ***t*-statistic** is defined by:

$$\tilde{t}_\beta = \beta^* / S_\beta \qquad (8\text{-}38)$$

where β^* is given by (8-23) and S_β by the denominator in (8-35)

Thus defined, t_β is the ratio of the least-squares slope estimate to the standard deviation of the error associated with that estimate. Because $\beta^* = \beta^* - 0$, it can also be interpreted as the number of standard deviations that β^* is away from zero. Observe that when $\beta^* - 1.96S_\beta > 0$ because the CI is completely positive, then $\beta^* > 1.96S_\beta \rightarrow \beta^*/S_\beta > 1.96 \rightarrow \widetilde{t_\beta} > 1.96$. Similarly, an entirely negative CI implies that $\beta^* < -1.96S_\beta$ and thus that $t_\beta < -1.96$. Thus whenever the confidence interval for β excludes 0 and the normality approximation holds, *the t-statistic will exceed 1.96 in absolute value.*

A similar argument yields a complementary point: if the 95% CI *does* contain 0, then $|t_\beta|$ must be less than 1.96. The upshot of all this is that the t-value tells us immediately what we need to know to perform the test about H_0. That test can be reformulated as follows:

When n is sufficiently large that W_β and β_0 are approximately normal, the test of H_0: $\beta_0 = 0$ in (8-38) becomes:

- Accept H_0 if $\left|\widetilde{t_\beta}\right| \leq 1.96$ (8-40)
- Reject H_0 if $\left|\widetilde{t_\beta}\right| > 1.96$

This is all mildly interesting, but the reader might be wondering why it is at all useful. In the 21st century, computer printouts of regression results routinely present the 95% confidence interval for β_0, so there no need to consult the t-statistic to see whether zero falls in that interval. Yet there are reasons that studying the t-statistic might be valuable. One is purely public relations: you might find yourself working with "old timers" who were taught to look at the t-statistic before taking a slope estimate seriously. You should at least understand what they are thinking so you can communicate with them. Furthermore, if the t-statistic takes a value like 11, that means that zero is eleven standard deviations below β^*. When that happens, zero is not merely absent from the range of plausible values for β_0: it is practically in a different solar system. Another advantage of the t-statistic is more practical: computer printouts often omit S_β but, knowing β^* and $\widetilde{t_\beta}$, we can immediately solve for S_β under the rule $S_\beta = \beta^*/\widetilde{t_\beta}$.

EXAMPLE 8.14 **Hot *t***

In the regression analysis "Heat Wave" in Example 10.14, β^* was 0.125 and its t-statistic was 7.72. Given that information, estimate the probability that β_0 was at least 0.15.

Solution

Because $\widetilde{t_\beta} = \beta^*/S_\beta = 7.72$ and $\beta^* = .125$, we see that $S_\beta = .125/7.72 = .0162$. Because $n = 27$, we can plausibly make the approximation that $\beta_0 \approx N(.125, .0162)$. We then find that $P(\beta_0 \geq 0.15) \approx .061$.

EXERCISE 8.16 Given the information in Example 8.14, does the hypothesis that $\beta_0 = 0$ for the New York temperature data seem absurd? Discuss briefly.

Not surprisingly, the t-statistic is related to the t-distribution that has previously appeared in our study of regression. If H_0: $\beta_0 = 0$ is correct, then t_β has a t-distribution with $n - 2$ degrees of freedom for any $n \geq 3$. We can use that fact in working out the p-values that accompany observed results, as the following example illustrates.

EXAMPLE 8.15 ***t* Turns Hostile**

In an OLS regression with 12 data points on $Y = \alpha_0 + \beta_0 x + \varepsilon$, suppose that $\beta^* = 1.6$ and $S_\beta = 0.8$. What is the p-value of this outcome when H_0: $\beta_0 = 0$ is tested against the alternative H_a: $\beta_0 \neq 0$?

Solution

Under H_0, the t-statistic $\widetilde{t_\beta} = \beta^* / S_\beta$ should follow a t-distribution with $n - 2$ degrees of freedom. Because $\beta^* = 1.6$ and $S_\beta = 0.8$, $t_\beta = 2$; and the relevant question in the p-value calculation becomes: What is the probability of an outcome at least as hostile to H_0 as $t_\beta = 2$? Given H_a, and the fact that $n = 12$, the question can be restated as: What is the probability of an absolute outcome of 2 or greater for a t-distributed random variable with $12 - 2 = 10$ degrees of freedom. A visit to the t-table reveals that:

$$p\text{-value} = P\left(\left|\widetilde{t_\beta}\right| \geq 2 \,\middle|\, t_\beta \text{ is } t\text{-dist with 10 degrees of freedom}\right) = .0734$$

EXERCISE 8.17 Find a 95% confidence interval for $\widetilde{t_\beta} = \beta^* / S_\beta$ when $n = 14$ and $\beta_0 = 0$. If $S_\beta = 3$ and $\widetilde{t_\beta}$ falls in that confidence interval, what can we say about β^*?

At this point, we should recall an earlier caveat about what it means to accept the null hypothesis H_0: $\beta_0 = 0$. Accepting H_0 in that context means that 0 is a *plausible* candidate for the true value of β_0, and thus that those initially skeptical that x affected Y have every right to remain so in the face of the test results. But that is *not* to say that it has been shown unequivocally that β must equal 0. If 0 is deemed "acceptable" because it falls in the 95% CI, then every other value in the interval is acceptable by the same standard. As we know, the same data can be compatible with a full range of possibilities.

In the abstract, it is easy to see that "it is possible that $\beta_0 = 0$" is a weaker statement than "it is *impossible* that β_0 is anything other than 0." Yet in practice, this distinction is blurred with surprising frequency. Sometimes an explanatory variable is discarded from a regression model simply because its t-statistic falls below 1.96. That exclusion—which is equivalent to setting $\beta_0 = 0$—overstates what has actually been determined. If there are strong prior grounds for suspecting that a variable is influential, deleting it solely because its t-statistic is low is neither necessary nor especially wise.

EXERCISE 8.18 In the Philadelphia-delinquent regression in Example 8.11, the printout revealed that $\beta^* = 0.325$ and (unbeknownst to you until now), that $\widetilde{t_\beta} = 10.75$. The printout also said that the 95% confidence interval for β_0 ranged from 0.260 to 0.389. Show that the listed confidence interval is the 95% CI implied by β^* and t_β. Assume that the computer used the t-distribution for β_0 rather than a normal approximation.

The R^2 Measure

So far, our discussion about linear regression has assumed that all its various axioms are satisfied perfectly. Yet, like any statistical enterprise, a regression analysis is but an exercise in approximation. The widespread recognition of this point has led to an equally widespread yearning for some simple measure of how "good" a regression is. A very popular indicator on the subject is the one designated R^2, the ***coefficient of determination***.

The literal definition of R^2 contrasts sharply with the more grandiose image the statistic sometimes acquires. Given a set of data points (x_i, y_i), one can imagine at least two linear models that generated them:

$$\text{Model I: } Y = c_0 + \xi$$

$$\text{Model II: } Y = \alpha_0 + \beta_0 x + \varepsilon$$

where α_0, β_0, and c_0 are constants and ε and ξ satisfy the standard assumptions (homoscedasticity, zero-mean normality, independence).

The second of these models, of course, is the one we have been considering through most of this chapter. The first posits that the observed variations of the Y_i's in the data set are just random oscillations around the true value of c_0. Geometrically, the true y-to-x relationship would be horizontal.

One can estimate the parameters of both of these models under the least-squares criterion. We already have the formulas for α^* and β^*. As for c^*, we turn to the quadratic loss function $L(c)$ that follows:

$$L(c) = \sum_{i=1}^{n}\left(y_i - c\right)^2 = \sum_{i=1}^{n} y_i^2 - 2c\sum_{i=1}^{n} y_i + nc^2$$

Setting $dL(c)/dc$ equal to 0 yields: $c^* = \overline{Y}$

This formula is eminently plausible: If one really believes that the Y_i's reflect random scatter about some common value, it makes sense to estimate that value by their arithmetical average $\overline{Y}$.

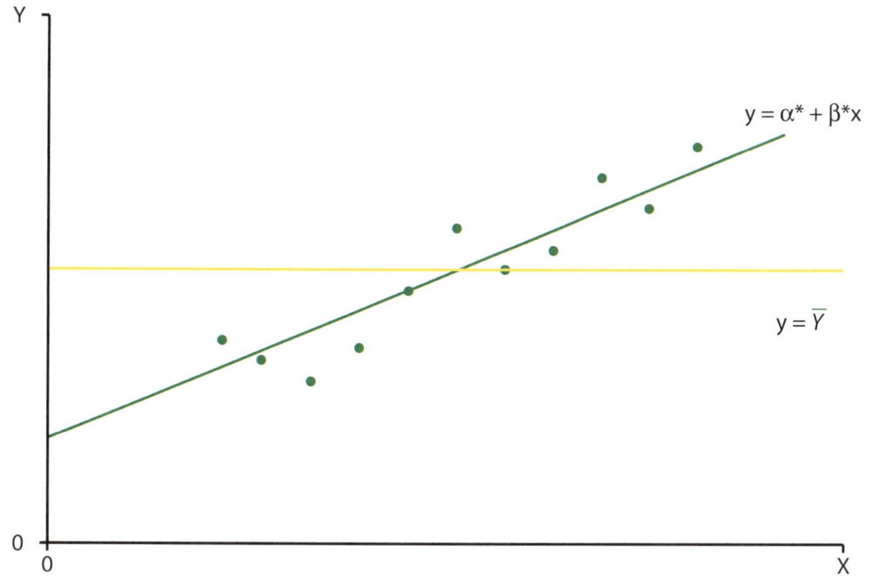

FIGURE 8.9 Generally a tilted line achieves lower residuals than a horizontal one.

Figure 8.9 presents the least-squares lines for a hypothetical set of data points. Note that the tilted line seems to fit the points better; this is unsurprising because, while model I is constrained to be horizontal, model II can take on any slope and any intercept.

To quantify the improvement of Model II over Model I, we might compare their sums of squared residuals (SSR's), which are the sums of their squared "vertical misses." If we denote these sums as SSR(I) and SSR(II), then [SSR(I) – SSR(II)]/SSR(I) is the *proportional reduction* in the SSR when the "best" horizontal line is replaced by the "best" unrestricted line. R^2 is *defined* as this percentage reduction:

R^2, the coefficient of determination in a least-squares regression analysis, is the *proportional reduction* in the sum of squared residuals when the OLS regression line $y = \alpha^* + \beta^* x$ is fitted to available data points rather than the horizontal line $y = \overline{Y}$.

The quantity [SSR(I) – SSR(II)]/SSR(I) can be simplified:

If SSR(I) and SSR(II) are respectively, the sum of squared residuals around $y = \overline{Y}$ and $y = \alpha^* + \beta^* x$, then R^2 is given by:

$$R^2 = 1 - \text{SSR(II)}/\text{SSR(I)} \tag{8-41}$$

Given that neither SSR(I) nor SSR(II) can be negative, it is obvious that $R^2 \leq 1$. And it is almost as obvious that R^2 cannot be negative in this setting. The line $y = \alpha^* + \beta^* x$ will never have a greater SSR than $y = \bar{Y}$ because the parameters $\alpha = \bar{Y}$ and $\beta = 0$ are among those considered in the SSR optimization that yielded $y = \alpha^* + \beta^* x$. Thus, SSR(II) is at most SSR(I): if the line with intercept α^* and slope β^* cannot improve the fit over the horizontal one, it can match the fit by sitting directly atop $y = \bar{Y}$. Because SSR(II) ≤ SSR(I), it follows that SSR(II)/SSR(I) ≤ 1 and hence that $R^2 = 1 - $ SSR(II)/SSR(I) ≥ 0.

It must be the case that $0 \leq R^2 \leq 1$.

EXAMPLE 8.16 **One Off**

Give an example of three data points for which $R^2 = 1$.

Solution

Sure. Consider (1,1), (3,3) and (5,5). The horizontal line $y = \bar{Y}$ is $y = 3$ here. The OLS regression line is $y = x$, which achieves a perfect fit to the data.

To find SSR(I), we note that the residual for (1,1) against the line $y = 3$ is −2 (i.e., 1 minus 3), the residual at (3,3) is 0, and the residual at (5,5) is 2 (i.e., 5 − 3). Thus, SSR(I) = $(-2)^2 + 0^2 + 2^2 = 8$. Because $y = x$ passes through all three points, SSR(II) = 0. Thus, the sum of squared residuals drops by 100% when we move from model I to model II, and $R^2 = 1 - 0/8 = 1$.

EXERCISE 8.19 What is the R^2-value associated with OLS regression and the three data points (1,1), (3,3), and (11,5)? (Yes. You have the find the least-squares line before you can find residuals around it.)

EXAMPLE 8.17 **It's Nothing**

What is the R^2-value for OLS regression on the points (1,1), (3,3), and (5,1)?

Solution

Here $\bar{Y} = 1.67$, so SSR(I) = $(-.67)^2 + (1.33)^2 + (-.67)^2 = 2.67$.

For the OLS line, we have:

$$\beta^* = \frac{\sum_{i=1}^{3} y_i (x_i - \bar{X})}{\sum_{i=1}^{3} (x_i - \bar{X})^2} = \frac{1 * -2 + 0 + 1 * 2}{\sum_{i=1}^{3} (x_i - \bar{X})^2} = 0$$

But then $\alpha^* = \bar{Y} - \beta^* \bar{X} = \bar{Y} = 1.67$.

In other words, the regression line is also $Y = 1.67$. It follows that SSR(II) is also 2.67, and therefore:

$$R^2 = 1 - 2.67/2.67 = 0.$$

EXERCISE 8.20 Suppose that the three points of interest are (1,1), $(3,y_2)$, and (5,1). If $y_2 > 1$ and SSR(I) = 24, what is the value of y_2? What is R^2 in this situation?

When considering R^2-values between 0 and 1, one should remember that proportional drops in the *squares* of the residuals are larger than those in their absolute values. If each residual fell by half, for instance, R^2 would be .75 and not .5. Thus the improvement over the horizontal model might well seem greater after an R^2 computation than after an inspection of a picture like Figure 8.9.

Some Misconceptions about R^2

R^2 is not especially difficult to understand, and a statistical package that performs OLS will do the calculations for us (which, not surprisingly, get less pleasant when there are more than three data points). But a curious gap has arisen between what the statistic actually means and what many people think that it means. The issue is important enough to warrant a bit of discussion.

The R^2-statistic is often believed to reflect the *degree of success* of a regression analysis. When the computer printouts of regression results are scanned, R^2's near 1 can produce elation while those near 0 can provoke despair. But such strong reactions could be misplaced, for a model's real value and its R^2 statistic can easily part company.

Consider, for example, the situation depicted in Figure 8.10. The tilted least-squares line entails far smaller squared residuals than its horizontal counterpart; hence the calculated R^2 would be rather high. But even the line with nonzero slope has fairly large residuals, and the data would evidently be better summarized by (say) a parabola.

The underlying problem is that R^2 is a *relative* and not an absolute measure of accuracy and that, in many settings, the "baseline" model I against which comparisons are made is something of a "straw man." In such conditions, even a large amount of improvement over model I could still leave us far short of excellence.

At the same time, a low R^2 need not imply that a model lacks credibility. Figure 8.11 presents a regression line with both very low residuals and very low R^2; these are quite compatible outcomes when the horizontal line itself fits the data very well. It is one thing to be disappointed that x seems irrelevant to y, but quite another to impugn the regression analysis that accurately conveyed the information. To do the latter because of the low R^2 would be to blame the messenger for the content of its message.

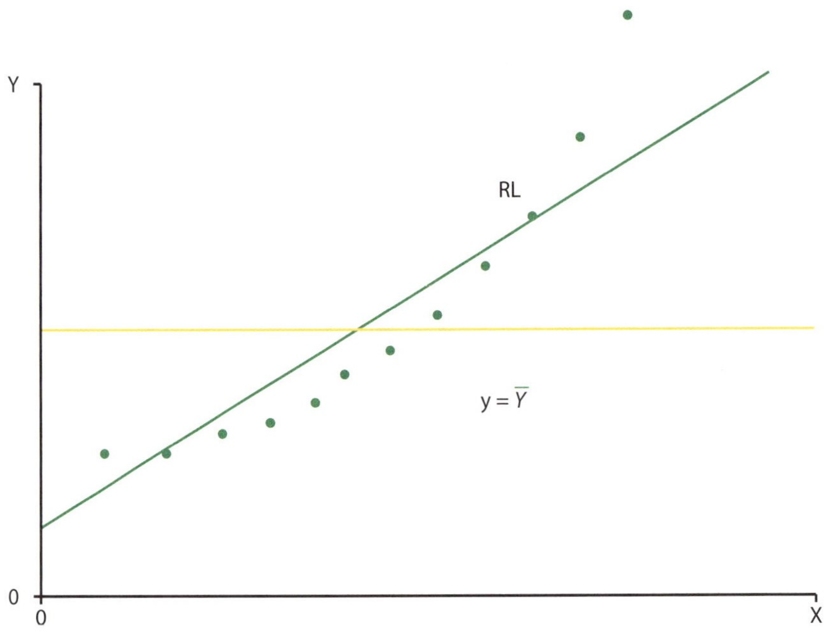

FIGURE 8.10 The regression line attains a high R^2 but a less-than-ideal description of the data.

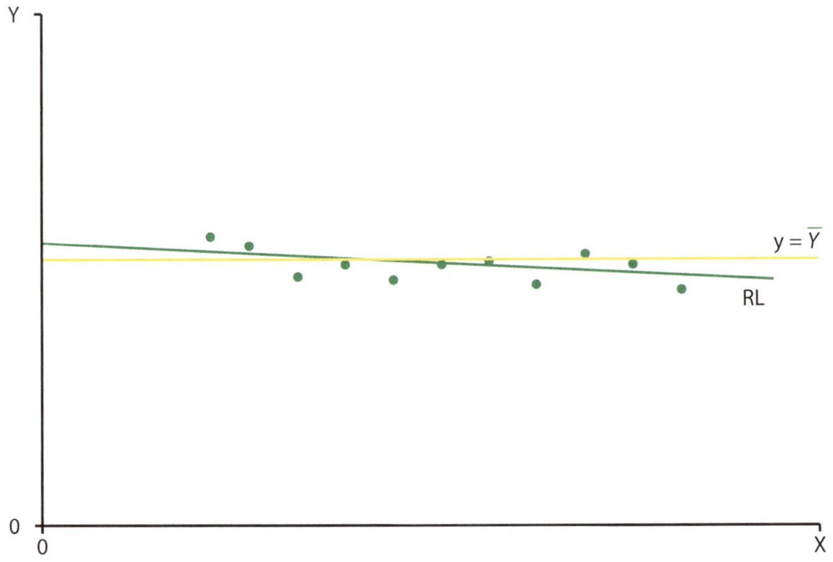

FIGURE 8.11 If a horizontal line fits the data points well, R^2 is rather likely to be low.

In other words, if the aim is successful modeling, a high R^2 is not necessarily good and a low R^2 is not necessarily bad. Of course the statistic has some value: if one treats the horizontal line as "ground zero" in describing the behavior of Y, R^2 suggests how far linear regression has pushed things above that level.

In the Y vs. x case, however, R^2 is not inherently more informative than the other statistics available. For example, the mean-squared residual (or the square root of that quantity) provides a direct indication of the fit between the regression line and the data points, while relation between β^* and zero—summarized by the t-statistic t_β—suggests the strength of the case for preferring a tilted line to a horizontal one. Viewed at the same time as other indicators about a linear regression, R^2 can increase our intuitive "feel" for the results. But if seized upon as the preeminent indicator, R^2 can diminish our understanding rather than raise it.

Two More Points about R^2

R^2 and the Correlation Coefficient

We have noted that β^* depends in part on $\widehat{\rho_{xy}}$ the correlation between x and y. For that reason, the residuals in OLS regression relate to $\widehat{\rho_{xy}}$, which means that R^2 does so. But what is not obvious is the simplicity of the relationship between $\widehat{\rho_{xy}}$ and R^2:

> In y-vs.-x OLS regression, R^2 is the **square** of the correlation coefficient $\widehat{\rho_{xy}}$ between x and y. (8-42)

The derivation of (8-42) requires the usual ingredients: algebra and patience (Exercise 8.36). The fact that R^2 tells us nothing new beyond the correlation coefficient adds to the argument that it is no "be all and end all" statistic about regression analysis.

R^2 and Regression Through the Origin

R^2 is not taken too seriously when the regression line must pass through (0,0). In that setting, it is perfectly possible that the SSR's around the line $y = \beta^* x$ will exceed those around $y = \overline{Y}$: consider, for instance, the points (3,3), (4, 3), and (5,3.01) in which SSR around $y = 3.003$ is roughly .00007, while the SSR around $y = .72x$ (the least-squares line through the origin) is about 2. There, the use of (8-41) would yield an R^2 of about −30,000. Because the range of R^2 in the zero-intercept case is essentially $(-\infty, 1)$, the R^2 statistic is both harder to interpret and less respected (although some statistical packages persist in reporting it).

EXAMPLE 8.19 **Walk the Talk**

Go back to some of the OLS regression analyses we considered earlier in the chapter, retrieve their R^2-values from the regression printout, and discuss the results briefly.

Solution

In the "Fare Game" in Example 8.12, the R^2-value in the regression output was 0.385. That statistic means that the OLS model cut the sum of squared residuals by 38.5% compared to a baseline model ($y_i = \bar{Y}$) that takes the extreme position that the fare charged on a given route has no effect on its passenger traffic. Given that the dependent variable—participation rate—varied hugely on the 35 routes considered, there was ample opportunity to achieve a big percentage reduction in SSR when fare was used as an explanatory variable. That the actual reduction was modest enhances the sense that the model fell short of a complete explanation why some routes have far higher participation rates than others. And it increases the worry that the omitted factors compromised our attempt to isolate the effect of a single variable: average fare paid.

In the regression "Heat Wave" in Example 8.13, the R^2-value was 0.704. That statistic means the regression line cuts squared residuals by 70.4% compared to the horizontal line at 54.1°, the average New York temperature over the 126 year-period considered. Here the horizontal line reflecting no time trend fits the data pretty well, because the upward drift in the raw data is quite gradual. Improving over that line by 70% therefore seems impressive, especially given the year-to-year fluctuations in mean temperature that no simple model of the overall trend could hope to capture.

EXERCISE 8.21 Go back to the regression in Example 8.11 about Philadelphia juvenile delinquents, and compute the R^2-value from the information available there (i.e., do not go to a statistical package; create your own little spreadsheet and apply the key formulas). Then briefly discuss the result.

8.5 Two More Regression Examples

For substances that can cause health problems, the likelihood that they will do so often depends on the amount to which a person was exposed. Regression analysis is sometimes used to approximate a *dose-response* curve, which relates the probability of harm to the dose of the toxin experienced. Now we illustrate the calibration of such a curve against actual data.

EXAMPLE 8.20 **A Dose-Response Model**

The 1945 atomic bombing of Hiroshima caused life-threatening health problems even for those individuals who survived the initial blast. One of these consequences was an increased risk of leukemia caused by gamma and neutron radiation. The dose of radiation exposure was estimated for large numbers of people in or near Hiroshima, as were the doses for the people known to have developed leukemia over the period from 1950 to 1971 (when the bulk of leukemia cases tied to the explosion would have materialized). Table 8.5 summarizes the relevant data as reported in a paper published in the journal *Radiation Research.*

The data for the analysis are:

TABLE 8.5 Leukemia Over 1950–71 Among People Exposed to Radiation From the Atomic Bombing of Hiroshima in 1945

Exposure Range	Effective Dose (Rads)[a]	Number of People Exposed[b]	Number of Acute Leukemia Cases
0–1	0	28,463	20
1–49	11	23,453	25
50–99	70	2573	7
100–199	139	1619	12
200–399	281	981	15
400+	525	477	12

Source: T. Ishimaru, M. Otake, and M. Ischimaru (1979), "Dose-response Relationship of Neutrons and Gamma Rays to Leukemia Incidence among Atomic Bomb Survivors in Hiroshima and Nagasaki by Type of Leukemia, 1950–71, *Radiation Research* 77(2), pp. 377–394

[a] Effective dose is apparently the average among people in the corresponding exposure range, which can be used in a linear-regression model.

[b] The number exposed was estimated by the author from data in the paper, with a modest degree of approximation.

Given these data:

(i) Discuss modeling the dose-response relationship with a linear equation of form: $Y = \alpha_0 + \beta_0 x + \varepsilon$
where x = dose received and Y = leukemia risk

(ii) What is the OLS linear regression equation that arises from the Hiroshima data?

(iii) What supplementary details about the analysis do we learn from the regression output?

(iv) What caveats should be attached to the regression results to reduce the danger that they will be misused?

Solution

(i) While there are debates among radiologists, many believe that the leukemia dose-response curve is linear in the dose received. A simple model that embodies this viewpoint is:

$$Y = \alpha_0 + \beta_0 x + \varepsilon$$

where Y = chance of suffering leukemia over a period 5–26 years after exposure

x = dose of radiation received (in rads)

In this formulation, α_0 is the risk of leukemia over the observation period for reasons unrelated to the atomic bombing. The ε-term reflects three distinct factors:

- Even if a given dose of radiation has a 1% chance of causing leukemia, the number of people who get leukemia out of (say) 1000 exposed to that dose might not be exactly ten; more plausibly, the number will be the outcome of a binomial process with $n = 1000$ and $p = .01$. Thus, there will be some random variation in the observed leukemia rate around the true probability.
- The radiation exposure of particular individuals is not known exactly: it depends on the distance and direction from the center of the blast, but also on possible shielding based on exactly where the person was. Actually, the uncertainty here is really part of x rather than y, but treating it as part of ε is reasonable as an approximation.
- The age and gender distribution among those exposed might vary with the distance from "ground zero," and these differences might contribute to differences in susceptibility to radiation-induced leukemia

The hope is that these phenomena can be approximately modeled as creating zero-mean variation around the true dose-response curve. But we need recognize at the outset that the regression exercise cannot yield something as precise as a law of Physics. Rather, it might offer a useful rule-of-thumb for the relationship between radiation dose and leukemia risk.

(ii) Use of ordinary-least squares regression analysis on the data in Table 8.5 generates the approximate line:

$$y = .0005 + .000048x$$

(iii) Among other things, the regression output includes the following information about the results:

95% confidence interval for α_0: −.0009 to .0019

95% confidence interval for β_0: .000042 to .000054

Estimate of σ_ε (i.e., S_ε) = .0009

$R^2 = 0.993$

These various results about β_0 suggest that someone exposed to 100 rads of radiation suffers an incremental risk of acute leukemia of approximately 100 * .000048 = .0048 = 1 in 208. A low estimate of the increased risk (based on the bottom of the 95% confidence interval) is .0042 = 1 in 238, while a high estimate is .0054 = 1 in 185.

The high R^2-value means that the regression line fits the data far better than the horizontal line associated with the model that radiation dose has nothing to do with acute-leukemia risk. Indeed, it cut squared residuals compared to that "no effect" model by a formidable 99.3%.

(iv) One important caveat is that the results can only be applied with confidence in the range of doses for which data were used in calibrating the regression model. We do not know whether the risk continues to grow linearly as x grows above 500 rads. Perhaps, for example, the risk increases faster than linearly beyond a certain dose.

Another point of importance is that all the individuals in the study were sustained virtually their entire radiation exposure all at once. People who get the same cumulative dose x over a longer period of time might have a different risk of suffering leukemia than is suggested by the regression equation.

Furthermore, the fact that a linear model provided a good fit to the Hiroshima data does not prove in itself that the dose-response pattern is linear. Perhaps a quadratic model also yields an acceptable fit, as does a cubic model. Our discussion in Chapter 5 about the range of acceptable hypotheses continues to be relevant.

EXERCISE 8.22 Analogous to Table 8.5, Table 8.6 presents dose-response information about leukemia risk among survivors of the atomic bombing of Nagasaki, which took place in August 1945 three days after the bombing of Hiroshima.

TABLE 8.6 Leukemia Over 1950–71 Among People Exposed to Radiation From the Atomic Bombing of Nagasaki in 1945

Exposure Range	Effective Dose (Rads)[a]	Number of People Exposed[b]	Number of Acute Leukemia Cases
0–1	0	4547	2
1–49	10	10021	9
50–99	71	1282	0
100–199	144	1368	4
200–399	268	1008	8
400+	525	349	7

Source: Ishimaru, Otake, and Ichimary (1979), cited above

(i) Replicate for Nagasaki the analyses corresponding to (i)–(iii) in Example 8.20

(ii) Compare the results of the Nagasaki analysis with those of the Hiroshima regression. Do you believe the two analyses yielded consistent results about the dose response pattern?

The Capital Asset Pricing Model

Under a major financial theory, there is no such thing as a "good buy" on the stock market. The price of a company's stock at any time already incorporates in the appropriate way all relevant information about the company's prospects. Thus, whether a particular firm will outperform the full stock market is only a matter of chance.

On the other hand, the theory does allow some systematic differences across stocks. Some are intrinsically **volatile**, and will exaggerate overall market movements in either direction (e.g., when the market goes up 1%; they rise more than 1%; when the market drops 1%, however, they drop more). Other stocks are unusually **stable**, and swing less than the market as a whole. Through analyzing historical data, one can estimate where a given stock falls on the stable/volatile spectrum.

More precisely, we can define the *net return* Y on a given stock over a period by:

$$Y = (P_f + d - P_b)/P_f - T \qquad (8\text{-}43)$$

where P_b = price per share at the beginning of the period

P_f = price per share at the end of the period

d =dividend per share over the period

T = prevailing interest rate on a US Treasury Bond

The first term on the right of (8-43) measures the proportional gain to the investor who bought the stock at the start of the period and sold it at the end. The subtracted quantity T reflects the gain that the investor could have achieved through a risk-free investment in a bond (assuming the continued solvency of the US government). The quantity Y, therefore, reveals the net reward for accepting the risks of the stock purchase. Not surprisingly, Y is sometimes negative.

The Capital Asset Pricing Model (CAP-M) relates what happens to a particular stock to what happens to the market as a whole. Its basic equation takes the familiar form:

$$Y = \alpha_0 + \beta_0 x + \varepsilon \qquad (8\text{-}44)$$

Where Y = net return for the stock under study:

x = net return on the entire stock market

ε is a zero-mean normally distributed term reflecting "random variation"

Under CAP-M, the slope β_0 in is often interpreted as a *volatility multiplier* and the key descriptor of the stock's price behavior. If $\beta_0 > 1$, then the stock tends to amplify market price movements, while, if $1 > \beta_0 > 0$, it appears to dampen them. If $\beta_0 < 0$, then the stock tends to move opposite the direction of the full market (which is not always a bad thing).

The quantities α_0 and ε play subsidiary roles in CAP-M. The constant α_0 allows for the possibility of systematic price movements unrelated to those of the market.

Many theorists believe that α is *almost always zero,* because all stocks "sway in rhythm" although to varying amplitudes. The "random noise" term ε reflects the fact that unforeseen developments can affect some companies (and their stock prices) more than the market as a whole, and "unforeseen" developments are typically modeled as random effects.

Typically, x and Y values are gathered over a past period, and least-squares regression is used to estimate β_0 in (8-44).

EXAMPLE 8.21 **A Stable Energy Source?**

Five years of monthly-returns data for Exxon from 1989–93 were examined in conjunction with corresponding data for the full New York Stock Exchange. OLS regression on the $12 * 5 = 60$ data pairs (x_i, y_i), led to the CAP-M equation:

$$Y = .0027 + .490x$$
$$(0.72) \quad (4.66)$$

The numbers in parenthesis were t-statistics and, in the primitive conditions of the mid-1990's, 95% confidence intervals for parameters did not appear in the printout.

(i) Is zero in the 95% confidence interval for α_0? For β_0?
(ii) What are the 95% confidence intervals for the two parameters?
(iii) Are the results consistent with the view of some theorists that α_0 must be zero?
(iv) Do the results affirm that Exxon stock showed less volatility during 1989–93 than the stock market as a whole?

Solution

(i) Because there were 60 data points, we can assume that the distributions for α_0 and β_0 were both approximately normal. Therefore, zero is in the confidence interval for a parameter if and only if its t-statistic is below 1.96 in absolute value. By that standard, we see that zero is in the confidence interval for α_0 but not for β_0.
(ii) The standard deviation S_α assigned to α_0 is obtained from the formula:

$$\widetilde{t_\alpha} = \alpha^* / S_\alpha$$

Therefore $S_\alpha = \alpha^* / \widetilde{t_\alpha} = .0027/0.72 = .00375$.
The 95% confidence interval for α_0 is approximately $(\alpha^* - 1.96S_\alpha, \alpha^* + 1.96S_\alpha)$, which is $(-.0047, .0101)$.
For β_0, the analogous calculation yields a 95% CI of (.284, .696)
(iii) Because zero is comfortably within the 95% CI for the intercept, these theorists would see in the results an affirmation of their views. The results, of course, do not prove that α_0 must be zero.

(iv) Superficially, the β-estimate of about 1/2 might suggest that, whatever the market does, Exxon does about half as much (e.g., that when the market's return is 2%, Exxon's is 1%.) Exxon, therefore, emerges from the period as something of a paradigm of stability.

We gain some insight, however, if we recall that the least-squares estimate of the slope can be stated under (8-25) as:

$$\beta^* = \widehat{\rho_{xy}}\left(\frac{\widehat{\sigma_y}}{\widehat{\sigma_x}}\right) \tag{8-25 again}$$

where $\widehat{\rho_{xy}}$ is the calculated coefficient of correlation of x and y, while $\widehat{\sigma_y}$ and $\widehat{\sigma_x}$ are the standard deviations of, respectively, the various y and x measurements.

This equation has an interesting interpretation. In CAP-M, $\widehat{\sigma_y}$ reflects the *historical volatility* of returns for the stock in question (as measured by their standard deviation). The ratio $\widehat{\sigma_y}/\widehat{\sigma_x}$ compares the stock's volatility to that of the market as a whole. $\widehat{\rho_{xy}}$, in turn, measures the degree of *simultaneity* between movements of the particular stock and the entire market. Under CAP-M, $\widehat{\rho_{xy}}$ is often believed to be close to 1; the only reason for imperfect correlation would be those occasional random events that make life and investing more interesting.

Yet it is important to heed what (8-25) implies about what β^* really means. When $\beta^* = 0$, there are two possible explanations:

(i) $\widehat{\sigma_y}$ is essentially zero, meaning that the stock is inert and motionless

(ii) $\widehat{\rho_{xy}} = 0$, which means that the movements of the stock are uncorrelated with those of the market.

In the latter case, it is conceivable that the $\widehat{\sigma_y}/\widehat{\sigma_x}$ ratio is large despite the zero slope estimate. β^* is zero not because the stock lacks strong volatility, but rather because its large swings are unrelated in time to swings in the market. For that reason, x is of little help in estimating the concurrent value of y. The least squares procedure reacts to this circumstance by offering a horizontal regression line.

The quantity β^*, in other words, is not really a volatility measure for the stock. It reflects *not just the volatility of the stock but also the likelihood that the stock and the market move the same way over time.* To say that a stock with a low β is a "safe" stock might be silly: its volatility might be several times as large as the market.

One can, however, make some statements on the basis of (8-25):

(i) If β^* exceeds one, then the volatility of the stock must exceed that of the market. $\widehat{\rho_{xy}}$ after all, cannot exceed one, so $\widehat{\sigma_y}/\widehat{\sigma_x}$ must do so if the product of the two quantities is to exceed one).

(ii) If β^* is negative, then the coefficient of correlation of x and y must be negative, and vice-versa. So a negative β^* does mean that the stock and the market tend to move in opposite directions, much as a positive β^* means a tendency to move in the same direction.

In the Exxon case, It turned out that β^* was about 0.5 because $\widehat{\rho_{xy}}$ was about 0.5 and $\widehat{\sigma_y}/\widehat{\sigma_x}$ was about one. Exxon was therefore not less volatile than the market as a whole: when the market rose by 4%, Exxon did not obediently rise by 2%. It typically moved up or down to about the same extent as the market, though not precisely at the same times.

Of course, given both the factors in the formula for β^* are of interest, one should presumably look at each of them. $\widehat{\rho_{xy}}$ is immediately reported in regression printouts. Then the ratio $\widehat{\sigma_y}/\widehat{\sigma_x}$ can immediately be calculated under (8-25) as $\beta^*/\widehat{\rho_{xy}}$. When these two factors are both examined, we might find that two stocks with the same β^*-value have actually behaved very differently.

EXERCISE 8.23 If a stock β_0-estimate of 2 in a CAP-M regression, what are the upper and lower limits of $\widehat{\sigma_y}/\widehat{\sigma_x}$ over the period? (Hint: One or both of these limits might not be finite.)

The discussion about α_0 in the last example raises a question: *Has regression through the origin been superseded by general linear regression?* Why not always work with the general regression of form $Y = \alpha_0 + \beta_0 x + \varepsilon$ rather than perform regression with a zero intercept, on the grounds that, if the true line does in fact pass through the origin (i.e., if $\alpha_0 = 0$), then we will find that out by getting an α^* close to zero and a confidence interval for α_0 that includes zero?

This argument is not senseless, but it can do us considerable harm when data points are few. Suppose, for example, that in reality $Y = 2x + \varepsilon$, where ε is $N(0,1)$. Suppose too that we observe the three data points (3, 7), (4,7), and (5,8), corresponding to $\varepsilon_1 = 1$, $\varepsilon_2 = -1$, and $\varepsilon_3 = -2$. If we perform OLS, we will get $\beta^* = 1/2$. Regression through the origin, however, yields $\beta^* = 1.78$, far closer to the actual slope. The upshot is that, if we are confident that x and Y are related by a line that passes through (0,0), then forfeiting that insight can cost us.

8.6 Linear Regression in Situations that are Initially Nonlinear

It is sometimes possible to use linear regression to estimate parameters in models that are not themselves linear. Much like algebraic substitution in Calculus, a change of variables might convert the original model to an equivalent linear one.

For example, suppose that y is believed to vary with the square root of x rather than x itself. Including the usual random error term ε, we might posit the model:

$$Y = \alpha + \beta\sqrt{x} + \varepsilon$$

If we write $w = \sqrt{x}$ and replace the coordinates (x_i, y_i) by $\left(\sqrt{x_i}, y_i\right) = (w_i, y_i)$ then we can work with the linear model:

$$Y = \alpha + \beta w + \varepsilon$$

If we assume homoscedasticity, we can then use OLS to provide information about α and β.

Linearity Via Logarithms

Now we turn to Minerva's effort to model the relationship between public transport usage and population density. Any approximate formula that relates the two variables should presumably pass through (0,0): when population density is zero, there is no one to transport. But, as Minerva argued, it seems plausible that the use of public transport grows might grow faster than linearly as population density grows. To allow for that possibility, one might try to model the relationship with a formula of the form:

$$V = c_0 x^{\beta_0}(1 + \varepsilon) \tag{8-45}$$

where V = percentage of residents who use public transit to commute to work

x = the city's population density in residents per square mile

ε = a random-error term reflecting special characteristics of the city that move its use of public transport away from the trend-line value of $c_0 x^{\beta_0}$.

For example, if $\beta_0 = 2$, (8-45) would reflect a quadratic relationship between population density and public-transport use. If $\beta_0 = 1/2$, by contrast, usage with grow only with the square root of density (i.e., slowly than linearly). The linear model through the origin is the special case of (8-45) with $\beta_0 = 1$, so the formula does not preclude the possibility that the relationship is linear despite any initial conjecture that $\beta_0 > 1$.

In (8-45), random variability is modeled by the multiplicative factor $1 + \varepsilon$ rather than an additive term ε. When $\varepsilon = .02$ in (8-45), for example, the variability yields a V-value that is 2% higher than the trend line value $c_0 x^{\beta_0}$. Note that the multiplicative factor equals one when $\varepsilon = 0$ so that, as before, $\varepsilon = 0$ means no random deviation from the trend line.

Readers will not have failed to notice that (8-45) is not the equation of a straight line. On the other hand, if we take logarithms (to base e) of both sides of (8-45), we reach:

$$\ln V = \ln c_0 + \beta_0 \ln x + \ln(1 + \varepsilon) \tag{8-46}$$

The Taylor series for ln $(1 + \varepsilon)$ yields the approximation ln $(1 + \varepsilon) \approx \varepsilon$ if ε is small. Thus, if we assume that ε is zero mean normal in (8-44) with a small standard deviation σ_ε that applies at all data points, then $\xi = \ln(1 + \varepsilon)$ is itself approximately $N(0,\sigma_\varepsilon)$. We can then rewrite (8-44) as:

$$Y = \alpha_0 + \beta_0 z + \xi \tag{8-47}$$

where $Y = \ln V$, $\alpha_0 = \ln c_0$, $z = \ln x$, and $\xi = \ln(1 + \varepsilon) \approx \varepsilon \approx N(0, \sigma_\varepsilon)$

After the logarithmic transformation, in other words, we are in the domain of linear of regression, in which (x_i,v_i) is replaced by $(\ln x_i, \ln v_i) = (z_i,y_i)$. The key exponent β_0 in the original equation has become the slope in the new coordinate system. On to data!

EXAMPLE 8.22

Car Less?

Suppose that Minerva gathered the data in Table 8.7, which concerns US cities that have populations over 100,000 and in which at least 10% of commuters to work use public transport. (The cities are listed by the percentage of commuters who took bus and/or rail transit to work in 2006, in descending order.) It presents their actual percentages of public-transport commuters as well as the population densities She converted these data via a logarithmic transformation into the form (8-47), and performed OLS regression on the new variables.

(i) Given what she said at the start of this chapter, what would she have expected about β_0, which she was calling Minerva's constant?
(ii) What was the estimate of β_0 arising from the regression analysis?
(iii) What can be said about the precision of this estimate?

TABLE 8.7 Percentage of Commuters Who Use Public Transport vs. Population Density for Some US Cities in 2006

CITY	% Transit	Population Density
New York	54.24	26403
Jersey City	46.62	16094
Wash DC	38.97	9316
Boston	31.6	13321
San Francisco	30.29	16634
Philadelphia	26.43	11234
Arlington VA	26.28	7995
Yonkers NY	25.47	10848
Chicago	25.38	12750
Newark	24.04	11495

CITY	% Transit	Population Density
Pittsburgh	21.14	6019
Alexandria VA	20.55	8451
Baltimore	19.55	8058
Seattle	17.79	6717
Berkeley CA	17.36	6362
Daly City CA	17.27	13709
Oakland	16.72	7126
Buffalo	15.62	7205
Richmond CA	15.55	3450
Hartford	15.5	7026
Atlanta	14.85	4019
Edison NJ	14.82	3184
Paterson NJ	13.85	17674
East LA	13.75	16697
Minneapolis	13.19	6970
Portland OR	12.64	3939
Cleveland	12.22	6166
Miami	12.16	10161
Stamford CT	11.86	3102
San Juan	11.26	5139
Bridgeport CT	11.24	8720
Cincinnati	11.17	4249
Honolulu	11.08	4337
Los Angeles	10.97	7876
Concord CA	10.85	4041
Rochester	10.78	6133
St. Louis	10.28	5622
New Haven	10	6558

Source: American Community Survey, US Bureau of the Census

Population densities are measured in people per square mile.

Solution

(i) Minerva had anticipated that transit use would grow faster than linearly with population density. Thus, she expected that β_0, the exponent in (8-45) and the slope in (8-47), would exceed one.

(ii) Taking logarithms of the two columns in Table 8.7 yields the ingredients for the linear regression package, which generates the following results:

$$Y = -1.99 + 0.54z$$

- The 95% confidence interval for α_0 extends from –3.94 to –.032
- The 95% confidence interval for β_0 extends from .322 to .758
- The unbiased estimate of σ_ε^2 is $(.343)^2$.

Well, that's a surprise. Minerva had thought that β_0 would exceed one, but the least-squares estimate β^* is 0.54, which suggests that the usage of public transport varies with roughly the *square root* of population density. And the confidence interval for β_0 only extends as high as .758, so even the simple linear model ($\beta_0 = 1$) is inconsistent with the data if the model is to be believed. Of course, we perform data analyses because are prior conjectures are just that: conjectures. We should be ready to give them up when confronted with empirical evidence that works against them.

If we look carefully at Table 8.7, the result about β_0 was more-or-less foreordained by the underlying data. Public transport usage in the table varied by a factor of about five from low to high, from 10% to around 50%. But population densities among the cities varied by a greater factor of about eight, from about 3000 to 25,000. Thus, β_0 must somewhat "compress" the variation in the explanatory variable to match that of the dependent variable. Values of β_0 below one offer a mechanism for such compression.

Somewhat disturbing is that fact that S_ε is 0.343. Given that $(1 + \varepsilon)$ is a multiplicative factor in (8-45), the implication is that the formula is routinely off by 25% or more compared to the actual rate of public transport use, and is sometimes off by 50% of more. (These deviations are in percentages of the estimate and not in raw percentage points: a deviation of 25% against an estimate of 24% usage yields a residual of 25% $*$ 24% = 6%.) When the ε's are so large, we worry that the city-specific factors not included in the model might not behave like zero-mean random noise. If that is the case, then the slope estimation error might be larger than the confidence interval for β_0 suggests.

EXERCISE 8.24 When Minerva submitted her work to her thesis advisor, the advisor praised her for a skillful use of regression analysis. The advisor was concerned, however, about the raw data used in the analysis, because it was restricted to cities in which at least 10% of commuters use public transport. That circumstance raises the danger of a selection bias that could yield misleading results. A city with high population density and (say) only 3% usage of public transport could never make it into the data set. And if the omitted cities with low transit usage had low population densities, then the relationship between usage and density might emerge stronger if all cities were considered. The advisor therefore suggested that Minerva expand the data set to all US cities with populations of least 100,000.

The advisor was making an important point: if the data used in a regression analysis are inappropriate for some reason, then the statistical package that performs the regression is unable to point out the problem. But is it plausible that β^* would exceed one if cities with high population density and low transit usages were added to the data set? Or if a few cities with population density around 1500 and transit usages around 2% were added? Briefly discuss.

Final Remark

This concludes our introduction to linear regression analysis. But we are not finished with the linear regression model, for we have scarcely touched upon questions about the appropriateness of a linear model in a particularly setting. For example, we have not fully considered the possibility that the relationship between Y and x is nonlinear. Nor have we dealt with an obvious issue: What if there are two or more explanatory variables that could affect the value of Y? But before moving on, the reader might enjoy some further exercises, which aim to sharpen an intuitive understanding of regression modeling. Or, if not enjoy the exercises, at least benefit from them.

The Takeaway Bar

Glossary

Linear regression analysis strives to estimate which straight line a set of data points would fall along were it not for the presence of zero-mean random vertical deviations.

The **dependent variable** in a linear regression analysis is on the left of the equation, and its value is assumed to be determined (except for zero-mean statistical "noise") by nothing more than the **explanatory variable** on the right. Geometrically, the dependent variable is usually plotted against the y-axis, while the explanatory variable is plotted against the x-axis.

Linear regression through the origin assumes that the data were generated by a straight line with a y-intercept of zero.

General Linear Regression assumes that the data were generated by a straight line, but not necessarily a line with a y-intercept of zero

The **residual** in a linear regression analysis is the difference in y-coordinates between a data point and the regression line at the same x-value. It is positive if the point is above the line and negative if the line is above the point.

Homoscedasticity in linear regression analysis is the assumption that the zero-mean random errors (ε_i's)—which cause points to stray vertically from the "true" line that generated them—all have the same standard deviation.

Least-squares is a method for estimating the slope and (in general linear regression) the intercept of the regression line that summarizes the y to x relationship in a data set. It chooses as parameter estimates those values that minimize the sum of squares of residuals between the data points and the regression line. As used in general linear regression, the method is usually called **ordinary least squares**.

In the context of regression analysis, the **best linear unbiased estimator** of a parameter is a linear combination of values of the dependent variable (i.e., of the Y_i's) that yields an unbiased parameter estimate and, in addition, that achieves the lowest mean squared estimation error among all unbiased estimators of that parameter based on linear combinations of the Y_i's.

The **R^2-statistic** in general linear regression analysis is the proportional reduction in the sum of squared residuals (SSR) when the data are fit by the least-squares line $y = \alpha^* + \beta^* x$ instead of the least-squares horizontal line $y = \bar{Y}$. It falls between 0 and 1 in general linear regression. R^2 is generally not calculated for regression through the origin, because the statistic could enter the negative domain.

The **t-statistic** in linear regression analysis is the ratio of a least-squares parameter estimate to the standard deviation assigned to that estimate.

The **CAP-M financial model** is a linear regression model meant to summarize the relationship between a stock's price movements and those of the stock exchange of which it is a part. The slope "beta" in that regression is the product of volatility of the stock's movements relative to the market as a whole and the correlation between the stock's movements and those of the market.

Formulas

Linear Regression Model Through the Origin:

$Y = \beta_0 x + \varepsilon$ (ε's at different data points are independent and $N(0, \sigma_\varepsilon)$

Least squares slope estimate:

$$\beta^* = \sum_{i=1}^{n} x_i y_i \Big/ \sum_{i=1}^{n} x_i^2$$

Residual at i^{th} data point:

$$r_i = y_i - \beta^* x_i$$

Slope estimation error:

$$W_\beta = \beta^* - \beta_0$$

If σ_ε^2 is *known*, then:

- W_β is normally distributed.
- W_β has mean 0 and standard deviation $\sigma(W_\beta)$ that follows:

$$\sigma(W_\beta) = \sigma_\varepsilon \Big/ \sqrt{\sum_{i=1}^{n} x_i^2}$$

If σ_ε^2 is *unknown*, then:

- an unbiased estimator S_ε^2 of σ_ε^2 is given by:

$$S_\varepsilon^2 = \frac{\sum_{i=1}^{n} r_i^2}{n-1}$$

- $\sigma(W_\beta)$ can be estimated by $S(W_\beta)$ which follows:

$$S(W_\beta) = S_\varepsilon \Big/ \sqrt{\sum_{i=1}^{n} x_i^2} = \sqrt{r_i^2 \Big/ \left((n-1) * \sum_{i=1}^{n} x_i^2\right)}$$

- The quantity $t_\beta = W_\beta/S(W_\beta)$ follows a t-distribution with $n-1$ degrees of freedom
- For sufficiently large n (at least a few dozen), $W_\beta \approx N(0, S(W_\beta))$
- Confidence intervals for β_0 can be calculated based on β^* and the distribution of W_β
- Probabilistic statements about the location of β_0 can arise from the fact that $\beta_0 = \beta^* + W_\beta$ and the assumption that the distribution of W_β once β^* is known is the same as its distribution before β^* was known

General Linear Regression Model

$Y = \alpha_0 + \beta_0 x + \varepsilon$ (ε's at different data points are independent and $N(0, \sigma_\varepsilon)$

Least-Squares Parameter Estimates:

$$\beta^* = \frac{\sum_{i=1}^{n} y_i \left(x_i - \overline{X}\right)}{\sum_{i=1}^{n} \left(x_i - \overline{X}\right)^2}$$

$$\alpha^* = \overline{Y} - \beta^* \overline{X}$$

where $\overline{Y} = \sum_{i=1}^{n} y_i / n$ and $\overline{X} = \sum_{i=1}^{n} x_i / n$

This model yields the Ordinary Least Squares Regression equation:

$$y = \alpha* + \beta^* x$$

Residual at i^{th} data point:

$$r_i = y_i - (\alpha^* + \beta^* x_i)$$

In ordinary least squares regression analysis:

$$\sum_{i=1}^{n} r_i = 0 \quad \text{and} \quad \sum_{i=1}^{n} r_i x_i = 0$$

Slope estimation error:

$$W_\beta = \beta^* - \beta_0$$

Intercept estimation error:

$$W_\alpha = \alpha^* - \alpha_0$$

If σ_ε^2 is *known*, then:

- W_β and W_ε are both normally distributed.
- Each has mean 0, and their standard deviations follow:

$$\sigma\left(W_\beta\right) = \sigma_\varepsilon\sqrt{1/\sum_{i=1}^{n}\left(x_i - \bar{X}\right)^2}$$

$$\sigma\left(W_\alpha\right) = \sigma_\varepsilon\sqrt{\left(\frac{1}{n} + \left(\bar{X}\right)^2\right) / \left(\frac{(n-1)\sum_{i=1}^{n}\left(x_i - \bar{X}\right)^2}{n}\right)}$$

If σ_ε^2 is *unknown*, then:

- an unbiased estimator S_ε^2 of σ_ε^2 is given by:

$$S_\varepsilon^2 = \sum_{i=1}^{n} r_i^2/(n-2)$$

- $\sigma(W_\beta)$ can be estimated by $S(W_\beta)$ which follows:

$$S\left(W_\beta\right) = S_\varepsilon\sqrt{1/\sum_{i=1}^{n}\left(x_i - \bar{X}\right)^2}$$

- σ_α can be estimated by $s(W_\alpha)$ which follows:

$$S\left(W_\alpha\right) = \sigma\left(W_\alpha\right) = S_\varepsilon\sqrt{\left(\frac{1}{n} + \left(\bar{X}\right)^2\right) / \left(\frac{(n-1)\sum_{i=1}^{n}\left(x_i - \bar{X}\right)^2}{n}\right)}$$

- The quantity $t_\beta = W_\beta/S(W_\beta)$follows a t-distribution with $n - 2$ degrees of freedom
- The quantity $t_\alpha = W_\alpha/S(W_\alpha)$ follows a t-distribution with $n - 2$ degrees of freedom
- For sufficiently large n (at least a few dozen), $W_\beta \approx N(0, S(W_\beta))$ and $W_\alpha \approx N(0, S(W_\alpha))$
- Confidence intervals for β_0 can be calculated based on β^* and the distribution of W_β
- Ditto for α_0
- Probabilistic statements about the location of β_0 can arise from the fact that $\beta_0 = \beta^* + w_\beta$ and the assumption that the distribution of W_β once β^* is known is the same as its distribution before β^* was known
- Ditto for α_0

t-statistics:

The t-statistics for α^* and β^* are defined by:

$$\widetilde{t_\beta} = \beta^*/S\left(W_\beta\right) \quad \widetilde{t_\alpha} = \alpha^*/S\left(W_\alpha\right)$$

Note that $\widetilde{t_\beta} = \beta^*/S\left(W_\beta\right) \neq \left(\beta^* - \beta_0\right)/S\left(W_\beta\right)$

Given β^* and $\widetilde{t_\beta}$, $S(W_\beta)$ can be determined. Ditto for $S(W_\alpha)$

R^2 Statistic:

The R^2-statistic is given by:

$$R^2 = 1 - \frac{\sum_{i=1}^{n}\left(y_i - \left(\alpha^* + \beta^* x_i\right)\right)^2}{\sum_{i=1}^{n}\left(y_i - \bar{Y}\right)^2}$$

R^2 measures the proportional decline in sum-of-squared residuals (SSR) when the least-squares regression line replaces the horizontal line at $y = \bar{Y}$. R^2 ranges from 0 to 1.

Some Parting Thoughts

- In practice, the "random error" term (ε) in linear regression is rarely meant as just random noise tied to imprecise measurements; rather, "random error" is something of a catchall term for all factors left out of the regression model. We cannot always be certain, however, that the omitted factors really function as nothing beyond zero-mean noise.
- Even though ordinary least-squares regression seems to treat all data points equally (giving equal weight to the squared residuals at all points), in practice the points further from $x = 0$ in regression through the origin and from $\bar{X}$ in general linear regression have a disproportionate (and often highly disproportionate) influence on slope estimates. That circumstance is not inherently troubling, but it does mean we should be sure that the far-off points belong in the analysis and are correctly recorded.
- If the 95% confidence interval for a parameter in linear regression includes the value zero, it means that we lack clear evidence that x affects Y. But that does not mean that we are sure that x has no effect on Y, or that zero deserves greater weight than other values in the confidence interval.
- If the values of the explanatory variable used to fit a regression line are limited to a particular range, it is often perilous to assume that the regression equation is accurate outside that range.
- It is simplistic to believe that, in linear regression, the higher the R^2, the better the model. If one believes that $\beta_0 = 0$, then the ideal regression result would be a horizontal line with low residuals and an R^2-value of zero.
- The uncertainty statistics about a parameter in linear regression analysis (like confidence intervals) are calculated on the assumption that the linear model used is exactly correct, and the only reasons for uncertainty are the zero-mean "random errors" that are unavoidable. If the assumptions underlying the analysis

are wrong, then the parameter estimates may be subject to greater error than the regression calculations suggest.

Chapter 8 EZ Pass

1. In least-squares regression on $Y = \alpha_0 + \beta_0 x + \varepsilon$, the residuals measure:
   (A) the horizontal distances between individual points and the regression line
   (B) the vertical distances between individual points and the regression line
   (C) the shortest distances between individual points and the regression line
   (D) the distances between the best fit horizontal line to the data and the regression line
2. If zero is in the 95% confidence interval for β_0 in a linear regression analysis on $Y = \alpha_0 + \beta_0 x + \varepsilon$ then:
   (A) the analysis shows that x has no effect on Y
   (B) the analysis shows that a horizontal fits the data just as well as the regression line (in terms of SSRs)
   (C) the analysis offers no compelling evidence that x influences Y
   (D) all of these
3. The regression line of the form $Y = \alpha^* + \beta^* x$ through the points (2,5), (3,8), and (4,q) had a R^2-value of 1. Given that information, the value of q:
   (A) must be 11
   (B) must be 5
   (C) must be negative
   (D) cannot be determined
4. In a least-squares regression on $Y = \alpha_0 + \beta_0 x + \varepsilon$ under standard assumptions, the variance of the *intercept* estimation error:
   (A) must equal that of the slope estimation error
   (B) must exceed that of the slope estimation error
   (C) must fall below that of the slope estimator error
   (D) none of these
5. If a regression least-squares regression analysis on $Y = \alpha_0 + \beta_0 x + \varepsilon$ yields $\alpha^* = 0$, then a least-squares regression on $Y = \beta_0 x + \varepsilon$ would;
   (A) definitely yield the same line
   (B) definitely yield a different line
   (C) possibly but not definitely yield the same line
   (D) none of these
6. If $\beta^* = 0$ in a CAP-M regression, then what can we conclude about the standard deviation of the stock's price movement over the observation period?
   (A) The standard deviation was zero
   (B) The standard deviation was equal to that of the market as a whole
   (C) The standard deviation was equal to the stock's correlation with the overall market
   (D) none of these

Further Exercises

EXERCISE 8.25 For each of the following situations in which linear regression might be contemplated, tell whether you would expect the true relationship of Y to x to pass through the origin:

	Dependent Variable	Explanatory Variable
(i)	Typing Errors per Minute	Typing Speed in Words per Minute
(ii)	Score on an Exam	Hours of Study the Night Before
(iii)	Home Runs this Year	Home Runs Last Year
(iv)	Bear Population	Amount of Oxygen in the Air

EXERCISE 8.26 In a regression through the origin, with the two points $(1,y_1)$ and $(2, y_2)$:

(i) Express the equation of the regression line in terms of y_1 and y_2.
(ii) Assuming that the regression line does not pass through both $(1,y_1)$ and $(2,y_2)$, is it possible that the sum of the residuals is zero?
(iii) As y_2 changes, so does the slope of the regression line in (i). If y_2^* is the point *on the regression line* when $x = 2$, then $y_2^* = 2\,\beta^*$. As y_2 changes while y_1 is held constant, what is the derivative of $y_2^* = 2\,\beta^*$ with respect to y_2? (Hint: Express y_2^* in terms of y_2 and the constant y_1.) This derivative relates to the concept of *leverage* in regression analysis.)

EXERCISE 8.27 If $W_{\hat{\beta}} = \hat{\beta} - \beta_0$ in Example 8.2 and the x_i's are not all the same, is $\sigma^2\left(W_{\hat{\beta}}\right)$ always greater than $\sigma^2(W_\beta)$ for $W_\beta = \beta^* - \beta_0$?

EXERCISE 8.28 A simpleton decides to use a (well, simple) linear regression model to quantify the relationship between the area of a circle and its radius. He can measure the radius exactly at various data points, but his estimate of the area is subject to zero-mean homoscedastic measurement error. The data points he achieves—where x_i is the radius and Y_i the estimated area of the circle—are:

(1, 3.26)	(8, 200.78)	(15, 708.17)
(2, 13.09)	(9, 254.93)	(16, 805.39)
(3, 28.53)	(10, 313.64)	(17, 908.21)

(4, 49.61)	(11, 379.3)	(18, 1018,59)
(5, 80.13)	(12, 451.81)	(19, 1134.08)
(6, 114.72)	(13, 531.67)	(20, 1256.09)
(7, 154.78)	(14, 615,66)	

(i) Suppose that the simpleton performs the regression analysis $Y = \alpha_0 + \beta_0 x + \varepsilon$ on these data. What will be his least squares parameter estimates? (Yes. Do it yourself).
(ii) What is the R^2-value for his analysis?
(iii) Suppose that it occurs to the simpleton that he should really perform linear regression through the origin on $Y = \beta_0 x + \varepsilon$. Now what are his parameter estimates?
(iv) How does the SSR under this regression through the origin compare with the SSR under the original model?
(v) Suppose that Mendel, on observing the simpleton at work, gently suggests that the simpleton should perform the regression $Y = \beta_0 x^2 + \varepsilon$ on the data. (Mendel doesn't have the heart to tell the simpleton the formula that everyone knows.) The first data point would be $(1, 3.26^2)$, etc. What are the results of this analysis?
(vi) Does the 95% CI for β_0 in this last analysis include the value π?
(vii) What is the moral of this story?

EXERCISE 8.29 Minerva believes that rainfall levels in a certain community follow a sinusoidal pattern of variation, under the general model:

$$Y_t = \alpha_0 + \beta_0 \sin((\pi/2)t) + \varepsilon \text{ where } t = \text{time}$$

She gathers 12 data points as follows at $t = 0 \to t = 11$, and will perform OLS regression with $\sin((\pi/2)t)$ as the explanatory variable.

(0, 28.2)	(1, 31.3)	(2, 27,7)	(3, 25.6)	(4, 28.0)	(5, 30.8)
(6, 27.5)	(7, 24.9)	(8, 28.4)	(9, 31.4)	(10, 27.5)	(11, 24.7)

(Data are of form (t, Y_t), not $\sin((\pi/2)t), Y_t)$)

(i) What are the results of the regression?
(ii) Do the results support the assumption that $\sin((\pi/2)t)$ is a reasonable explanatory variable? Briefly discuss.
(iii) Suppose that Mendel, on learning of this endeavor, suggests to Minerva that she should use $\cos((\pi/2)t)$ as her explanatory variable. If Minerva performs this new regression analysis, what results does she get?
(iv) What is the moral of this story?

EXERCISE 8.30 Suppose that it is hypothesized that, in industry A, worker satisfaction Y_A on a 0–1 scale varies with hourly wages (x_A) according to the formula:

$$Y_A = \alpha_A + \beta_A x_A + \varepsilon_A$$

where ε_A is $N(0, \sigma_A)$
(We assume that the workers considered perform similar work.)
In industry B, the comparable equation is:

$$Y_B = \alpha_B + \beta_B x_B + \varepsilon_B$$

where ε_B is $N(0, \sigma_B)$

It is desired to test the hypothesis $H_0 : \beta_A = \beta_B$ based on the results of linear regression analysis. Suppose that β^*_A and β^*_B are the least squares slope estimates in the two regressions, and that $S_{\beta A}$ and $S_{\beta B}$ are, respectively, the standard deviations of slope estimation error in the industry A and industry B regressions. Suppose further that 100 data points are obtained in each industry for randomly-chosen workers.

(i) Write an expression for the probability distribution of $\beta_A - \beta_B$ given the regression results (and assuming that the assumptions of OLS regression are satisfied). (Hint: What is the distribution of β_A given β^*_A and $S_{\beta A}$?)
(ii) Propose a test of H_0 analogous to a difference-of-means test, with a 5% risk of accidental rejection. Be specific. Not surprisingly, such a test is called a *difference-of-slopes* test.
(iii) A labor analyst suggests that if H_0 is accepted, then workers at the same hourly wage are equally happy in both industries. Give at least two reasons that this assessment might be wrong.

EXERCISE 8.31 Suppose that the results of the two regressions in Exercise 8.30 are:

$$\begin{aligned} Y_A &= \underset{(3.44)}{0.15} + \underset{(5.50)}{.02x_A} \\ Y_B &= \underset{(4.45)}{0.18} + \underset{(2.88)}{.018x_B} \end{aligned}$$

where x_A and x_B are measured in dollars
(The numbers in parenthesis are t-statistics for the corresponding parameter estimates.)

(i) Perform the difference in slopes test you devised in Exercise 8.30. What is the result?
(ii) What is the p-value of the result in testing H_0 against H_a: $\beta_A \neq \beta_B$?
(iii) Neglecting zero-mean fluctuations and taking the parameter estimates as literally correct, for what range of salaries are workers in industry B happier

than their counterparts in industry *A* with the same salary? (For example, where are workers who make $11 per hour happier?)

(iv) Suppose that the data used in the regressions concerns workers with hourly wages between $9.00 and $13.00. Why might the results at either workplace be misleading for workers there with wages outside this range?

EXERCISE 8.32 Suppose that the Hiroshima data from Example 8.20 on leukemia risk versus radiation dose are analyzed via the quadratic model: $Y = \alpha_0 + \beta_0 x^2 + \varepsilon$.

(i) What are the least-squares parameter estimates?

(ii) What is the R^2 value for the regression?

(iii) Suppose that Minerva asserts that this model is inferior to the linear model because it has a lower R^2-value. What do you think of that assessment?

EXERCISE 8.33 In a particular data set, the standard deviation of the *y*-measurements is 6.5. A linear regression analysis for the model $Y = \alpha_0 + \beta_0 x + \varepsilon$ is performed on the data, and it achieves an R^2-value of 0.75. Do you have enough information to determine the SSR of the regression? If so, what is the SSR? If not, what additional information would you need about the regression to determine the SSR?

EXERCISE 8.34 A Standardized Testing Service (STS) contends that, when a student who took its Math Exam takes it for a second time, the *change* in score from the first time is normally distributed with mean 20 and standard deviation 15. This is true for all scores in the range 500–700, regardless of the specific score on the first test.

To investigate STS's claim, Ezekiel gathers data for 7500 students who took the Math Exam twice and originally scored in the range 500–700. He will perform an ordinary-least-squares regression analysis of the form:

$$Y = \alpha + \beta x + \varepsilon$$

where Y = *change* in Math score between first and second tests

x = Math score on the first test

α, β, and ε are the usual suspects

In (i)–(iii), assume that STS's assessment is correct.

(i) Approximately what should be the parameter estimates α^* and β^*?

(ii) The R^2-value for the regression should be:

(A) close to zero

(B) close to 1/2

(C) close to one
(D) close to two

(iii) Estimate the average squared residual in this regression analysis.

Now suppose that, in reality, the change in score is normally distributed with mean equal to 5% of the original Math score and standard deviation 15.

(iv) Approximately what parameter estimates should Ezekiel reach?

(v) Would this pattern threaten the homoscedasticity assumption in Ezekiel's regression analysis? Briefly explain.

EXERCISE 8.35 When DVD's still existed, a linear regression analysis was performed on certain DVD sales versus time, and the least-squares line was:

$$S = 8.88 - .065t,$$

where S = sales and t = time in months since some origin five years earlier.

In all, there were 60 data points, one for each month.

Suppose that Q_t was the *absolute* value of the regression's residual for time t, and that a least-squares regression of Q_t vs. t yielded the line:

$$\begin{array}{cccc} Q_t = & 0.008 & + & .008t \\ & (0.35) & & (2.55) \end{array}$$

The numbers in parenthesis are t-statistics.

Do the results of the second regression analysis cast doubts about those from the first? Discuss briefly but wisely.

EXERCISE 8.36 Showing considerable patience and dexterity with algebra, prove that R^2 is the square of the coefficient of correlation between x and y in the data set used for regression.

EXERCISE 8.37 The Professor has been playing Easy Sudoku, and he is not great at it. An accompanying spreadsheet records his time-to-completion for 124 rounds, which are listed in the order in which he played them over several sad months at websudoku.com. (These are actual data.)

(i) Using linear-regression analysis, fit the data with the ordinary-least squares linear regression line $y = \alpha^* + \beta^* x$, where y = time to completion, and x = game number. The assumption is that the true model is $Y = \alpha + \beta x + \varepsilon$, where ε follows a zero-mean normal distribution with the same variance at all x-values.

(ii) What is β^*, the least squares estimate of β?
(iii) What is the 95% confidence interval for β under the regression assumptions?
(iv) What is the 99% confidence interval for β? (You can use a normality approximation.)
(v) What is the R^2-statistic for the regression?
(vi) What is the t-statistic associated with the slope estimate?
(vii) What do the regression results imply about the hypothesis that the Professor has not improved at all over the 124 games?
(viii) What is the approximate probability based on the regression results that the Professor did actually improve over the 124 games?
(ix) What aspect of the game Sudoku could the random error term ε represent?

EXERCISE 8.38 The Congressman likes to fly from Washington to his country club in his private jet. Based on data about his last 50 trips, the relationship between the airplane's fuel consumption and ground temperature when he left Washington is estimated from the ordinary-least-squares regression equation:

$$Y = \underset{(23.3)}{116.2} + \underset{(2.72)}{1.24x} \quad R^2 = .688$$

where Y= fuel consumed, x = ground temperature in Washington on departure, and the numbers in parenthesis are t-statistics.

(i) Is 0 in the confidence interval for the slope of x?
(ii) Under the usual regression assumptions, what is the probability that fuel consumption actually drops as the temperature goes up? (Hint: What slope-values would correspond to such a drop?)

EXERCISE 8.39 An electric utility company gathers the following data about average daily power in a certain region use versus average daily temperature for 24 half-months of a particular year

Period	Avg Temp (°F)	Avg Power Use (Million Kilowatt-Hours)
1/1–1/15	30.8	42.5
1/16–1/31	25.9	44.8
2/1–2/14	32.8	40.9
2/15–2/28	33.6	40.8

3/1–3/15	40.5	38.7
3/16–3/30	45.6	36.0
4/1–4/15	50.3	34.5
4/16–4/30	53.1	33.6
5/1–5/15	58.0	32.0
5/16–5/31	60.2	31.6
6/1–6/15	63.4	33.9
6/16–6/30	68.5	43.5
7/1–7/15	77.2	50.4
7/16–7/31	79.8	53.8
8/1–8/15	74.0	48.9
8/16–8/31	68.3	42.8
9/1–9/15	61.1	33.5
9/16–9/30	54.3	33.9
10/1–10/15	49.5	35.4
10/16–10/31	44.8	37.8
11/1–11/15	39.2	38.5
11/16–11/30	36.5	39.9
12/1–12/15	34.0	40.9
12/16–12/31	32.1	41.1

(i) Suppose an OLS regression is performed on the annual data, for the relationship:

$$Y = \alpha_0 + \beta_0 t + \varepsilon$$

where Y = average power consumption and t = average temperature

What are the OLS parameter estimates, and the average squared residual?

(ii) Make a plot of Y vs. t for these data. Does the pattern suggest that the linear regression model is not realistic in this setting? Briefly discuss.

(iii) Suppose instead that two separate linear regressions are performed, one on the data from April through September, and the other for the data from January through March plus October through December. Find the OLS results. What do you make of the findings? Do the slope estimates differ to a statistically significant extent?

(iv) Why might the pattern of electrical use versus time follow the pattern shown above? (Can you imagine why power use could be highest in the summer?)

(v) What is the moral of this story?

CHAPTER 9

Further Regression Analysis

Challenger, RIP

The temperature was near freezing at Cape Canaveral, Florida, on the morning of January 28, 1986. It was near launch time for NASA's Space Shuttle Challenger, and some engineers were worried that its crucial O-rings might fail at such a low temperature. But NASA managers very much hoped not to postpone the launch. They noted that, while no previous space shuttle had lifted off on such a cold day, past shuttles had done so at a wide variety of temperatures, and O-ring failures had occurred at high temperatures as well as low ones. (There were five O-rings, and only the failure of all five would doom the shuttle. Previous failures had affected no more than three O-rings.)

In the intense discussion that morning, NASA managers asked the doubters to "take off their engineering hats and put on their management hats." When they then asked whether any participants had strong objections to the launch, they were met with silence. The Challenger took off at 11:37 AM.

To be continued

Here we extend Chapter 8's treatment of linear regression in several ways. We discuss:

- Using a linear regression model to make predictions
- Model validity tests for a linear regression model
- Logistic regression as an alternative to linear regression
- Multiple regression models involving more than one explanatory variable

However, we end the chapter with three examples of regression models that suffered major flaws. The aim in presenting them is not to suggest that regression analysis is unreliable, but rather to stress the degree of careful thought that it requires of successful users.

9.1 Regression and Prediction

So far, we have used regression analysis to make inferences about which straight line produced a given set of data points. Generally speaking, though, we are interested less in finding out about the points we already have than about others we presently lack. Whether directly or indirectly, whether explicitly or implicitly, the real aim of regression is often prediction rather than simply description.

Various examples in Chapter 8 support this last contention. The leukemia regression analysis about Hiroshima residents was meant primarily to estimate the dangers that radiation poses to other individuals exposed to similar doses. The trend line about New York temperatures suggests what will happen in the near future should recent trends continue. And the CAP-M regressions and their "beta" estimates are often used to create an investment portfolio with a particular level of future risk and expected return.

Like those arising from any other source, predictions based on regression analyses range from general to highly specific. From the Philadelphia equation $Y = .908 + .325x$, one could assert that the higher a delinquent's early arrest rate, the higher on average will his later one be. One could also assert that for a youth apprehended a fifth time whose past average arrest rate was two per year, his expected future rate is .908 + .325(2) = 1.558. Of these two forecasts, the latter is obviously far more precise and, for that reason, more vulnerable. Problems that would leave untouched the overall thrust of the Philadelphia finding could raise major qualms about its detailed ramifications.

In this section, we will consider only *numerical* predictions arising from a linear regression model. Our particular focus will be on the degree of accuracy that attends such forecasts. (We use the words "forecasts" and "predictions" interchangeably.) As in Chapter 8, we start with regression through the origin because it is simpler to deal with.

Prediction with Regression Through the Origin

Suppose that there are n data points (x_i, y_i) that are governed by the linear relationship:

$$Y_i = \beta_0 x_i + \varepsilon_i \tag{9-1}$$

where ε_i is $N(0,\sigma_\varepsilon)$ and different ε_i's are independent

The least squares estimate of β_0, namely β^*, has been calculated based on n data points. A new data point also governed by (9-1) will arise at $x = x_{n+1}$, and Y_{n+1} will be estimated in advance by Y^*_{n+1} which follows: $Y^*_{n+1} = \beta^* x_{n+1}$ What can we say about the probability distribution of $W_{n+1} = Y^*_{n+1} - Y_{n+1}$, the prediction error associated with the estimate ?

We can write $W_{n+1} = \beta^* x_{n+1} - (\beta_0 x_{n+1} + \varepsilon_{n+1}) = (\beta^* - \beta_0)x_{n+1} - \varepsilon_{n+1}$

Because β^* is an unbiased estimate of β_0 and ε_{n+1} has mean zero, we see that

$$E(W_{n+1}) = E((\beta^* - \beta_0)x_{n+1}) - E(\varepsilon_{n+1}) = 0$$

As for $\sigma^2(W_{n+1})$, we can write:

$$\sigma^2\left(W_{n+1}\right) = x_{n+1}^2\sigma_\beta^2 + \sigma_\varepsilon^2 \tag{9-2}$$

where σ_β^2 is the variance of the least-squares slope estimation error.

Here we are using the fact that ε_{n+1} and β^* are independent random variables.

If n is sufficiently large, then we can approximate σ_ε^2 well by S_ε^2 which we met in Chapter 8, and which follows:

$$S_\varepsilon^2 = \frac{\sum_{i=1}^n \left(y_i - \beta^* x_i\right)^2}{n-1} = \frac{\sum_{i=1}^n r_i^2}{n-1}$$

where r_i is the ith residual

We also indicated in Chapter 8 that, when n is large, σ_β^2 can be approximated by S_β^2 which is given by:

$$S_\beta^2 = \sum_{i=1}^n r_i^2 \Big/ \left((n-1)\left(\sum_{i=1}^n x_i^2\right)\right)$$

Therefore (9-2) implies that:

Under the model $Y_i = \beta_0 x_i + \varepsilon_i$, the approximate variance $S^2(W_{n+1})$ of the prediction error when OLS regression on n data points is used to estimate the dependent variable at a new point $x = x_{n+1}$ is given by:

$$S^2\left(W_{n+1}\right) = \left(\sum_{i=1}^n r_i^2/(n-1)\right)\left(1 + x_{n+1}^2/\sum_{i=1}^n x_i^2\right) \tag{9-3}$$

We can go further to assert that, when n is sufficiently large, W_{n+1} has approximately a normal distribution. We assume that this distribution is unchanged once β^* has been determined, meaning that having the specific slope estimate does not change the distribution of the slope estimation error $\beta^* - \beta_0$. We can then say that say that $Y_{n+1} = Y^*_{n+1} + W_{n+1}$ is approximately normally distributed with mean $\beta^* x_{n+1}$ and standard deviation $S(W_{n+1})$.

If $Y_i = \beta_0 x_i + \varepsilon_i$ and β^* is the least-squares slope estimate based on n data points, then:

If a new measurement is to be made at $x = x_{n+1}$ and n is sufficiently large, then $Y_{n+1} \approx N\left(\beta^* x_{n+1}, S\left(W_{n+1}\right)\right)$ where $S(W_{n+1})$ is indicated by (9-3). (9-4)

If we study (9-3), we see that the variance of Y_{n+1} depends on x_{n+1} through a proportionality with its square. That formula implies that predictions about a future data point become less accurate as the point moves *further away from the origin*. If we think about it, that outcome makes perfect sense. If β^* deviates from β_0, then the regression line gets further and further from the true line as x_{n+1} grows in absolute value. The prediction error is $(\beta^* - \beta_0)x_{n+1} + \varepsilon_{n+1}$, which means that, apart from the zero-mean "noise" tied to ε_{n+1}, the absolute error grows linearly with $|x_{n+1}|$. Not surprisingly, the mean squared prediction error is least if $x_{n+1} = 0$, when the only contributor to that error is ε_{n+1}.

EXAMPLE 9.1

Gas Leak

An American state energy agency estimates each month the average statewide price of a gallon of gasoline, based on a survey of a particular set of 100 gas stations viewed as representative. This month, it has heard from 98 of these stations, and their average price per gallon increased by 12 cents over the previous month. It uses least-squares regression relating this month's price Y at a station to last month's price x to get the relationship: $Y = 1.04x + \varepsilon$, where $S_\varepsilon = .01$ (dollars) and $S_\beta = .005$. The agency must release its estimate now without the information from the two remaining stations. But it knows that, last month, the price of gas was \$2.985 per gallon at one of the stations and \$3.25 per gallon at the other.

(i) What estimates might the agency make of this month's price per gallon at the two missing stations?

(ii) What is a 95% confidence interval for the price at each station, based on the regression results?

(iii) What is the probability that the estimate in (i) for the more expensive gas station is correct to within two cents ?

(iv) Why shouldn't the agency simply add 12 cents to last month's gas price per gallon at the two stations to estimate this month's price?

Solution

(i) Using the regression equation with its slope of 1.04, the agency would multiply last month's price by 1.04 to get \$3.104 at the first station and \$3.38 at the second.

(ii) Following (9-2), the agency would estimate the variance of the price as $(2.985)^2 S_\beta^2 + S_\varepsilon^2$ at the first station and $3.25^2 S_\beta^2 + S_\varepsilon^2$ at the second. Because $S_\beta^2 = (.005)^2$ and $S_\varepsilon^2 = .01^2$, we have variances of .00032 and .00036 at the two stations. The corresponding standard deviations are, respectively .0179 and .0191.

Assuming approximate normality, the approximate 95% confidence intervals for price per gallon are: $(3.104 - 1.96 * .0179, 3.104 + 1.96 * .0179) =$

(\$3.069, \$3.139) at the first station and (3.38 – 1.96 * .0191, 3.38 + 1.96 * .0191) = (\$3.343, \$3.417) at the second station

(iii) At the station that charged \$3.25 last month, this month's price has an approximate normal probability distribution with mean \$3.38 and standard deviation .0191. The probability the price is within two cents of \$3.38 (i.e., is between \$3.36 and \$3.40) is about .705.

(iv) Twelve cents is the average price increase at the 98 stations, but the regression implies that price increase was larger at the stations that were more expensive one month ago. Thus, if the two missing stations were comparatively expensive last month (or, for that matter, comparatively cheap), then it makes sense to use that information in estimating their price increases. It turns out that the 12-cent approximation gives almost the same results as the regression estimates, but that doesn't alter the general point that using an average is risky if it is applied to something that is different from average.

EXERCISE 9.1 Estimate the probability that the price increase at the \$2.985 station in Example 9.1 was greater than the average dollar increase for the 98 stations that reported. (Hint: Reread the question if this last phrase seems frightening.)

Prediction with the Standard Linear Model

Now suppose that $Y_i = \alpha_0 + \beta_0 x_i + \varepsilon_i$, where the various ε_i's are independent and $N(0,\sigma_\varepsilon)$ A new data point will come up at $x = x_{n+1}$ and we need to predict the value of Y_{n+1}. How reliably can Y_{n+1} be estimated by Y^*_{n+1} that follows:

$$Y^*_{n+1} = \alpha^* + \beta^* x_{n+1}$$

where $\alpha^* + \beta^*$ arise from OLS regression using the n original (x_i,y_i)'s?

If we again let $W_{n+1} = Y^*_{n+1} - Y_{n+1}$, we can write:

$$W_{n+1} = \left(\alpha^* - \alpha_0\right) + \left(\beta^* - \beta_0\right) x_{n+1} - \varepsilon_{n+1}$$

Because $E(W_{n+1}) = E(\alpha^* - \alpha_0) + E(\beta^* - \beta_0)x_{n+1} - E(\varepsilon_{n+1})$ and because all three random variables on the right have mean zero, it follows that $E(W_{n+1}) = 0$. The least-squares estimator of Y_{n+1} is therefore unbiased.

Finding the variance of Y_{n+1} is somewhat more demanding. Hold on.

If $\alpha^* - \alpha_0$, $\beta^* - \beta_0$, and ε_{n-1} were independent random variables, we could invoke the "variance of sum is sum of variances" rule for $\sigma^2(W_{n-1})$, which would be pleasant because we have estimates for all three of these variances (see (8-33), (8-34) and

(8-32)). But do we have such independence? Writing $W_\alpha = \alpha^* - \alpha_0$ and $W_\beta = \beta^* - \beta_0$, we can express W_{n+1} as:

$$W_{n+1} = W_\alpha + W_\beta x_{n+1} - \varepsilon_{n+1}$$

We comfortably say that ε_{n+1} is independent of W_α and W_β, for the new data point had nothing to do with the least-squares estimation. But a simple picture casts doubt on the idea that W_α and W_β are independent.

EXAMPLE 9.2

Negative Reinforcement

Draw a simple picture with three data points that suggests that W_α and W_β are negatively correlated when $\bar{X} > 0$.

Solution

In Figure 9.1, the three measurement errors (ε_i's) "conspire" to cause β^* substantially to underestimate β_0, yielding a negative W_β But the picture makes clear that the errors also induces an α^* value that overestimates α_0, producing a positive W_α. Thus, we might suspect a negative correlation between W_α and W_β.

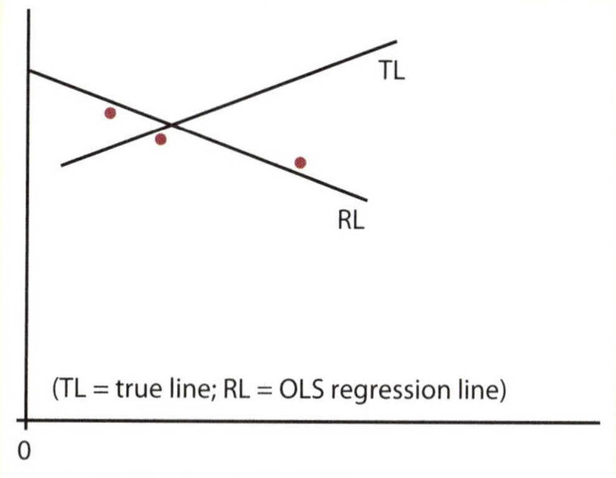

FIGURE 9.1 When $\bar{X} > 0$, Underestimation of the Slope Tends to Accompany Overestimation of the y-intercept

EXERCISE 9.2 Draw a picture similar to Figure 9.1 that suggests that, when $\bar{X} < 0$ 0, W_α and W_β are positively correlated.

Under the circumstances, we cannot assume that W_α and W_β are independent. That circumstance does not render finding $\sigma^2(W_{n+1})$ impossible: we discussed in

Chapter 1 how to find the variance of a sum of correlated random variables. We can do so here without too much conceptual difficulty, because both α^* and β^* are linear combinations of the original Y_i's (i.e., $\alpha^* = \sum_{i=1}^{n} c_i Y_i$ and $\beta^* = \sum_{i=1}^{n} d_i Y_i$, where the c_i's and d_i's are (different) constants identified in (8-23) and (8-24)). We present the details at the web site that accompanies this book, where we find that:

$$\operatorname{cov}\left(\alpha^*, \beta^*\right) = -\overline{X}\sigma_\varepsilon^2 / S_x^2$$

where $S_x^2 = \sum_{i=1}^{n}\left(x_i - \overline{X}\right)^2 / n$, and $\operatorname{cov}(\alpha^*,\beta^*) = E(\alpha^*\beta^*) - E(\alpha^*)E(\beta^*)$ and then that:

$$\sigma^2\left(W_{n+1}\right) = \sigma_\varepsilon^2\left(1 + \frac{1}{n} + \frac{\left(x_{n+1} - \overline{X}\right)^2}{S_x^2}\right) \tag{9-5}$$

Note that the variance of the prediction error is minimized in (9-5) when $x_{n+1} = \overline{X}$, meaning when the new data point has an x-coordinate at the average of those for the prior points. The intuitive reason for that outcome is that errors in slope estimation take their greatest toll as x_{n+1} recedes further from $\overline{X}$ and the cumulative effect of the difference between $\hat{\beta}$ and β_0 gets more and more consequential.

For sufficiently large n, we know that $\sigma_\varepsilon^2 \approx S_\varepsilon^2$, where $S_\varepsilon^2 = \sum_{i=1}^{n} r_i^2 / n - 2$. Furthermore, we can approximate W_{n-1} as normally distributed:

For sufficiently large n and the standard linear model, the OLS prediction error W_{n+1} for an $n + 1^{st}$ point at $x = x_{n+1}$ is approximately $N(0,S_{n+1})$, where S_{n+1} is given by:

$$S_{n+1} = \sqrt{S_\varepsilon^2\left(1 + \frac{1}{n} + \frac{\left(x_{n+1} - \overline{X}\right)^2}{S_x^2}\right)} \tag{9-6}$$

We would follow this statement with another:

For sufficiently large n and the standard linear model, the value of Y_{n+1} for an $n + 1^{st}$ point at $x = x_{n+1}$ is approximately $N(\alpha^* + \beta^* x_{n+1}, S_{n+1})$ where S_{n+1} is given by:

$$S_{n+1} = \sqrt{S_\varepsilon^2\left(1 + \frac{1}{n} + \frac{\left(x_{n+1} - \overline{X}\right)^2}{S_x^2}\right)} \tag{9-7}$$

and α^* and β^* are OLS parameter estimates based on the original n data points.

At the risk of being repetitious, let's again go over the logic under which we move from a probability distribution for W_{n+1} to a probability distribution for Y_{n+1}. If we knew all the x_i's and σ_ε^2 in advance of the least squares analysis (on the first n data points), that information would be enough to work out the distribution of W_{n+1}. (Note that $\sigma^2(W_{n+1})$ in (9-4) does not depend on the actual values of α_0 and β_0.) We might conclude, for example, that W_{n+1} has a 95% chance of falling between −3 and 3. We make the assumption that actually carrying out OLS and determining α^* and β^* does not change that prior accuracy assessment: we still think there is a 95% chance that the prediction error is between −3 and 3. Thus, we assign probability 95% to the event that Y_{n+1} falls within three of the calculated Y^*_{n+1} and, proceeding in this way, we develop a probability distribution for where Y_{n+1} will be.

When the number of data points is not "sufficiently large," (meaning, roughly speaking, is less than two dozen), we can take refuge in the t-distribution:

Under the standard linear model and OLS, the ratio t_{n+1} defined by

$$t_{n+1} = \left(Y_{n+1} - \left(\alpha^* + \beta^* x_{n+1}\right)\right)/S_{n+1} \tag{9-8}$$

follows a t-distribution with $n - 2$ degrees of freedom

At all sample sizes, therefore, we can make probabilistic statements about Y_{n+1}.

EXAMPLE 9.3

Central Heating

In Example 8.13, we analyzed the time trend in mean temperatures at New York's Central Park, using five-year periods starting in 1876. Suppose that it is believed that the standard linear model that we calibrated with OLS continues to be accurate for the next ten years. Let Y_{28} be the mean Central Park temperature over 2012–16 (the 28th five-year period since the start of record keeping), and let Y_{29} be the five-year mean for 2017–21.

(i) Based on the regression results, what are the predicted values of Y_{28} and Y_{29}?
(ii) What are 95% confidence intervals for Y_{28} and Y_{29}?
(iii) Are the prediction errors for Y_{28} and Y_{29} independent? Discuss briefly.
(iv) Sensibly estimate the probability that Y_{28} will exceed Y_{29}.

Solution

(i) The regression equation in Example 8.13 for Central Park mean temperatures was:

$$Y = 52.16 + .125t$$

where t was measured in five-year increments and 2007–11 was $t = 27$.
The predicted value of Y_{28} for 2012–16 is $52.16 + 28 * .125 = 55.66$°F

The predicted value of Y_{29} for 2012–16 is 52.16 + 29 * .125 = 55.79°F

(ii) Because the t-values were the integers from 1 to 27, we have $\bar{t} = 14$ and $S_t = \sqrt{\sum_{t=1}^{27}(t-14)^2/27} = 7.79$. The Central Park regression printout reveals that $S_\varepsilon^2 = 0.427$. Assuming that we have enough data for normality, we can use (9-7) to say that:

$$S_{28} = \sqrt{.427\left(1+\frac{1}{27}+(28-14)^2/7.79^2\right)} = 1.35$$

Thus, $Y_{28} \approx N(55.66, 1.35)$. Similarly, $Y_{29} \approx N(55.79, 1.42)$

Therefore: The 95% CI for Y_{28} is

$$(55.66 - 1.96 * 1.35, 55.66 + 1.96 * 1.35) = (53.01, 58.31)$$

In the same way, the 95% CI for Y_{29} is (53.00, 58.58)

Both these prediction intervals are widened because t = 28 and t = 29 are far from the middle of the time line used in regression (which was at t = 14, in the late 1940's).

(iii) By no stretch of the imagination are the prediction errors independent. Both predictions involve the same α^* and β^*, and thus both are subject to the same slope and intercept estimation error.

(iv) Under the linear model we are using:

$$Y_{28} = \alpha_0 + 28\beta_0 + \varepsilon_{28} \text{ and } Y_{29} = \alpha_0 + 29\beta_0 + \varepsilon_{29}$$

Therefore, $Y_{28} - Y_{29} = \varepsilon_{28} - \varepsilon_{29} - \beta_0$

We know that $\beta^* = .125$, and the regression printout tells us that $t_\beta = \beta^*/S_\beta = 7.72$. (Return to Chapter 8 if you've forgotten how the t-statistic was defined there.) Thus, $S_\beta = .125/7.72 = .016$. From (8-38) and (8-35), we therefore approximate $\beta_0 \approx N(.125, .016)$. Furthermore, ε_{28} and ε_{29} are independent (and independent of β_0), and each has the approximate distribution $N(0, \sqrt{.427}) = N(0, .653)$ because $S_\varepsilon^2 = .427$. We can therefore treat $Y_{28} - Y_{29}$ as approximately normal with mean –.125 and a standard deviation of $\sqrt{.016^2 + .427 + .427} = .924$.

Because $Y_{28} - Y_{29} \approx N(-.125, .924)$, $P(Y_{28} > Y_{29}) = P(Y_{28} - Y_{29} > 0) \approx 0.45$.

What this result is saying is that the "random component" of five-year average temperature is large compared to the slow upward trend, so it is not unusual for a given period to be cooler than the preceding one. Indeed, that happened ten times among the 27 periods over 1876–2011.

EXERCISE 9.3 Estimate the probability that the mean temperature over 2012–16 will be the highest of all the five-year periods from 1876–2016. Return to Example 8.13 for the mean temperatures in the individual periods from 1876–2011. (The first period 1876–81 had six years, but don't be petty.)

Some Cautionary Remarks about Prediction

While we have formulas for using OLS regression for predictive purposes, we should be careful when doing so. The confidence intervals for the forecasts are predicated on the assumption that the underlying linear model is correct for the new data point (as well as the previous ones). If the model is not correct, then the calculated confidence intervals are not reliable.

Before applying a forecast to a new data point at $x = x_{n+1}$, we should consider several points:

- Does x_{n+1} fall within the range of $(x_1,\ldots,x_n)$? If not, we should recognize that even a regression equation that does well in summarizing the y vs. x relationship in a certain domain of x-values might fall off sharply in accuracy beyond that range. For example, a time-series regression that worked well in the past might fare miserably in the future. A formula about annual gains in height that works well for six-year olds might work considerably less well for 96-year-olds.
- Even if x_{n+1} apparently falls within the range, are there circumstances that distinguish the new data point from the ones used in the regression? For example, the Philadelphia delinquency regression in Example 8.11 was based on before/after arrest rate-comparisons among various groups with 16 youths apiece. The formula estimated the *average* later arrest for each such group, but did not take account of the variations of later arrest rates among the 16 youths in a given group. Thus, applying the regression formula to an individual youth would omit a potentially-large component of uncertainty about how he will behave. Similarly, applying the Hiroshima dose-response results to individuals who sustained their radiation exposure gradually rather than at a single instant may involve a level of inaccuracy that the regression analysis could not capture.
- As noted, the linear regression model might be inaccurate for the original n data points, let alone any later ones. It could be, for example, that a dose-response relationship is quadratic rather than linear. Ignoring issues of model validity does not make them go away, and an inaccurate model is unlikely to yield accurate forecasts. All the more reason we now turn to some ways that one might test a linear regression model to assess its validity.

9.2 Considerations of Model Validity

Any regression model depends on a series of strong assumptions. For example, we have routinely assumed that:

- the "random errors" in the regression analyses are independent
- these errors are zero-mean normal

- these errors are homoscedastic
- the dependent variable is linearly related to the explanatory one.

If some or all of these assumptions are inaccurate, the calculated results are compromised to an unknown and possibly enormous extent. For that reason, there is understandable concern about whether a particular regression model is "good enough" to be taken seriously.

Alas, there is no single definitive test on that subject; as we have discussed, the popular R^2-statistic falls far short of a magic bullet. But it is possible to explore whether the results of a regression analysis are consistent with the assumptions that generated those results. If not, then the analysis has the unfortunate property that it is self-contradictory. A series of informative tests about internal consistency exist, based on examining the residuals of the regression analysis. Here we discuss one of the simpler tests—the runs test—but we consider several others at the course website.

Before turning to those tests, however, we should consider the ε_i's, the true measurement errors that arose in the y-coordinates of the (x_i, Y_i)'s. Consider the standard regression model:

$$Y_i = \alpha_0 + \beta_0 x_i + \varepsilon_i$$

where ε_i is $N(0,\sigma_\varepsilon)$ and different ε_i's are independent

If the model is an accurate description of reality, then each ε_i has a 50% chance of being negative and 50% of being positive. (The chance that ε_i is exactly zero is itself zero.) Moreover, the signs of successive ε_i's are independent: the chance that ε_4 is negative is 50% regardless of the sign of ε_3. Suppose that the ε_i's were (somehow) determined and that their *signs* were recorded, starting with ε_1 and continuing through ε_n. (We assume that x_i's are listed in ascending order, with x_1 the lowest, etc.) We might get a sequence like the following for $n = 12$.

$$+ \; - \; - \; + \; + \; + \; - \; + \; - \; + \; + \; -$$

We could ask the question: how many runs of outcomes of the same sign do we have in the sequence? For the data above, we have eight: +, − −, + + +, −, +, −, + +, −. We could move on to the general question: If the ε_i's are independent selections from $N(0,\sigma_\varepsilon)$, then what is the probability distribution of the number of runs? We can find that distribution if we recognize that, starting with ε_2, each of the ε_i's independently has a 50% chance of switching signs compared to its predecessor. The total number of switches, therefore is the outcome of a binomial process with $n - 1$ trials and $p = 1/2$. (We say $n - 1$ because ε_1 has no predecessor with which to differ.) Moreover, each such switch starts a new run. It follows that the number of runs is one plus the number of switches. (We add one because ε_1 always starts the first run.) Putting things together, the number of runs is *one plus the outcome of a binomial process with $n - 1$ trials and $p = 1/2$.*

EXAMPLE 9.4

Figure Eight

What is the probability of getting exactly eight runs among the 12 ε_i's in a linear regression model that is correctly specified?

Solution

For $n = 12$, the number of runs X is one selection from the distribution $1 + Z$, where Z is $B(11, 1/2)$. Thus, $P(X = 8) = P(Z = 7 | Z \text{ is } B(11, 1/2)$. A visit to the binomial table informs us that $P(X = 8) = .161$.

EXERCISE 9.4

(i) What are the maximum and minimum values X could take on in Example 9.4?

(ii) True or False: The probability that X takes its maximum value is the same as the chance that it takes its minimum value.

Testing Model Validity Via the Residuals

Of course, we would never know the ε_i's in an actual situation even when the model $Y_i = \alpha_0 + \beta_0 x_i + \varepsilon_i$ applies perfectly. What we do see are the residuals around the regression line, which is an imperfect proxy for the true line. Can we assume that the general properties of the ε_i's carry over to the r_i's, meaning that the number of runs will be one plus the outcome of a binomial process with $p = 1/2$?

In a word, no. We saw in Chapter 8 that, in OLS regression, $\sum_{i=1}^{n} r_i = 0$ and $\sum_{i=1}^{n} r_i x_i = 0$, where r_i is the ith residual. Given these properties, we showed that, for $n = 3$, knowing one residual and the x_i's not only affected the distribution for the other two residuals but revealed precisely what they were. (So much for the independence assumption.) And while all the ε_i's could be of the same sign, the same could never be true of the residuals when their sum must be zero.

But such dramatic differences between the r_i's and the ε_i's are only consequential when data points are few. As the number of points gets larger—and, presumably, the regression line gets closer to the true line—the constraints on the r_i's become progressively less important. Extensive simulation has confirmed that, with a reasonably large number of data points, the residuals should behave approximately like the ε_i's if the linear regression model is correctly specified (i.e., is accurate in all its assumptions).

What do we mean by "reasonably large?" A simple rule of thumb is that, if n is large enough that the t-distribution for slope estimation error can be approximated as normally distributed, then the residuals should exhibit the general characteristics of the ε_i's. For that reason, testing whether the residuals do exhibit these characteristics gives us one vantage point for making assessments of model validity.

A Runs Test on the Residuals

We have seen that the number of runs we see as we consider the ε_i's from left to right should be one plus the outcome of a binomial random process. With a few dozen data points, we can have the same expectation about the residuals. Thus, we might devise a runs test as follows:

A Runs Test for Residuals

- Focusing solely on the signs of individual residuals, find the number of runs with the same sign as one moves from r_1 to r_n.
- Find the p-value of that number under the hypothesis H_0 that each residual is equally likely to be positive or negative (i.e., $P(\text{positive}) = P(\text{negative}) = 1/2$), against a specified alternative H_a.
- Accept H_0 if the p-value of the result is larger than some pre-specified threshold; otherwise, reject H_0 (9-9)

Why did we speak of some "pre-specified threshold" rather than .05? If we are conducting the runs test as one of several tests on the residuals, we might want to choose α-values for the individual tests (where α here refers to this risk of accidental rejection, not the intercept α_0) so that their sum is .05 or some other desired value.

EXAMPLE 9.5

Run for Philadelphia

In the regression analysis about Philadelphia juvenile delinquents, the signs of the residuals (starting with the group with lowest early arrest rate) were:

$$+ \; - \; - \; - \; - \; + \; + \; - \; - \; - \; - \; + \; + \; + \; - \; -$$

Given this information, perform a runs test of H_0: *All residuals independently are equally likely to have positive or negative sign* against alternative H_a: *H_0 is false*

Solution

There are a total of six runs in this sequence. If each residual independently had a 50% chance of being positive and 50% negative, then the number of runs X should be 1 + Z, where Z is $B(15, .5)$. The actual number of runs was X = 6, corresponding to Z = 5. To find the p-value associated with Z = 5, we need identify those values of Z at least as hostile to H_0 as Z = 5. Because the "ideal" outcomes under H_0 are Z = 7 and Z = 8, the following outcomes are at least as hostile to H_0 versus H_a are:

$$Z = 5 \qquad Z < 5 \qquad Z = 10 \qquad Z > 10$$

Thus p-value = $P\big(Z \le 5 \mid Z \text{ is } B(15,1/2)\big) + P\big(Z \ge 10 \mid Z \text{ is } B(15,1/2)\big)$

From the binomial table, we find that the p-value is .151 + .151 = .302. Thus, while we had fewer runs than the expected number (namely, 1 + 7.5 = 8.5), the deficit does not go beyond the realm of common fluctuations.

EXERCISE 9.5 Perform a runs test on the residuals in the Central Park temperature regression. (You can deduce the signs of the individual residuals from Figure 8.8.)

What Does Failing a Runs Test Mean?

If the residuals do not pass a runs test because runs are too few, that does not necessarily mean that the fault lies with the ε_i's rather than some other aspect of the model. In Figure 9.2 below, we see that the scarcity of runs was apparently caused by the failure of the linearity assumption, in consequence of which the best linear fit to the data systematically underpredicted at low and high x-values and systematically overpredicted in the middle of the x-range. Had the relationship been modeled as (say) $Y_i = \beta_0 x_i^2 + \varepsilon_i$, then the residuals might well have behaved like outcomes from a zero-mean normal process. It helps to look at a scatterplot of y vs. x when the regression line is present, and we can routinely do so with computer packages.

When the number of runs is considerably larger than we would expect under the binomial "coin toss" distribution, the reason might be a picture something like Figure 9.3.

Such a rigid pattern in sign shifts does not seem random, any more that it would seem random to get heads and tails in perfect alternation in 100 tosses of a fair coin. Calculations with the binomial distribution might make clear that would be far-fetched to dismiss the pattern as a coincidence.

Of the two patterns depicted above, the one in Figure 9.2 is often considered the more troubling. The excess in sign changes in Figure 9.3 is curious, but it does not

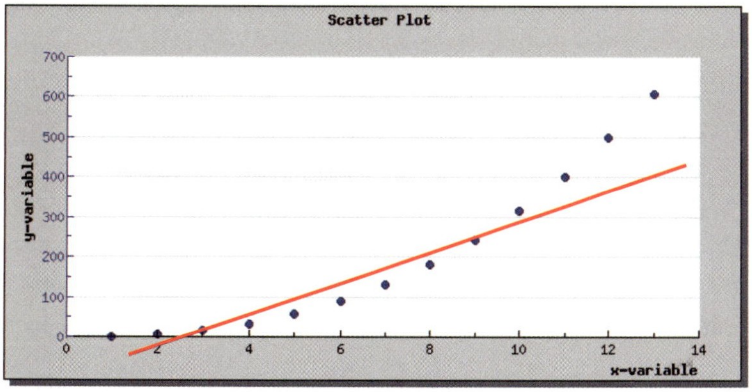

FIGURE 9.2 When the y vs. x relationship is nonlinear, sign changes for successive residuals might be surprisingly few.

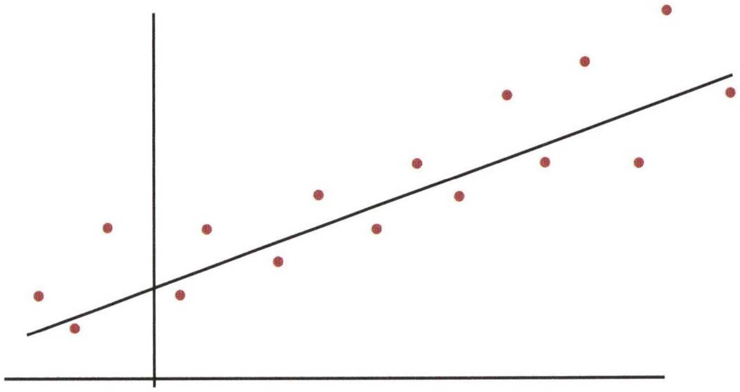

FIGURE 9.3 When consecutive residuals change signs with monotonous regularity, the number of runs is unusually high.

necessarily reduce our belief that the regression line is a reasonable first approximation to the y-to-x relationship. The scarcity of sign changes in Figure 9.2, on the other hand, arises in a picture that suggests that the y-to-x relationship might be nonlinear, which casts doubt on the fundamental validity of the regression equation. For that reason, people often conduct a one-sided runs test on the residuals, in which the rejection region is at the lower end of the distribution of runs. The test therefore takes the form:

A One-Sided Runs Test for Residuals

- Find the number of runs of consecutive residuals with the same sign as one moves from r_1 to r_n.
- Find the p-value of that number under the hypothesis H_0 H_0: *All residuals independently are equally likely to have positive or negative sign* against the alternative H_a: *Successive residuals will more likely than not have the same sign.*
- Accept H_0 if the p-value of the result is larger than some pre-specified threshold; otherwise, reject H_0 (9-10)

EXERCISE 9.6 Perform a one-sided runs test on the Philadelphia data, against the alternative H_a: *Successive residuals will more likely than not have the same sign.*

There are procedures beyond the sign test for exploring whether the residuals are consistent with the axioms of the linear regression model. Some tests consider whether, collectively, the histogram of the residuals resembles that of a zero-mean normal distribution. Others probe whether the absolute sizes of the residuals are correlated with the explanatory variable x, which would violate the homoscedasticity assumption. We discuss several such procedures at the website that accompanies this volume. While the regression results are not worthless if the key assumptions are not literally satisfied, major problems with the residuals are often symptoms of major problems with the model. There are potential remedies for many of these difficulties, and those are also discussed at the website.

But recognizing that linear regression might be inappropriate in a given setting increases the desire for alternatives to linear regression. Now we turn to logistic regression models, which are especially helpful in synthesizing available data to make probability estimates.

9.3 Logistic Regression

Some Preliminaries

A few points and definitions will make our journey through logistic regression easier. First of all, consider the function e^z, where z is some real number. The maximum value of e^z is infinity, which arises when z get boundlessly positive. But the minimum value of e^z is not negative infinity but rather zero: as z gets boundlessly negative, e^z is the number one divided by some exploding positive number. So far so good?

Now consider the function $f(z)$ defined by

$$f(z) = \frac{e^z}{1+e^z}$$

Note that $f(z)$ cannot exceed one because the numerator is smaller than the denominator. As $z \to \infty$, however, $f(z)$ approaches one because adding one to e^z has only a trivial effect. As $z \to -\infty$, e^z approaches zero and thus so does $f(z)$. $f(z)$ is thus bounded by one from above and zero from below, much as a probability is.

One last introductory item: the definition of odds. If $P(E)$ is the probability of event E, then $O(E)$, the *odds* of E is defined by:

$$O(E) = \frac{P(E)}{1-P(E)}$$

The odds (also called the odds ratio) is thus the probability that E will happen divided by the probability that it will not. Some comparisons of $P(E)$ and $O(E)$ yield the following chart:

P(E)	*O(E)*
0	0
0.25	0.33
0.5	1
0.75	3
0.9	9
0.99	99

The formula for $O(E)$ implies that $O(E)$ rises when $P(E)$ does: as $P(E)$ increases, the numerator of $O(E)$ increases while the denominator goes down. And the $O(E)$ formula can be rewritten after some algebra as:

$$P(E) = 1/\left(1+\frac{1}{O(E)}\right)$$

Thus, when $O(E)$ goes up (and $1/O(E)$ goes down), $P(E)$ goes up too. In short, $O(E)$ and $P(E)$ move in the same direction: when one of them increases, so does the other.

One last point. Suppose that $P(E)$ can be estimated by a function like $f(z) = e^z/(1 + e^z)$ above, where z is a number reflecting some background conditions (e.g., the outside temperature, the age of the customer). Then the probability that E does *not* happen would follow:

$$P(\text{not } E) = 1 - P(E) = 1 - \frac{e^z}{1+e^z} = \frac{1+e^z}{1+e^z} - \frac{e^z}{1+e^z} = \frac{1}{1+e^z}$$

Because $(1 + e^z)$ is the denominator of both $P(E)$ and $1 - P(E)$, that quantity disappears when we take the odds ratio $O(E)$, yielding:

$$O(E) = \frac{P(E)}{1-P(E)} = e^z$$

Now, let's go logistic.

The Main Act

In usual linear regression analysis, the basic model takes the form:

$$Y_i = \alpha_0 + \beta_0 x_i + \varepsilon_i$$

In logistic regression, the expression (without the random error term) becomes:

$$P(E \text{ given } x) = \frac{e^{\alpha_0+\beta_0 x}}{1+e^{\alpha_0+\beta_0 x}} \tag{9-11}$$

As we have seen, an expression of this form must fall in the range (0,1) regardless of the values of the parameter α_0 and β_0.

The event E in question (e.g., customer purchases a new product) might be assumed to relate to one or more of the person's characteristics: perhaps age, gender, income, and purchase history. An explanatory variable related to such factors might become x, and the parameters might be estimated so as to approximate the observed probabilities of purchase within a data set. For example, the product under study might be a hybrid car, and the aim is to predict the probability that a person in a certain age range who now drives a Toyota Corolla will purchase a hybrid when he next buys a car. In the data set used to estimate the parameters of the logistic regression model, there might be eight Toyota Corolla drivers at age 36 who recently bought a new car. (Here the explanatory variable x would be age: all the drivers in question would start with Toyota Corollas so that would not be a variable in the regression model (9-11).) If three of these drivers switched to hybrids at that purchase, then α^* and β^* would strive to offer a probability estimate close to 3/8 for 36 year-olds. They would similarly try to mimic the purchase probabilities for Toyota Corolla drivers at other ages.

The parameter estimation process would not be least squares; it would involve instead the maximum likelihood method that we met in Chapter 3. (We will shortly illustrate with an example how the technique is used.) The observed probabilities that the formula seeks to match are themselves subject to sampling error: while 3/8 of the 36 year-olds considered bought a hybrid car, the fraction might not be the same for all Toyota Corolla drivers at that age. In consequence, the parameters are fitted to observed probabilities that often are themselves imprecise estimates of the true probabilities. For such reasons, we recognize that the maximum-likelihood parameters are themselves imprecise, and we follow a familiar approach in offering 95% confidence intervals for the parameters to accompany the point estimates.

What about the odds of event E given x? Under (9-11), the probability that E *fails* to happen follows:

$$P(\text{not } E) = 1 - P(E) = 1 - \frac{e^{\alpha_0+\beta_0 x}}{1+e^{\alpha_0+\beta_0 x}} = \frac{1}{1+e^{\alpha_0+\beta_0 x}}$$

Canceling the common denominator in $P(E)$ and $1 - P(E)$ as we did earlier, we reduce the odds ratio $O(E)$ for event E to:

$$O(E \text{ given } x) = e^{\alpha_0+\beta_0 x}$$

This last expression allows an interpretation of the coefficient of x. Suppose that x increases by one to $x + 1$. Then we can write:

$$O(E \text{ given } x+1) = e^{\alpha_0+\beta_0(x+1)}$$

It follows that:

$$\frac{O(E \text{ given } x+1)}{O(E \text{ given } x)} = e^{\beta_0}$$

Put another way, the new odds ratio is the old one *multiplied by* e^{β_0}.

What can we say about this multiplier e^{β_0}? Clearly $e^{\beta_0} > 1$ for positive values of β_0, $e^{\beta_0} = 1$ when $\beta_0 = 0$, and $e^{\beta_0} < 1$ when β_0 is negative. (Right?) A positive β_0, in other words, means that the odds ratio increases when x increases by one, *regardless of the original value of* x. And when the odds ratio $O(E)$ goes up, so must $P(E)$. The upshot is that a positive β_0 means that the probability of E goes up as x does. A negative β_0 means that increases in x reduce $P(E)$; when $\beta_0 = 0$, event E could not care less about the value of x.

If instead x_1 decreases by one, then we have:

$$\frac{O(E \text{ given } x - 1)}{O(E \text{ given } x)} = e^{-\beta_0}$$

Here the reduction in x *decreases* the odds of E when $\beta_1 > 0$, increases the odds when $\beta_1 < 0$, and leaves the odds unchanged when $\beta_1 = 0$.

As in linear regression, we do not know the exact value of β_0; rather, we have a maximum-likelihood estimate β^* that arises from the curve-fitting exercise. Yet the intuitive interpretation of β^* is the same as suggested in the last paragraphs:

Sign of β^*	Effect of Increasing x	Effect of Decreasing x
Positive	$P(E)$ and $O(E)$ increase	$P(E)$ and $O(E)$ decrease
Zero	$P(E)$ and $O(E)$ unchanged	$P(E)$ and $O(E)$ unchanged
Negative	$P(E)$ and $O(E)$ decrease	$P(E)$ and $O(E)$ increase

Suppose that β^* is positive but the 95% confidence interval for β_0 includes zero. Then it is not outlandish to assert that $\beta_0 = 0$, meaning that we lack conclusive evidence that x affects $P(E)$. As we have said before, that outcome does not demonstrate that x is irrelevant: failing to prove that x matters is different from proving for sure that x does not matter. If, however, the confidence interval does not include zero, that circumstance would mean that zero is not in the "range of plausible values" for β_0. If the confidence interval is entirely positive and the logistic regression model is reliable, then the evidence is considerable that the odds ratio (and thus the probability) increases when x does.

Maximum Likelihood Estimation in Logistic Regression

As noted, the usual approach for estimating the logistic-regression parameters α_0 and β_0 is the method of maximum likelihood that we first met in Chapter 3. Here we illustrate how the method proceeds, although in practice the logistic regression package performs the calculation without our active involvement.

Suppose that the data set consists of yes/no questions about individuals, and some information about each one that is reflected by the number x. If the answer is Yes for a person i with x-coordinate x_i and No for a person j with coordinate x_j, then logistic regression with parameters α and β would imply that:

- The probability of getting Yes when $x = x_i$ is given by:

$$P\left(\text{Yes} \middle| x = x_i\right) = e^{\alpha+\beta x_i} / \left(1 + e^{\alpha+\beta x_i}\right)$$

- The probability of getting no at $x = x_j$ would follow:

$$P\left(\text{No}\middle|x = x_j\right) = 1/\left(1 + e^{\alpha+\beta x_j}\right)$$

Assuming independence, the probability of this joint outcome follows:

$$P\left(\text{Yes at } x = x_i \text{ and No at } x = x_j\right) = \frac{e^{\alpha+\beta x_i}}{\left(1 + e^{\alpha+\beta x_i}\right)} \frac{1}{\left(1 + e^{\alpha+\beta x_j}\right)}$$

If these were the only two observations, then the maximum likelihood estimates of α and β would be those that maximize the probability above.

More generally, the maximum likelihood estimates of α and β—which we call α^* and β^*—maximize the joint probability of all the observed yes/no outcomes. These quantities become the point estimates of, respectively, α_0 and β_0. Note that α^* and β^* are *not* least-squares estimates of the parameters: here we are maximizing rather than minimizing, and doing so for a function of α and β that is decidedly not quadratic. It's time for an example.

EXAMPLE 9.6

Work Makes Right?

In a Psychology experiment, a Professor is interested in how study improves performance on an especially difficult exam question. The exam is a vocabulary test; the questions are all multiple-choice with *four* possible answers. The Professor focuses on one question about a very obscure word. She takes it as a given that those who did not study at all would have no clue about the answer, and would have to guess randomly. The actual test outcomes were:

Of the ten students who did not study, three got the correct answer.

Of the ten students who studied for one hour, six got the correct answer.

Of the ten students who studied for two hours, eight got the correct answer.

Of the ten students who studied for three hours, eight got the correct answer.

Of the ten students who studied for four hours, nine got the correct answer.

The Professor wants to summarize this pattern by a logistic regression model of the form:

$$P\left(\text{correct answer}\middle|x \text{ hours of study}\right) = e^{\alpha_0+\beta_0 x}/\left(1 + e^{\alpha_0+\beta_0 x}\right)$$

(i) What value of α_0 would yield $P(\text{correct}|\text{no study})$ equal to the success rate under random guessing?

(ii) Using that value for α_0, what would be the likelihood function for these data in terms of β?

(iii) What value of β (i.e., β^*) maximizes the likelihood function?
(iv) In general terms, how well does the logistic formula for the chance of a correct answer given x hours of study correspond to the data for the 50 students?
(v) Can you imagine other values of β that also give a statistically good-fit to the data?
(vi) Why do you think you were asked question (v)?
(vii) Suppose that Minerva is about to take the test, and that she has studied for 1.5 hours. Based on the formula, approximately what is the probability she will get this question correct?

Solution

(i) With four choices, the probability of guessing the answer at random is 1/4. If those who didn't study will always guess at random, then:

$$P\left(\text{correct}\middle|\text{no study}\right) = P\left(\text{correct}\middle|x = 0\right) = e^{\alpha_0}/\left(1 + e^{\alpha_0}\right)$$

Equating that probability to 1/4 yields:

$$\frac{e^{\alpha_0}}{1+e^{\alpha_0}} = \frac{1}{4} \rightarrow e^{\alpha_0} = \frac{1}{3}$$

$$\text{Thus, } \alpha_0 = \ln\left(\frac{1}{3}\right) = -1.10$$

(ii) Insisting that α_0 must be –1.10 is like performing regression through the origin, in that one parameter is fixed at a particular value. Under this constraint, α_0 enters the logistic formula as $e^{\alpha_0} = 1/3$, so we will use 1/3 in what follows rather than something with the symbol α_0.
Among the ten students who did not study, the number who got the question correct is treated as the outcome of a binomial process with $n = 10$ and $p = 1/4$. Thus familiar formulas tell us that:
- *P*(three correct answers among ten students who did not study):

$$\binom{10}{3}\left(\frac{1}{4}\right)^3\left(\frac{3}{4}\right)^7$$

With similar reasoning and using the logistic formula with β_0 estimated as β, we write:
- *P*(six correct answers among ten students who studied one hour):

$$\binom{10}{6}\left(e^{\beta}/3\right)^6/\left(1+e^{\beta}/3\right)^{10}$$

This expression arises because the probability of a correct answer among those who studied one hour is $(e^{\beta}/3)/(1 + (e^{\beta}/3))$ while the probability of a wrong answer is $1/(1 + (e^{\beta}/3))$

- P(eight correct answers among ten students who studied two hours):

$$\binom{10}{8}\left(e^{2\beta}/3\right)^8/\left(1+e^{2\beta}/3\right)^{10}$$

- P(eight correct answers among ten students who studied three hours):

$$\binom{10}{8}\left(e^{3\beta}/3\right)^8/\left(1+e^{3\beta}/3\right)^{10}$$

- P(nine correct answers among ten students who studied four hours):

$$\binom{10}{9}\left(e^{4\beta}/3\right)^9/\left(1+e^{4\beta}/3\right)^{10}$$

(iii) The overall probability of the 50 outcomes is the product of the five expressions above. Perhaps the easiest way to find β^* is to conduct a grid-search, evaluating the exponential product in a spreadsheet for various values of β. Doing so here yields the estimate $\beta^* = 0.97$. Thus, the regression formula becomes:

$$P\left(\text{correct}\middle|x \text{ hours of study}\right) = \left(\frac{1}{3}\right)e^{.97x}/\left(1+\left(\frac{1}{3}\right)e^{.97x}\right)$$

(iv) Evaluating this expression at various values of x, we reach the estimates:

$P(\text{correct}|x = 0) = .250$

$P(\text{correct}|x = 1) = .468$

$P(\text{correct}|x = 2) = .699$

$P(\text{correct}|x = 3) = .860$

$P(\text{correct}|x = 4) = .942$

As to whether the observed outcomes at each x value are consistent with these estimates, the answer is easily yes. At $x = 1$, for example, the actual number who got the answer correct was six. In a binomial process with $n = 10$ and $p = .468$, the mean is 4.68, and thus the observed outcome was 1.32 away from the mean. Calculations for B(10, .468) show that probability of getting an observed outcome at least 1.32 away from the mean is 52%. We get similar outcomes at other values of x.

(v) Well, one way to "imagine" other values of β that fit the data well is to evaluate $P(\text{correct}|x$ hours of study) for other values. If we try $\beta = 1.1$, for example, we find:

$P(\text{correct}|x = 0) = .250$

$P(\text{correct}|x = 1) = .500$

$P(\text{correct}|x = 2) = .751$

$P(\text{correct}|x = 3) = .900$

$P(\text{correct}|x = 4) = .964$

$$\text{where } P(\text{correct}|x \text{ hours of study}) = \left(\frac{1}{3}\right)e^{1.1x} / \left(1 + \left(\frac{1}{3}\right)e^{1.1x}\right)$$

These probabilities are also easily consistent with the data for the 50 students.

(vi) In OLS regression, we estimated confidence intervals for β_0 because the random "errors" (ε's) in the data could create differences between β^* and β_0. Here the maximum likelihood method yields a formula that (ideally) matches the data well, but other values of β near β^* might also do well. We benefit if we have confidence intervals that identify the range of β-values that are nearly as consistent with the data as β^* itself.

(vii) If Minerva studies 1.5 hours, then the logistic regression equation suggests that she has a probability of $(1/3)e^{.97*1.5}/\left(1+(1/3)e^{.97*1.5}\right)$ of getting the right answer, which is 58.8%.

EXERCISE 9.7

Suppose that the Professor realizes that she misgraded one of the papers of a volunteer who had studied for one hour, finding that a person she had marked as wrong actually got the question correct.

(i) With this amendment to the data pattern, how does the likelihood function change?

(ii) Would the original parameter estimates $\alpha_0 = -1.10$ and $\beta^* = 0.97$ still provide a good fit to the data? Briefly discuss.

EXAMPLE 9.7

Challenger, RIP

As we discussed at the start of this chapter, concerns were expressed just before the space shuttle Challenger was launched that the temperature at Cape Canaveral was dangerously low. The shuttle had five O-rings, and it could not survive if all five failed on departure. On previous launches, some of the O-rings had failed (though not all), and the failure rate seemed at least somewhat related to the temperature at liftoff time. It was 36°F when the Challenger was to be launched.

(i) Treating the five O-rings as *independent* and using logistic regression on the data about previous space shuttle launches, approximate the relationship between temperature at launch and the chance of O-ring failure. (The data appear below.)

(ii) Discuss whether the regression formula seems plausible when its risk estimates are contrasted with actual O-ring failure rates.

(iii) Using the formula, estimate the probability that a given O-ring would fail at 36°F?

(iv) Assuming that different O-rings are independent in terms of failure, estimate the probability that all five would fail at 36°F.

(v) Why is there uncertainty about whether the logistic regression model is reliable at 36°F?

(vi) Do the problems in (v) justify disregarding the estimate in (iv)? Briefly discuss.

Solution

(i) As Table 9.1 below shows, there were 24 previous launches, at temperatures ranging from 53°F to 81°F. On the first one, the temperature was 66°F and no O-rings failed. Thus, there were five different data points of the form (66,0), each one meaning that the O-ring in question had an actual failure probability of zero. The second launch took place at 70°F and one of the O-rings failed. For that launch, the five points generated were (70,0), (70,0), (70,1), (70,0), and (70,0). (The overall probability was 20%, but we list the points separately to reflect the sample size of five at 70°F. For each point, the observed failure probability was either zero or one.) The 24 launches generated a total of 24 * 5 = 120 data points.

The logistic regression equation we use to describe the key relationship is:

$$P(\text{O-ring fails given temperature } t) = \frac{e^{\alpha_0+\beta_0 t}}{1+e^{\alpha_0+\beta_0 t}}$$

If we perform the logistic regression on all 120 data points that were available before launch time for the Challenger, we reach the estimate $\beta^* = -.147$. (Why not perform the regression yourself? Use the logistic regression module of your statistical package, or find a logistic regression program on the web.) The reported 95% confidence interval for β_0 extends from –.054 to –.239: importantly, it does not contain the value 0. Thus, under the model, there is statistically significant evidence that the risk of failure does relate to temperature at liftoff.

Because $e^{-.147} = .863$, the parameter estimate $\beta^* = -.147$ implies that odds of failure drops by about 14% (i.e., $e^{\beta^*} \approx .86$) when the temperature goes up by one degree Fahrenheit. It also implies that, when the temperature *drops* by one degree Fahrenheit, the odds of failure go up by a factor of $e^{.147} = 1.16$, or by about 16%. That outcome does not mean the probability of failure goes up by 16% for each one-degree drop in temperature: the probability formula is more complicated than that for the proportional change in odds. But because the odds ratio is going up, we can be sure that the failure probability is also increasing.

The regression package estimates the constant α_0 by $\alpha^* = 7.41$. Thus, its formula for failure probability at temperature t follows:

$$P(\text{fail at temperature } t) = e^{7.41-.147t}/\left(1+e^{7.41-.147t}\right)$$

TABLE 9.1 O-Ring Failures on Space Shuttle Launches Prior to 1/28/86

Temperature (°F)	Number of O-Ring Failures (out of 5)
66	0
70	1
69	0
80	0
68	0
67	0
72	0
73	0
70	0
57	1
63	1
70	1
78	0
67	0
53	3
67	0
75	0
70	0
81	0
76	0
79	0
75	2
76	0
58	1

(ii) Evaluated at various temperatures, the regression formula yields the following chart:

Temperature (Fahrenheit)	Estimated O-Ring Failure Probability
80°	.013
73°	.035
66°	.092
59°	.220
53°	.406

Because the formula implies a failure probability of 40.6% at 53°, we would expect on average about two failures among the five O-rings at that temperature (i.e., .406 * 5). We actually saw three on the 53° launch day, but that deviation from the mean would hardly be unusual for a binomial process with $n = 5$ and $p = .406$. Against the raw data used to calibrate the formula, the logistic regression curve modestly overestimates the observed risk for temperatures in the 60's, and underestimates for temperatures of 70°F and above. Again, however, these deviations between predicted probabilities and actual frequencies are consistent with binomial random fluctuation, and do not render the model unsuitable as an *approximation* to the key relationship.

The logistic regression model is reacting to is the following general pattern in the data:

Temperature Range	Overall Failure Rate in O-Ring Data
Below 60°	5/15 = 33.3%
60° or Higher	5/105 = 4.8%

Viewed this way, the data suggest that that the model was not being reckless in depicting a statistically-significant relationship between launch temperature and failure risk.

(iii) Evaluating the regression formula at $t = 36°F$, we reach the estimate that an O-ring has a 90% chance of failing at that temperature.

(iv) If each O-ring has a 90% chance of failing $t = 36°F$ and their individual outcomes are independent, then the chance is$(.9)^5 = 59\%$ that *all five* would fail.

(v) Because we have no data for temperatures below 53°F, we cannot be sure that the logistic regression model applies at lower temperatures. We cannot be sure that any regression model is reliable beyond the range of data from which it was calibrated.

(vi) The regression model reflects a real pattern in prior data: a sharp and rapid rise in failure rates as the temperature falls below 60°F. Taken literally, it implies that the space shuttle would be more likely than not to crash if launched at 36°F. Even though crash-risk estimate of 59% is not exact, on what grounds would one ignore the red flag that the analysis does is frantically waving? If the correct risk estimate is 40%, would anyone argue for launch at that risk level?

No such regression was performed on the data, and the Challenger was cleared for liftoff. Moments later, it crashed to earth killing everyone on board.

9.4 Multiple Regression Analysis

Despite all the attention we have lavished on it, the case in which Y depends solely on x is in fact rather rare. The sales of a product typically depend not just on advertising but also on, among other things, price, quality, the state of the economy, and the activities of competitors. And though the risk of cancer does depend on the radiation dose absorbed, it relates too to age, gender, and exposure to other carcinogens. A major reason for emphasizing the Y vs. x case is pedagogical: geometric notions are easier to visualize in two dimensions than five, and concepts grasped in the simpler case often apply without great revision in the more complex one.

A general model under which Y depends linearly on the m explanatory variables $x_i, \ldots, x_m$ is:

$$Y_i = \beta_0 + \beta_1 x_{1i} + \beta_2 x_{2i} + \ \ldots\ \beta_m x_{mi} + \varepsilon_i \tag{9-12}$$

where ε_i represents "measurement error" in Y_i, which is assumed to be zero-mean normal, and x_{ki} is the value of x_k for the ith individual.

Some of the x_j's might have been numerical at the outset, while others could constitute *indicator variables* devised to reflect some qualitative condition. For example, x_j might be 1 if a person is female and 0 if male, or x_j could be 1 if a restaurant accepts American Express cards and 0 if it does not. It is typically assumed that the ε_i's are independent and share the common distribution $N(0,\sigma_\varepsilon)$.

As in linear regression analysis, we can use the data points to make estimates of the constant β_0 in (9-12) as well as the coefficients (slopes) of the individual explanatory variables. (Note that β_k represents the change in y when x_k is increased by one and the other x_j's are held constant.) The usual process from which parameters estimates arise is called **multiple regression analysis**. In essence, multiple regression brings to $m + 1$ dimensional space the notions that animate linear regression in two-dimensional space.

Three-Dimensional Least Squares

We focus first on the case when Y is a function of only two explanatory variables. In that case, we can reduce the general model above to:

$$Y_i = \beta_0 + \beta_1 x_{1i} + \beta_2 x_{2i} + \varepsilon_i$$

The presumption is that we have n data points ($n > 3$) with which to estimate the three parameters (β_0, β_1, and β_2). We might interpret the data points geometrically as follows: Y reflects height above the ground, while x_1 measures position on an East-West axis and x_2 on a North-South axis.

As readers who have studied Analytic Geometry will be aware, $Y = \breve{\beta}_0 + \breve{\beta}_1 x_1 + \breve{\beta}_2 x_2$ is the equation of a *plane*, for any constants $\breve{\beta}_0$, $\breve{\beta}_1$, and $\breve{\beta}_2$. When $\breve{\beta}_1 = \breve{\beta}_2 = 0$, that plane is parallel to the ground at a height $\breve{\beta}_0$ above it. When $\breve{\beta}_0 = 0$, $\breve{\beta}_1 > 0$, and $\breve{\beta}_2 = 0$, the plane is at ground level when $x_1 = 0$, and moves further above the ground as x_1 gets more positive and deeper underground as x_1 gets more negative More generally, $\breve{\beta}_0$ connotes the height of the plane when x_1 and x_2 are both 0, while $\breve{\beta}_1$ and $\breve{\beta}_2$ engender the precise pattern of tilt.

There is no reason why least-squares analysis cannot proceed in a three-dimensional setting. As the quantities $\breve{\beta}_0$, $\breve{\beta}_1$, and $\breve{\beta}_2$ are varied, so is the orientation of the plane that they describe. And at any given orientation, one can compute the vertical distances between the n Y_i-values and the plane. (At the ith point, the vertical separation is $Y_i - \left(\breve{\beta}_0 + \breve{\beta}_1 x_{1i} + \breve{\beta}_2 x_{2i}\right)$.) Except in very special cases that we will discuss, a unique set of coefficients $\left(\widehat{\beta_0}, \widehat{\beta_1}, \widehat{\beta_2}\right)$ achieves the lowest sum of squares of the n residuals. These coefficients constitute by definition the least-squares estimates of their respective parameters.

Finding the least squares parameters in multiple regression is conceptually similar to what is done in OLS linear regression. When there are n parameters, one simultaneously sets a series of n partial derivatives equal to zero, a process that yields a set of n linear equations in n unknowns. Their solution yields the least-squares estimates.

Because multiple regression is so universally performed by computers in the 21st century, it is unnecessary for us to work through the details to be able to conduct multiple regression analyses. With that explanation (or excuse), we will not present the actual formulas here that express the $\hat{\beta}_j$'s in terms of the Y_i's and the x_{ji}'s. But while least-squares calculations are feasible in three dimensions, that alone does make them desirable. In the Y vs. x case, the least-squares estimators had a number of attractive properties Unless at least some of these properties survive the transplant to higher dimensions, the appeal of least-squares withers away. Happily, that survival occurs.

Similarities with the Two-Variable Case

"Nothing has changed; it's all the same."

— *The Beatles*

When y is regressed against x alone under the requisite conditions, the least-squares solution has the following virtues:

- α^* and β^* are minimum-variance unbiased estimators and maximum likelihood estimators of, respectively, α_0 and β_0 in $Y_i = \alpha_0 + \beta_0 x_i + \varepsilon_i$. *(Don't be confused by our use of β_0 as a slope in linear regression and the constant term in multiple regression. We are following familiar notation in both instances.)*
- Given α^* and β^*, one can readily compute an unbiased estimate of σ_ε^2, the variance of the common distribution of measurement error.

- The probability distributions of $\beta^* - \beta_0$ and $\alpha^* - \alpha_0$ are known under the assumptions of the model. One can therefore find confidence intervals for α_0 and β_0, and assess the strength of the evidence that these parameters differ from zero.
- One can readily determine the R^2-statistic, which suggests how much closer the least-squares line is to the data points than is the best-fit horizontal line.

Fortunately, it can be shown that all of these advantages are retained when the number of variables (dependent plus explanatory) goes from two to three. To elaborate on this statement, we consider a hypothetical example.

EXAMPLE 9.8

Threesome

Suppose that there are 100 data points and that least-squares analysis yields the computer printout below:

$$Y_i = 3.26 + 1.38x_{1i} - 1.97x_{2i}$$
$$(9.85) \quad (3.32) \quad (-1.35)$$
$$S_\varepsilon^2 = 9.09 \qquad R^2 = .874$$

where the numbers in parenthesis are t-statistics for the parameter estimates above them.

How should this output be interpreted?

Solution

The item-by-item demystification of this printout goes as follows:

(i) The least-squares estimate of β_0 is 3.26; of β_1, 1.38; of β_2, −1.97. These numbers also serve as maximum likelihood and unbiased estimates.

(ii) An unbiased estimate of the measurement-error variance σ_ε^2 is 9.09.

(iii) The t-statistics are defined as before and interpreted the same way. As there are 100 data points and $t_{\beta_1} = 3.32$, for example, the slope estimation error $W_1 = \widehat{\beta_1} - \beta_1$ is approximately normally distributed with mean zero, and the 95% CI for β_0 does not include 0. Moreover, because $\widehat{\beta_1} = 1.38$ while $t_{\beta_1} = \widehat{\beta_1}/S_{\beta_1} = 3.32$, the standard deviation of W_1 follows $S_{\beta_1} = 1.38/3.32 = .42$. And, finally, because $W_1 \approx N(0,.42)$ while $\widehat{\beta_1} = 1.38$, we can make inferences about the probability that β_1 falls in specific ranges under the reasoning discussed in Chapter 8.

(iv) $\widehat{\beta_2}$ and t_{β_2} allow the corresponding inferences about β_2, as do $\widehat{\beta_0}$ and t_{β_0} about β_0.

(v) Among horizontal models of the form $Y_i = c + \varepsilon_i$, the one closest to the data under the least-squares criterion chooses $c = \bar{Y}$ (as before). That $R^2 = .874$ means that the sum of squared residuals under the multiple-regression model is 87.4 % less than the corresponding sum for the best-fit horizontal plane.

Regression analysis in the three-variable (Y_i, x_{1i}, x_{2i}) case can proceed with as few as four data points. As before, an n larger than (say) two dozen or so continues to permit a zero-mean normality approximation for the parameter estimation errors. For small numbers of data points, we can use the fact that the distribution of $t_{\beta i} = W_i/S_{\beta_1}$ is, once again, in the t-family; this time, though, it has $n - 3$ rather than $n - 2$ degrees of freedom.

EXERCISE 9.8 In Example 9.8, estimate the probability that β_2 is in fact positive even though $\widehat{\beta_2}$ was −1.97.

A New Complication

Example 9.8 indicates that we could hardly ask for a more straightforward generalization of the two-dimensional theory. But behind the facade of continuity lurk some new elements in the internal computer calculations. Because these elements can affect the way the regression results should be evaluated, it is important to discuss them. To frame the issue, we return to an example that first arose in Chapter 7 about Analysis of Variance.

Suppose that a company wants to estimate the relationship between its sales of a product and two variables: the amount of advertising on the product's behalf (x_1) and the price charged for the product (x_2). It arranges to vary both price and advertising from district to district, and gathers n data points of the form (Y_i, x_{1i}, x_{2i}) where Y_i = observed sales in district i ($i = 1,\ldots,n$), x_{1i} = advertising per capita in district i, and x_{2i} = unit price in district i. (We assume that it is actually possible to control advertising and price to achieve the experimental conditions.) Then it desired to use the data to conduct OLS regression analysis on the model:

$$Y_i = \beta_0 + \beta_1 x_{1i} + \beta_2 x_{2i} + \varepsilon_i$$

where ε_i's are independent and $N(0, \sigma_\varepsilon)$

Prior to the regression analysis, however, a disconcerting pattern appears in the data. When (x_{1i}, x_{2i}) pairs are plotted, they cluster in two of the four possible quadrants on the high/low dimension (Figure 9.4):

Suppose that a cursory review of data makes clear that sales were higher in the quadrant with low price/high advertising than in the quadrant with high price/low advertising. Given that pattern, was the increase in sales caused by lower price alone? Higher advertising alone? Some weighted average of high advertising and low price? If so, what was the weighting factor?

The reader who answered "I don't know," has answered correctly. Because price and advertising were so strongly correlated in the data set (negatively so), trying to

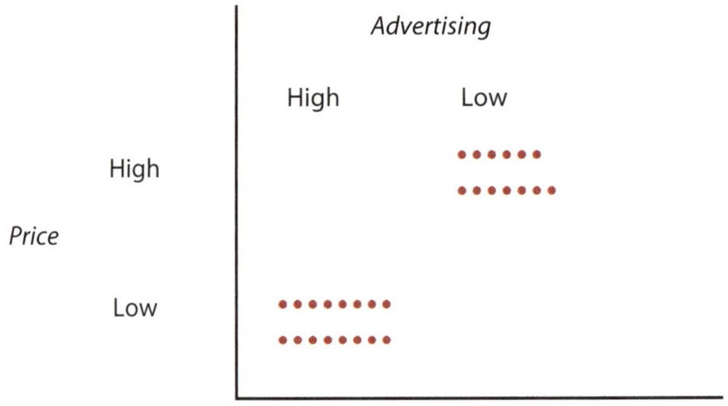

(Each ° represents a different district.)

FIGURE 9.4 The available data concern only two of four possible price/advertising combinations.

partition their joint effect between them is a hapless venture. If the aim of the experiment was to distinguish between sales stimulation caused by advertising and stimulation caused by low prices, then the endeavor was largely a waste of time.

Could the company have avoided this mess? Sure, if it had arranged to get data points in all four quadrants. Had it done so, watching the change in sales between the "low advertising/low price" cluster to the "high advertising/low price" one would have suggested advertising's impact when price is held constant. One would gain further insight by comparing the "low advertising/high price" and "high advertising/high price" clusters. Likewise, the company could have made low price vs. high price comparisons with advertising level held constant, and thus learned about price sensitivity among potential users of the product.

EXERCISE 9.9 Suppose that the data pattern just discussed had instead been:

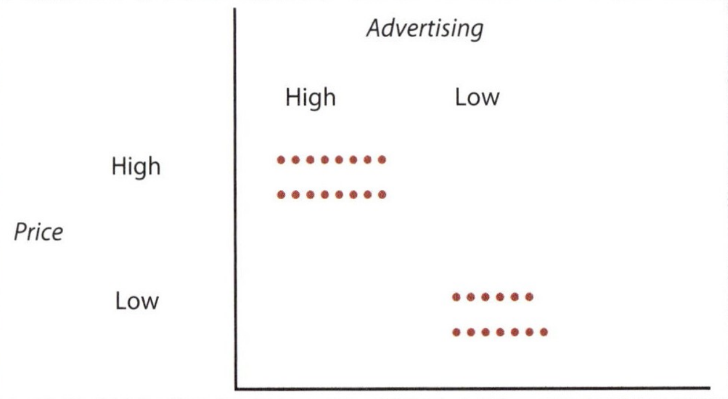

If sales were virtually the same at all data points, would that outcome indicate that neither advertising nor price affects sales? Are there other hypotheses about price and advertising that are consistent with these data? Briefly discuss.

Let's talk more generally about the issue just raised. Consider the data set $\{(Y_i, x_{1i}, x_{2i})\}$ for which the least squares fit is attained by the plane $Y = 2x_1 + 3x_2$. Suppose that, within this data set, $x_2 = 5x_1$ at each available point. Under these conditions, the plane $Y = 2x_1 + 3(5x_1) = 17x_1$ will generate exactly the same residuals as $Y = 2x_1 + 3x_2$.; hence it too must achieve the lowest sum of squares. So, for that matter, does the plane $Y = 2(x_2/5) + 3x_2 = 3.4x_2$, and so does $Y = x_1 + (x_2/5) + 3x_2 = x_1 + 3.2x_2$. In other words, not one but an enormous number of planes fulfills the least-squares requirement, and all of them give a different overall assessment of how x_{1i} and x_{2i} affect the value of Y_i.

Such a multiplicity of least-squares solutions is unavoidable when x_1 and x_2 are perfectly correlated in the data set; however, that adverse circumstance seldom arises. Far more frequently, x_1 and x_2 are strongly though not perfectly correlated, in which case least-squares regression *does* produce a unique solution. But the dilemma so visible in the extreme case does not disappear as soon as one departs it.

For example, suppose that in the situation just described, the relationship $x_2 = 5x_1$ is almost but not precisely satisfied. Then $Y = 2x_1 + 3x_2$ might actually triumph over all other planes in the least-squares competition. But $y = 17x$ and $y = 3.4z$ could have sums of squares that are only barely higher and, given the presence of random errors (the ε_i's), such "runners up" would seem nearly as likely as the "winning" plane to reflect the true causal relationship. No one, therefore, could say with strong confidence that β_1 is really 2 as opposed to 0 or 17, or that β_2 is really 3 as opposed to 0.

The generic problem is that correlations between explanatory variables could gravely threaten the accuracy of the least-squares parameter estimates. (Such a difficulty is unknown in the two-dimensional case, where there are no other explanatory factors with which x can be correlated.) To put the issue more technically, the variance of $W_1 = \widehat{\beta_1} - \beta_1$ depends not just on traditional factors like n and S_ε^2, but also on the degree of *association* between x_1 and x_2. Independence between these variables is most conducive to reliable slope estimation; a strong correlation (whether positive or negative) is least so.

A Tough Case

The ramifications of these comments are perhaps made clearer by a few pictures. Imagine that the pairwise relations among Y vs. x_1 and Y vs. x_2 are as depicted in Figure 9.5.

Bearing in mind the random dislocations in the y-coordinates, consider two important questions. Does only one of the explanatory variables affect Y and, if so, which? Or do both x_1 and x_2 exert influence and, if so, in what proportions? The reader who finds these questions difficult should sympathize with the plight of the computer, which must assign standard deviations for both $W_1 = \widehat{\beta_1} - \beta_1$ and W_2.

Given the real possibility that x_2 alone governs the movements of Y, the 95% CI for β_1 would presumably contain 0. But, by the same reasoning, 0 is also within the 95% CI for β_2. A person skimming the regression printout, therefore, might surmise that *neither* x_1 nor x_2 has significant effect on y. Such an interpretation, however, would

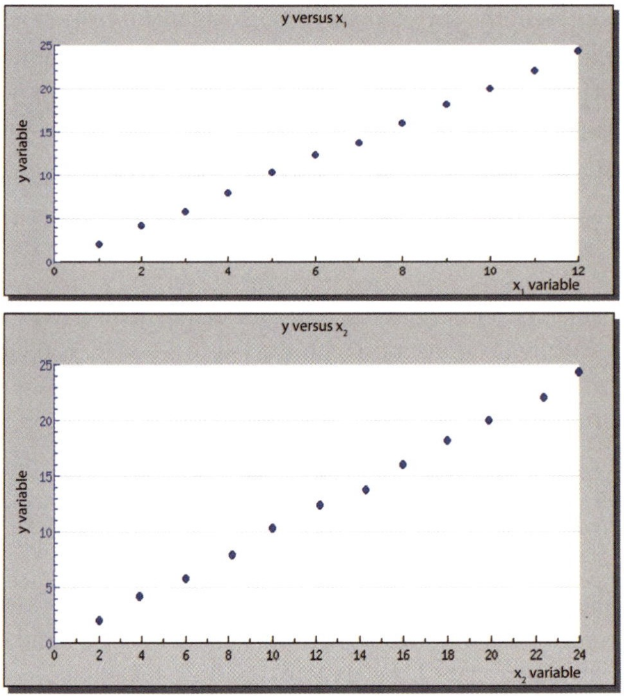

FIGURE 9.5 Two relationships involving Y, x_1 and x_2. (Note that the horizontal axis has a different scale in the two graphs.)

be most unfortunate: Of the four possible dependence relationships—both x_1 and x_2 affect y, x_1 alone does, x_2 alone does, and neither variable does—the last is surely the least consistent with Figure 9.5

We have entered a situation in which the whole is different than the sum of its parts; statements that are accurate about individual parameters can, taken together, give a misleading description of the overall pattern. In multiple regression analysis, having zero in the confidence intervals for all slopes fails to distinguish between two quite different cases:

(i) It seems plausible that none of the explanatory factors affects the dependent variable.

(ii) Though it seems likely that some (though perhaps not all) of the variables matter, strong correlations among them make it unclear precisely which factors are important.

Situation (ii) is generally referred to as **collinearity**.

Such ambiguity raises an obvious question: is there supplementary information that can help us distinguish (i) from (ii)? The answer, luckily, is yes. Many regression packages present the correlations among explanatory variables; when there are two such factors, we learn $\widehat{\rho_{12}}$, the correlation coefficient for x_1 and x_2 within the data set. Roughly speaking, a high $\widehat{\rho_{12}}$ would raise the possibility of the Figure 9.5 scenario, while a low $\widehat{\rho_{12}}$ would work against it.

However, this correlation coefficient is only partially illuminating. A high $\widehat{\rho_{12}}$ does not automatically mean that x_1 and/or x_2 affects y; it is possible, after all, that x_1 and x_2 are strongly correlated and yet both are irrelevant to Y. A more definitive resolution of the problem arises from the R^2-statistic which, whatever its limitations in the linear Y vs. x case, has a useful role to play in the present one. We will discuss the issue shortly in connection with the F-test of a regression model.

Many-dimensional Multiple Regression

The general multiple regression model is:

$$Y_i = \beta_0 + \beta_1 x_{1i} + \beta_2 x_{2i} + \ \ldots \beta_m x_{mi} + \varepsilon_i$$

Not too much changed when we moved from regression in two variables to regression in three, and going from three factors to a larger number entails even fewer conceptual adjustments. Least-squares analysis, t-statistics, S_ε^2 and R^2, all get generalized in the natural way.

What the Computer Says

The output we might get about an OLS multiple regression analysis might well include the following:

- The least squares estimates of the intercept(s) and slopes
- The t-statistic assigned to each of the estimates
- The 95% confidence intervals for each parameter
- The R^2-value for the regression analysis
- An F-test result
- An unbiased estimate of σ_ε^2

We will not go through the specific formulas in detail, because we rarely use them directly. We might, however, note that:

An unbiased estimate S_ε^2 of σ_ε^2 follows:

$$S_\varepsilon^2 = \sum_{i=1}^{n} \frac{r_i^2}{n-q} \tag{9-13}$$

where $q = m + 1$ = the number of parameters in the regression model (including the intercept β_0) and r_i is the ith residual

Equation (9-13) generalizes the results we saw in Chapter 8 for linear regression through the origin ($q = 1$) and general linear regression ($q = 2$).

Many regression printouts include a simple *correlation matrix* for the variables:

	Y	x_1	x_2	$\cdots$	x_m
Y	$\widehat{\rho_{yy}}$	$\widehat{\rho_{y1}}$	$\widehat{\rho_{y2}}$	$\cdots$	$\widehat{\rho_{ym}}$
x_1	$\widehat{\rho_{1y}}$	$\widehat{\rho_{11}}$	$\widehat{\rho_{12}}$	$\cdots$	$\widehat{\rho_{1m}}$
x_2	$\widehat{\rho_{2y}}$	$\widehat{\rho_{21}}$	$\widehat{\rho_{22}}$	$\cdots$	$\widehat{\rho_{2m}}$
		$\cdots$	$\cdots$		
x_m	$\widehat{\rho_{my}}$	$\widehat{\rho_{m1}}$	$\widehat{\rho_{m2}}$	$\cdots$	$\widehat{\rho_{mm}}$

where $\widehat{\rho_{ij}}$ is the correlation in the data between x_i and x_j while $\widehat{\rho_{iy}}$ is the correlation between x_i and y. (Note that $\widehat{\rho_{ii}}$ is always one, while $\widehat{\rho_{ij}} = \widehat{\rho_{ji}}$, meaning that all useful information in the matrix appears above its diagonal.)

The correlation matrix is increasingly helpful as m get larger because, as m goes up, so does the danger of "concealed significance." The more explanatory factors there are, the more numerous are the possibilities for pairwise correlations. And when two highly correlated variables both have confidence intervals including zero, it could be because the computer is unable to partition their joint effect between them. What is called collinearity when there are two explanatory variables is called **multicollinearity** when there are several.

The regression output might also present the variance-covariance matrix for the regression, which takes the form:

	β_0	β_1	β_2	$\cdots$	β_m
β_0	$S^2(\beta_0)$	$\text{cov}(\beta_0,\beta_1)$	$\text{cov}(\beta_0,\beta_2)$	$\cdots$	$\text{cov}(\beta_0,\beta_n)$
β_1	$\text{cov}(\beta_1,\beta_0)$	$S^2(\beta_1)$	$\text{cov}(\beta_1,\beta_2)$	$\cdots$	$\text{cov}(\beta_1,\beta_m)$
β_2	$\text{cov}(\beta_2,\beta_0)$	$\text{cov}(\beta_2,\beta_1)$	$S^2(\beta_2)$	$\cdots$	$\text{cov}(\beta_2,\beta_n)$
		$\cdots$	$\cdots$		
β_m	$\text{cov}(\beta_m,\beta_0)$	$\text{cov}(\beta_m,\beta_1)$	$\text{cov}(\beta_m,\beta_2)$	$\cdots$	$S^2(\beta_m)$

where $S^2(\beta_j)$ is the estimated variance of β_j (when β_j is treated as a random variable with a distribution based on the data) and $\widehat{\text{cov}(\beta_i,\beta_j)}$ is the estimated covariance of the random variables β_i and β_j. (Because $\widehat{\text{cov}(\beta_i,\beta_j)} = \widehat{\text{cov}(\beta_j,\beta_i)}$, the elements below the diagonal of the matrix repeat the information above the diagonal.

In the discussion about regression and prediction at the start of this chapter, we noted that the estimation errors for the slope and intercept are correlated in linear regression (as are α_0 and β_0, when they are treated as random variables). The variance-covariance matrix generalizes the earlier treatment to multiple regression analysis with the $m + 1$ random variables ($\beta_0 \ldots \beta_m$). The matrix is useful for making predictions about new data points, and also for describing the behavior of combinations of the β_i's. (We'll illustrate this last point shortly.)

Does Anything Matter?

The regression output also addresses a key question raised above: Is there evidence that at least *some* of the x_{ij}'s affect Y_i, even if we're not sure which they are? There is a helpful procedure tied to the F probability distribution and ultimately to our friend R^2.

R^2 is the reduction in the sum of squared residuals when the $m + 1$-dimensional regression line replaces the simple line $y = \bar{Y}$. We might intuitively expect that a test of the hypothesis H_0 given by:

$$H_0\colon \beta_1 = \beta_2 = \ldots \beta_m = 0$$

could take the form:

$$\begin{cases} \text{Accept } H_0 \text{ if } R^2 \le c \\ \text{Reject } H_0 \text{ if } R^2 > c \end{cases} \quad (9\text{-}14)$$

This test has appeal because, even if none of the explanatory variables matter, having m extra parameters to match to the data (i.e., the $\widehat{\beta_j}$'s) should yield some improved fit compared to the line $y = \bar{Y}$. Thus, a small positive R^2 might not signify real improvement, just meaningless tweaking. But once R^2 exceeds a certain threshold, we sense that it reflects a real relationship between Y_i and one or more of the x_{ij}'s. In fact, a test of form (9-14) is applied, though it actually involves the F-ratio defined by:

$$F = \left(\frac{n - m - 1}{m}\right)\left(\frac{R^2}{1 - R^2}\right) \quad (9\text{-}15)$$

Note that, as R^2 increases, so does F in (9-15).

If all the slopes in the multiple regression equation are zero (i.e., if none of the "explanatory" variables explain anything about Y), then F in (9-15) follows an F-distribution with $(m, n - m - 1)$ degrees of freedom. (We met the F-distribution, you might recall, in connection with Analysis of Variance.) Therefore:

An F-test of $H_0\colon \beta_1 = \beta_2 = \ldots \beta_m = 0$ takes the form:

$$\begin{array}{ll} \text{Accept } H_0 & \text{if } F \le d \\ \text{Reject } H_0 & \text{if } F > d \end{array} \quad (9\text{-}16)$$

Here F is given by (9-15) and d is some high percentile (e.g., the 95th) of the F-distribution with $(m, n - m - 1)$ degrees of freedom.

The statistical package typically offers us the p-value of the calculated F-statistic, and we might reject H_0 if the p-value falls below (say) 0.05. We typically see something in the output like:

F	Pr > F
59.567	< 0.0001

What this display means is that the F-statistic for the OLS regression line was 59.57, and that the p-value of the test of H_0 (against the alternative H_a that at least one of the β_i's is nonzero) was below 1 in 10,000. Thus, we might reasonably conclude that the β_i's collectively matter, though we might not know how to divide the credit among them.

EXAMPLE 9.9

Statin Island?

Suppose that a multiple regression analysis is performed on data about a random sample of 200 men from Nassau County on Long Island, New York, in which cholesterol level is the dependent variable while age and weight are the explanatory variables. The results are as follows:

$$\begin{aligned} Y = \ & 94 \quad + \ .22x_1 \ + \ .29x_2 \\ & (9.33) \ (0.57) \quad (0.44) \end{aligned}$$

where Y = cholesterol level, x_1 = age in years, and x_2 = weight in pounds (The numbers in parenthesis are t-statistics.)
$R^2 = .067$. The F test on H_0: $\beta_1 = \beta_2 = 0$ yields a p-value of .27.

(i) In general terms, what does the regression equation say?
(ii) What do the R^2-value and F test results imply?

Now suppose that someone suggests that the analysis should take into account whether the men are taking statins to reduce their cholesterol levels. The information is obtained for these men, and a 0–1 indicator variable for statin use is added to the model. The revised regression equation is:

$$\begin{aligned} Y = \ & 55 \quad + \ .96x_1 \ + \ .52x_2 \ - \ 54x_3 \\ & (4.88) \ (3.65) \quad (3.20) \quad (9.55) \end{aligned}$$

where $x_3 = \begin{cases} 1 \text{ if a man takes a statin} \\ 0 \text{ if a man does not} \end{cases}$

$R^2 = .73$ The F test on H_0: $\beta_1 = \beta_2 = \beta_3 = 0$ yields a p-value below .0001

(iii) What does the revised regression equation suggest?

(iv) What about R^2 and F?

(v) What is the moral of this story?

Solution

(i) The regression equation suggests a modest tendency for cholesterol levels to increase with age and weight. However, the low t-statistics reflect enough uncertainty in the parameters that neither estimate differs significantly from zero.

(ii) The R^2 result suggests that the regression model only slightly reduces squared residuals compared to the model that estimates each man's cholesterol level by the average level across all ages and weights. The F-test works against the notion that collinearity between age and weight is the reason that both 95% confidence intervals for the slope included zero.

(iii) The revised regression equation suggests that, for men in the same category of statin use, there is a statistically significant increase in cholesterol levels with age and weight. Men who take statins benefit from a 54-point drop in the cholesterol levels they would otherwise have.

Here statin use is reflected by a 0–1 **indicator variable**, a frequently-used device in multiple regression analysis. Because $x_3 = 1$ for statin users and zero for others, the estimated cholesterol level for a man of age x_1 and weight x_2 follows:

$$Y = \begin{cases} 55 + .96x_1 + .52x_2 & \text{if he takes no statin} \\ 55 + .96x_1 + .52x_2 - 54 & \text{if he does take a statin} \end{cases}$$

(iv) The rise in R^2 from .067 to .73 suggests a dramatic decline in squared residuals when statin use is considered (Both R^2 calculations used as their baseline the same quantity, namely, $\sum_{i=1}^{n}(Y_i - \bar{Y})^2$, where Y_i is the ith man's cholesterol level.) And the new F-test result implies that the hypothesis that the true slopes are zero for all three explanatory variables is preposterous.

(vii) It's not a great idea to omit important variables from a regression model. In this case, the omission led to the misleading conclusion that cholesterol levels scarcely vary with age or weight, when the reason for the small variation could be that higher cholesterol levels among older and heavier men led many of them to take statins that sharply cut their cholesterol levels.

EXERCISE 9.10 Given the discussion above, what would you expect the signs in the correlation matrix involving the four variables (Y, x_1, x_2, and x_3) to be? Which correlations would you expect to be positive, which negative, and which close to zero? Are there some correlations that you cannot deduce from the discussion above?

EXAMPLE 9.10 A Visit to Parry Sound

Multiple regression analysis can help in Meteorology by providing approximate weather conditions at particular places based on known conditions at surrounding locations. Here we illustrate how the approach might work, using data from the Canadian province Ontario.

The aim is to approximate the temperature at noon at Parry Sound, Ontario, given the temperatures at that time in four other Ontario locations: North Bay, Sault Ste. Marie, Timmons and Toronto. Of course, we actually know the temperature at noon at Parry Sound, so there is no need to approximate it. In a real situation, meteorologists might go to a location of interest on a few occasions to measure the temperature there, and then devise a formula based on surrounding temperatures that comes close to matching the actual one. They would use that formula in the future in lieu of true readings at the place of interest.

To obtain data for the regression analysis, we record temperatures at the five Ontario locations at noon every ten days in 2011, starting on January 5. That effort yields 36 data points—three for each month—and covers all four seasons of the year. The raw data appear below in Table 9.2.

It seems reasonable to assume a linear relationship between the dependent variable—temperature Y_i at Parry Sound on the ith given day—and the concurrent temperatures (x_{ji}'s) at the four surrounding weather stations. The formula would take the form:

$$Y_i = \beta_0 + \beta_1 x_{mi} + \beta_2 x_{si} + \beta_3 x_{ti} + \beta_4 x_{ni} + \varepsilon_i \tag{9-17}$$

where the subscripts m, s, r, and n refer, respectively, to Timmons, Sault Ste. Marie, Toronto, and North Bay

(i) Perform a least-squares multiple regression analysis to calibrate (9-17) with the data below.

(ii) Then discuss the findings.

Solution

Given these data, the OLS regression equation of form (9-17) is:

$$Y_i = 2.399 - 0.141x_{mi} + 0.052x_{si} + 0.054x_{ti} + 0.963x_{ni}$$

Among the implications of this equation is that the Parry Sound temperature would be estimated at 2.399°C on a day that the temperatures were zero at the four surrounding points. We also see that the temperature at North Bay plays a much larger role in the estimation formula than those at the other locations, given that its slope estimate is near one while the other slope estimates are near zero.

The correlation matrix for the variables—which presents the various coefficients of correlation is:

TABLE 9.1 Noon temperatures Celsius at the five Ontario locations (all in same time zone)

2011 Date	Timmins	Sault Ste. Marie	Toronto	North Bay	Parry Sound
5-Jan	−11.1	−4.7	−1.5	−10.7	−6.5
15-Jan	−11.5	−5	−2.5	−9.9	−5.5
25-Jan	−8.7	−4.7	−1	−9.4	−8.1
5-Feb	−1.7	−0.3	−1.9	−2.9	−2
15-Feb	−12.2	−6.8	−7.6	−13.2	−8.6
25-Feb	−16.2	−10.4	−0.7	−7.5	−2.5
5-Mar	−13.3	−3.7	7.5	−3.6	0.1
15-Mar	5.4	2.2	5.1	0.7	3.4
25-Mar	−13.2	−7.5	−3.6	−9.8	−5.7
5-Apr	−4.6	0.8	3.5	−2.5	−0.2
15-Apr	0	3.4	3.3	0.8	6.2
25-Apr	14.2	14.5	10.3	10.5	9.9
5-May	13.5	16.7	15.1	12	12.1
15-May	4.1	12.8	9.1	8.7	12
25-May	12.2	12.4	14.5	12.4	15.7
5-Jun	14.6	16.8	25.3	16.7	17.7
15-Jun	26.4	23.9	25.9	24.8	21.8
25-Jun	14.5	15.6	19.5	16.9	15.7
5-Jul	25.9	22	26.1	23	23.2
15-Jul	25.3	24.2	26.8	24.8	27
25-Jul	21.6	21.6	25.4	21	22.5
5-Aug	29.4	28.4	26.3	25.9	27.1
15-Aug	25.4	21.5	25.5	24.5	25.9
25-Aug	13.3	17.7	23.6	19.8	22
5-Sep	6.9	11.6	16.5	10.1	12.8
15-Sep	6.6	8.7	14	7.8	11.2
25-Sep	18.4	17.1	22.3	18.8	21.3
5-Oct	11.7	16.6	17.9	13.3	16.5
15-Oct	6.2	7.6	10.7	8.4	9.5
25-Oct	4.2	8.2	11.9	4.8	8.3
5-Nov	6.9	8.5	6.7	4.7	6.8
15-Nov	5.9	7.3	10	8.1	10.5
25-Nov	6.6	10.2	13	6.4	9.9
5-Dec	−14.5	−2.1	6.2	−3.6	0.9
15-Dec	0.5	3.8	13.3	7.9	10.6
25-Dec	−1.4	−0.5	4.1	−1.1	2.7

Variables	Timmins	Sault Ste. Marie	Toronto	North Bay	Parry Sound
Timmins	**1.000**	0.975	0.908	0.963	0.942
Slt Ste. Mar	0.975	**1.000**	0.939	0.976	0.962
Toronto	0.908	0.939	**1.000**	0.972	0.971
North Bay	0.963	0.976	0.972	**1.000**	0.991
Parry Sound	0.942	0.962	0.971	0.991	**1.000**

The correlation coefficient between (say) Toronto and North Bay answers the question: over the course of the year, did high temperatures (and low temperatures) at the two locales tend to occur together? For that reason, it is no surprise that all the correlations in the table are high. High temperatures at Toronto occur during the summer, when temperatures are also high throughout Ontario. More generally, the temperature paths at the different sites go up and down together over the course of the year, so they are strongly correlated. Understanding the high correlations, however, does not diminish the danger that will result in unreliable slope estimates because of multicollinearity.

The regression output includes the following information about uncertainty in the slope and intercept estimates.

Source	Value	Standard error	*t*	*P*-value	Lower bound (95%)	Upper bound (95%)
Intercept	2.399	0.723	3.320	0.002	0.925	3.873
Timmins	–0.141	0.095	–1.482	0.148	–0.335	0.053
Slt Ste. Marie	0.052	0.131	0.394	0.696	–0.216	0.319
Toronto	0.054	0.115	0.472	0.640	–0.180	0.289
North Bay	0.963	0.163	5.920	< 0.0001	0.631	1.295

Several points arise from this table. The p-values in the table pertain to the hypothesis that the associated parameter estimate is actually zero. They reflect the probability under that hypothesis that slope estimation error would yield a parameter estimate at least as far from zero as the estimate that actually arose. Thus, the p-value of 0.002 for the hypothesis that the intercept $\beta_0 = 0$ implies that we should take the positive intercept-estimate seriously. Among the explanatory variables, only North Bay has a slope whose 95% confidence interval that does not include zero. (North Bay also has by far the largest absolute slope estimate.) Actually, the statistical significance of the North Bay parameter is encouraging: multicollinearity was clearly present here, but it did not preclude getting a statistically significant finding. The larger point is that, so long as the explanatory variables are not perfectly correlated, a strong enough pattern can "break through" the confusion that multicollinearity can cause about which variables have real influence.

The printout also informs us that $R^2 = .984$ But this outcome is not breathtaking: the baseline in the R^2 calculations is average squared spread in noon temperatures at Parry Sound over the course of the year (i.e., $\sum_{i=1}^{36}(Y_i - \bar{Y})^2$, where $\bar{Y}$ is the average of the 36 recorded temperatures). That variation is substantial, and even a simple model that treats summer as warmer than winter can fit the data much better.

EXERCISE 9.11 Assuming that slope estimation error for the Toronto variable is normally distributed with mean zero and the standard deviation (standard error) shown above, estimate the p-value for the hypothesis that Toronto's true slope is zero in the Parry Sound regression model. Do you come close to the estimate in the printout?

EXAMPLE 9.11

A Simpler Model for Parry Sound?

An obvious temptation given the regression results above is to reduce the prediction formula to a bivariate regression analysis of the form:

$$Y_i = \alpha_0 + \beta_0 x_{ni} + \varepsilon_i \tag{9-18}$$

The argument is that the other three locations had low slope estimates and confidence intervals for their slopes that contained the value zero. Maybe they constitute meaningless clutter and complexity when the North Bay temperature is the only serious predictor.

If we give in to that temptation, what is the revised model about Parry Sound?

Solution

Running the regression (9-18) against the 2011 data, we get the following least squares equation:

$$Y = 3.09 + .902\, x_n$$

$$R^2 = .981$$

Source	Value	Standard error	t	P-value	Lower bound (95%)	Upper bound (95%)
Intercept	3.090	0.287	10.778	< 0.0001	2.507	3.673
North Bay	0.902	0.021	42.445	< 0.0001	0.858	0.945

Here the intercept estimate has increased from about 2.4 to 3.1, while the slope of the North Bay variable has slightly decreased, from 0.963 to 0.902. Note that the standard errors assigned to the parameter estimates have dropped sharply compared to those in the original regression analysis, presumably because multicollinearity has disappeared as a source of uncertainty.

In light of these findings, we might reasonably argue that (9-18) offers a better rule for estimating Parry Sound temperature than (9-17). Simplicity has its charms, and a one-variable formula is simpler than one with four variables. And eliminating three of the four variables only reduced R^2 by .003, from .984 to .981.

EXERCISE 9.12 Using the data in Table 9.1, perform a linear regression analysis of the form:

$$Y_i = \alpha_0 + \beta_0 x_{ti} + \varepsilon_i \tag{9-19}$$

where x_{ti} is the noon temperature at Timmins on the ith day and Y_i the concurrent temperature at Parry Sound. Discuss the results briefly but incisively. What do you make of the fact that x_{ti} achieves a statistical significance in (9-19) that it failed to achieve in (9-17)?

Adjusted R^2

Actually, the case for modeling for Parry Sound temperature using one explanatory variable rather than four is even stronger than we might realize. It is true that the four-variable model in Example 9.10 had a higher R^2 than the North Bay model in Example 9.11, namely, .984 rather than .981. But is it really fair to compare those R^2-values? The original model had five parameters (including the intercept) it could use to optimize fit to the data, while the simpler model had only two.

We should recognize that, as defined, R^2 will almost surely increase if an another explanatory variable is added to a regression model. (R^2 cannot possibly decrease in that circumstance, because there is always the option of setting the new variable's slope at zero and thus achieving the same sum-of-squares as existed before its inclusion.) And, chances are, the additional variable will allow a bit of fine-tuning that increases R^2 *even if the new variable is actually irrelevant.*

For that reason, attention has been drawn to the question of whether R^2 has meaningfully improved as a model includes more variables. An approach pioneered by Thiel that is discussed at the website has won favor. His adjusted R^2 statistic takes account of the number of explanatory variables, and can actually decrease if added variables do not sufficiently reduce the sum of squared residuals. This adjusted R^2-statistic is routinely listed in regression-analysis output.

EXAMPLE 9.11 **Did North Bay Win?**

How does adjusted-R^2 for the Parry Sound model with North Bay alone compare with that in the model based on four surrounding locations in Ontario?

Solution

We need merely compare the printouts for the two regressions to answer this question. R^2_{adj} was .981 (to the nearest thousandth) in the North Bay model, almost unchanged from its original R^2 value. R^2_{adj} was .982 for the four-variable model, down from its original R^2 of .984. But .982 is greater than .981—however infinitesimally—so we cannot declare that adding the three other locations to North Bay was utterly worthless. Yet the improvement was so slight that we might reasonably treat it as not worth the added complexity.

9.5 Three Troubled Regressions (all based on true stories)

To end this chapter, we offer brief overviews and critiques of three separate regressions. All these regressions suffer some problems, but they are not riddled with troubles from beginning to end. Their study provides some useful insights about how multiple regression should be properly performed.

Discussing the flaws in these analyses does not suggest that regression is a disreputable technique. The advice columnist Ann Landers once said that recognizing a difficulty is halfway to solving it; for some of the problems we will discuss, the proportion is even greater than 1/2. The attack is not on regression analysis *per se* but on the thoughtless use of regression. Simply to work through the first regression that comes to mind can be dangerous; the path of least resistance is sometimes the road to ruin. If the examples ahead make that point, they will have served their purpose.

ILLUSTRATION 9.A **The Future Demand for Cookies**

A producer of cookies was faced with a decision on whether to expand its baking capacity. Cookie consumption had been rising in recent years and, even after per capita figures were compiled to weed out the effect of population growth, an upward trend was still apparent. Without some new facilities, demand would soon exceed supply if the upsurge were to persist.

To affirm the authenticity of the visual pattern in the data, it was decided to calibrate a regression model of the form:

$$Y_t = \alpha_0 + \beta_0 t + \varepsilon_t$$

where Y_t = per capita cookie consumption at t; t = time in years as measured from some baseline time 0; ε_t = zero-mean random variation around the cookie-consumption trend line

No one was suggesting that time on the clock could itself affect cookie consumption; rather, time was serving as a surrogate for continuing changes in dietary patterns. A significant and positive estimate of β_0 would be construed as reflecting real growth over time in the demand for cookies.

Least-squares regression was performed and the slope estimate β^* was positive. Moreover, the 95% confidence interval for β_0 did not include zero, R^2 was very high, and the residuals behaved as expected. (Draw yourself a picture consistent with this pattern.) This outcome was thoroughly congenial to the proposals for additional ovens.

In subsequent years, however, a problem became evident: consumption did *not* rise at roughly β^* per year but instead embarked upon a path of steady decline. The regression analysis, in other words had provided a grossly misleading description of the process it was meant to portray.

What had gone wrong? Though there are many conceivable villains, the "silent killer" was in fact demographic: the model had corrected for changes in the *size* of the population but not in its age distribution. (Given that cookies are particularly popular among the young, that was a perilous omission.) In the years used for the regression, the post-World War II baby boom was steadily raising the proportion of people in the "cookie-eating" age range; in the *post-regression period*, by contrast, the end of the boom was reducing that proportion year by year. Per capita cookie sales were declining because their prime consumers were dropping in number.

This example highlights the hazard of **omitted variables**. If a factor that affects the dependent variable is excluded from a regression, some variable correlated with the absentee might erroneously get credit for the latter's influence. It is far from guaranteed that R^2, a confidence interval, or some other indicator will send up a flare to warn that something is awry.

Discussion

Before leaving this defective regression, we might consider how it could have been modified. One is tempted to suggest its replacement by the alternative regression:

$$Y = \alpha_0 + \beta_0 f + \varepsilon$$

where f = the *fraction* of people in the "prime age range" for cookie consumption

This least-squares analysis would presumably have generated a positive and significant β^*. That outcome—coupled with the widely-available knowledge that f-values were about to start *falling*—would have given fair warning that per-capita sales of cookies were destined to decline rather than rise.

We must be wary, however, of the eternal hazard of confusing hindsight with foresight. To exclude the factor t from the regression is to deny even the possibility that cookies were experiencing a genuine rise in popularity. In retrospect, that assessment might be correct, but it would be far better for that conclusion to arise as a *consequence* of the analysis than as one of the starting premises. To refuse t even the chance of an explanatory role is to risk an omitted-variables problem just as surely as to leave out f.

These concerns bring us to the three-dimensional regression:

$$Y = \alpha_0 + \beta_0 t + \gamma_0 f + \varepsilon$$

The author considers this regression the most defensible of the three considered. If f affected per capita cookies sales over the observation period but t did not, the hope would be that β^* would be close to 0 and γ^* significantly higher than 0.

There is, of course, no certainty that this last hope would be realized. Clearly implicit in the preceding discussion is the point that, within the data available for regression, f and t were highly correlated. Confronted with such collinearity, the computer might well report that zero is in the 95% CI's for both β_0 and γ_0. Coupled with the high R^2, that regression results would imply that at least one of the two explanatory factors matters, but it would tell us little beyond that point.

But such an outcome should not be viewed as a calamity. A highly equivocal finding would imply that β_0 and/or γ_0 is positive but that neither can be estimated reliably; hence the sales growth was caused by some unknown combination of changing diets and transient demographics. But by this open confession of inadequacy, the regression might stimulate the more specialized studies that would resolve the matter once and for all. Surely that is a happier end to the analysis than a quick, definitive conclusion that is wrong.

Obviously, one cannot include in a regression *every* variable that might conceivably be relevant. But when a factor has a reasonable chance of being important, to exclude it from the modeling is to risk substantial distortion. And, as this example suggests, the mistake can go undetected until it results in serious harm.

EXERCISE 9.13 How do the consequences of the omitted-variable problem in Illustration 9.A compare with those of the omitted variable problem in Example 9.9, the regression about cholesterol levels?

ILLUSTRATION 9.B

Unemployment Benefits and the Incentive to Work

A business analyst argued in a prestigious publication that high unemployment benefits *cause* high unemployment, the reason being that they reduce the pressure on the jobless to find work. He was hardly the first to espouse this view but, unlike many others, he tried to back it up statistically. First he posited the model:

$$U = \alpha_0 + \beta_0 x + \varepsilon$$

where U = rate of unemployment in the period of study; x = an index of the generosity of unemployment benefits (related to their duration, mean monetary value, etc.)

Then he gathered data from several Western nations in a given year, and estimated the model's parameters via ordinary least-squares (OLS) regression. When he obtained a positive and significant value of β^*, he advanced that finding as evidence for his thesis.

Though his empirical approach is in no sense irrational, neither is it wholly persuasive. He treated U as the dependent variable and x the explanatory factor; arguably, however, it should be the other way around. When the economy is booming and thus U is low, the compensation offered the jobless might be minimal; the feeling might be that anyone wanting a job can find one and hence that to raise x above subsistence would be to subsidize sloth. But when the economy is in real trouble—as evidenced by a high U-value—the sense might grow that the jobless are the victims of forces beyond their control. To augment their suffering with harsh deprivation might seem cruel; hence x might increase *in response to* the growth in U.

It might, in other words, be more plausible *to regress x on U rather than U on x.* The general issue we are discussing, of course, is the direction of causality. OLS regression requires the user to answer that question at the start of the work; the variable on the left is the *dependent* variable, while those on the right are explanatory. The only possible outcomes are a significant are a significant casual relationship in the assumed direction (right to left) or no clear relationship at all. Should the cause-and-effect assessment reverse the true pattern, there is no reason to expect the regression results to make that problem apparent. If after all, a Y-on-x regression gives a "good fit" to the data, it seems rather likely that an X-on-y regression would do the same.

Especially when a regression analysis enters the area of policy debate, a direction-of-causality problem can be destructive. In the regression under discussion, if movements in x induce movements in U, then a cut in benefits will increase the number of jobholders. But if, historically, it is x that has followed U, then suddenly to cut benefits might do less to stimulate employment than to stimulate greater misery.

In this situation, some supplementary analyses could be revealing. The researcher restricted his attention to "cross-sectional" data about U and x values for several countries in one recent year. But given uncertainty about whether changes in x stimulate changes in U or vice-versa, "longitudinal" studies in given countries that trace the *timing* of movements in the two variables might help clarify the "leader-follower" relationship. Of course, we might be facing a "feedback" process in which U affects x and x also affects U. (A recession, for example, might lead to higher unemployment benefits, which then reduce the incentive to search for jobs and thereby prolong the recession.) With sufficient data, even a contingency like this could be investigated.

The "takeaway" from this example is that it is important to think hard about the cause-and-effect issue and not, in the understandable desire to get started on the regression analysis, to treat the issue nonchalantly.

ILLUSTRATION 9.C

Alleged Job Discrimination at the University

A female faculty member denied promotion at a state college filed suit against the school for sex discrimination. To buttress her claim that she was not an isolated victim, she included in her brief a multiple regression analysis that suggested that the school's female academics were systematically paid less than their male counterparts. The reliability of this regression became an issue in the courtroom proceedings.

The discussion here is not about the merits of the woman's lawsuit, of which the salary regression was only one part. But that salary analysis provides an unusually interesting setting to review an underestimated threat to the use of multiple regression model: the hazard of functional form. It is for this methodological reason that we pursue the statistical evidence into the courtroom.

A Salary Regression

In essence, the analysis treated seniority and gender as the only two determinants of a faculty member's salary. The relationship among the three variables was represented by the linear model:

$$S_i = \alpha_0 + \beta_0 x_i + \gamma_0 z_i + \varepsilon_i \tag{9-20}$$

where S_i = annual salary in dollars for the ith faculty member

x_i = faculty member i's total number of years of college teaching

z_i = a gender-indicator variable = $\begin{cases} 1 \text{ if faculty member } i \text{ is a woman} \\ 0 \text{ if faculty member } i \text{ is a man} \end{cases}$

ε_i = a zero-mean random variation term

One can imagine other variables that might affect salary patterns, such as field of study or individual merit. The first of these factors was deliberately excluded from the analysis: it was argued that, if salary were lower in areas of study in which women were concentrated, that might be a *consequence* of sex discrimination rather than some innocent explanation for any sex-related differences. Merit-pay differentials across individuals were assumed to be reflected in the ε_i's. The argument was that, if differences in merit were uncorrelated with seniority or gender, then performance-related pay variations could, in a macroscopic sense, be treated as fluctuations around the overall trend line.

Given the definition of z, a "two-track" salary pattern is implied by (9-20). Because $z = 0$ for men, a male academic is paid under the general formula:

$$S_M = \alpha_0 + \beta_0 x + \varepsilon$$

And because $z = 1$ for women, a female faculty member is compensated according to:

$$S_F = \alpha_0 + \beta_0 x + \gamma_0 + \varepsilon$$

Because the ε's for both genders are assumed to have means of zero, $S_F - S_M$ has an expected value of γ_0 at a fixed value of x. The parameter γ_0 is therefore interpreted as the *average difference in salary between a female and a male of equal seniority.* Under the assumptions of the regression, a significant negative estimate of γ_0 would be construed as evidence of salary bias against women.

To calibrate (9-20), OLS regression was performed on data for the entire faculty. (Each member contributed one data point of the form (S_i, x_i, z_i).) The least-squares estimate of γ_0 was $\gamma^* = -\$443$, and the 95% confidence interval for γ_0 was completely in the negative range. On the basis of this outcome, the brief described \$443 as "the price of being a woman" at the state college in question. Actually, \$443 would be an annual price; the cumulative penalty over a career of several decades would reach many thousands of dollars.

Discussion of the Model

In assessing the credibility of the evidence this regression produced, we might start by considering the strengths of the work. Though its decisions would not universally be accepted, the analysis did attempt seriously to explain its choice of explanatory variables. Moreover, the work shows a clear understanding of a crucial point: Because male faculty are often older on average than their female colleagues, simply comparing mean salary by gender could yield spurious evidence of wage discrimination. The analysis was designed to attribute to gender bias only that portion of salary discrepancy that differences in seniority cannot explain.

If there is any respect in which the model seems implausible, it is in the specification of how S depends on x. In most American organizations, annual wage increases are computed as a percentage of the current salary. Yet (9-20) assumes that salaries go up by the same dollar amount every year, namely β_0. That pattern would imply that, as one gets increasingly senior, the annual rises in pay are steadily dropping in percentage terms. To treat S as growing linearly with y, therefore, is to ascribe to the accused state college an unusual pay policy.

The plaintiff's brief offered no explanation for the assumption of linearity. That seems an unfortunate omission, but it is important to keep the problem in perspective. The focus of interest is whether S depends on z, not how S depends on x. An error in describing the S-to-x relationship is only relevant insofar as it distorts the value of γ^*. Whether such distortion is a real danger is not immediately clear, so we need pursue the matter.

Three Observations about Salaries

A mathematical model is only useful if it reflects the realities of the situation to which it is applied. In the present context, there are three general truths about university salaries that need be borne in mind. They apply to most American colleges and universities and, absent any information to the contrary, one might presume they apply to the defendant institution:

(i) Annual salary raises tend to be computed as a percentage of existing wages. Thus, the higher the present salary, the greater the increase in dollar terms. As a practical matter, this means that S grows faster when y is large than when it is small.

(ii) Given the "pyramid" structure of most universities, one would expect a disproportionate part of the faculty to be at *lower* experience levels.

(iii) Both the fraction of PhD's earned by women and the number of women desiring academic careers have grown steadily in recent years. Even in the absence of discrimination, therefore, the female members of a given faculty would probably tend to have lesser seniority than the males.

A Thought Experiment

In physics, one sometimes conducts "thought experiments" to explore the implications of a theory under a given set of conditions. In that tradition, we consider what would happen under the linear model (9-20) assuming that (i), (ii), and (iii) all obtain and further *there is no gender discrimination whatever in the university under scrutiny*. We particularly want to examine whether any forces exist in that setting that could push the value of γ^* from its proper value of 0.

The hypothetical data that underlie the "experiment" appear in Figure 9.6, where the ordinate is present salary and the abscissa, years of experience. Points

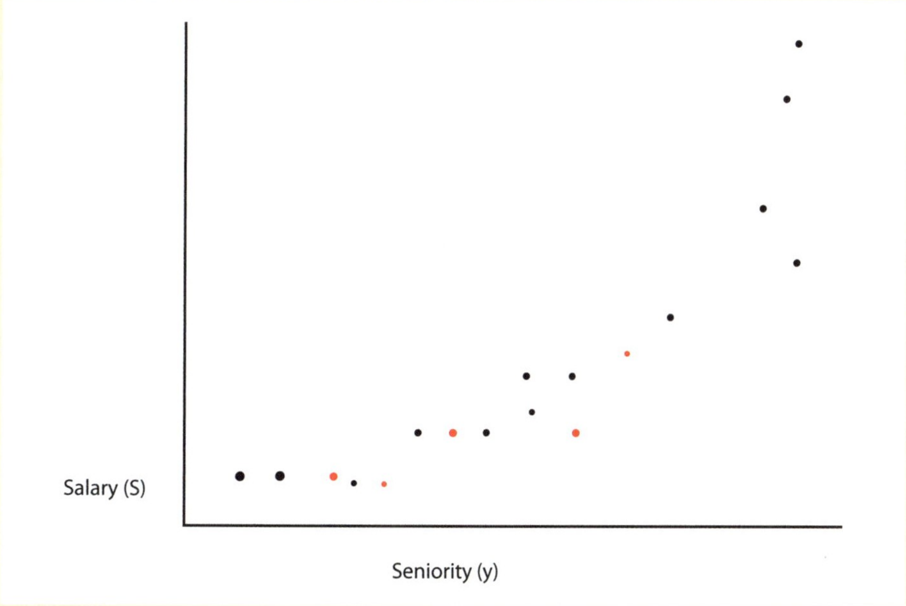

FIGURE 9.6 Salary vs. seniority at a university that does not discriminate by gender. (Red points are women; black points are men.)

in black are for men and those in red for women. Note that, consistent with (i), the points tend to climb faster as y goes up; because of (ii), there are more points at lower than higher x-values; because of (iii), the women are especially concentrated at the lower x's. Observe too that, because the school does not discriminate by gender, the "female" data points are not abnormally are not abnormally low given their y-values.

Imagine that, consistent with the hypothesis of no discrimination, one decides to calibrate (9-20) with γ_0 *set fixed* at 0. (In other words, one uses the simpler equation $S_i = \alpha_0 + \beta_0 x_i + \varepsilon_i$.) The results of the least-squares analysis are illustrated in Figure 9.7.

Here the regression line is doing its best to accommodate two divergent forces. Beside so many of the data points are situated in the low-x region, a "good fit" in that area is a major priority of the least-squares line. On the other hand, the "far off" x-values of the senior faculty exert disproportionate influence on the values of α^* and β^*. In consequence, the older faculty "tilt" the regression line upward, as is shown. Because of this tilt, the line tends to *overpredict* at the low x-values where the female faculty are especially concentrated.

But how does all this relate to the value of γ^*? Suppose that, in an attempt to reduce the sum of squared residuals, γ^* is allowed to float freely while α^* and β^* are held at their Figure 9.7 values. As noted, the value assigned γ^* is irrelevant to the prediction of men's salaries; hence the "male" residuals are unaffected whatever choice is made. But given an appropriate negative value of γ^*, the overprediction

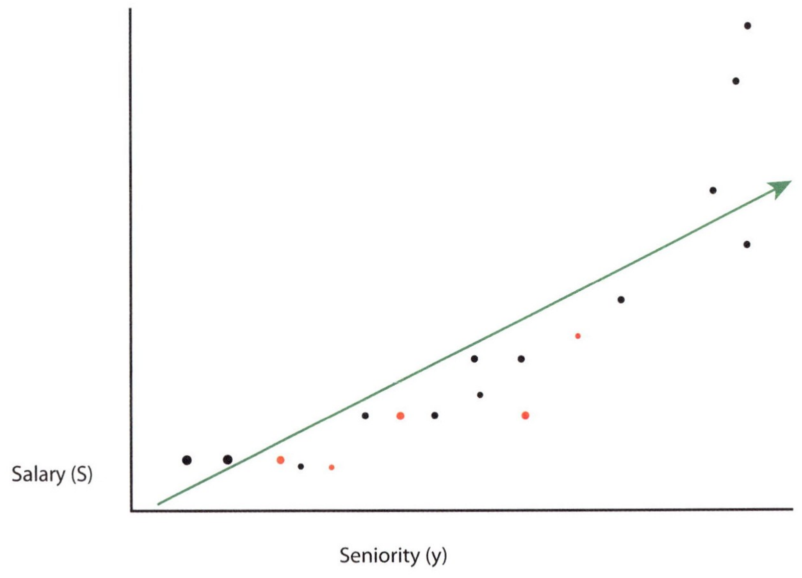

FIGURE 9.7 A simple regression line is fit to the data from the nonlinear salary schedule.

of S in the original female projections would be largely counteracted (Figure 9.8). With the female residuals thus reduced, the overall fit of the regression equation to the data would improve under the least-squares criterion.

What we have just presented is the first stage of a dynamic that could lead to a negative estimate of γ. But this negative value, we must remember, reflects not the presence of sex discrimination, but rather the attempt of the regression model to recover from its false assumption that S depends linearly on x. The thought experiment suggests, therefore, that technical failings predispose the analysis to suggest gender discrimination even when it is absent.

A Matching Test

Though these last arguments clearly cast doubt on the –$443 estimate, they need not discredit it totally. We have, after all, deduced only the *existence* of a bias, and said nothing of its actual magnitude. Should the numerical effect of the problem be far smaller than 443, the thrust of the original conclusion would scarcely be affected.

It turned out that, at the university that had been sued, one could investigate the issue quite easily. Nearly all the female faculty had male counterparts with exactly the same seniority (e.g., five years). In comparing salaries within such pairs, there is obviously no need to adjust for differences in x-values; one can concentrate on the raw differences themselves.

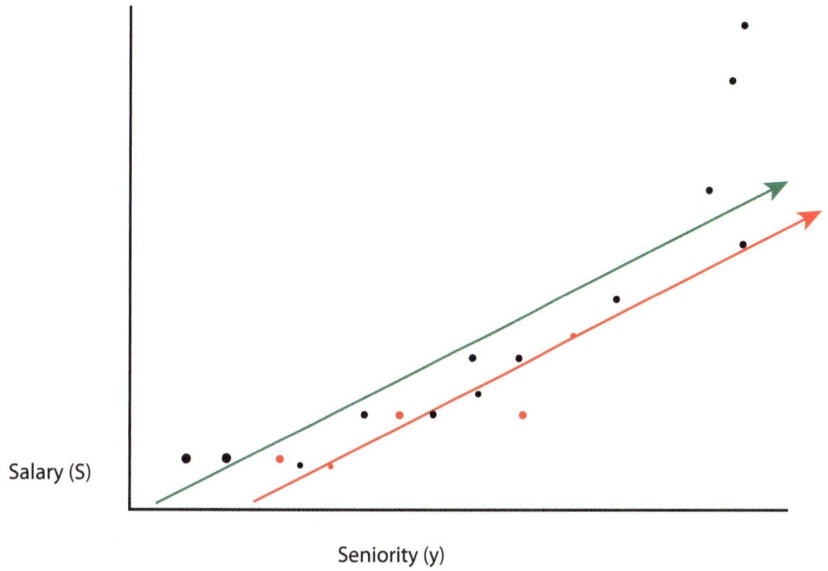

FIGURE 9.8 When γ* is chosen to lower the women's salary line, the overall sum-of-squared-residuals is reduced.

When such simple matchings were performed, the results did not corroborate those in the regression equation based on (9-20). It emerged that the woman had the higher salary in 60% of the comparisons. And, over all the pairings, the salaries for women averaged \$35 higher than those for men. This matched comparison suggests that the statistic \$443 was not just inflated by induced *in its entirety* by flaws in the regression model. The most conspicuous of the inadequacies is the linear functional form.

This example has a message: false assumptions of linearity can do terrible damage to regression results. The problem here was especially unfortunate because it was unnecessary. One could have performed a regression which allowed for fixed percentage annual salary raises, in which case the gender-coefficient would probably have been close to zero. But, unless specifically told otherwise, the regression package will assume a linear relationship among the variables.

EXERCISE 9.14 In Illustration 9.C, suppose that everything is the same about Figure 9.6 except that the women might be distributed differently by seniority (i.e., the same number of dots are red, but the red dots might move around within the full set of dots)

(i) Under what seniority-pattern for women might we expect γ^* to be essentially 0? (Answer with either words or pictures.)

(ii) Under what seniority-pattern for women might we expect γ^* to be positive?

Summing Up

Taken together, these illustrations have identified three hazards to the use of standard multiple regression:

- Omitted Variables
- Dubious Assumptions about Direction of Causality
- Improper Linearity Assumptions

We have indicated that each of these problems is potentially strong enough *on its own* to cause a greatly distorted answer to the question that spurred the regression analysis. And there are other hazards with similar potential to do mischief.

This exercise, however, has not been wholly negative. In the various cases, we have described (or at least alluded to) tractable modifications of the analysis that could have steered it out of harm's way. To cite the lapses we have observed to discredit regression would be like citing power failures to discredit electricity; whenever anything potent is used in an improper way, the results can be catastrophic. But, with a bit of foresight and common sense, we can substantially reduce the risk that we are led astray.

The Takeaway Bar

Glossary

A **prediction** made by a regression model is an estimate of the value of the dependent variable for a new data point that was not used in calibrating the regression model (i.e., in estimating its parameters)

A **sign test** of a regression model is a runs test that investigates whether the signs of consecutive residuals change in a random manner, with each residual independently having a 50% chance of being positive and 50% of being negative.

Logistic regression is a form of regression analysis that yields estimates of the dependent variable in the range (0,1), and which can therefore be used to estimate the probability of an event as function of explanatory variables.

A **Multiple Regression Analysis** is one in which there are two or more explanatory variables and the dependent variable is assumed linear in all of them.

Collinearity arises in a multiple regression analysis with two explanatory variables if these variables exhibit high correlation, a circumstance that makes it harder to estimate their slopes accurately.

Multicollinearity arises when there can be more than one strong pairwise correlation among the explanatory variables in a multiple regression analysis.

A **Basic F test** in a multiple regression analysis considers the hypothesis that the slopes of the explanatory variables are all zero, under which multicollinearity does not explain why the confidence intervals for these slopes include the value zero.

An **Indicator Variable** in a regression analysis is a 0/1 variable that reflects whether a data point has a certain characteristic.

An **Omitted Variable** in a regression analysis is a potentially important explanatory variable that is excluded from the model, an omission that can distort the slopes calculated for the variables that are included and can otherwise compromise the analysis.

Formulas

Linear Regression Analysis

Standard Linear Model

$$Y_i = \alpha_0 + \beta_0 x_i + \varepsilon_i$$

(ε_i's are independent and $N(0,\sigma_\varepsilon)$)

Prediction: If the OLS regression equation for n data points is used to estimate Y_{n+1} at the new point $x = x_{n+1}$ by $Y^*_{n+1} = \alpha^* + \beta^* x_{n+1}$ and the standard assumptions are satisfied at all points then:

$$E\left(Y^*_{n+1} - Y_{n+1}\right) = 0$$
$$\sigma^2\left(Y^*_{n+1} - Y_{n+1}\right) \approx S^2_{n+1}$$

where $S_{n+1} = \sqrt{S^2_\varepsilon\left(1 + \frac{1}{n} + \frac{\left(x_{n+1} - \bar{X}\right)^2}{S^2_x}\right)}$

where $S^2_\varepsilon = \sum_{i=1}^{n} r_i^2/n - 2.\quad S^2_x = \sum_{i=1}^{n}(x_i - \bar{X})^2/n.$

For sufficiently large n, $Y^*_{n+1} - Y_{n+1} \approx N(0, S_{n+1})$

Sign Test on Residuals:

If $Y_i = \alpha_0 + \beta_0 x_i + \varepsilon_i$ is true and n is sufficiently large, then X, the number of runs of consecutive residuals of the same sign should be essentially $1 + Z$, where Z is $B(n - 1, 1/2)$.

Logistic Regression with one Explanatory Variable:

- $O(E)$, the Odds of event E, follows: $O(E) = P(E)/(1 - P(E))$

General Logistic Model:

$$P(E \text{ given } x) = \frac{e^{\alpha_0 + \beta_0 x}}{1 + e^{\alpha_0 + \beta_0 x}}$$

In which case:

$$O(E) = e^{\alpha_0 + \beta_0 x}$$

e^{β_0} = Odds Multiplier when x increases by one

Multiple Regression Analysis:

Standard Model:

$$Y_i = \beta_0 + \beta_1 x_{1i} + \beta_2 x_{2i} + \ldots \beta_m x_{mi} + \varepsilon_i$$

(ε_i's independent and $N(0,\sigma_\varepsilon)$)

- Unbiased Estimate of σ^2_ε: $S^2_\varepsilon = \sum_{i=1}^{n} \frac{r_i^2}{n - m - 1}$.

F test of H_0: $\beta_1 = \beta_2 \ldots = \beta_m = 0$:

$$\text{Calculate } F = \left(\frac{n-m-1}{m}\right)\left(\frac{R^2}{1-R^2}\right)$$

$$\begin{cases} \text{Accept } H_0 \text{ if } F \le c \\ \text{Reject } H_0 \text{ if } F > c \end{cases}$$

where c is the $100(1-\alpha)$ percentile of the F-distribution with $(m, n-m-1)$ degrees of freedom and $\alpha = P(\text{reject } H_0 | H_0 \text{ true})$.

$$\text{Indicator Variable } z_i = \begin{cases} 1 \text{ if data point satisfies a certain condition} \\ 0 \text{ if data point does not satisfy the condition} \end{cases}$$

EZ Pass

1. Under the usual assumptions in a Y vs. x linear regression analysis, the slope estimation error and intercept estimation error:
   (A) are uncorrelated
   (B) are negatively correlated
   (C) are positively correlated
   (D) can be any of these
2. Suppose that $R^2 = 0.88$ in a multiple regression analysis with three explanatory variables and 100 data points, but that the confidence intervals for all three slopes include zero. If direction of causality is not an issue, it is reasonable to infer that:
   (A) None of the explanatory variables exerts influence on the dependent variable.
   (B) Exactly one of the explanatory variables exerts influence, but it is uncertain which one.
   (C) At most two of the explanatory variables exert influence, but the exact number is unknown
   (D) None of these inferences are reasonable.
3. If the odds of an event is 5, then the probability of the event is:
   (A) 1/5
   (B) 1/6
   (C) 4/5
   (D) 5/6
4. If the sign test on residuals in linear regression indicates unusually few runs and the deficit is statistically significant, then:
   (A) the residuals change sign less than half the time
   (B) the residuals change sign more than half the time
   (C) the dependent variable is uncorrelated with the explanatory variable
   (D) none of these
5. Mendel is uncertain whether y depends on x, x^2, or x^3, so he performs a multiple regression analysis. on the model $Y = \beta_0 + \beta_1 x + \beta_2 x^2 + \beta_3 x^3 + \varepsilon$. Assuming that $x > 1$ in his data set, then multicollinearity is:

(A) not present in this regression
(B) definitely an issue, though it might not create much extra uncertainty in the slope estimates
(C) definitely an issue, and it will create great uncertainty in the slope estimates
(D) only an issue if x is limited to the prime numbers

6. If an explanatory variable that matters is omitted from a multiple regression analysis, then the omission will be suggested by:
(A) the sign test of the residuals
(B) wider confidence intervals for the slopes of the explanatory variables included in the model
(C) a low adjusted R^2-value
(D) perhaps none of these

Further Exercises

EXERCISE 9.15 The study of 180 women and their female offspring yields the following relationship between the heights of mothers and daughters:

$$Y = 40.3 + .38x \quad R^2 = .568 \quad S_\varepsilon^2 = 9 \quad S_x^2 = 36 \quad \bar{X} = 66$$

where Y = the height of the daughter on her 25th birthday (in inches)
x = the mother's height on her own 25th birthday

A woman who was 68 inches tall on her 25th birthday has a daughter who just turned 25. Based on the regression results:

(i) Make an unbiased estimate of the daughter's height.
(ii) Find a 95% CI for that quantity.
(iii) Estimate the probability that the daughter is taller than her mother.
(iv) Find the value of Q such that, if a woman is Q inches tall on her own 25th birthday, the chance is 90% that her daughter will be at least six feet tall.

EXERCISE 9.16 Because Mendel is unsure whether the y-to-x relationship is linear, quadratic, or cubic, he decides to calibrate three separate regressions:

$$Y = \alpha_0 + \beta_0 x + \varepsilon$$
$$Y = \alpha_0 + \beta_0 x^2 + \varepsilon$$
$$Y = \alpha_0 + \beta_0 x^3 + \varepsilon$$

To resolve the functional form issue, he will treat the regression that attains the highest R^2 as reflecting the correct relationship.

(i) Referring to the relevant formulas, show that the regression with the lowest sum of squared residuals will appear the best under Mendel's R^2 criterion.
(ii) Briefly discuss the strengths and weaknesses of his approach.
(iii) Minerva proposes that, rather than perform three separate regressions, Mendel concentrate on the more general model:

$$Y = \beta_0 + \beta_1 x + \beta_2 x^2 + \beta_3 x^3 + \varepsilon$$

Show that each of Mendel's models is a special case of Minerva's.

EXERCISE 9.17

(i) Perform a *linear* regression analysis on the Challenger data, in which the dependent variable is the number of O-ring failures at a given launch, and the explanatory variable is the temperature.
(ii) What does this model imply about the chance that all five O-rings would fail at 36°?
(iii) Does this model fare well in a sign test?

EXERCISE 9.18 A logistic regression estimates the probability an aircraft engine acts abnormally during a flight as a function of flight length. It takes the form:

$$P(\text{abnormal during flight of length } x) = \frac{e^{\beta_0 + \beta_1 x}}{1 + e^{\beta_0 + \beta_1 x}}$$

where x = flight length in miles

(i) Are there *any* possible values of β_1 under which the chance of abnormality would be higher at both 200 miles and 1000 miles than at 500 miles? If so, give an example of such a value; if not, explain why not. (Hint: Think odds.)
(ii) Suppose that the estimated odds of abnormality under the equation are 2/1000 for a flight of 1000 miles and 3/1000 for a flight of 2000 miles. What are the odds of abnormality under the equation for a flight of 3000 miles? (Yes: you have enough information.)

EXERCISE 9.19 A logistic regression model involving several explanatory variables can take the form:

$$P\left(E \text{ given } x_1, x_2, \ldots, x_n\right) = e^{\beta_0 + \beta_1 x_1 + \beta_2 x_2 + \ldots + \beta_m x_m} / \left(1 + e^{\beta_0 + \beta_1 x_1 + \beta_2 x_2 + \ldots + \beta_m x_m}\right)$$

(i) In this formulation, what is $O(E|x_1,\ldots,x_m)$?

(ii) What is the odds multiplier when x_1 increases by one and the other x_i's are held constant?

(iii) Suppose that

$$P(E|x) = e^{\beta_0+\beta_1 x+\beta_2 x^2}/(1 + e^{\beta_0+\beta_1 x+\beta_2 x^2})$$

Why would it be problematic to say that the odds multiplier when x increases by one is e^{β_1}?

(iv) What is the actual odds multiplier in (iii) when $x = 3$ and x increases by one?

(v) Does this example suggest that multicollinearity can be a problem in logistic regression with several explanatory variables as well as in multiple regression analysis? Briefly discuss.

EXERCISE 9.20 Below are data by age and gender of 39 people who encountered a banner ad online during a short period, accompanied by information about whether they clicked through on the ad with intent to purchase the product. The indicator variable "click" is 1 if the person clicked through and 0 if not, while the indicator variable "gender" is 1 for women and 0 for men.

28,1,0 (age, gender, click)	33,1,1
36,0,0	41,0,0
25,0,1	38,0,0
44,1,0	50,0,0
33,0,0	32,0,1
29,1,1	36,0,1
55,0,0	40,1,0
39,0,0	29,0,0
27,1,0	27,1,1
52,0,0	45,0,0
34,1,0	39,0,0
39,1,1	35,0,0
41,0,0	49,1,0
35,0,0	46,0,1
33,1,1	40,0,0
39,0,0	38,0,1
41,0,0	34,0,0
37,1,0	46,0,0
28,0,1	39,0,0
44,0,0	

(i) Using a statistical package, perform a simple logistic regression analysis to create the formula that estimates the probability of clicking through as a function of age and gender. Write it out.

(ii) Assuming that we can make normality assumptions for the parameter estimates (i.e., the β_i's), find the 95% confidence intervals for the coefficients of the age variable and the gender variable.

(iii) Are your confidence intervals in (ii) consistent with the confidence intervals for the odds ratios that appear in the printout (e.g., for e^{β_1} rather than β_i)?

(iv) What do these confidence intervals suggest about the strength of evidence that the chance of clicking though depends on age? On gender?

(v) Under the formula, who has a higher chance of clicking though: a 38 year-old man or a 29 year-old woman?

EXERCISE 9.21 In a particular suburb, the selling price, Y, of a house depends solely on the spaciousness of the residence, x_1, and the size x_2 of its garden. To develop a linear formula for the linkage, an OLS multiple regression is performed on data about the 112 houses sold last year. The equation that emerges is:

$$Y = \underset{(9.33)}{94} + \underset{(0.57)}{.22x_1} + \underset{(0.44)}{.29x_2} \qquad R^2 = .863, \ \widehat{\rho_{12}} = .92$$

(x_1 and x_2 are measured in units of 1000 sq. ft.; Y is measured in units of \$10,000; the numbers in parenthesis are t-statistics)

(i) A typical house in the community has an x_1-value of 8.4 and an x_2-value of 13.8. Based on the regression, what would be its expected selling price?

(ii) What general interpretation of these regression results seems most sensible? Is it plausible that both house size and garden size are irrelevant to the selling price? (Assume that the underlying model is sound.)

EXERCISE 9.22 This exercise concerns the useful regression device called an *interaction term*. In some cases, it seems likely that "the whole exceeds the sum of its parts," in that the effect of two factors taken together far exceeds the sum of their individual effects. When x and z are nonnegative, a helpful model for reflecting such a phenomenon is:

$$Y = \beta_0 + \beta_1 x + \beta_2 z + \beta_3 xz + \varepsilon$$

where xz is called the interaction term.

Note that when x and z are both large, their product xz could be enormous.

Hoping to quantify how alcohol consumption and cigarette smoking can affect the risk of a particular cancer, suppose that some researchers gathered data about several groups with comparable age, ethnic and gender distributions. Suppose further that the OLS regression that summarized their results was:

$$\begin{array}{lllll} Y = & .44 & + .01x & - .01z & + .04xz \\ & (8.02) & (1.16) & (-1.48) & (3.01) \end{array}$$

$$\widehat{\rho_{xz}} = .15,\ R^2 = .758$$

where Y = annual incidence of the cancer under study (*cases per 1000 people*)

x = average alcohol consumption in ounces per day (Note: Alcohol ounces are the number of ounces in the drink multiplied by the percent alcohol.)

z = average cigarette smoking in packs per day

(i) Suppose that someone drinks two ounces of alcohol per day but does not smoke. From the regression equation, estimate his annual risk of the cancer under study.
(ii) Under the regression formula, roughly what proportion of his total risk is a consequence of his alcohol drinking?
(iii) What does the regression suggest about the circumstances in which cigarette smoking augments the danger of the cancer under study?
(iv) Under the regression formula, how much alcohol would a nonsmoker have to drink per day to suffer the same risk as someone who smokes one pack of cigarettes and drinks one ounce of alcohol daily?
(v) Does the regression present serious evidence that cigarette smoking in itself might reduce the risk of the particular cancer under study?

EXERCISE 9.23 Suppose that a college student's first year GPA is estimated via multiple regression analysis to relate to his high-school GPA and SAT score under the rule:

$$Y = \begin{array}{lll} .06 & + .97x & - .0001z \\ (2.17) & (3.20) & (-0.47) \end{array} \quad R^2 = .448,\ \widehat{\rho_{xz}} = .88$$

where Y = first-year college GPA; x = high school GPA; z = SAT score (out of 1600) (GPA's are measured on 0–4 scale)

(i) Mendel asserts that there is a high probability that a student with 1600 on the SAT's has a higher first-year GPA than another student who scored 1280. Does this regression equation contradict that statement? Discuss sensibly.
(ii) Suppose that Ezekiel contends that, because of the high correlation between x and z, multicollinearity precludes reliable slope estimates for x and z. Do the results above support that contention?

EXERCISE 9.24 A regression analysis across 72 American cities yielded the following relationship:

$$\underset{}{Y} = \underset{(16.88)}{1286.2} + \underset{(4.99)}{1.54x}$$

where Y = annual felony crime rate per 100,000 residents (t-statistics in parenthesis)
x = police officers per 100,000 residents

On the basis of these results, Mendel argued that increasing the size of the police force, far from reducing crime, would actually serve to stimulate it. Discuss some potential problems in Mendel's reasoning.

EXERCISE 9.25 In a study of worker dissatisfaction, the researcher performed a multiple regression analysis to see how Y, the time a cook stays at a given job, depends on x, her hourly wage, and z, the required number of hours of work per week. The results, based on a study of 67 workers, are summarized below.

$$Y = \underset{(8.29)}{24.2} + \underset{(2.88)}{1.23x} - \underset{(-2.26)}{.09z} \qquad R^2 = .768,\ S_\varepsilon^2 = 8.68$$

Y is measured in months; x in dollars; z in hours. The quantities in parenthesis are t-statistics.

(i) From its definition, what does the R^2 of .768 mean in this case?

(ii) Minerva announces that, given the equation above and the x and z values for a particular cook, she can predict to within one month how long the worker will stay at a job. Is that a reasonable statement given the information above? Briefly discuss.

(iii) Consider the statement: "Since the t-statistic for x has a larger absolute value than that for z, x has a greater influence on Y than does z." Discuss the validity of such an argument both in this case and in general.

INDEX

Note: This index generally presents important terms and concepts only when they are introduced. (There is no point recording every occasion that a phrase like "confidence interval" appears, and doing so could confuse the reader.) Readers should take note of the glossary and formula listings in the Takeaway Bars of individual chapters, which offer a good review of what happened in the chapter.